Remote Sensing of Leaf Area Index (LAI) and Other Vegetation Parameters

Remote Sensing of Leaf Area Index (LAI) and Other Vegetation Parameters

Special Issue Editors

Francisco Javier García-Haro
Hongliang Fang
Juan M. Lopez Sanchez

MDPI • Basel • Beijing • Wuhan • Barcelona • Belgrade

Special Issue Editors

Francisco Javier García-Haro
Universitat de València
Spain

Hongliang Fang
Chinese Academy of Sciences
China

Juan M. Lopez Sanchez
Universidad de Alicante
Spain

Editorial Office
MDPI
St. Alban-Anlage 66
4052 Basel, Switzerland

This is a reprint of articles from the Special Issue published online in the open access journal *Forests* (ISSN 1999-4907) from 2017 to 2019 (available at: https://www.mdpi.com/journal/forests/special_issues/LAI_remotesensing)

For citation purposes, cite each article independently as indicated on the article page online and as indicated below:

LastName, A.A.; LastName, B.B.; LastName, C.C. Article Title. *Journal Name* **Year**, *Article Number*, Page Range.

ISBN 978-3-03921-239-2 (Pbk)
ISBN 978-3-03921-240-8 (PDF)

Contents

About the Special Issue Editors

Francisco Javier García-Haro obtained his PhD (1997) in quantitative remote sensing from the University of Valencia. He is currently a Professor in the Department of Earth Physics and Thermodynamics, University of Valencia. His main research interests lie in canopy radiative transfer modeling and retrieval vegetation properties using satellite, including applications such as agro-meteorology, land and soil resources, agriculture, and forestry. He is responsible for the design and scientific validation of LSA SAF vegetation products from EUMETSAT satellites (https://landsaf.ipma.pt). His scientific production includes 60 papers, over 200 conference proceedings, and numerous technical reports. He is involved in several validation networks and exploitation programs of satellite missions and has received several research awards.

Hongliang Fang obtained his PhD (2003) in quantitative remote sensing from the Department of Geographical Sciences, University of Maryland, College Park. He is now a Professor in the Institute of Geographic Sciences and Natural Resources Research, Chinese Academy of Sciences (CAS). His main research interests lie in land surface biophysical parameter estimation, radiative transfer modeling, calibration and validation studies, and in situ measurements. He is an associate editor of the *IEEE Geoscience* and *Remote Sensing Letters* and *Acta Geographica Sinica*. He now serves as the leaf area index (LAI) focus area lead of the CEOS/WGCV/LPV. He is a senior member of IEEE and a member of AGU. (http://sourcedb.igsnrr.cas.cn/yw/zjrck/200910/t20091028_2638835.html).

Juan M. Lopez Sanchez obtained his PhD (2000) in electrical engineering from the Technical University of Valencia (UPV). He is currently a Full Professor in the Department of Physics, System Engineering and Signal Theory, University of Alicante. His main research interests lie in radar remote sensing technology and applications, with special attention to SAR interferometry and polarimetry for geophysics and agriculture. He is the leader of the SST research group at the University of Alicante since 2003. His scientific production includes 70 papers, over 110 conference proceedings, and numerous technical reports. He is a senior member of IEEE.

Preface to "Remote Sensing of Leaf Area Index (LAI) and Other Vegetation Parameters"

The monitoring of vegetation structure and functioning is critical to the modeling of terrestrial ecosystems and energy cycles. In particular, leaf area index (LAI) is an important structural property of vegetation used in many land surface vegetation, climate, and crop production models. Canopy structure (LAI, fCover, plant height, and biomass) and biochemical parameters directly influence the radiative transfer process of sunlight in vegetation, determining the amount of radiation measured by passive sensors in the visible and infrared portions of the electromagnetic spectrum.

Optical remote sensing methods build relationships exploiting in situ measurements and/or as outputs of physical canopy radiative transfer models. The increased availability of passive (radar and LiDAR) remote sensing data has fostered their use in many applications for the analysis of land surface properties and processes, thanks also to their insensitivity to weather conditions and the capability to exploit rich structural and texture information. Data fusion and multi-sensor integration techniques are pressing topics to fully exploit the information conveyed by both optical and microwave bands.

This Special Issue reviews the state-of-the-art in the retrieval of LAI, biomass, and other vegetation parameters using field, satellite, and airborne data, assimilation of remote sensing data with vegetation models, and its usage in wide variety of forest applications. It is composed of the following six sections:

- Field methods to measure LAI and other vegetation parameters

This section brings together innovative methods to measure LAI and other vegetation parameters. The section is composed of four chapters. The first chapter presents a smartphone-based method to measure conifer forest LAI. The second chapter presents a practical procedure to estimate the over and understory vegetation cover in Mexican forests using digital photography. The third chapter demonstrates the applicability of GPS and hemispheric photography-based methods to determine a forest signal absorption coefficient index. The final chapter in this section deals with the estimation of forest LAI and biomass patterns across Northeast China from digital aerial photograph data.

Potential and Limits of Retrieving Conifer Leaf Area Index Using Smartphone-Based Methods
Estimation of Vegetation Cover Using Digital Photography in a Regional Survey of Central Mexico
Development of a GPS Forest Signal Absorption Coefficient Index
Estimation of Forest Aboveground Biomass and Leaf Area Index Based on Digital Aerial Photograph Data in Northeast China

- Estimation of forest aboveground biomass (AGB)

The three chapters of this section are devoted to estimates of forest AGB using different data and methods. The first chapter deals with the use RapidEye optical data at the national level over Tanzania. The second chapter demonstrates the use of optical (Landsat, MODIS) and radar (ALOS-1 PALSAR, Sentinel-1) for mapping the AGB of a degraded Amazonian forest at the regional scale. The third chapter presents an approach for the estimation of forest vertical profiles and AGB using SAR (P-Band PolInSAR) images.

The Potential of High Resolution (5 m) RapidEye Optical Data to Estimate Aboveground Biomass at the National Level over Tanzania The Potential of Multisource Remote Sensing for Mapping the Biomass of a Degraded Amazonian Forest Forest Aboveground Biomass Estimation Using Single-Baseline Polarization Coherence Tomography with P-Band PolInSAR Data

- Unmanned aerial vehicle (UAV) sensors

This section is composed of three chapters. The first chapter evaluates UAV point cloud for estimating forest biophysical properties in managed temperate coniferous forests. The second chapter deals with individual tree detection from a UAV-derived canopy height model. The final chapter in this section presents an approach for the estimation of tree parameters using spectral correlation between UAV and Pléiades data.

Forest Structure Estimation from a UAV-Based Photogrammetric Point Cloud in Managed Temperate Coniferous Forests Individual Tree Detection from Unmanned Aerial Vehicle (UAV)-Derived Canopy Height Model in an Open Canopy Mixed Conifer Forest Estimation and Extrapolation of Tree Parameters Using Spectral Correlation between UAV and Pléiades Data

- LiDAR remote sensing

This section addresses new insights in the development, application, and benefits of terrestrial laser-scanning methods. The section is composed of three chapters. The first chapter addresses the automatic mapping of mapping of tree positions and tree diameter based on LiDAR and HP. The second chapter deals with the mapping of forest stands based on three-dimensional point clouds derived from terrestrial laser-scanning. The third chapter deals with the estimation of forest LAI and biomass patterns across Northeast China based on allometric scale relationship with large footprint LiDAR waveform data.

Effects of Tree Trunks on Estimation of Clumping Index and LAI from HemiView and Terrestrial LiDAR Automatic Mapping of Forest Stands Based on Three-Dimensional Point Clouds Derived from Terrestrial Laser-Scanning Estimation of Forest Biomass Patterns across Northeast China Based on Allometric Scale Relationship

- Validation of LAI products

This section addresses recent advances in the validation of LAI products. The first chapter evaluates the uncertainty and spatiotemporal consistency of global LAI and FPAR Products from VIIRS and MODIS Sensors. The second chapter calculates and validates LAI estimates generated from the USDA model in the Southeastern USA.

Analysis of Global LAI/FPAR Products from VIIRS and MODIS Sensors for Spatiotemporal Consistency and Uncertainty from 2012 to 2016 A Comparison of Simulated and Field-Derived Leaf Area Index (LAI) and Canopy Height Values from Four Forest Complexes in the Southeastern USA

- Forest applications

The last section includes a variety of forest applications for mapping and monitoring of forest disturbance, degradation, and regrowth using remotely sensed imagery. The first chapter assesses the extent of the forest cover and deforestation rates in the tropical forests in Paraguay over 17 years. The second chapter addresses the use of phenological metrics from MODIS data to monitor phenological phases and altitudinal variations in European beech-dominated stands. The last chapter in this section assesses the response of the photosynthetic activity of Mediterranean evergreen oaks using remote sensing physiological indices.

Assessing Forest Cover Dynamics and Forest Perception in the Atlantic Forest of Paraguay, Combining Remote Sensing and Household-Level Data Validation and Application of European Beech Phenological Metrics Derived from MODIS Data along an Altitudinal Gradient Assessment of the Response of Photosynthetic Activity of Mediterranean Evergreen Oaks to Enhanced Drought Stress and Recovery by Using PRI and R690/R630

Francisco Javier García-Haro, Hongliang Fang, Juan M. Lopez Sanchez

Special Issue Editors

Article

Potential and Limits of Retrieving Conifer Leaf Area Index Using Smartphone-Based Method

Yonghua Qu *, Jian Wang, Jinling Song and Jindi Wang

State Key Laboratory of Remote Sensing Science, Beijing Key Laboratory for Remote Sensing of Environment and Digital Cities, Institute of Remote Sensing Science and Engineering, Faculty of Geographical Science, Beijing Normal University, Beijing 100875, China; leonw63@mail.bnu.edu.cn (J.W.); songjl@bnu.edu.cn (J.S.); wangjd@bnu.edu.cn (J.W.)
* Correspondence: qyh@bnu.edu.cn; Tel.: +86-10-5880-2041

Academic Editor: Timothy A. Martin
Received: 17 May 2017; Accepted: 15 June 2017; Published: 19 June 2017

Abstract: Forest leaf area index (LAI) is a key characteristic affecting a field canopy microclimate. In addition to traditional professional measuring instruments, smartphone-based methods have been used to measure forest LAI. However, when smartphone methods were used to measure conifer forest LAI, very different performances were obtained depending on whether the smartphone was held at the zenith angle or at a 57.5° angle. To further validate the potential of smartphone sensors for measuring conifer LAI and to find the limits of this method, this paper reports the results of a comparison of two smartphone methods with an LAI-2000 instrument. It is shown that the method with the smartphone oriented vertically upwards always produced better consistency in magnitude with LAI-2000. The bias of the LAI between the smartphone method and the LAI-2000 instrument was explained with regards to four aspects that can affect LAI: gap fraction; leaf projection ratio; sensor field of view (FOV); and viewing zenith angle (VZA). It was concluded that large FOV and large VZA cause the 57.5° method to overestimate the gap fraction and hence underestimate conifer LAI. For the vertically upward method, the bias caused by the overestimated gap fraction is compensated for by an underestimated leaf projection ratio.

Keywords: conifer forest; leaf area index; smartphone-based method; canopy gap fraction

---

1. Introduction

Leaf area index (LAI), defined as a single leaf area per unit ground area [1], is an essential parameter for controlling mass and energy exchanges between the forest and the environment, and thus LAI affects many ecosystem processes [2]. LAI is also required when quantifying or modelling forest functioning, e.g., it is the primary factor related to the amount of light captured and the efficiency of light use, and then it is used to model the growth rate of trees [3,4] and in estimating biomass of forest in ground field experiment [5,6] or in remote sensing application [7].

The most reliable method for LAI measurement is destructive sampling, but this is time-consuming and suited only to low-growing and broad leaf plants. While for taller and needle-like leaf forests, such as conifer forests, it is difficulty to destroy forests to measure their leaf area. In this context, it will be very significant to research the method to measure LAI, especially that of a conifer forest, using indirect measurement. It is found that, compared with the manual destructive sampling method, optical sensors based indirect measurement on conifers have proved to be a more efficient in field scale [8–10] or in regional scale while remotely sensed data are considered [11,12].

The optical sensor-based indirect LAI measurement, which measures forest transmittance directly, and forest LAI were estimated using a canopy light attenuation model. In this context, an imaging or non-imaging sensor may be used depending on how forest transmittance is measured. For example,

LAI-2000 (Li-Cor Biosciences, Lincoln, NE, USA), TRAC (3rd Wave Engineering, Nepean, ON, Canada), AccuPar (Decagon Devices, Inc., Pullman, WA, USA) are representatives of the non-imaging method, whereas Can_Eye [13] and HemiView (Delta-T Devices, Cambridge, UK), which are normally used to process digital hemispherical photographs (DHPs) to calculate canopy transmittance, are popular imaging sensor-based methods. Due to greater availability of cameras for capturing images, imaging sensor-based methods are becoming increasingly popular. Among imaging sensor-based methods, a new development that uses a smartphone camera sensor rather than a general-purpose digital camera has emerged recently [14–19]. Building on the high performance-price ratio and multi-sensor integration of the smartphone, this approach has attracted much attention to measuring LAI using smartphone camera sensors.

Confalonieri et al. [15,16] have developed an android-based mobile application called PocketLAI that can be used to classify forest and background pixels (e.g., sky) in an image acquired at a zenith angle (VZA) of 57.5° and can then estimate LAI using the forest gap fraction at this zenith angle. Qu et al. [19] implemented a smartphone-based LAI measurement instrument called LAISmart. Unlike PocketLAI, which can capture images only at a VZA of 57.5°, LAISmart has no constraint on the observation angle. The operator determines the optimum viewing zenith angle depending on the canopy structure. Therefore, LAISmart can measure not only tall vegetation using upward-facing images, but also low vegetation using downward-facing images. To adapt to different vegetation heights, LAISmart provides a scalable support system. De Bei et al. [18] developed an iOS-based application called (APP)-VitiCanopy to extract LAI from vertically upward photography [20,21]. It produced a canopy clumping index besides the canopy LAI or effective LAI. At present, VitiCanopy is mainly applied to grapevines.

This paper compares the performance of smartphone-based methods to measure conifer LAI. Two of the three above mentioned methods were chosen to carry out the comparison: LAISmart and PocketLAI. The reason the VitiCanopy method was omitted was mainly due to the fact that when LAISmart is operated in upwards zenith mode, there will be no obvious difference from VitiCanopy if the classification accuracy of the image is kept at a comparable level. Another reason is that the effective LAI is the focus of this paper, and therefore there is no need to calculate clumping index which functioned in transforming efficient LAI to true LAI [22,23], although it is the unique feature of VitiCanopy.

Francone et al. [24] have achieved good verification results measuring maize and grassland with PocketLAI. However, in conifer forest, Orlando et al. [25] found that it was hard to obtain a satisfactory LAI using PocketLAI. In addition, Qu et al. [18] found that the LAI value obtained at 57.5° were much less than those from LAI-2000. Preliminary work carried out independently in different countries on similar conifer forest canopies has given indications that there may be some limitations of the smartphone-based method at a viewing zenith angle of 57.5° when a coniferous forest was measured [19,25]. However, all the above preliminary results were reached based on a small observed data set. For instance Orlando et al. [25] used twelve observed data on coniferous forests, whereas Qu et al. [19] obtained data only on two dates, so it's necessary to collect much more data to further validate the above research results.

As a result, this line of research is now entering a new stage where, despite the advantages of smartphone-based LAI measurement methods, it is time to perform more verifications and comparisons to ensure that this method is suitable for measuring coniferous forest LAI. The goal of this paper is to further verify the potential and limitations of different smartphone sensor-based methods for measuring coniferous forest LAI, to reveal the main factors influencing observational accuracy, and to provide a reference for quick, accurate acquisition of coniferous forest LAI data on a regional scale.

It should be noted on the term of LAI that, in general, not only leaves, but some other plant organs, such as trunks and branches can be seen in the digital images. Though the term plant area index (PAI) may be more appreciate than leaf area index in this context, to conform to the previous studies, we still use the term LAI herein.

2. Methodology and Data

2.1. Estimation of LAI Using a Smartphone Camera Sensor

As imaging methods, LAISmart [19] and PocketLAI [15] capture canopy photos using a smartphone camera sensor, then an automatic classification algorithm is used to separate image pixels into leaf and background to calculate canopy gap fraction. While calculating leaf area index from gap fraction, both methods share a common canopy light attenuation theory that describes how light transits from the top of the canopy to the sensors placed underneath. This theory can be formulated as the Beer-Lambert law [26]:

$$p(\theta) = e^{-G(\theta)/\cos(\theta)LAI}.$$ (1)

Then LAI can be obtained as:

$$LAI = -\ln P(\theta)\cos(\theta)/G(\theta)$$ (2)

where $P(\theta)$ is the canopy gap fraction at the sensor zenith θ and $G(\theta)$ is the projection of unit foliage area on the plane perpendicular to the view zenith θ.

The basic difference between the two methods is that PocketLAI uses the gap fraction at a fixed zenith of 57.5°, whereas LAISmart provides a flexible zenith option with no constraint to a fixed angle. However, previous work has suggested that LAISmart will produce fairly good results if it is operated using the upward zenith mode for measuring conifer canopy LAI. This paper follows the indications of previous work.

To calculate LAI using the gap fraction, the G-value should be determined for the canopy structure. LAISmart uses a G-value of 0.5 on the assumption of a spherical distribution of leaf angle [27], whereas PocketLAI uses the same value, but derived from the theory of inclined point quadrats [28].

Hence, LAISmart and PocketLAI can easily calculate LAI when the gap fraction is obtained from images segmented into vegetation and sky pixels:

$$LAI_0 = -2\ln p(0)$$ (3)

$$LAI_{57} = -2\cos(57.5)\ln p(57.5),$$ (4)

where LAI_0 and LAI_{57} are the estimated LAI from LAISmart and PocketLAI respectively.

Another instrument for measuring conifer LAI is LAI-2000, which uses a multi-angle gap fraction on blue band (450 nm) to estimate LAI, unlike the smartphone-based methods that use only one zenith angle. The LAI-2000 algorithm has been widely recognized in the literature, and interested users are encouraged to find detailed information on LAI-2000 in the reference [29]. As LAI-2000 has been validated on a wide range of vegetation types, this research has used the LAI value from the LAI-2000 instrument as references to compare the two smartphone methods.

2.2. Field Experiments

The experimental area was located around the National Field Observation and Research Station of Inner Mongolia-Great Xing'an forest Ecosystem of China (50.906°N, 121.502°E) in Genhe City of the Inner Mongolia Autonomous Region in northeastern China [8] (Figure 1a). The investigated forest area belongs to the CForBio (Chinese Forest Biodiversity Network) [30] and is selected as one of the priority areas for the conservation of perennial plants in China [31]. As the northernmost and coldest area in Inner Mongolia, Genhe is located in the cold and humid temperate forest climate region. Annual average temperature is approximately −5.3 °C, and precipitation is approximately 300 mm. The research site has a hilly topography with slight gradients (80% less than 15 degrees) and a mean altitude of approximately 950 m.

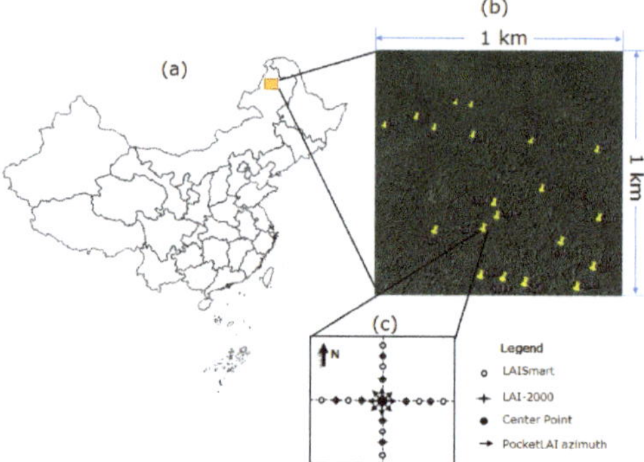

Figure 1. The research area (yellow square) located in China (**a**); the ESU (yellow symbols) distributed in the research area (**b**); and the sampling method of three instruments in an ESU (**c**).

Field experiments were carried out in a 1 km × 1 km area covered by conifer forest mainly composed of *larch (Larix gmelinii)* (Figure 1b). Stand measurements were not available during the experimental period of this research, but descriptive information about forests in the same region can be found in the published literature (Table 1) [32]. The actual data may differ slightly from published data, but will still be valid as a reference for this region since the Genhe region is covered by old-growth primary forest that has never suffered major human disturbance.

Table 1. Description on the forest in the Genhe old growth primary forest region [32].

Forest Type	Dominant Species	Forest Structure (Mean/std)				
		Density (stems/ha)	Da (cm)	Dmax (cm)	Ha (m)	Hmax (m)
Larix gmelinii **forest**	*L. gmelinii* (94%)	1135/515	15.1/4.6	41.3/11.6	13.5/3.4	26.9/4.0

Da, mean DBH (diameter at breast height); Dmax, maximum DBH; Ha, mean tree height; Hmax, maximum tree height.

Twenty essential sample units (ESUs), 20 m × 20 m in size, were established in the experimental area according the optimal sampling method [33]. For the measurement of the smartphone method, a smartphone with the CPU of Quad-Core Processer of 2.5 GHz, the camera of 13 million pixels and the aperture of F1.8 was configured. Both LAISmart and PocketLAI were installed in the same configured smartphone to ensure the images have the same optimal parameters except for different viewing angles. Different sampling methods were designed according to different instruments inside the ESUs (Figure 1c). For example, for LAISmart, the operator kept the smartphone sensor vertically upward and obtained eight observations at approximately regular intervals along an east-west route and a north-south route, giving 16 observed values for one ESU. For PocketLAI, the operator stood in the center of the ESU with the zenith angle of the smartphone at 57.5° and photographed along eight azimuths at 45° intervals, obtaining eight observed values. To obtain the reference LAI using LAI-2000, two LAI-2000 instruments were used. One was placed in the open space outside the forest for vertically upward observation to measure the total downward sky radiance. The logging frequency was configured as 1 min. The other instrument was held by an operator walking along the same line of LAISmart, but obtaining four pieces of canopy transmittance data in each line, thus there were eight LAI-2000 points in one ESU. A 180° lens cap was used while measuring under the conifer canopy. The measurements of the two smart APPs and the LAI-2000 instrument were completed within 5–10 min to ensure they were operated under the approximated sun illustration condition.

The field measurements were conducted on 26 May, 25 July, and 22 September 2016, an observation period that covered the growth, development, and withering of coniferous leaves. Note that in the first set of observations (26 May), only 10 ESUs were measured due to limited duration of field observation time, whereas 20 ESUs were measured on the other two dates. When the data were collected using the three instruments described above, cloudy weather conditions were selected to ensure strong scattered light and weak direct sunlight in the sky. The experimental observation dates, instruments, and numbers of data points are listed in Table 2.

Table 2. The instruments and the data number on every experiment date.

Date	Instruments and Data Number		
	LAISmart	**PocketLAI**	**LAI-2000**
26 May	10	/	10
25 July	20	20	20
22 September	20	20	20

Both smartphone sensor methods first captured the canopy images, after which the gap fraction was retrieved from each image. An automated segmentation method was used to classify all the pixels in blue band into sky or vegetation pixels [15,34]. The gap fraction was calculated as the ratio of the number of sky pixels to the total number of pixels. Then the LAI value was calculated using Equations (3) and (4).

To compare the two smartphone sensor methods, a regression model was established as a linear equation through the origin using estimated LAI with a reference LAI equal to LAI-2000. To evaluate the performance of the linear model, the coefficient of determination (R^2) was calculated in the software of Golden Grapher [35]. Certain other metrics were also calculated to evaluate the two smartphone sensor methods against LAI-2000 [36]. The performance of the models in reproducing observed data was quantified using the mean absolute error [37] (MAE, from 0 to $+\infty$, optimum 0), the relative root mean square error [38] (RRMSE, from 0 to $+\infty$, optimum 0), the coefficient of residual mass [39] (CRM, from $-\infty$ to $+\infty$, optimum 0), and the modeling efficiency [40] (EF; from $-\infty$ to 1, optimum 1).

3. Results

By taking LAI-2000 as a reference, the correlations between the measured values of the two smartphone sensors and the reference value were calculated, and the bias of the results obtained by the two smartphone methods was analyzed. As shown in the scatter diagram in Figure 2, the observed values of LAISmart were approximately evenly distributed on both sides of the 1:1 line (Figure 2a,b,d). This suggests that there is no obvious systematic deviation between the values obtained by LAISmart and LAI-2000. However, the values obtained by PocketLAI were almost all below the 1:1 line (Figure 2c,e), suggesting an obvious systematic deviation between the values obtained by PocketLAI and LAI-2000. In other words, the observed LAI values using PocketLAI were underestimated. This pattern can also be found in the statistical characteristics of the observed data (Table 3).

As shown in Table 3, the regression lines for LAISmart and LAI-2000 passing through the origin had slopes close to 1.0 (0.94, 0.97, and 0.93 respectively) and a mean value closer to LAI-2000 (biases were 0.07, 0.01, and 0.02 respectively). However, the regression lines for PocketLAI and LAI-2000 had slopes of 0.57 and 0.73 respectively, the maximum deviation of the mean value was 1.32, and the minimum was 1.0.

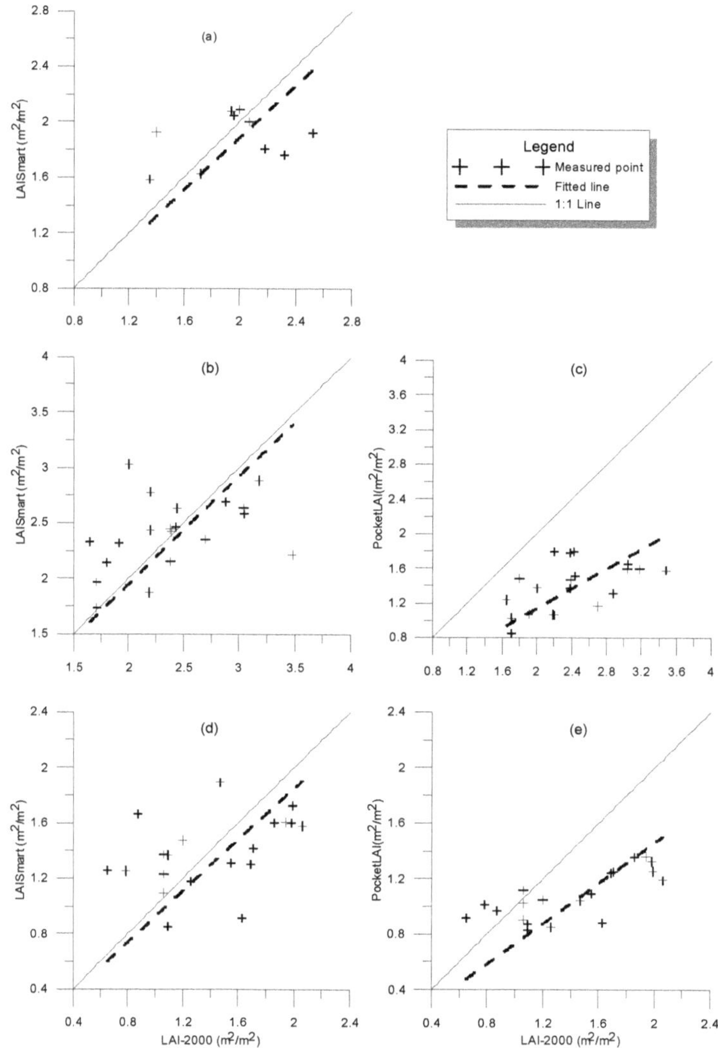

Figure 2. Scatter plot between LAI values of LAI-2000 and those of two smartphone methods (left column for LAISmart and right column for PocketLAI) on three dates (subplot (**a**) for 26 May, (**b**) and (**c**) for 25 July and (**d,e**) for 22 September).

Table 3. Statistics parameters of the measured and reference LAI values.

Date	Instruments	S	R^2	Averaged LAI		AB
				LAI-2000	**Smartphone**	
26 May	LAISmart	0.94	0.97	1.95	1.88	0.07
	PocketLAI	-		-	-	
25 July	LAISmart	0.97	0.96	2.39	2.40	0.01
	PocketLAI	0.57	0.96	2.39	1.39	1.00
22 September	LAISmart	0.93	0.93	1.40	1.38	0.02
	PocketLAI	0.73	0.96	1.40	1.08	0.32

S: Slope of a linear regression through the origin; R^2: Coefficient of determination of regress model; AB: the absolute bias of the measured LAI and the reference LAI

Table 3 also shows that all the regression models established between the results of both smartphone methods and the reference value from LAI-2000 had high coefficients of determination, suggesting that although the two methods had different performance with regards to magnitude, both methods could be used to identify the dynamic tendency of coniferous forest LAI.

To judge whether the two smartphone methods can produce a statistically identical result to a reference instrument (LAI-2000), a hypothesis test was carried out using Student's *t*-test for the above data. The basic principle of the test is that, for independent experiments on the same objects (here, conifer LAI), the different observation methods should lead to the same statistically significant result. In other words, the data obtained by the different sampling methods should come from the same population.

The Student's *t*-test results are listed in Table 4, where H = 0 indicates that the smartphone sensor results come from the same population as the LAI-2000 values and H = 1 indicates the opposite test result. The parameter *p* is the probability that H = 0 can be accepted, and the parameters CI_2.5, CI_97.5 define the 95% confidence interval for the observation bias of LAI-2000 and smartphones. The test results show that the observed results from LAISmart and LAI-2000 on the three days can be considered as being from the same population (H = 0). The probability of being unable to reject the hypothesis on the dates of 25 June and 22 September was greater than 85%. Moreover, the 95% confidence interval for the deviation between LAI-2000 and LAISmart was located approximately with zero as the center of a symmetric region, suggesting that there was little deviation between the two measures, a result that can be approximately considered as a zero deviation. Although, the results obtained on 26 May show a slightly lower probability of acceptance (0.6) and an asymmetrical confidence interval, they still could pass the Student's *t*-test (H = 0). However, for all the observed results from PocketLAI, Student's *t*-test indicated acceptance of H = 1, suggesting that these observed values could not be considered as coming from the same population as the LAI-2000 data. This means that there was an obvious difference in performance between PocketLAI and LAI-2000, which also was manifested by the confidence interval. These results indicate that the LAI obtained from LAI-2000 was significantly higher than that obtained from PocketLAI, because the lower bound (CI_2.5) of the 95% confidence interval for their deviation was greater than zero. Recalling the results in Table 3, the slopes of all the regression lines between PocketLAI and LAI-2000 were far less than 1.0, showing that in most cases, forest LAI was underestimated by PocketLAI.

The next step was to compare the observed data obtained by the two kinds of smartphones on all three dates with the values observed by LAI-2000. A scatter plot of these data is shown in Figure 3, and the regressive statistical parameters are shown in Table 5. The results are similar to the comparison made on separate dates, as shown in Table 3 and Figure 2. Both methods could measure the dynamic tendency with almost the same coefficient of determination, but the linear regression equation of PocketLAI had a lower slope.

Table 4. Results of Student's *t*-test results for the LAIs of smartphone sensors and those of LAI-2000.

Date	Method	H	*p*	CI_2.5	CI_97.5
26 May	LAISmart	0	0.60	−0.19	0.32
	PocketLAI	-	-	-	-
25 July	LAISmart	0	0.88	−0.25	0.22
	PocketLAI	1	0	0.78	1.21
22 September	LAISmart	0	0.85	−0.17	0.21
	PocketLAI	1	0	0.17	0.48

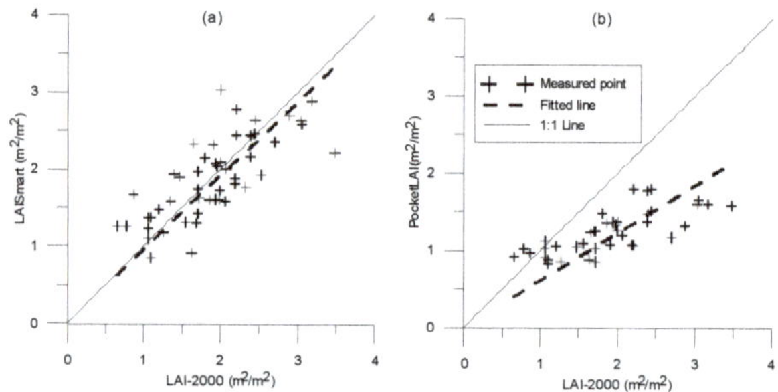

Figure 3. Scatter plot of the measurements of two smartphone sensors method with the data of LAI-2000 using all the data collected during the experiment period.

Table 5. Regression parameters and the evaluation metrics of the two smartphone sensor methods compared with references LAI.

	S	R^2	RMSE	RRMSE	MAE	EF	CRM
LAISmart	0.96	0.96	0.43	0.22	0.34	0.55	0.00
PocketLAI	0.61	0.95	0.84	0.44	0.69	−0.49	−0.35

MAE: mean absolute error; RRMSE: relative root mean square error; EF: modelling efficiency; CRM: coefficient of residual mass.

RMSE, RRMSE, and MAE, which indicate the accuracy of the two methods, show that LAISmart has much higher accuracy. As for the statistical indices of model efficiency (EF and CRM), in general, the Nash-Sutcliffe model efficiency coefficient (EF) confirmed the overall good performance of LAISmart, with a value of 0.55, whereas PocketLAI had a negative EF (−0.49). The value of E < 0 indicates that the reference mean is better than the mean of PocketLAI. The value of the LAISmart CRM equal to zero indicates that there is no difference between LAISmart and the LAI-2000 instrument, whereas the negative value of the PocketLAI CRM indicates that the LAI from the images at a zenith angle of 57.5° is lower than the reference value.

To visualize the difference in measurement results between these two kinds of smartphones and LAI-2000, their Q-Q plot is shown in Figure 4. A Q-Q plot is a type of probability plot that compares the probability distribution of two data sets by plotting their quantiles against each other. If the two datasets being compared have similar distributions, the points will lie approximately on the line $y = x$. If the distributions are linearly related, the points will lie on a line, but not necessarily the $y = x$ line.

The Q-Q plot in Figure 4a basically coincides with the $y = x$ line, although there was a slight bias as the LAI value approached 3.0 m²/m², suggesting that the measured values from LAISmart and LAI-2000 can be considered as coming from the same sample population. However, because the PocketLAI line in Figure 4b obviously deviated from the $y = x$ line, it is hard to conclude that the data from PocketLAI obey the same distribution as those from LAI-2000. This result again confirms the result from the Student's *t*-test, as shown in Table 4.

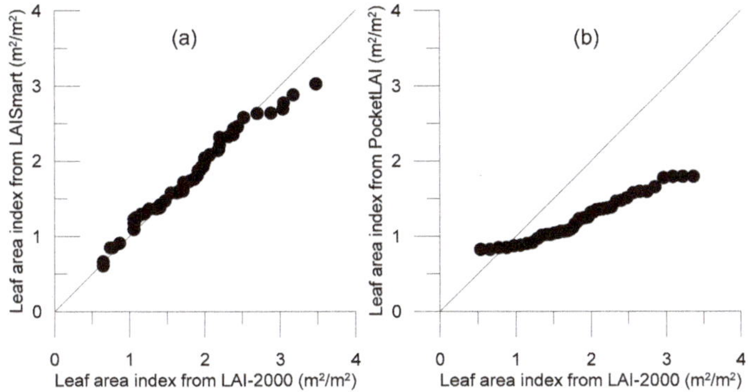

Figure 4. The Q-Q plot of two smartphone sensor methods with the LAI-2000 instrument. The similar quantile-quantile plot around $y = x$ line of LAISmart and LAI-2000 (**a**) indicates that they share a same distribution, while the PocketLAI measurement can't be taken as a same distribution of LAI-2000 as there is much bias to the $y = x$ line (**b**).

4. Discussion

A comparison between the measurement results of the two smartphone sensor methods with LAI-2000 indicates that although the measured values obtained by both methods can reflect dynamic changes in coniferous forest LAI with a high determinate coefficient of regress model, in terms of measurement magnitude accuracy, the LAISmart results were closer to those from LAI-2000.

As the research area is covered by old-growth coniferous stands, the LAI values are kept in a stable level inter-annually without human activity. Li et al. [41] reported the measured LAI in 27 ground points of coniferous forest in the site near to the research area was 2.24 m^2/m^2. Wen et al. [42] reported a higher LAI of 3.3 m^2/m^2 estimated from MISR data, unfortunately, they didn't provide the ground truth directly, but they found that the estimated LAI was higher than the ground measurement with a bias of 0.285 m^2/m^2, as a result when considering the estimation error and the clumping index of 0.76 they used, the ground LAI value can be deduced as 2.29 m^2/m^2. Both reported LAI values are measured on the dates from July to August in different years, though it was not the identical growth date as we have reported here, as the reference values, their measurement were very closer to the mean value (2.40 m^2/m^2) of LAISmart at the end of July (Table 3).

Since the two smartphones had identical hardware configurations and almost identical sampling methods in the experiment, the difference in their results could only have been caused by a classification algorithm or a shooting zenith angle.

The classification algorithm mainly influenced the canopy gap fraction extracted from the digital image. The algorithm used in both smartphone methods used the blue band as the feature to extract forest and skylight pixels. Therefore, the classification algorithm did not greatly affect these two methods, although even a difference in some specific detail of the programming implementation might have caused a slight bias.

Given that there was not much difference between the classification algorithms, the main factor that can influence gap fraction accuracy is the camera's auto-exposure mode. Beckschäfer et al. [43] found that auto-exposure might overestimate the canopy gap fraction, but for forests with low canopy density, the gap fraction overestimation caused by auto-exposure would be less than 5%. In these experiments, the maximum LAI as measured by LAI-2000 was 3.48 m^2/m^2, which can be considered as a low-density canopy most of the time. Hence, it can be inferred that at the beginning or end of the experiment, corresponding to leaf emergence and defoliation stages, the accuracy of LAI produced by smartphone methods might be better.

Another reason for gap fraction overestimation could have been the global optimum threshold method for automatic image segmentation [44]. Due to gap fraction overestimation, the estimated value of LAI was lower than the real LAI. As shown by the results presented in this paper, both methods produced some underestimation. However, underestimation was more severe in PocketLAI than in LAISmart, mostly because canopy gap fraction was overestimated. However, Qu et al. [19] found that the underestimation of LAI caused by gap fraction overestimation was inversely proportional to the real gap fraction. Hence, for sparse discrete forests with a large gap fraction, overestimation of the canopy gap fraction caused by the classification method would not lead to an obvious loss of LAI estimat45ane et al. [45] also reached a similar conclusion that "none of the more complicated classification methods yielded results that greatly differed from a simple global binary threshold classification" after they carefully investigated four complicated algorithms and a simple global optimum threshold method.

The shooting angle (camera zenith angle) affected LAI extraction accuracy in two main ways: first, the degree of closeness between the G-value used in the model and the real G-value at a specific shooting angle; and second, the proportion of leaf elements that might be influenced by shooting angle under large FOV (field of view). These two aspects will be discussed separately in the following section.

Firstly, the zenith angle of the sensor affects the G-value used in the two smartphone methods.

Figure 5 shows the G functions (dashed-dotted lines) measured by LAI-2000 on the three days, as well as the average G functions (grey line) on the three dates. Although, G is not related only to the viewing zenith angle, but also to the leaf inclination angle, for a certain forest type (conifers here), the influence of leaf inclination angle can be ignored. Hence, in this paper, it has been assumed that G is related only to the viewing zenith angle.

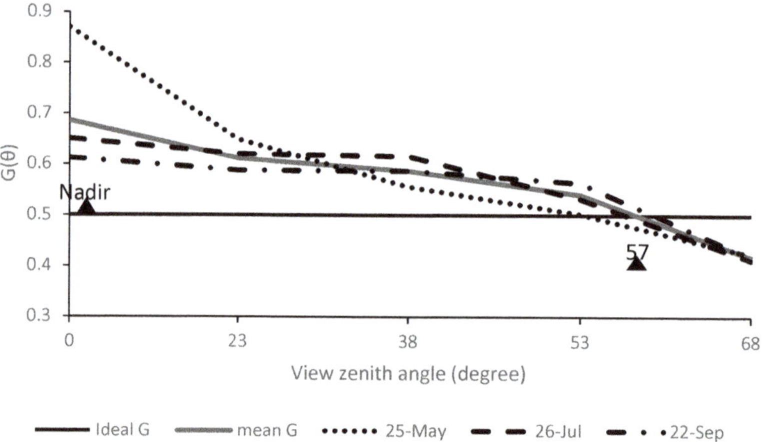

Figure 5. Foliage projection coefficient (G) derived from LAI-200 instrument at three dates. The mean G values of the measured LAI G values on the three dates is plotted as grey line, and the ideal G = 0.5 also is plotted as solid black line. The averaged G for LAISmart and PocketLAI are plotted as a solid triangle.

Because the FOV of the camera in a mobile phone is basically around 70° [19], the mean of the G-value of the first three angles in LAI-2000 was taken to be the G-value corresponding to the LAISmart viewing zenith angle, and the mean of the last two angles was taken to be the G-value corresponding to the PocketLAI viewing zenith angle of 57.5°. The calculated G-values of the two methods are shown as the solid triangle in Figure 5.

As shown in Figure 5, when the image was taken at a small or moderate viewing angle (corresponding to the observation angle in LAISmart), the G-value (0.59) of the coniferous forest was greater than the assumed value ($G = 0.5$). However, when the zenith angle was larger (corresponding than the observation angle of 57.5° in PocketLAI), the real G-value (<0.49) was slightly less than the assumed value. Note that the G-value corresponding to the viewing angle in PocketLAI should be much lower than the calculated value because G decreased with increased zenith angle, as shown in Figure 5. With the impact of the camera FOV, the real zenith angle in PocketLAI was greater than 68°, therefore the corresponding G-value should be less than 0.49.

According to Equation (2), the estimated value of LAI is inversely proportional to G-value, therefore it can be easily inferred that when the canopy gap fraction is kept constant, the error in G estimation might bring about a positive deviation in LAISmart and a negative deviation in PocketLAI. Under the combined impact of the positive deviation caused by the G-value and the negative deviation caused by gap fraction error, the value measured by LAISmart was close to that measured by LAI-2000. On the other hand, the value estimated by PocketLAI was much lower than the real value because the error sources in PocketLAI were additive for the two factors.

Secondly, the camera zenith angle and FOV jointly determined the proportion of leaf elements entering the field. At an FOV of 70°, the actual viewing angle in LAISmart should be less than 35°, whereas the maximum viewing angle in PocketLAI is approximately equal to 90°. A large FOV might lead to a dramatic decrease in gap fraction, and this is a major reason why the observed values in PocketLAI were generally lower than those produced by LAISmart and LAI-2000. This phenomenon is similar to the result of the 3D canopy simulation performed by Woodgate et al. [44]. They found that as the viewing angle increased, the proportion of trunk (woody) components in the visual field increased. Particularly when the viewing angle was greater than 75°, the proportion of trunks sharply increased, and the sharp increase in proportion of trunks greatly increased the ratio of gaps in pictures, making the estimated LAI value lower than the real value. This could partially explain why the LAI was always very much lower when Orlando et al. [25] observed coniferous trees with a height greater than 2.0 m.

Therefore, when a digital image-based (e.g., DHP) method is segmented into multi-viewing-angle sub-image, a small zenith bin size is often used [44] to ensure a good correspondence between the gap fraction and the given sensor's viewing angle. However, another question must then be answered: why can a more accurate result be achieved in LAISmart when the same wide-FOV sensor is used? The answer is provided by the fact that when the LAISmart sensor is operated with vertically upward observation, the forest leaf component still occupies a dominant proportion of the camera FOV, and hence the larger FOV has less effect on the canopy gap fraction than in PocketLAI, which is operated at a large viewing zenith angle.

5. Conclusions

In this research, coniferous canopy LAI in northeastern China was measured using two smartphone sensor-based methods, LAISmart and PocketLAI, and their performance was validated using a sophisticated commercial instrument (LAI-2000). The greatest difference between the two smartphone-based methods is the camera viewing zenith angle: the former is close to 0° and the latter 57.5°. The conifer LAI data collected by the two smartphone applications were compared with those obtained by the LAI-2000 instrument, and the factors that limit the accuracy of the retrieved LAI were analyzed. The following conclusions can be drawn:

(1) Considering the performance of LAISmart and PocketLAI, when the coniferous forest is observed, the vertically upward observation, such as the 0° zenith used in the LAISmart method, can help in obtaining observed values that are closer to those of the LAI-2000 instrument.

(2) The two smartphone-based methods show different behaviors with regards to accuracy of the estimated leaf area index when there is uncertainty in the canopy gap fraction and the leaf projection coefficient (G). For LAISmart, the effects of an overestimated canopy gap fraction may be compensated for by the underestimated G-value when estimating LAI; for PocketLAI,

the effects of an overestimated canopy gap fraction and an underestimated *G*-value on LAI are cumulative.

(3) The large field of view (FOV) of the smartphone camera sensors have different effects on these two methods. For LAISmart, a large FOV did not obviously decrease the leaf component in the field, whereas with PocketLAI, too many trunks and canopy gaps were included under the combination of the sensor's large zenith angle and large FOV. Thus, it is difficult for PocketLAI to measure coniferous forest canopy LAI accurately.

Compared with a professional canopy analyzer or a digital single-lens reflex (SLR) camera, the smartphone has some advantages such as light weight, low price, and high environmental adaptability. According to current experimental data and analytical results, the LAISmart method can produce acceptable coniferous forest LAI, and it may be a promising prospective for forest management to reduce intensive field labor and to improve the working efficiency. Furthermore, by combining the camera and gyro of the smartphone, besides LAI, more interested parameters e.g., canopy height, stand basal area, and stand volume can be calculated through the method of photogrammetry [46]. When connected to the internet cloud server as stated by the concept of smart forest [47], the smartphone method may highly facilitate the data collecting in the forest survey application.

Acknowledgments: This work was supported in part by the National Basic Research Program of China (2013CB733403) and the Natural Science Foundation of China (41671333/41531174). We also thank to Yu Bo, Tian Xiaodan of Beijing Normal University for the work collecting ground LAI-2000 data, Zhang Qiuliang of Inner Mongolia Agricultural University, China as well as Zhang Yongliang from Forestry Bureau of Genhe, Inner Mongolia, China for they providing many convenient conditions while we conducted the field experiment in Genhe forest Region. All the supporting material in this paper, including original LAISmart and PocketLAI data together with the data of LAI-2000 instrument can be available on the request by email to corresponding author Y. Qu (qyh@bnu.edu.cn).

Author Contributions: Yonghua Qu and Jindi Wang conceived and designed the experiments; Jinling Song and Jian Wang performed the experiments; Yonghua Qu analyzed the data and wrote the paper.

Conflicts of Interest: The authors declare no conflict of interest.

References

1. Chen, J.M.; Black, T.A. Defining leaf area index for non-flat leaves. *Plant Cell Environ.* **1992**, *15*, 421–429. [CrossRef]
2. Thimonier, A.; Sedivy, I.; Schleppi, P. Estimating leaf area index in different types of mature forest stands in Switzerland: A comparison of methods. *Eur. J. For. Res.* **2010**, *129*, 543–562. [CrossRef]
3. Chaturvedi, R.K.; Raghubanshi, A.S.; Singh, J.S. Relative effects of different leaf attributes on sapling growth in tropical dry forest. *J. Plant Ecol.* **2014**, *7*, 544–558. [CrossRef]
4. Li, Y.; Kröber, W.; Bruelheide, H.; Härdtle, W.; von Oheimb, G. Crown and leaf traits as predictors of subtropical tree sapling growth rates. *J. Plant Ecol.* **2017**, *10*, 136–145. [CrossRef]
5. Timilsina, N.; Beck, J.L.; Eames, M.S.; Hauer, R.; Werner, L. A comparison of local and general models of leaf area and biomass of urban trees in USA. *Urban For. Urban Green.* **2017**, *24*, 157–163. [CrossRef]
6. Forrester, D.I.; Tachauer, I.H.H.; Annighoefer, P.; Barbeito, I.; Pretzsch, H.; Ruiz-Peinado, R.; Stark, H.; Vacchiano, G.; Zlatanov, T.; Chakraborty, T.; et al. Generalized biomass and leaf area allometric equations for European tree species incorporating stand structure, tree age and climate. *For. Ecol Manag.* **2017**, *396*, 160–175. [CrossRef]
7. Heiskanen, J. Estimating aboveground tree biomass and leaf area index in a mountain birch forest using ASTER satellite data. *Int. J. Remote Sens.* **2006**, *27*, 1135–1158. [CrossRef]
8. Qu, Y.; Han, W.; Fu, L.; Li, C.; Song, J.; Zhou, H.; Bo, Y.; Wang, J. LAINet—A wireless sensor network for coniferous forest leaf area index measurement: Design, algorithm and validation. *Comput. Electron. Agric.* **2014**, *108*, 200–208. [CrossRef]
9. Leblanc, S.G.; Fournier, R.A. Hemispherical photography simulations with an architectural model to assess retrieval of leaf area index. *Agric. For. Meteorol.* **2014**, *194*, 64–76. [CrossRef]

10. Nilson, T.; Anniste, J.; Lang, M.; Praks, J. Determination of needle area indices of coniferous forest canopies in the NOPEX region by ground-based optical measurements and satellite images. *Agric. For. Meteorol.* **1999**, *98–99*, 449–462. [CrossRef]

11. Fernandes, R.A.; Miller, J.R.; Chen, J.M.; Rubinstein, I.G. Evaluating image-based estimates of leaf area index in boreal conifer stands over a range of scales using high-resolution CASI imagery. *Remote Sens. Environ.* **2004**, *89*, 200–216. [CrossRef]

12. Wang, R.; Chen, J.M.; Pavlic, G.; Arain, A. Improving winter leaf area index estimation in coniferous forests and its significance in estimating the land surface albedo. *ISPRS J. Photogramm. Remote Sens.* **2016**, *119*, 32–48. [CrossRef]

13. CAN-EYE. Available online: http://www6.paca.inra.fr/can-eye/ (accessed on 15 June 2017).

14. Fuentes, S.; Bei, R.D.; Pozo, C.; Tyerman, S. Development of a smartphone application to characterise temporal and spatial canopy architecture and leaf area index for grapevines. *Wine Vitic. J.* **2012**, *27*, 56–60.

15. Confalonieri, R.; Foi, M.; Casa, R.; Aquaro, S.; Tona, E.; Peterle, M.; Boldini, A.; De Carli, G.; Ferrari, A.; Finotto, G.; et al. Development of an app for estimating leaf area index using a smartphone. Trueness and precision determination and comparison with other indirect methods. *Comput. Electron. Agric.* **2013**, *96*, 67–74. [CrossRef]

16. Confalonieri, R.; Francone, C.; Foi, M. The PocketLAI smartphone app: An alternative method for leaf area index estimation. In Proceedings of the 7th International Congress on Environmental Modelling and Software: Bold Visions for Environmental Modeling, San Diego, CA, USA, 18 June 2014; iEMSs: Toulouse, France, 2014; pp. 288–293.

17. Aquino, A.; Millan, B.; Gaston, D.; Diago, M.P.; Tardaguila, J. vitisFlower®: Development and Testing of a Novel Android-Smartphone Application for Assessing the Number of Grapevine Flowers per Inflorescence Using Artificial Vision Techniques. *Sensors* **2015**, *15*, 21204–21218. [CrossRef] [PubMed]

18. De Bei, R.; Fuentes, S.; Gilliham, M.; Tyerman, S.; Edwards, E.; Bianchini, N.; Smith, J.; Collins, C. VitiCanopy: A Free Computer App to Estimate Canopy Vigor and Porosity for Grapevine. *Sensors* **2016**, *16*, 585. [CrossRef] [PubMed]

19. Qu, Y.; Meng, J.; Wan, H.; Li, Y. Preliminary study on integrated wireless smart terminals for leaf area index measurement. *Comput. Electron. Agric.* **2016**, *129*, 56–65. [CrossRef]

20. Macfarlane, C.; Hoffman, M.; Eamus, D.; Kerp, N.; Higginson, S.; McMurtrie, R.; Adams, M. Estimation of leaf area index in eucalypt forest using digital photography. *Agric. For. Meteorol.* **2007**, *143*, 176–188. [CrossRef]

21. Fuentes, S.; Palmer, A.R.; Taylor, D.; Zeppel, M.; Whitley, R.; Eamus, D. An automated procedure for estimating the leaf area index (LAI) of woodland ecosystems using digital imagery, MATLAB programming and its application to an examination of the relationship between remotely sensed and field measurements of LAI. *Funct. Plant Biol.* **2008**, *35*, 1070–1079. [CrossRef]

22. Qu, Y.; Fu, L.; Han, W.; Zhu, Y.; Wang, J. MLAOS: A Multi-Point Linear Array of Optical Sensors for Coniferous Foliage Clumping Index Measurement. *Sensors* **2014**, *14*, 9271–9289. [CrossRef] [PubMed]

23. Chen, J.M.; Menges, C.H.; Leblanc, S.G. Global mapping of foliage clumping index using multi-angular satellite data. *Remote Sens. Environ.* **2005**, *97*, 447–457. [CrossRef]

24. Francone, C.; Pagani, V.; Foi, M.; Cappelli, G.; Confalonieri, R. Comparison of leaf area index estimates by ceptometer and PocketLAI smart app in canopies with different structures. *Field Crops Res.* **2014**, *155*, 38–41. [CrossRef]

25. Orlando, F.; Movedi, E.; Paleari, L.; Gilardelli, C.; Foi, M.; Dell'Oro, M.; Confalonieri, R. Estimating leaf area index in tree species using the PocketLAI smart app. *Appl. Veg. Sci.* **2015**, *18*, 716–723. [CrossRef]

26. Nilson, T. A theoretical analysis of the frequency of gaps in plant stands. *Agric. Meteorol.* **1971**, *8*, 25–38. [CrossRef]

27. Goudriaan, J. The bare bones of leaf-angle distribution in radiation models for canopy photosynthesis and energy exchange. *Agric. For. Meteorol.* **1988**, *43*, 155–169. [CrossRef]

28. Wilson, J.W. Estimation of foliage denseness and foliage angle by inclined point quadrats. *Aust. J. Bot.* **1963**, *11*, 95–105. [CrossRef]

29. Li-COR. *LAI-2000 Plant Canopy Analyzer Instruction Manual*; Li-COR: Lincoln, NE, USA, 1992.

30. Feng, G.; Mi, X.; Yan, H.; Li, F.Y.; Svenning, J.-C.; Ma, K. CForBio: A network monitoring Chinese forest biodiversity. *Sci. Bull.* **2016**, *61*, 1163–1170. [CrossRef]

31. Zhang, M.-G.; Slik, J.W.F.; Ma, K.-P. Priority areas for the conservation of perennial plants in China. *Biol. Conserv.* **2016**, in press. [CrossRef]

32. Wang, X.; Fang, J.; Tang, Z.; Zhu, B. Climatic control of primary forest structure and DBH-height allometry in Northeast China. *For. Ecol. Manag.* **2006**, *234*, 264–274. [CrossRef]

33. Zeng, Y.; Li, J.; Liu, Q.; Qu, Y.; Huete, A.R.; Xu, B.; Yin, G.; Zhao, J. An Optimal Sampling Design for Observing and Validating Long-Term Leaf Area Index with Temporal Variations in Spatial Heterogeneities. *Remote Sens.* **2015**, *7*, 1300–1319. [CrossRef]

34. Otsu, N. A Threshold Selection Method from Gray-Level Histograms. *IEEE Trans. Syst. Man Cybern.* **1979**, *9*, 62–66. [CrossRef]

35. Golden Grapher 10.0. Available online: http://www.goldensoftware.com (accessed on 15 June 2017).

36. Bregaglio, S.; Orlando, F.; Forni, E.; De Gregorio, T.; Falzoi, S.; Boni, C.; Pisetta, M.; Confalonieri, R. Development and evaluation of new modelling solutions to simulate hazelnut (*Corylus avellana* L.) growth and development. *Ecol. Model.* **2016**, *329*, 86–99. [CrossRef]

37. Shaeffer, D.L. A model evaluation methodology applicable to environmental assessment models. *Ecol. Model.* **1980**, *8*, 275–295. [CrossRef]

38. Jørgensen, S.E.; Kamp-Nielsen, L.; Christensen, T.; Windolf-Nielsen, J.; Westergaard, B. Validation of a prognosis based upon a eutrophication model. *Ecol. Model.* **1986**, *32*, 165–182. [CrossRef]

39. Loague, K.; Green, R.E. Statistical and graphical methods for evaluating solute transport models: Overview and application. *J. Contam. Hydrol.* **1991**, *7*, 51–73. [CrossRef]

40. Nash, J.E.; Sutcliffe, J.V. River flow forecasting through conceptual models part I—A discussion of principles. *J. Hydrol.* **1970**, *10*, 282–290. [CrossRef]

41. Li, H.; Xiao, K.; Fan, W.; Wen, Y.; Ma, Z.; Sun, S. Determination and Estimation of the Forest Leaf Area Index with Remote Sensing in Daxing'an Mountain. *J. Northeast For. Univ.* **2013**, *41*, 66–70.

42. Wen, Y.-B.; Chang, Y.; Fan, W.-Y. Algorithm for leaf area index inversion in the Great Xing'an Mountains using MISR data and spatial scaling for the validation. *J. Beijing For. Univ.* **2016**, *38*, 1–10.

43. Beckschäfer, P.; Seidel, D.; Kleinn, C.; Xu, J. On the exposure of hemispherical photographs in forests. *iForest-Biogeosci. For.* **2013**, *6*, 228–237. [CrossRef]

44. Woodgate, W.; Jones, S.D.; Suarez, L.; Hill, M.J.; Armston, J.D.; Wilkes, P.; Soto-Berelov, M.; Haywood, A.; Mellor, A. Understanding the variability in ground-based methods for retrieving canopy openness, gap fraction, and leaf area index in diverse forest systems. *Agric. For. Meteorol.* **2015**, *205*, 83–95. [CrossRef]

45. Macfarlane, C. Classification method of mixed pixels does not affect canopy metrics from digital images of forest overstorey. *Agric. For. Meteorol.* **2011**, *151*, 833–840. [CrossRef]

46. Itoh, T.; Eizawa, J.; Yano, N.; Matsue, K.; Naito, K. Development of Software to Measure Tree Heights on the Smartphone. *J. Jpn. For. Soc.* **2010**, *92*, 221–225. [CrossRef]

47. Smart Forests Sensor Network. Available online: https://smartforests.org (accessed on 15 June 2017).

 *forests*

Article

Estimation of Vegetation Cover Using Digital Photography in a Regional Survey of Central Mexico

Víctor Salas-Aguilar [1,*], Cristóbal Sánchez-Sánchez [2], Fabiola Rojas-García [2], Fernando Paz-Pellat [3], J. René Valdez-Lazalde [2] and Carmelo Pinedo-Alvarez [4]

[1] Programa Mexicano del Carbono, Texcoco 56230, Mexico
[2] Postgrado en Ciencias Forestales, Colegio de Postgraduados, Texcoco 56230, Mexico; crisdansanchez@gmail.com (C.S.-S.); fabiosxto1981@gmail.com (F.R.-G.); jrene.valdez@gmail.com (J.R.V.-L.)
[3] Postgrado en Hidrociencias, Colegio de Postgraduados, Texcoco 56230, Mexico; ferpazpel@gmail.com
[4] Facultad de Zootecnia y Ecología, Universidad Autónoma de Chihuahua, Chihuahua 33820, Mexico; cpinedo@uach.mx
* Correspondence: vsalasaguilarl@gmail.com; Tel.: +52-595-120-2056

Received: 9 September 2017; Accepted: 9 October 2017; Published: 15 October 2017

Abstract: The methods for measuring vegetation cover in Mexican forest surveys are subjective and imprecise. The objectives of this research were to compare the sampling designs used to measure the vegetation cover and estimate the over and understory cover in different land uses, using digital photography. The study was carried out in 754 circular sampling sites in central Mexico. Four spatial sampling designs were evaluated in three spatial distribution patterns of the trees. The sampling designs with photographic captures in diagonal form had lower values of mean absolute error (MAE < 0.12) and less variation in random and grouped patterns. The Carbon and Biomass Sampling Plot (CBSP) design was chosen due to its smaller error in the different spatial tree patterns. The image processing was performed using threshold segmentation techniques and was automated through an application developed in the Python language. The two proposed methods to estimate vegetation cover through digital photographs were robust and replicable in all sampling plots with different land uses and different illumination conditions. The automation of the process avoided human estimation errors and ensured the reproducibility of the results. This method is working for regional surveys and could be used in national surveys due to its functionality.

Keywords: automated classification; sampling design; adaptive threshold; over and understory cover

1. Introduction

Vegetation cover is used in studies of the aerosphere, pedosphere, hydrosphere, and biosphere, as well as their interactions [1]. Remote sensing (RS) technology, particularly the development of vegetation indices, has allowed researchers to estimate vegetation cover at a regional and global scale [2].

Validation of high and medium resolution satellite products is a critical aspect of its usefulness in operational approaches [3]. The feasibility and precision of RS must be verified before data can be applied [4]. One way of validating and re-scaling RS products is the use of field measurements, especially the application of digital photography [5,6].

The use of digital photography to estimate the understory cover (shrub and herbaceous—nadir angle) and overstory cover (arboreal—zenith angle) has been advocated in recent years as one of the most accurate methods to estimate these variables [7,8]. According to Liang et al. [1] and White et al. [9], this technique is the most reliable and can be easily employed to extract vegetation cover information in different physiographic conditions.

In relation to shrub and herbaceous cover (understory), robust segmentation algorithms have been developed to discriminate bare soil and vegetation regardless of the type of vegetation present [10,11] and type of luminosity (presence of shadows) [12]. On the other hand, vegetation cover estimations have also progressed in terms of the classification methods [13,14], automation [15], and classification when there is a mixture of vegetation-sky pixels [16].

The advantage of using a non-destructive method such as digital photography to estimate the vegetation cover is that it can be related to the biophysical variables of an ecosystem at a lower cost and time [17,18]. However, few operational studies (local, regional, and national surveys) contemplate measuring this variable for lack of knowledge on how to pursue it efficiently [19] or for the inconsistency of associating forest attributes to a sampling site with an estimated vegetation cover that may exceed the site surface [20].

A disadvantage of the mentioned methods is that sampling sites in which the experiments are made are generally small and homogeneous (low slope and similar land use conditions and plant architecture) [21]. Therefore, the application of these techniques in operational inventories must be planned to efficiently provide the information for which they are developed, and this condition includes considerations of the accuracy of estimates, costs of data collection and processing, and the speed of the process from the planning stage to the presentation of results [22].

At present, there is no agreement among researchers on how to define the optimum sampling design to derive the leaf area index and vegetation cover in field measurements [23]. In a sampling plot, the vegetation cover and its spatial distribution may vary when considering the effects of management [24].

A rapid, reliable, and economical way to compare vegetation cover sampling designs is by predicting the crown diameter through allometric ratios [25] and by estimating the spatial patterns of trees in the sample site. The advantage is that the crown diameter and spatial clustering of trees can be projected into a geographic information system [26], avoiding the intensive work of conducting and comparing them directly in a survey campaign [27].

Due to the lack of an accurate vegetation cover estimation method for forest surveys in México, the objectives of the present study were: (1) to compare the sampling patterns used to measure the vegetation cover; and (2) to estimate the overstory (trees) and understory (shrub and herbaceous) cover in sampling sites of the State of Mexico, Mexico, using a practical procedure and easily reproducible method.

2. Materials and Methods

The State of Mexico is located in the southern part of the southern plateau of Mexico, between parallels 18°22′ and 20°17′ North and meridians 98°36′ and 100°37′ West, in an area of 22,333 km^2. In this region and particularly around the Valley of Mexico, there are specific environmental and historical conditions that have resulted in great biological and cultural diversification along mountain ranges, basins, rivers, and forests.

Ceballos et al. [28] consider that the vegetation of the State of Mexico is represented by three main ecosystems with variations: temperate-cold (temperate forests), semi-warm and sub-humid warm (low deciduous forests), and arid zones (arid and semi-arid vegetation).

The study was carried out from January to September 2015; 754 circular sampling sites of 1000 m^2 were established and distributed in eight forest regions of the State of Mexico (Figure 1) [29].

In each region, we collected information on the type of vegetation cover and land use. The classification of vegetation was established according to the land use and vegetation chart, Series IV, scale 1:250,000 [30], and was verified in the field.

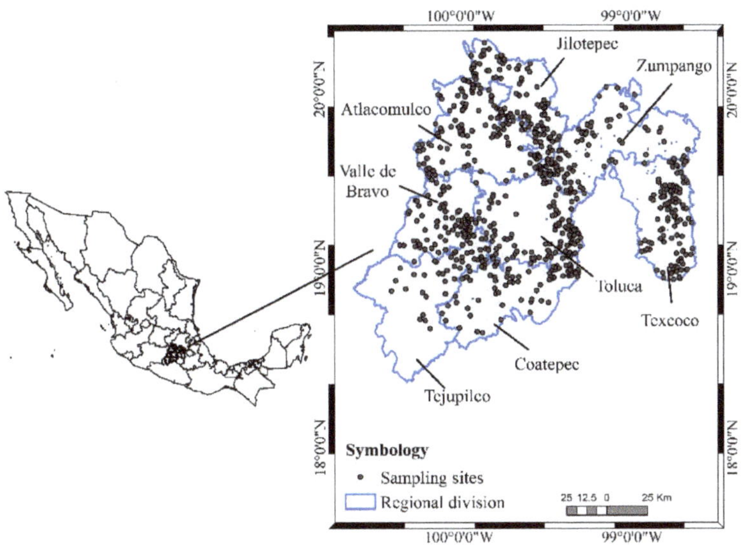

Figure 1. Spatial location of the sampling sites, within the forest regions of the State of Mexico.

2.1. Spatial Projection to Evaluate Sampling Designs

The regional survey was planned as a complement to the National Forest and Soil Inventory (NFSI) in a simplified way, where the height and crown diameters were not measured as done in the NFSI. These variables were planned to be estimated from the state and national surveys.

Before the survey phase, a pre-survey of 30 sites was carried out in the Texcoco forest region to evaluate the spatial pattern of trees in four sampling designs of vegetation cover. Comparisons were made between VALERI [31] and SLAT [32] designs, along with two alternative samples: CBSP (carbon and biomass sampling plots) and RM (regular mesh). The VALERI design is composed of 13 samples, SLAT of 15 samples, CBSP of 21 samples, and RM of 37 samples (Figure 2a).

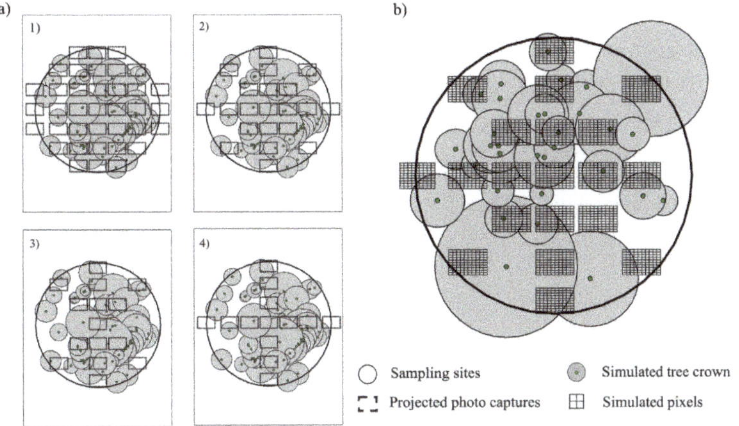

Figure 2. (**a**) Photographic captures by sampling design: (1) RM, (2) CBSP, (3) SLAT, (4) VALERI; (**b**) Projected photo capture areas (pixels) in a CBSP sampling design.

Due to the difficulties of knowing the real vegetation cover within a sampling site, the comparison of sampling designs was performed within a geographic information system. Initially, we recorded the central coordinates of each site (as planned in the regional survey), the distance of the trees to the central point, and their azimuth. Then we calculated the location of trees using the central location of the plot and the azimuth and distance to the central point of the respective tree.

Given that no sampling of the tree crowns diameter was recorded in the survey, an allometric relationship was established between the crown diameter and diameter at breast height [25]. The data of this function were obtained from the National Forest and Soil Inventory [33]. The function is the following: DC = 0.1553 + 0.1859 (Dn) (R^2 = 0.79, p < 0.001), where DC is the tree crown diameter and Dn is the diameter at breast height. The linear model is generalized because it comprised all of the timber species found in the survey. The estimated DC allowed us to construct the crown influence area projection of the trees, assuming a circular shape.

The spatial patterns of the trees were evaluated using the Average Nearest Neighbor (ANN) equation [34]. If the pattern of the tree distribution is completely random ANN = 1, if ANN < 1 the trees are grouped, and if ANN > 1 the tree mass is regular (dispersed). The ANN analysis was performed within the ArcGIS (10.3, Esri, Redlands, CA, United States).

2.2. Projected Photographic Captures within the GIS

A single photographic capture area of 16.38 m^2 was established to determine the projected cover per site and type of sampling (the procedure for estimating the area is described below). In each area we built a grid (10 × 10) with the purpose of simulating the pixels of a photographic camera (Figure 2b).

The total observed cover was calculated by dividing the area of overlapping crowns between the sizes of the plot (1000 m^2). On the other hand, the estimated cover resulted from the following equation:

$$\sum_{i=0}^{n} \frac{NPSi}{NTP} \tag{1}$$

where *NPSi* is the number of projected pixels (grid) intersecting with the tree crown area and *NTP* is the total number of pixels per sampling design.

As a quantitative measure of the error, the mean absolute error (*MAE*) was estimated:

$$MAE = N^{-1} \sum_{I=1}^{N} |Oi - Ei| \tag{2}$$

where *O* is the observed value of the total projected cover, *E* is the estimated value (Equation (1)), and *N* is the number of captures per sampling design.

2.3. Field Sampling

Sampling sites were targeted to include vegetation succession and degradation among land uses in Central Mexico [29]. Information was collected on sites with and without anthropogenic intervention.

2.4. Photo Features

The photographic images were taken at a resolution of 5184 × 3456 pixels in JPG format. We used a Canon EOS Rebeld T5^{RM} camera. The camera lens was adjusted to a range of 18 to 55 mm focal length and an ISO 200 with aperture and exposure in automatic mode.

2.5. Taking Photos at the Sampling Sites

We applied the CBSP sampling design with 21 captures to nadir and zenith, according to Figure 3a. The lines represent the transects within the sampling site (L1–L4) and each letter represents a photographic capture. Figure 3b shows the photograph taken at zenith, where the distance between the camera and the ground is 1.5 m.

Figure 3c shows the process of shooting understory (shrub and herbaceous strata), where the interference of the personnel in the photograph was avoided using a stick of five meters long; in this case, the distance between the camera and the ground was 3 m. Two levels of bubble were used to control the angle in which the photographs were taken, one near the operator and the other one stuck to the side of the camera.

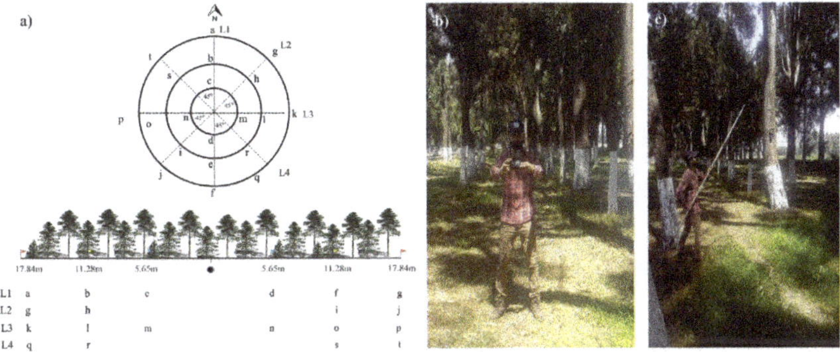

Figure 3. Photographic capture process at sampling sites. (**a**) CBSP design, the letters correspond to the distance of capture from the center of the sampling site; (**b**) photographic capture to zenith; (**c**) photographic capture to nadir.

The purpose of the CBSP sampling was to capture the largest possible physical area with the fewest samples. To do so, the visual field angle of the lens (θ) was adjusted depending on the size of the sensor (n) and its focal length (f):

$$\theta = 2 \times \tan^{-1}\left(\frac{n}{2 \times f}\right) \tag{3}$$

The real area covered by a photograph depends on three variables: sensor size (n_{ij}), focal length of the lens (f), and distance of the lens to the object (h). In the case of nadir, h is the distance between the camera lens and the ground. For zenith, h corresponds to the distance between the lens and the tree crowns. Equation (4) defines the calculation of the real area of the photograph:

$$Gij = \frac{n_{ij} \times h}{f} \tag{4}$$

where G is the actual length of the object in the horizontal (i) and vertical (j), where the horizontal distance of the n_i sensor size of the camera used was 22.3 mm and the vertical distance n_j was 14.9 mm.

The value of f for the nadir photographs was set at 18 mm because h was established at 3 m. By solving Equation (4), we estimated a real area to nadir of 9.2 m^2.

In the case of zenith photographs, a larger real area was required to be representative at the sampling site. The minimum average height of the tree crowns in the forested areas of the region was 4 m. At this point, the value of θ must be adjusted to reach the largest surface, so the value of f was set at 18 mm. Then, by solving Equation (4), the real area at zenith is 16.38 m^2.

In heterogeneous forests, such as the study area, the height of the trees can vary in short distances, so in order to maintain the area captured independently of the height of the tree, the value of f can be adjusted by multiplying the average height of the trees at the point of capture by the constant 4.5 ($f = 4.5 \times h$). If the height value was four meters, the camera was placed as close as possible to the ground; in the case of exceeding six meters in height, the camera was placed at a fixed distance of 1.5 m on the ground.

2.6. Estimation of over and Understory Cover

The processing of images to estimate the vegetation cover is different in nadir and zenith projections; in the first case a robust classifier is needed to distinguish the shade of the vegetation, whereas the second one requires a methodology that distinguishes the cover of the canopy in contrast to the sky (atmosphere). Due to the large number of photographs that needed to be processed (24,182 photographs), a code was written in the Python 2.7^{RM} language (Python Software Foundation (PSF): Wilmington, DE, USA) to optimize the process (the software can be requested from the authors). The following sections describe the methodology used.

2.6.1. Estimation of Overstory Cover

The photographs were taken in the morning and in the afternoon, before the sun surpassed 130° of azimuth or after 230° to avoid confusion due to the brightness of the leaves in association with the sky. We used the SunEarthTool tool (http://www.sunearthtools.com/dp/tools/pos_sun.php) to identify the appropriate times to take the photographs.

The methodology of Fuentes et al. [15] was adjusted within the Python language for image processing. The images were converted to vector format in order to separate the three color channels (R, G, B). The blue channel (B) was used to filter the clouds from the image because it gives the best contrast between the cover of the foliage and the sky with the presence of clouds. The adaptive threshold method was used to classify the image [35].

The method consists of dividing the image into sub-images. The threshold (M) of the sub-image is calculated using the mean or median Gaussian methods. In this case, the median was used as a threshold to perform the separation. The size of the blocks used to divide the image was 200 × 200 pixels. The sum of the proportions of the number of pixels with vegetation in each block to the total number of pixels of the photograph was the cover of the canopy per photograph. Figure 4 shows an outline of the threshold calculation using this method. The overstory cover includes branches and the upper stems of trees.

Figure 4. Outline of thresholds in blocks by the adaptive threshold method; the left part of the histogram corresponds to pixels with vegetation, the right side of the histogram corresponds to pixels without vegetation. The *M* value is the threshold for each block.

2.6.2. Estimation of Understory Cover

The classification of green vegetation and soil was achieved by calculating a threshold within a two-dimensional space. The images were transformed in the color space L*a*b* [10]. The green-red component a* was used to distinguish the vegetation from the bare soil, where the values skewed to the left of the histogram indicated green pixels (vegetation) and those skewed to the right showed pixels in red (bare soil). The assumption of the methodology is that the distribution of this component tends to be a bimodal Gaussian distribution.

$$F(x) = \frac{W_1}{\sqrt{2\pi}\sigma_1} e^{\left(\frac{(x-\mu_1)^2}{2\sigma_1^2}\right)} + \frac{W_2}{\sqrt{2\pi}\sigma_2} e^{\left(\frac{-(x-\mu_2)^2}{2\sigma_2^2}\right)} \tag{5}$$

where μ_1 and μ_2 are the green vegetation and soil average, respectively; and σ_1 and σ_2 are the standard deviations of green vegetation and soil, respectively. The value W_1 is a weighting of the pixels in green and W_2 is the respective weighting for soil. The image is scaled to values of 0–255.

Threshold Adjustment

Regardless of the land use, the value of the pixels is between 75 and 150 in all photographs; to make an optimal adjustment, an initial threshold was set, which was obtained in the middle of the range 75–150 ($T0 = 112$). The optimal value of the threshold ($T1$) occurred when the functions of Equation (5) were equal (Figure 5). In this case, the error of omission of vegetation and soil classification, represented by areas S1 and S2, is minimal. The following equation was used to solve $T1$:

$$AT1^2 + BT1 + C = 0 \tag{6}$$

where:

$$
\begin{aligned}
A &= \sigma_1^2 - \sigma_2^2 \\
B &= 2 \times \left(\mu_1\sigma_2^2 - \mu_2\sigma_1^2\right) \\
C &= \sigma_1^2\mu_2 - \mu_1\sigma_2^2 + 2\sigma_1^2 2\sigma_2^2 ln(\sigma_2 W1/\sigma_1 W2)
\end{aligned}
\tag{7}
$$

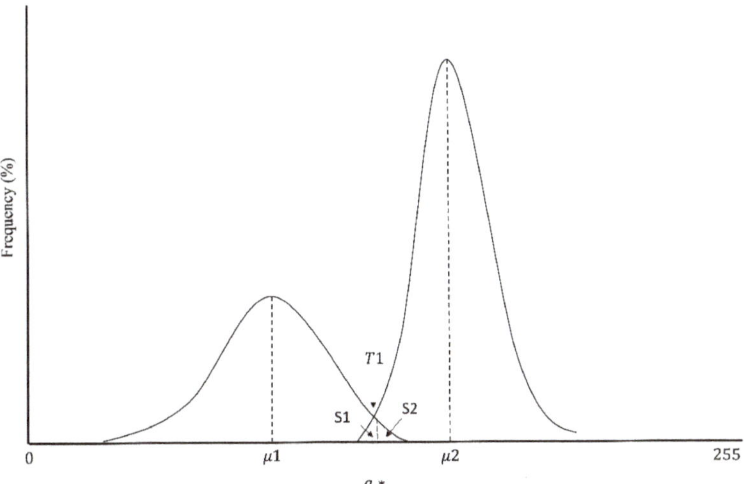

Figure 5. Identification of the optimal $T1$ threshold in a bimodal distribution.

In extreme situations where bimodality is not evident, that is, photographs where there is only vegetation or bare soil, we applied the algorithm proposed by Liu et al. [10].

2.6.3. Calculation of Total Vegetation Cover

Total vegetation cover (*TVC*) was calculated as follows (Figure 6):

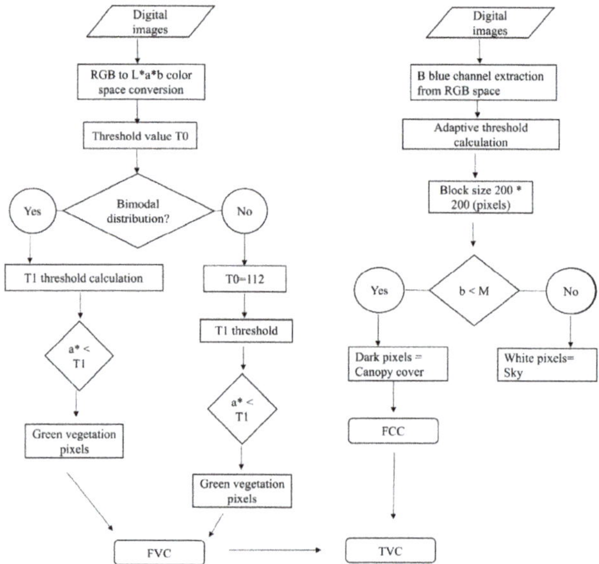

Figure 6. Flowchart for estimating the cover of vegetation vertical strata at sampling sites in the State of Mexico.

$$TVC = \frac{\sum_{i=0}^{n} FCC}{NPT} + \left(1 - \frac{\sum_{i=0}^{n} FCC}{NPT}\right) \times \left(\frac{\sum_{i=0}^{n} FCV}{NPT}\right) \tag{8}$$

where *FCC* is the proportion of the number of pixels classified as aerial (overstory) cover, *FCV* is the fraction of the vegetative cover of the understory (lower stratum), and *NPT* is the total number of pixels contained in the image. The sum of *TVC* in all images of the CBSP design was considered the total cover per plot. Figure 6 presents the flowchart of the process of classification.

2.6.4. Accuracy in Cover Classification

The accuracy of the cover estimates obtained using the proposed methodology was calculated through a comparison of these values using a visual classification of the images within the ENVI 5.0[RM] program. We considered two classes to distinguish the colors in the photographs. In understory, all pixels in green were considered as leaves, and the rest were classified as bare soil. In overgrowth, all pixels corresponding to leaves, stems, and branches were classified as cover, and the rest of the pixels were classified as sky. As mentioned in [11], visual classification is considered as the real values of cover in the image and those are compared with the automated threshold proposed in this research.

Images of 12 zenith plots (252 images) and 11 plots to nadir (231 images) were used. The plots represented different land uses. The accuracy of the implemented classifier (*AC*) was evaluated using the following formula [11].

$$AC = 100 \times \left(1 - \frac{\sum_{i=1}^{n} \left|\frac{A-B}{A}\right|}{N}\right) \tag{9}$$

where A represents the number of pixels with a real presence of vegetation in the reference image (visual classification) and B represents the number of pixels classified as having vegetation in the applied methods. An average accuracy was obtained in each plot evaluated, where 100% corresponds to a classification without errors.

3. Results

3.1. Sampling Design

The observed (total area of projected crowns) and estimated (projected photographic captured area) values of the projected cover (%) per type of sampling design and spatial pattern of the trees are shown in Table 1. In the pattern analysis of the trees, 10 plots corresponded to a grouped (clustered) pattern, 15 to a random pattern, and five plots belonged to a dispersed cluster. The calculation of the estimated area is explained in Equation (1).

Table 1. Estimated and observed cover values (%) per type of sampling and spatial pattern of trees.

Design	Spatial Pattern	Observed (%)	Estimated (%)	MAE	Coefficient of Variation
	Grouped	87.1	80.9	0.175	0.24
RM	Random	64.0	58.1	0.173	0.34
	Dispersed	34.9	29.0	0.171	0.06
	Grouped	87.1	85.3	0.091	0.09
CBSP	Random	64.0	62.5	0.097	0.33
	Dispersed	34.9	32.9	0.089	0.16
	Grouped	87.1	86.9	0.079	0.22
SLAT	Random	64.0	62.2	0.119	0.38
	Dispersed	34.9	33.7	0.117	0.20
	Grouped	87.1	34.2	0.153	0.23
VALERI	Random	64.0	31.7	0.179	0.41
	Dispersed	34.9	29.2	0.155	0.22

RM: regular mesh; CBSP: Biomass and carbon sampling plots; SLAT: tree and land use sampling; VALERI: Remote sensing ground validation instrument.

The CBSP design showed the least error in two of the three types of spatial patterns. The second design that showed minor error was SLAT, which indicates that the sampling design with photographic captures in diagonal form exhibits better results. The RM and VALERI designs had the highest and lowest number of samples, respectively. However, their errors were the highest (MAE > 0.15). These results practically discard them from being considered in operational sampling.

The random spatial pattern showed a higher coefficient of variation (CV) in the four sampling designs due to the design geometry. The dispersed pattern was the second one with the highest CV in the CBSP, SLAT, and VALERI designs. Within the grouped pattern, the CBSP sample recorded the lowest variation.

3.2. Segmentation of Images

The use of the Python program allowed us to make the segmentation threshold selection consistent. The number of captured photos in the sampling makes it impractical to analyze the photographs in a supervised way or in semi-automated processes (photo by photo). The developed program classifies a sampling plot of 42 photographs at zenith and nadir in 30 s. The processor used has 2.6 GHz and 8 GB of memory.

3.3. Classification of Overstory Images

Figure 7 shows the classification of zenith images in four cover conditions. Figure 7a and its classification 7b represent the photograph with the highest cover in the whole 97% sampling (Oyamel fir forest). Figure 7c, d represent 50% cover (Oyamel fir forest secondary vegetation). Figure 7g shows how the classifier correctly discriminates foliage from clouds (secondary vegetation of Pine Forest). Finally, Figure 7i presents a correct classification with minimum cover (secondary vegetation of Pine Forest).

Figure 7. Classification of zenith images, with different cover conditions. Images (**a,c,f,h**) correspond to the original photograph. Images (**b,d,g,i**) correspond to the adaptive threshold classifier.

3.4. Classification of Undestory Images

Figure 8 presents a sample of four photographs in different land use and different illumination and vegetation cover conditions, as well as their respective classification and histogram. Figure 8a,b are presented within a plot of land with secondary vegetation of Oyamel fir forest; the threshold in this capture was a* = 119 and the area under the curve with vegetation (V) was 43%. Figure 8c,d correspond to herbaceous vegetation within an Oyamel fir forest; in this particular vegetation the illumination was reduced because of the high overstory cover, with an understory cover of 62%.

Figure 8f,g show the image within a Rainfed Agriculture area and we observed that the classifier adequately discriminated between bare soil vegetation and shadows; the threshold in this image was a* = 122 and the vegetation cover was 35%. Figure 8i,h are presented within a plot without vegetation; the classifier was able to detect the minimal green cover found in the photograph. In this case, as the distribution of the histogram was unimodal, the threshold was set at a* < 105 and the vegetation cover was 0.8%.

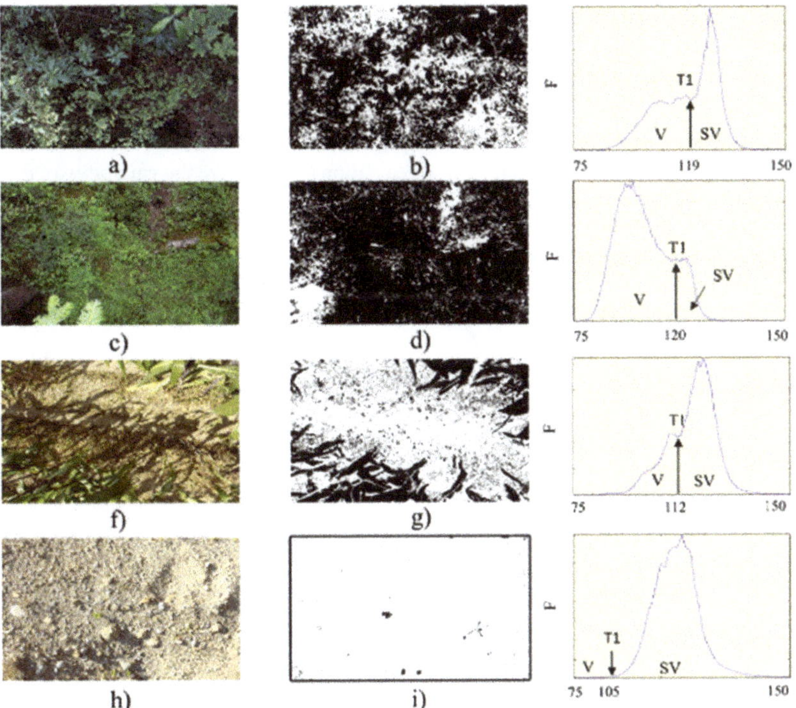

Figure 8. Classification of images in different land uses to nadir. Images (**a,c,f,h**) correspond to the original photograph. Images (**b,d,g,i**) correspond to the classified images. The third column shows the distribution of the histogram according to the cover.

3.5. Accuracy of the Classification

Table 2 presents the comparison of the classified cover (%) by the supervised classification method (visual classification) and the zenith method (Estimated). The accuracy was high in all land uses with an average of 93%. In relation to the coefficient of variation CV (representing the variability of the sampling design), we observed that this variability increased as the estimated cover of the different land uses declined. In primary forests (BQP, BQ, BA, BPQ), the variation of the sampling design was low; insofar as there was disturbance in the land use (VSa, VSA, VSh) the variation increased because the static designs were sensitive to the opening of the canopy.

Table 2. Accuracy evaluation of the visual interpretation of the images in 12 zenith sampling plots (overstory cover).

Land Use	Description	Classified (%)	Estimated (%)	AC	CV
BQP	Oak-pine forest	84.0	83.0	99.0	0.06
BQ	Oak forest	86.0	84.0	96.0	0.09
BA	Oyamel fir forest	82.0	78.0	94.0	0.10
BPQ	Pine-oak forest	74.0	75.0	96.0	0.12
VSa/BQ	Secondary shrub vegetation of oak forest	52.0	48.0	90.0	0.13
VSA/BPQ	Secondary arboreal vegetation of pine-oak forest	69.0	67.0	94.0	0.17
VSa/BQP	Secondary shrub vegetation of oak-pine forest	71.0	69.0	96.0	0.21
VSa/BQP	Secondary shrub vegetation of oak-pine forest	50.0	49.0	96.0	0.22
VSA/BA	Secondary arboreal vegetation of oyamel fir forest	68.0	62.0	91.0	0.24
VSA/BP	Secondary arboreal vegetation of pine forest	22.0	22.0	91.0	0.28
VSh/BP	Secondary herbaceous vegetation of pine forest	16.0	15.0	92.0	0.66
VSh/BPQ	Secondary herbaceous vegetation of pine-oak forest	31.0	35.0	87.0	0.68

Table 3 presents the evaluation of the vegetative cover to nadir. The average accuracy was 94%. As in zenith, the estimated cover maintains a negative correlation with CV. For this classification, the greatest cover is found in secondary herbaceous vegetation (VSh) where the variability of the sampling is lower; as the vegetation transits to mature forest, the CV is high due to the fact that the understory cover is random. In sites with non-vascular vegetation (bryophytes), the classifier overestimates the percentage of vascular vegetation because it associates the green color with this type of cover.

Table 3. Accuracy evaluation of the visual interpretation of the images in 12 sampling plots at nadir (understory cover).

Land Use	Description	Classified (%)	Estimated (%)	AC	CV
VSh/BPQ	Secondary herbaceous vegetation of pine-oak forest	82.10	85.00	89.0	0.06
VSh/BP	Secondary herbaceous vegetation of pine forest	78.43	79.00	96.0	0.10
TA	Rain-fed agriculture	27.37	28.00	90.0	0.16
VSh/BA	Secondary herbaceous vegetation of oyamel fir forest	84.00	83.50	98.0	0.19
VSa/BQ	Secondary shrub vegetation of oak forest	32.70	32.00	94.0	0.51
PC	Cultivated grassland	47.23	49.33	97.0	0.84
BP	Pine forest	29.30	27.67	95.0	0.86
VSA/BPQ	Secondary arboreal vegetation of pine-oak forest	37.65	34.42	92.0	0.87
BQP	Pine-oak forest	30.69	29.87	96.0	0.88
BQ	Oak forest	6.58	6.00	94.0	0.94
VSa/BQP	Secondary shrub vegetation of oak-pine forest	5.63	6.95	87.0	0.98

4. Application of CBSP Design to the Regional Survey

After validating that the CBSP design was robust and precise enough to be used in a regional survey, it was implemented in the campaign through a replication of the procedure used in the pre-survey evaluation in the 754 sampling plots.

It is identified that the application of the sampling design allowed us to capture photographs in an easy and fast way in the majority of land uses.

Figure 9 shows the average total vegetation cover of the main land uses in the regions of the State of Mexico. The cover data are presented from highest to lowest and from mature forest to agriculture. Mature forests have a tendency of 50–90% cover in the eight regions, and there was higher average coverage in the region of Toluca (70–90%) (Figure 9a); in wooded areas with secondary tree vegetation (VSA) and secondary shrub and herbaceous vegetation (VSa and Vsh), the cover ranges from 20 to 90%. This is because the limit of tree vegetation in these plots is a mixture of perennial and deciduous cover, therefore presenting a seasonal change of vegetation cover because of the weather pattern. In this case, the region with the highest coverage was Zumpango (Figure 9b) and that with the lowest coverage was Coatepec (Figure 9f).

With regard to cover in agricultural areas (RA, TA), the development of cover over time follows a spatial pattern associated with the time of planting and growth. Cover starts at <10% in all regions and the percentage increases up to 100%, as in the Toluca and Texcoco regions (Figure 9a,c).

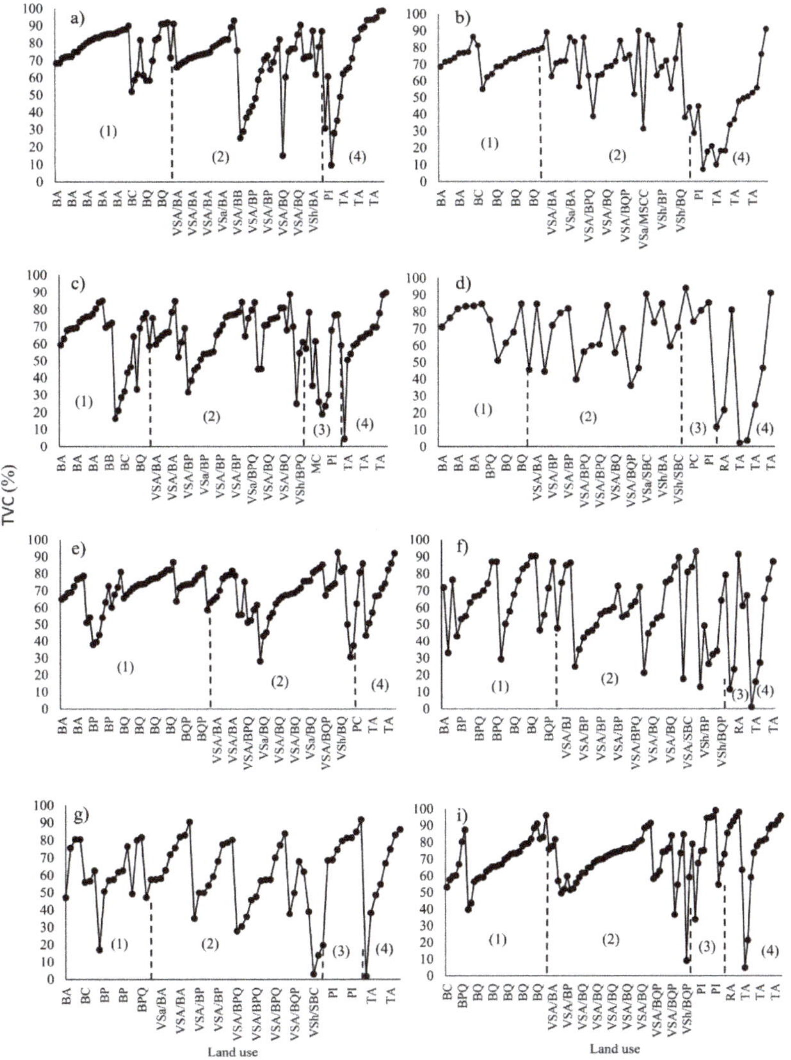

Figure 9. Average total vegetation cover by land use in the eight regions of the State of Mexico: (**a**) Toluca Region; (**b**) Zumpango Region; (**c**) Texcoco Region; (**d**) Tejupilco Region; (**e**) Atlacomulco Region; (**f**) Coatepec Region; (**g**) Valle de Bravo Region; (**i**) Jilotepec Region. (1) Mature forest; (2) disturbed forest; (3) grassland; (4) agricultural area.

5. Discussion

The use of digital photography for vegetation cover estimations is an easy, low-cost, and potentially suitable approach for hard-to-reach places. These properties give this method an advantage over direct (destructive) and indirect (fisheye cameras) methods [13]. In Mexico, operational surveys such as the National Forest and Soils Survey [33] generate vegetation cover estimations that rely on the technician criteria in the field. This research proposes an accurate survey design method which is potentially suitable for the forest sector in the country.

The advantage of using field data is its strength for the validation of satellite information, as a way to use the cover values at greater scales [6]. On the other hand, a disadvantage of field photography data is the requirement of several photographs in order to produce a reliable estimate. However, there are software methods for the automation of these processes, such as the one proposed in this research.

5.1. Sampling Designs Comparison

One important consideration in field vegetation cover measurements is the need to determine an appropriate sampling design [31]. The methodologies for measuring vegetation cover are ambiguous in reference to which method should be used to ensure a correct estimation within a sampling plot. The projection made in the GIS allowed us to observe differences in vegetation cover estimations with a simple scheme. The results provide important information for choosing a sampling design and reducing the costs of the collection and processing of field data [22], considering the large amount of samples of this survey.

Martens et al. [36] show that the spatial patterns influence the height, cover, and distribution of vegetation in their different strata. Our results showed that, independently of the spatial pattern of the survey sites, sampling designs that captured diagonal photographs (CBSP and SLAT) exhibited the least estimation error.

The advantages of these designs are the low number of photographs needed (42 phothographs) and their easy field implementation. This ratifies the vegetation cover estimation operability using this method and how it can be related to grouping indices to evaluate the effect of the forest management of zones without disturbance on disturbed zones [24].

5.2. Sampling Survey

Forest surveys in several parts of the world, including Mexico, are carried out in circular plots of 1000 m² (17.85 m radius) [22]. In this research, we adopted this design to evaluate the biomass and carbon storage in different land uses within the State of Mexico. The CBSP design proved to be feasible for its implementation in different land uses and spatial patterns of the trees; this simple design allowed the application in distant sampling points and rugged terrain, which makes it an operative method to capture vegetation cover with digital photographs.

An example of the sampling operability is that bubble levels were used to stabilize the camera at each sampling point, instead of using tripods to fix it. This technique helped to reduce the time to take the photographs and presented no considerable error when compared to tripod shots [37].

The number of samples and their arrangement was another variable to be considered for the sampling to be operative and related to other variables measured in the plot (i.e., biomass, carbon), so captures were fixed at the ends of the sites [20]. The CBSP design obtained the best results when estimating the vegetation cover. The efficiency of its application and smaller errors in the cover estimation turns it into a practical design for this type of application, besides saving storage and time when processing the images.

The spatial distribution of trees within the site is an important element for the dynamics of the forest ecosystem [26]. In the case of overstory cover, we observed that the applied sampling design showed a negative trend between the estimated cover and its coefficient of variation (CV). The primary forests presented high and compact canopy covers (low CV); when reducing it, the canopy cover tends to be dispersed (high CV). In the case of understory cover, the opposite occurs. In disturbed areas (secondary vegetation), the cover is larger and compact; when this cover is reduced, the pattern is dispersed.

The real area of a photograph is another important aspect that has been little explored in vegetation cover sampling design. Researchers generally use a fixed lens viewing angle (35–40°) to estimate the canopy cover, so a greater opening angle would be measuring the closure of the canopy. Jennings et al. [38] describe in detail the difference between these two concepts. In this study, we observed that in real situations the height of the trees in a sampling site is heterogeneous. For this reason, we proposed to adjust the focal length of the camera at the point of capture by a constant as this ensures that one is able to approximate and make repeatable the capture area independently of the architecture of the trees.

5.3. Automated Classification of Images

The automation of vegetative cover classification is a process that avoids the error of the human component and ensures the possibility of reproducing the results [11]. The efforts to automate image classification have focused on programs such as MATLAB [12,15,39], WinScanopy [40], or Photoshop [41], among others. But although the former meets the requirement of making the batch processing of the images operative, it has the disadvantage of having an additional cost. The other programs have the disadvantage of processing the images photo by photo, which discarded them for the analysis of cover in this study. The Python^{RM} program was chosen for the versatility of the specialized libraries available, and for being a free access program. The written code enabled us to process in batch the images of the sampling in a time similar to that described in programs like MATLAB.

5.3.1. Overstory Cover

In the classification of digital photographs, binary methods (global thresholds) are generally used to estimate canopy cover [42]. The colors of the classified images have gray tonalities for vegetation and white for the sky. According to Chityala and Pudipeddi [35], the accuracy in the classification for the global threshold methods is low. The problem is trying to find the maximum variance between two logical groups of segments within the whole image, which can cause confusion in the classification by overexposure in the camera or image capture at inappropriate times. In this research, we propose the use of the adaptive threshold in the blue space of the image [15]. The method is based on the same principles of the global threshold, but the segmentation statistics are done at the block level within the image, allowing greater accuracy in the overall classification.

The accuracy of the classification is high and related to the comparison of binary algorithms performed by Ghatthorn and Beckschäfer [42]. The sampling was directed to avoid the effect of the sun (the captures were made before noon and at near sunset); however, in the sampling, there were circumstances that caused the photographs to exceed the proposed range, such as the VSh/BPQ land use plot (Table 2), which showed the lowest precision (87%) [16]. Nonetheless, the development of methodologies applied to this problem are beyond the scope of this research.

5.3.2. Understory Cover

There are many methods to extract the vegetation cover fraction [11,12,43] so the degree of accuracy is an important factor in the efficiency of field measurements. For example, supervised classification has a high accuracy but low efficiency, while unsupervised classification has a high efficiency but low precision due to commission and omission errors [1].

The algorithm proposed by Liu et al. [10] has the property of being simple, easy to automate, and has a high degree of precision. When comparing it with the supervised classification in different sampling plots, the results in terms of precision were high (>87%); the main problem of the misclassification was the confusion of the vascular vegetation with bryophytes, which in the color space *a detects a green color that is difficult to discern. In conditions of low illumination due to the effect of the canopy, the algorithm had no problems in correctly classifying vegetation and shade [12], and with a single component predomination (bare soil) in the photograph, the classifier generated good results (Figure 8i).

The two methods proposed to evaluate the over and understory vegetation cover were robust and replicable in all sampling plots. The reason for estimating the foliar projective cover rather than the leaf area index is that the former is a more adequate variable to characterize vegetation [44] since the projective foliar cover captured with digital photographs contains information on individual plants and their spatial distribution. With this perspective and considering the validation of the proposed sampling and plot area, we will contemplate the validation of biophysical variables calculated with remote sensors in future work of the research group.

6. Conclusions

The over and understory vegetation cover was estimated with digital photographs in sampling plots of the State of Mexico. The high efficiency and precision of the classification methods indicate that they are robust for discerning vegetation in different land uses and illumination conditions.

The use of digital photography reduces the ambiguity of vegetation cover estimations in regional and national surveys. The proposed method is easily reproducible in heterogeneous land and vegetation conditions.

The automation of the process using a free programming language avoided human errors and ensured the reproducibility of the results at a low cost.

The sampling methods using the capture of diagonal-angled photographs were the best way to obtain less biased information when taking digital photographs at a circular sampling site. The simulation showed that the CBSP design has a smaller error when considering the spatial distribution of trees within the sampling site. Its easy field management, the number of photographs per site, and its precision make it an operative design. One additional advantage of the proposed field survey is that the real area covered by the photograph is independent of the height of trees. This guarantees representability and avoids image superposition in the sampling site.

Mature forests have a high and compact overstory vegetation cover, which tends to be reduced in secondary forests. The greater cover of the understory is found in secondary forests, where it is denser. The cover of the undergrowth declines in mature forests.

The application of this method for regional and national surveys is recommended.

Acknowledgments: We appreciate the financial support given by the Programa Mexicano del Carbono and Protectora de Bosques del Estado de México (PROBOSQUE) which allowed us to conduct this study.

Author Contributions: Víctor Salas-Aguilar and Fernando Paz-Pellat contributed to the initial proposal of the methodology. Víctor Salas-Aguilar designed and developed the software to process the data in python for implementation on a larger scale. Fabiola Rojas-García, Cristóbal Sánchez-Sánchez, J. René Valdez-Lazalde, and Carmelo Pinedo-Alvarez conducted the research methods. All authors discussed the structure and commented on the manuscript at all stages.

Conflicts of Interest: The authors declare no conflict of interest.

References

1. Liang, S.; Li, X.; Wang, J. *Advanced Remote Sensing: Terrestrial Information Extraction and Applications*; Academic Press: Cambridge, MA, USA, 2012.
2. Gutman, G.; Ignatov, A. The derivation of the green vegetation fraction from NOAA/AVHRR data for use in numerical weather prediction models. *Int. J. Remote Sens.* **1998**, *19*, 1533–1543. [CrossRef]
3. Li, Y.; Wang, H.; Li, X.B. Fractional vegetation cover estimation based on an improved selective endmember spectral mixture model. *PLoS ONE* **2015**, *10*, e0124608. [CrossRef] [PubMed]
4. Mu, X.; Hu, M.; Song, W.; Ruan, G.; Ge, Y.; Wang, J.; Huang, S.; Yan, G. Evaluation of sampling methods for validation of remotely sensed fractional vegetation cover. *Remote Sens.* **2015**, *7*, 16164–16182. [CrossRef]
5. Li, F.; Chen, W.; Zeng, Y.; Zhao, Q.; Wu, B. Improving estimates of grassland fractional vegetation cover based on a pixel dichotomy model: A case study in Inner Mongolia, China. *Remote Sens.* **2014**, *6*, 4705–4722. [CrossRef]
6. Mu, X.; Huang, S.; Ren, H.; Yan, G.; Song, W.; Ruan, G. Validating GEOV1 fractional vegetation cover derived from coarse-resolution remote sensing images over croplands. *IEEE J. Sel. Top. Appl. Earth Observ. Remote Sens.* **2015**, *8*, 439–446. [CrossRef]
7. Zhou, Q.; Robson, M.; Pilesjo, P. On the ground estimation of vegetation cover in Australian rangelands. *Int. J. Remote Sens.* **1998**, *19*, 1815–1820. [CrossRef]

8. Chianucci, F.; Cutini, A. Estimation of canopy properties in deciduous forests with digital hemispherical and cover photography. *Agric. For. Meteorol.* **2013**, *168*, 130–139. [CrossRef]

9. White, M.A.; Asner, G.P.; Nemani, R.R.; Privette, J.L.; Running, S.W. Measuring fractional cover and leaf area index in arid ecosystems: Digital camera, radiation transmittance, and laser altimetry methods. *Remote Sens. Environ.* **2000**, *74*, 45–57. [CrossRef]

10. Liu, Y.; Mu, X.; Wang, H.; Yan, G. A novel method for extracting green fractional vegetation cover from digital images. *J. Veg. Sci.* **2012**, *23*, 406–418. [CrossRef]

11. Coy, A.; Rankine, D.; Taylor, M.; Nielsen, D.C.; Cohen, J. Increasing the accuracy and automation of fractional vegetation cover estimation from digital photographs. *Remote Sens.* **2016**, *8*, 474. [CrossRef]

12. Song, W.; Mu, X.; Yan, G.; Huang, S. Extracting the green fractional vegetation cover from digital images using a shadow-resistant algorithm (SHAR-LABFVC). *Remote Sens.* **2015**, *7*, 10425–10443. [CrossRef]

13. Macfarlane, C.; Hoffman, M.; Eamus, D.; Kerp, N.; Higginson, S.; McMurtrie, R.; Adams, M. Estimation of leaf area index in eucalypt forest using digital photography. *Agric. For. Meteorol.* **2007**, *143*, 176–188. [CrossRef]

14. Chianucci, F.; Chiavetta, U.; Cutini, A. The estimation of canopy attributes from digital cover photography by two different image analysis methods. *iFor. Biogeosci. For.* **2014**, *7*, 255–259. [CrossRef]

15. Fuentes, S.; Palmer, A.R.; Taylor, D.; Zeppel, M.; Whitley, R.; Eamus, D. An automated procedure for estimating the leaf area index (LAI) of woodland ecosystems using digital imagery, MATLAB programming and its application to an examination of the relationship between remotely sensed and field measurements of LAI. *Funct. Plant Biol.* **2008**, *35*, 1070–1079. [CrossRef]

16. Macfarlane, C. Classification method of mixed pixels does not affect canopy metrics from digital images of forest overstorey. *Agric. For. Meteorol.* **2011**, *151*, 833–840. [CrossRef]

17. Tausch, R.; Tueller, P. Foliage biomass and cover relationships between tree-and shrub-dominated communities in pinyon-juniper woodlands. *Great Basin Nat.* **1990**, *50*, 121–134.

18. Muukkonen, P.; Mäkipää, R. Empirical biomass models of understorey vegetation in boreal forests according to stand and site attributes. *Boreal Environ. Res.* **2006**, *11*, 355–369.

19. Luna, J.A.N.; Hernández, E.H. Relaciones morfométricas de un bosque coetáneo de la región de El Salto, Durango. *Ra Ximhai* **2008**, *4*, 69–82.

20. Williams, M.S.; Patterson, P.L.; Mowrer, H.T. Comparison of ground sampling methods for estimating canopy cover. *For. Sci.* **2003**, *49*, 235–246.

21. Muir, J.; Schmidt, M.; Tindall, D.; Trevithick, R.; Scarth, P.; Stewart, J. *Field Measurement of Fractional Ground Cover: A Technical Handbook Supporting Ground Cover Monitoring for Australia*; Australian Bureau of Agricultural and Resource Economics and Sciences (ABARES): Canberra, Australia, 2011.

22. Matern, B. *Recopilación de Notas Sobre Técnicas de Muestreo Usadas en Inventarios Forestales*; SARH-INIFAP Pub. Especial: Distrito Federal, Mexico, 1993.

23. Gobron, N.; Verstraete, M. Remote sensing and geoinformation processing in the assessment and monitoring land degradation and desertification state of art and operational perspectives. In *Assessment of the Status of the Development of the Standards for the Terrestrial Essential Climate Variables: Fraction of Absorbed Photosynthetically Active Radiation (FAPAR)*; GTOS Secretariat Food and Agriculture Organization of the United Nation (FAO): Rome, Italy, 23 April 2009.

24. Corral-Rivas, J.J.; Wehenkel, C.; Castellanos-Bocaz, H.A.; Vargas-Larreta, B.; Diéguez-Aranda, U. A permutation test of spatial randomness: Application to nearest neighbour indices in forest stands. *J. For. Res.* **2010**, *15*, 218–225. [CrossRef]

25. Hemery, G.; Savill, P.; Pryor, S. Applications of the crown diameter–stem diameter relationship for different species of broadleaved trees. *For. Ecol. Manag.* **2005**, *215*, 285–294. [CrossRef]

26. LeMay, V.; Maedel, J.; Coops, N.C. Estimating stand structural details using nearest neighbor analyses to link ground data, forest cover maps, and Landsat imagery. *Remote Sens. Environ.* **2008**, *112*, 2578–2591. [CrossRef]

27. Shaw, J.D. *Models for Estimation and Simulation of Crown and Canopy Cover*; General Technical Report (GTR); US Forest Service: Washington, DC, USA, 2005.

28. Ceballos, G.; List, R.; Garduño, G.; López, R.; Muñozcano, M.; Collado, E.; San Román, J. *La Diversidad Biológica del Estado de México*; Estudio de Estado; Biblioteca Mexiquense del Bicentenario: Ventura, CA, USA, 2009.

29. Programa Mexicano del Carbono (PMC). Manual de Procedimientos Inventario de Carbono+. In *Estudio de Factibilidad Técnica Para el Pago de Bonos de Carbono en el Estado de México*; Programa Mexicano del Carbono: Texcoco, Mexico, 2015; p. 69.

30. INEGI Datos Vectoriales Escala 1:250,000 de Uso de Suelo y Vegetación. Available online: http://www.inegi.org.mx/go/contends/recant/mussel/ (accessed on 24 May 2017).

31. Baret, F.; Weiss, M.; Allard, D.; Garrigues, S.; Leroy, M.; Jeanjean, H.; Fernandes, R.; Myneni, R.; Privette, J.; Morisette, J. VALERI: A network of sites and a methodology for the validation of medium spatial resolution land satellite products. *Remote Sens. Environ.* **2005**, *76*, 36–39.

32. Kuhnell, C.A.; Goulevitch, B.M.; Danaher, T.J.; Harris, D.P. Mapping woody vegetation cover over the state of Queensland using Landsat TM imagery. In Proceedings of the 9th Australasian Remote Sensing and Photogrammetry Conference, Sydney, Australia, 24 July 1998; pp. 20–24.

33. CONAFOR (Comisión Nacional Forestal). *Inventario Nacional Forestal y de Suelos Informe de Resultados 2004–2009 National Forest and Soils Survey, Results Report 2004–2009*; CONAFOR: Zapopan, Mexico, 2012; p. 212.

34. Clark, P.J.; Evans, F.C. Distance to nearest neighbor as a measure of spatial relationships in populations. *Ecology* **1954**, *35*, 445–453. [CrossRef]

35. Chityala, R.; Pudipeddi, S. *Image Processing and Acquisition Using Python*; CRC Press: Boca Raton, FL, USA, 2014.

36. Martens, S.N.; Breshears, D.D.; Meyer, C.W. Spatial distributions of understory light along the grassland/forest continuum: Effects of cover, height, and spatial pattern of tree canopies. *Ecol. Model.* **2000**, *126*, 79–93. [CrossRef]

37. Origo, N.; Calders, K.; Nightingale, J.; Disney, M. Influence of levelling technique on the retrieval of canopy structural parameters from digital hemispherical photography. *Agric. For. Meteorol.* **2017**, *237*, 143–149. [CrossRef]

38. Jennings, S.; Brown, N.; Sheil, D. Assessing forest canopies and understorey illumination: Canopy closure, canopy cover and other measures. *Forestry* **1999**, *72*, 59–74. [CrossRef]

39. Korhonen, L.; Heikkinen, J. Automated analysis of in situ canopy images for the estimation of forest canopy cover. *For. Sci.* **2009**, *55*, 323–334.

40. Pekin, B.; Macfarlane, C. Measurement of crown cover and leaf area index using digital cover photography and its application to remote sensing. *Remote Sens.* **2009**, *1*, 1298–1320. [CrossRef]

41. Lee, K.-J.; Lee, B.-W. Estimating canopy cover from color digital camera image of rice field. *J. Crop Sci. Biotechnol.* **2011**, *14*, 151–155. [CrossRef]

42. Glatthorn, J.; Beckschäfer, P. Standardizing the protocol for hemispherical photographs: Accuracy assessment of binarization algorithms. *PLoS ONE* **2014**, *9*, e111924. [CrossRef] [PubMed]

43. Liu, J.; Pattey, E. Retrieval of leaf area index from top-of-canopy digital photography over agricultural crops. *Agric. For. Meteorol.* **2010**, *150*, 1485–1490. [CrossRef]

44. Poblete-Echeverría, C.; Fuentes, S.; Ortega-Farias, S.; Gonzalez-Talice, J.; Yuri, J.A. Digital cover photography for estimating leaf area index (LAI) in apple trees using a variable light extinction coefficient. *Sensors* **2015**, *15*, 2860–2872. [CrossRef] [PubMed]

Article

Development of a GPS Forest Signal Absorption Coefficient Index

William Wright [1,*], Benjamin Wilkinson [2] and Wendell Cropper Jr. [2]

[1] Department of Geography and Environmental Engineering, United States Military Academy, West Point, NY 10996, USA

[2] School of Forest Resources and Conservation, University of Florida, Gainesville, FL 32611, USA; benew@ufl.edu (B.W.); wcropper@ufl.edu (W.C.J.)

* Correspondence: william.wright@usma.edu; Tel.: +1-845-938-2063

Received: 14 March 2018; Accepted: 20 April 2018; Published: 25 April 2018

Abstract: In this paper GPS (Global Positioning System)-based methods to measure L-band GPS Signal-to-Noise ratios (SNRs) through different forest canopy conditions are presented. Hemispherical sky-oriented photos (HSOPs) along with GPS receivers are used. Simultaneous GPS observations are collected with one receiver in the open and three inside a forest. Comparing the GPS SNRs observed in the forest to those observed in the open allows for a rapid determination of signal loss. This study includes data from 15 forests and includes two forests with inter-seasonal data. The Signal-to-Noise Ratio Atmospheric Model, Canopy Closure Predictive Model (CCPM), Signal-to-Noise Ratio Forest Index Model (SFIM), and Simplified Signal-to-Noise Ratio Forest Index Model (SSFIM) are presented, along with their corresponding adjusted R^2 and Root Mean Square Error (RMSE). As predicted by the CCPM, signals are influenced greatly by the angle of the GPS from the horizon and canopy closure. The results support the use of the CCPM for individual forests but suggest that an initial calibration is needed for a location and time of year due to different absorption characteristics. The results of the SFIM and SSFIM provide an understanding of how different forests attenuate signals and insights into the factors that influence signal absorption.

Keywords: canopy closure; global positioning system; hemispherical sky-oriented photo; signal attenuation; geographic information system

---

1. Introduction

The Global Positioning System (GPS) constellation is primarily used for position, navigation, and timing purposes. However, the scientific community has used the signals transmitted from GPS satellite vehicles (SVs) for applications in many different research fields. Some GPS signal studies include GPS performance, wireless communication reliability, and the combination of GPS signal-to-noise ratios (SNRs) with light detection and ranging (LiDAR) data to measure signal loss in forests [1–5]. The L1 frequency of the GPS system broadcasts at 1575.42 MHz and is attenuated by vegetation. Developing a method to predict with confidence the degree to which GPS signals are affected by forest structure provides useful information on L-band scattering and absorption. This work is important to understanding GPS performance and to scientific studies that employ other microwave signals, such as satellite communications, air-to-ground communications, cellular phones, and synthetic aperture radar (SAR). It is also relevant to studies that explore forest growth modeling and use light interception predictions [6–12].

Both in previous studies and in conventional SAR remote sensing applications, forest vegetation is generally assumed to be uniformly-distributed stratified media [13]. This builds on Beer's Law and suggests that the zenith angle of the microwave source is a key factor governing the scattering of radio waves in a particular forest stand. However, ecologists have long recognized that forest structure is far from uniform.

In the literature, there are many different techniques used to model signal loss through forest structure. For the research presented here, the most relevant model is the Canopy Closure Predictive Model (CCPM) described in [14]. The CCPM model consists of capturing two primary components: one based on the atmospheric attenuation, and the second based on attenuation by the forest canopy based on Beer's Law. When considering Beer's Law, if we consider the forest as a uniform slab of vegetation, the absorption of a signal exhibits a linear dependence between the signal propagation path length through the media, an absorption coefficient, and the concentration of medium yielding:

$$L = \alpha dc \tag{1}$$

where L is attenuation, d is the path length, c is the concentration of the media, and α is an absorption coefficient [1]. The CCPM was developed for a managed pine forest and the Beer's Law component consists of the product of the sine of the zenith angle and the canopy closure value. The CCPM makes the assumption that the concentration and absorption parameters of Beer's Law can be combined into just the canopy closure (CC) of the forest, where CC is the percent of pixels classified as canopy in a window of interest inside a hemispherical sky-oriented photo (HSOP).

As such, there is a need to determine the concentration of the forest through the path of signal propagation [5]. Our hypothesis is that while zenith angle may be the dominant factor in attenuation, other independent parameters leading to variations in signal strength will be observable, and that the inclusion of HSOP-derived CC data at 1-degree zenith angles can precisely measure the degree to which GPS signals from individual SVs are affected by forest canopy.

The goal of this study is to estimate the values of L1-band GPS signals in multiple diverse forests using observations from GPS and CC values derived from HSOPs. The objectives are to (1) develop an atmospheric attenuation model for GPS SNR values, (2) develop an overall canopy closure predictive model (CCPM) independent of study site, and (3) create an adjustment index for each study site that can be applied to the CCPM in order to allow for refinement of predictions based on forest absorption characteristics.

2. Materials and Methods

2.1. Study Site

The data used in this study were collected in 15 different forests throughout the United States. Figure 1 depicts the location of each forest and Table 1 provides forest details, including the date of each data collection, the average and standard deviation of both the diameter at breast height (DBH) and tree height, and a brief description.

Table 1. Forest study sites and description.

ID	City Vicinity	Tree Type	Date (ddmmyy)	HT/STD (m)	DBH/STD (m)	Notes
1	West Point, NY	Oak/Hickory	110515	23.4/3.4	0.30/0.18	Military
		100% Deciduous	100815			Reservation
			241015			
			170216			
2	IMPAC	100% Pine				Managed Forest
	Gainesville, FL	Control Plot	110215/250815	5.77/1.4	0.09/0.05	Fertilization
	Gainesville, FL	Weeded Plot	110215/250815	8.21/0.64	0.12/0.04	Research plots
		Fertilized and Weeded	110215/250815	9.05/0.67	0.13/0.06	
3	Hogtown Forest	80% Deciduous	050216	20.4/2.47	0.46/0.08	Uplands Natural Mixed Forest
	Gainesville, FL	20% Coniferous				Loblolly Woods Nature Park
4	Charleston, SC	90% Pine, 10% Deciduous	230516	24.0/3.1	0.36/0.05	Francis Marion National Forest
5	Alexandria, LA	90% Pine, 10% Deciduous	190616	23.2/4.1	0.56/0.11	Kisatchie National Forest

Table 1. *Cont.*

ID	City Vicinity	Tree Type	Date (ddmmyy)	HT/STD (m)	DBH/STD (m)	Notes
6	Cold Spring, TX	80% Pine, 20% Deciduous	200616	19.5/4.7	0.52/0.13	Sam Houston National Forest
7	Georgetown, TX	Ceder Elm and Live Oak with Ash Juniper	220616	6.3/1.1	0.42/0.11	North Fork of San Gabriel River
8	Cloudcroft, NM	Ponderosa Pine	230616	23.3/3.2	0.41/0.12	Lincoln National Forest
9	Flagstaff, AZ	Ponderosa Pine	250616	19.2/6.8	0.41/0.07	
10	Guadalupe, CA	Eucalyptus	020716	28.2/3.3	0.42/0.14	
11	San Luis Obispo	Agrifolia	030716	6.9/1.5	0.22/0.09	Military Base
12	Davenport, CA	75% Redwood, 25% Douglas Fir and Tanoak	050716	54.0/6.3	1.20/0.56	California Polytechnic Research Center
13	Davenport, CA	80% Tanoak, 25% Douglas Fir	050716	18.7/1.4	0.28/0.08	California Polytechnic Research Center
14	Tahoe NF	Ponderosa Pine	070716	26.5/2.2	0.53/0.16	University of California, Berkley Sagehen Experimental Forest
15	Nederland, CO	Aspen	090716	8.4/2.6	0.20/0.06	

Note: Where HT is Tree Height, DBH is Diameter at Breast Height and STD is Standard Deviation.

Figure 1. A map showing all the forest study sites used in this research.

The vast majority of data were collected during the summer of 2015 (Table 1). Due to personnel availability constraints, weather conditions varied between each location. In each case, best efforts were made to collect data in the morning or during times with mostly-cloudy conditions to avoid sun glare on the images. During this data collection period, California had a lack of winter precipitation and was in drought conditions. In contrast, the gulf coast region had higher precipitation than usual.

2.2. GPS Signal Observations

To obtain a measurement of signal loss, GPS L1-band SNR observations were collected both in the open and inside each forest. Four Topcon Hiper Lite global navigation satellite system (GNSS) receivers were used, with three receivers set up inside each forest and one receiver positioned in an open area within 1 km of the others. Comparing SNR values observed from the GPS receiver in the open to those in the forest provides the signal attenuation observed at a specific site at a specific time. The three receivers that were set up inside each forest were positioned at random locations and recorded at least 60 min. of observations. The observations included multiple National Marine

Electronic Association (NMEA) messages at a rate of 1 Hz. The recorded messages included: time, GPS SV SNR values, GPS SV zenith angle, and the azimuth of each SV with respect to the GPS receiver. We collected data from an average of 10 GPS SVs, resulting in 36,000 observations per GPS receiver, per data collection. As such, given 19 data collections, each with four GPS receivers, the data used in this research includes over 2.5 million GPS SV observations. It is important to note that for each GPS receiver setup, we calculated the mean SNR for each SV at 1-degree increments from the horizon and used these values in the modeling process.

A control experiment was conducted in January 2015, where all four GPS receivers were set up within 20 m of each other in the open. No statistical difference between each GPS receivers' SNR observations was observed [15].

2.3. Hemispherical Sky-Oriented Photos and Image Processing

A single HSOP image was taken at each GPS receiver setup location in each forest with the camera directed straight up, and the top of the photo oriented north. The resulting photos are circular, with zenith in the center of the image and the horizon on the outer edge. An example is shown in Figure 2. The camera and lens combination used to collect these images consisted of an IPIX fisheye lens mounted on a Coolpix P6000 Nikon camera (Nikon Ltd., Tokyo, Japan).

Figure 2. A hemispherical sky-oriented photo taken during the spring data collection at West Point, New York, with the global positioning system (GPS) satellite vehicle positions (red circles) plotted inside the image.

Figure 3 shows the frequency distribution of the number of GPS SV observations recorded during the spring data collect at West Point, NY.

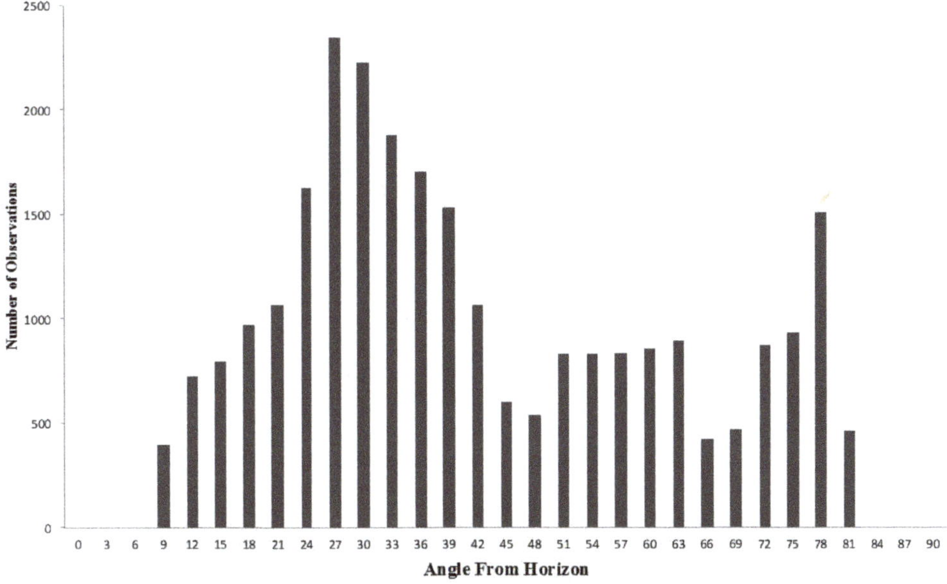

Figure 3. The frequency distribution of GPS observations at different angles from the horizon during the data collection in the spring at West Point, New York.

Image processing of the collected HSOP images was conducted using ArcGIS software (Environmental Systems Research Institute (ESRI), CA, USA). ArcGIS allows for the establishment of spatial reference, the delineation of each photo into one-degree rings associated with each angle from the horizon, and the ability to convert each image into a binary, black and white, image, where the sky is white and forest structure black. Tools within ArcGIS allow for the isolation of the blue channel of each HSOP for the creation of the binary image. This is beneficial because the blue channel is better suited to distinguish clouds and sun glare [16–21] than the red and green channels. When creating the binary image, the Natural Breaks function was used to determine appropriate threshold values. Additionally, each histogram and binary image was visually inspected for accuracy. During this process, each Red, Green and Blue (RGB)histogram and corresponding open-sky threshold values were examined to ensure there were no abnormalities. The resulting binary images were then compared to the original images to ensure proper classification. Figures 4 and 5 are the resulting binary HSOPs from the Intensive Management Practice Assessment Center (IMPAC), a managed forest in Gainesville, Florida, during needle minimum and needle maximum. The three plots in each of these forests were unique in that the species and spacing of the trees were the same. The difference between the plots resulted in different fertilization processes resulting in different DBH and tree heights between the plots.

During image processing, the percentage of pixels classified as canopy at specific angles from the horizon inside each specific forest was calculated. These CC values serve as the concentration of forest media at specific angles inside the forest. Instrumental to our modeling process is the calculation of CC fractions for each angle from zenith. This was conducted using the zonal statistics tool within ArcGIS for each 1° ring, as shown in Figure 6. Using the zonal statistics tool, the corresponding CC value for each SV location was extracted for use in statistical modeling.

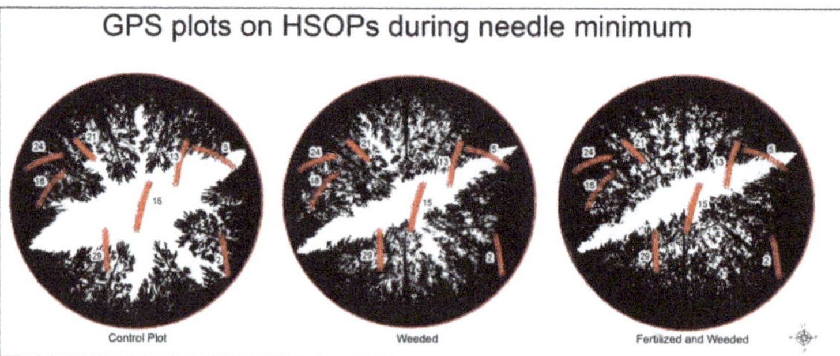

Figure 4. The resulting binary Hemispherical Sky-Oriented Photos (HSOPs) taken at the Intensive Management Practice Assessment Center managed forest taken during the needle minimum period with the GPS satellite vehicle positions (red circles) plotted inside the image.

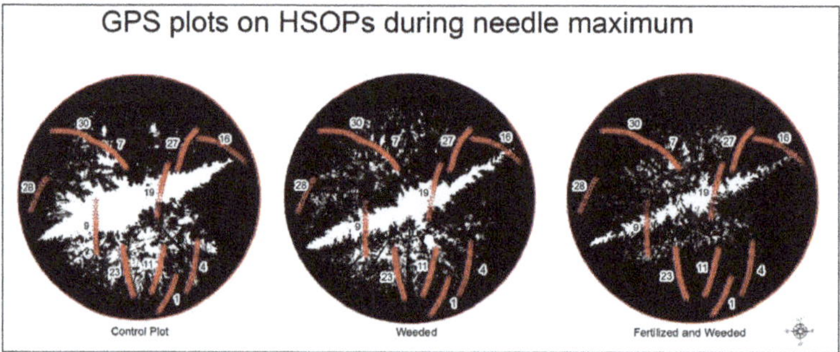

Figure 5. The resulting binary images taken at the Intensive Management Practice Assessment Center managed forest taken during the needle maximum period with the GPS satellite vehicle positions (red circles) plotted inside the image.

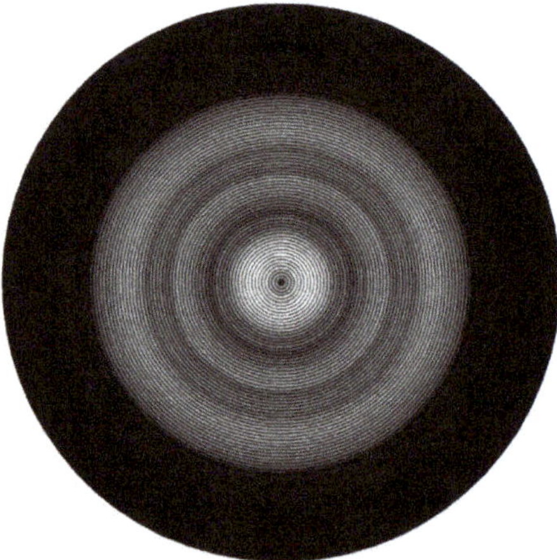

Figure 6. The 1°-ringed hemispherical sky-oriented photo with a gray scale range of values associated with the canopy closure values. Dark colored rings indicate a high level of canopy closure and white rings indicate less forest media.

3. Results

The models presented below are the results of regression modeling. This process consisted of initial data exploration of many different variables other than those presented. During the analysis, we identified ways to linearize the relationships within each model. Four models are presented and include an atmospheric model, a model optimized using HSOPs, and two models using dummy variables for each forest to generate an absorption index associated with the different forests.

The first analysis conducted used all the open receiver GPS observations to create an overall GPS SNR atmospheric model. In this work, we built on the previous work where the natural log of the angle from the horizon (*lnel*) of the GPS SV was the key parameter in the modeling process [21]. The resulting overall GPS L-band SNR atmospheric model (SAM) has a Root Mean Square Error (RMSE) of 2.01 dB and an adjusted R^2 of 0.81. The SAM equation applies to all observations where no vegetation is present over the GPS receiver. The SAM equation is:

$$SNR_{open} = 7.79 \; (lnel) + 18.85 \tag{2}$$

When exploring how different forests influence GPS signal, we incorporated the use of the CCPM. The CCPM approach to model SNR incorporates GPS observations from all forest study sites. The CCPM uses two variables. The first variable is *lnel*, as in the SAM equation, which linearizes the problem and is also vital in modeling the atmospheric component of the observed SNR. The other variable in the CCPM is the Beer's Law component, the product of CC and the sine of the zenith angle. Table 2 shows the results of the CCPM for each individual forest, with the equations taking the following form:

$$SNR = a + B_1 \; (lnel) + B_2 \; CC \; \sin(ZA) \tag{3}$$

where *lnel* is the natural log of the angle of the SV above the horizon, ZA is the zenith angle, and SNR units are in decibels.

Table 2. Canopy Closure Predictive Model Results with coefficients a, B_1, and B_2 are in reference to Equation (3) and the root Mean Square Error (RMSE).

ID	City Vicinity	a	B_1	B_2	RMSE	Adj R^2
1	West Point, NY	18.85	7.79	−5.53	3.28	0.60
2	IMPAC	19.32	7.79	−5.49	2.78	0.71
3	Hogtown Forest, Gainesville, FL	25.05	5.26	−6.02	3.02	0.66
4	Charleston, SC	27.87	4.25	−7.00	3.10	0.64
5	Alexandria, LA	25.77	4.87	−5.24	3.03	0.60
6	Cold Spring, TX	26.89	5.61	−16.35	2.80	0.59
7	Georgetown, TX	23.94	6.25	−8.02	3.71	0.61
8	Cloudcroft, NM	25.71	5.99	−6.99	3.77	0.60
9	Flagstaff, AZ	21.70	6.70	−0.50	3.33	0.57
10	Guadalupe, CA	28.83	5.08	−8.81	2.66	0.66
11	San Luis Obispo	26.26	5.79	−29.39	3.75	0.60
12	Davenport, CA	27.50	5.46	−6.03	2.83	0.72
13	Davenport, CA	27.50	3.15	−14.49	3.02	0.70
14	Tahoe NF	30.03	4.56	−8.59	3.24	0.70
15	Nederland, CO	31.04	4.79	−12.99	2.83	0.74

The last model we developed incorporated all aspects of the CCPM and also included dummy variables associated with each different forest site. The resulting dummy variables simply provide a Y-intercept shift for the expected SNR value based on a particular forest. The resulting dummy variable coefficients for each specific forest provide absorption indexes that help establish an understanding of how each forest affects the reception of GPS signals. The resulting model is termed the SNR forest index model (SFIM). The SFIM had an RMSE of 3.24 dB and an adjusted R^2 of 0.65, with the SFIM equation is as follows:

$$SNR = 7.79\,(lnel) - 0.26CC\,\sin(ZA) + 18.85 + I \tag{4}$$

where I is the index value.

Inspection of the SFIM equation shows that the Beer's Law component has a very small influence on the model. This result was unexpected initially. However, it agrees with the same points as discussed with respect to the overall CCPM model applied to multiple forest sites. Based on the lack of influence by the Beer's Law component, it was removed from the model to generate the simplified SFIM or SSFIM. The resulting SSFIM equation is:

$$SNR = 7.83\,(lnel) + 18.73 + I \tag{5}$$

The SSFIM resulted in the same RMSE and adjusted R^2 as the SFIM. Table 3 shows the results of the absorption index value (I) for each forest applied to the SFIM and the SSFIM.

Table 3. Absorption indexes of the forest study sites showing the results of the signal to noise forest index model (SFIM) and simplified signal to noise forest index model (SSFIM).

ID	City Vicinity	Tree Type	HT/STD (m)	DBH/STD (m)	SNR Index (dB) SFIM		SNR Index (dB) SSFIM
1	West Point, NY	Oak/Hickory 100% Deciduous	23.4/3.4	0.30/0.18	Fall	−3.74	−3.75
					Spring	−5.43	−5.44
					Summer	−5.54	−5.55
					Winter	−4.37	−4.38
2	IMPAC	100% Pine					
	Gainesville, FL	Needle Minimum	See Table 1	See Table 1		−3.31	−3.32
	Gainesville, FL	Needle Maximum	See Table 1	See Table 1		−4.30	−4.31
3	Hogtown Forest Gainesville, FL	80% Deciduous 20% Coniferous	20.4/2.47	0.46/0.08		−5.87	−5.89
4	Charleston, SC	90% Pine, 10% Deciduous	24.0/3.1	0.36/0.05		−7.68	−7.69
5	Alexandria, LA	90% Pine, 10% Deciduous	23.2/4.1	0.56/0.11		−6.48	−6.49
6	Cold Spring, TX	80% Pine, 20% Deciduous	19.5/4.7	0.52/0.13		−6.03	−6.06
7	Georgetown, TX	Cedar Elm and Live Oak with Ash Juniper	6.3/1.1	0.42/0.11		−5.15	−5.16
8	Cloudcroft, NM	Ponderosa Pine	23.3/3.2	0.41/0.12		−3.12	−3.12
9	Flagstaff, AZ	Ponderosa Pine	19.2/6.8	0.41/0.07		−2.83	−3.35
10	Guadalupe, CA	Eucalyptus	28.2/3.3	0.42/0.14		−5.02	−5.02
11	San Luis Obispo	Agrifolia	6.0/1.5	0.22/0.09		−3.10	−3.11
12	Davenport, CA	75% Redwood, 25% Douglas Fir and Tanoak	54.0/6.3	1.20/0.56		−10.78	−10.80
13	Davenport, CA	80% Tanoak, 25% Douglas Fir	18.7/1.4	0.28/0.08		−8.05	−8.07
14	Tahoe NF	Ponderosa Pine	26.5/2.2	0.53/0.16		−3.98	−3.98
15	Nederland, CO	Aspen	8.4/2.6	0.20/0.06		−4.99	−5.00

Note: Where HT is Tree Height, DBH is Diameter at Breast Height and STD is Standard Deviation.

4. Discussion

In the first portion of this study, we presented the SAM. The SAM methodology is simplistic and, as such, future work associated with atmospheric modeling has been considered. Factors such as humidity, barometric pressure, clouds versus open sky, and fog could all be potential components in atmospheric modeling. However, many of these factors change rapidly and would require a very substantial series of photos and measurements of the different variables over short periods of time, thus we used the approach outlined above to develop the SAM.

A primary objective of this research was to determine the parameters that influence signal attenuation. As such, during the modeling process, many other variables were considered to include the interaction of these variables. Parameters such as the leaf area index (LAI) (as calculated from gap light analyzer software), the density of the trees, average diameter at breast height, and average tree height (to name a few) were considered. However, the parameters that make up the CCPM and SSFIM proved the optimum method.

Many previous studies modeled forest structure. Larsen and Kershaw explained the evolution of different canopy structure models as the assumption of uniformity of foliage was removed [22]. Building on this work, Oker-Blom modeled forests with individual cylinder or parabolic crowns as trees with a uniformly distributed LAI density [23]. This work allowed for areas with no foliage and areas with overlap. The overlapping areas would cause a clumping effect. Other statistical models such as the Poisson, negative binomial, and Markov models predict the likelihood of a ray of light passing completely through forest canopy [24]. A great advantage of the models presented in this research is their simplicity. When comparing a single layer canopy to a triple layer canopy, for example, we simply obtain different CC values. In a triple layer dense canopy, there will be higher CC values compared to a single layer forest canopy.

The results shown in Table 2 suggest that for each individual forest, the CCPM performs well. However, when applying all the observations from each forest as a whole, the Beer's Law component was not found to be statistically significant at 90% confidence. This goes against our initial hypothesis. However, based on the results for dummy variable modeling for seasons at West Point, it is not surprising. The West Point seasonal study found that applying dummy variables for the seasons helped adjust the overall model [21]. This adjustment likely had to do with the health of the canopy. For example, during the fall season the CC values derived from HSOPs included foliage that was dry. However, these leaves were counted the same as leaves during the spring or summer that were healthy. Applying the same logic for different forests, each study site has different conditions. Some sites had received recent record rainfall while other sites were in drought-like conditions. Additionally, when comparing many different species of forests there are numerous factors that could influence the absorption component associated with Beer's Law. Most importantly, Beer's Law has both a concentration and an absorption component and the CCPM attempts to capture both using just CC. Therefore, it is justifiable that individual forests, sharing many of the same attributes, are successfully modeled individually using the CCPM, but as a conglomerate, the model does not work as well. As a result, the use of the CCPM is effective but would require a calibration prior to implementing for a specific forest, meaning that when photos are taken in a forest to get the CC values, GPS SNR observations should also be collected. If a GPS calibration is not feasible, a user of this work may benefit from a different approach, such as the SSFIM.

When comparing the results of the SFIM to the SSFIM, we identified that the SSFIM provides an equation that removes the need for photography. With both models having the same RMSE and adjusted R^2, ultimately, there is no need to go through the added complexity of taking photos. Rather, a user can reference Tables 1 and 3 to identify a forest with similar attributes and gain an insight into how signals may be attenuated in a particular forest site. The challenge with this concept is identifying what the key similarities are between forests. Would species play the largest role, or would tree height and DBH have a larger influence? In a similar vein, there is a seasonal influence on attenuation, as shown in Table 3 at the West Point study site. As such, a larger index is needed with more forest types. Further research is also needed to determine the variability of absorption within a single forest based on rainfall, foliage conditions, and other factors to ensure a good prediction of signal attenuation. All of these are just some of the questions that could be addressed in future work.

5. Conclusions

In this study, four different GPS SNR models were presented: one model predicts GPS SNR in the open and three models provide methods to predict GPS SNR in forest conditions. Previous work shows how applying predicted SNR values can easily be applied to determine estimated attenuation. While we expected all models to perform well, we were initially surprised that the CCPM did not perform well when modeling all forests together. However, when considering that the CCPM uses only CC to describe the Beer's Law component of signal loss, these results reflect that absorption variation is significant between different tree species and environmental conditions. The SSFIM accounts for this variation nicely according to its associated results.

This work specifically investigates GPS signal attenuation in different forest conditions. However, gaining a better understanding of techniques to model GPS signal attenuation will lead us to understand how other signals belonging to other technologies may be influenced. Technologies dependent on different cell phone frequencies, satellite communications, Bluetooth, and AM or FM radio transmissions are just a few of the different signals that could benefit from the presented predictive models. This work could also benefit forest growth research that uses light interception predictions.

When this research began, it was our desire to build on the knowledge of how GPS signals are attenuated in different forest environments. There was no desire or requirement to limit our modeling efforts to any specific technologies, only the desire to try as many different techniques available to us and identify the optimal modeling approach. The use of HSOPs in the modeling process proved fruitful from the beginning of our work. The historic use of HSOPs to estimate LAI led us to using HSOPs in our modeling process. We explored the use of LAI values derived from the HSOPs during the modeling process. However, for each photo there is just one LAI value. In contrast, using the HSOPs to calculate CC values at particular angles from zenith became an additional consideration and proved effective. The results of the research suggest that the only needed measurements are HSOPs and a calibration of the model using GPS observations from a specific forest. Another approach to estimate signal loss in a forest is to reference a given forest to the SSFIM index. Finding forests within the index that have similar attributes would guide the user towards selecting an appropriate model absorption coefficient.

Author Contributions: W.W. and W.C.J. conceived and designed the experiments; W.W. performed the experiments; W.W. and B.W. analyzed the data and wrote the paper.

Acknowledgments: This work was supported by the National Geospatial Intelligence Agency under Grant Number NIB8G15107GS71. We are also grateful to the Forest Biology Research Cooperative and Timothy A. Martin and Eric J. Jokela for their assistance in providing access to the experimental IMPAC study site. A special thanks to the following sites: Camp San Luis Obispo, California; California Polytechnic's Swanton Pacific Ranch Research Center in Davenport, California; and the University of California Berkeley's Sagehen Research center near Truckee, California.

References

1. Wright, W.C.; Liu, P.-W.; Slatton, K.C.; Shrestha, R.L.; Carter, W.E.; Lee, H. Predicting L-Band Microwave Attenuation through Forest Canopy Using Directional Structuring Elements and Airborne Lidar. In Proceedings of the Geoscience and Remote Sensing Symposium (IGARSS 2008), Boston, MA, USA, 6–11 July 2008; pp. III-688–III-691.

2. Lee, H.; Slatton, K.C.; Roth, B.; Cropper, W.P. Prediction of forest canopy light interception using three-dimensional airborne LiDAR data. *Int. J. Remote Sens.* **2009**, *30*, 189–207. [CrossRef]

3. Lee, H.; Kampa, K.; Slatton, K.C. Segmentation of ALSM point data and the prediction of subcanopy sunlight distribution. In Proceedings of the Geoscience and Remote Sensing Symposium (IGARSS 2005), Seoul, Korea, 25–29 July 2005; p. 4.

4. Lee, H.; Slatton, K.C.; Roth, B.; Cropper, W., Jr. Adaptive clustering of airborne LiDAR data to segment individual tree crowns in managed pine forests. *Int. J. Remote Sens.* **2010**, *31*, 117–139. [CrossRef]

5. Holden, N.; Martin, A.; Owende, P.; Ward, S. A method for relating GPS performance to forest canopy. *Int. J. For. Eng.* **2001**, *12*, 51–56.

6. Bode, C.A.; Limm, M.P.; Power, M.E.; Finlay, J.C. Subcanopy Solar Radiation model: Predicting solar radiation across a heavily vegetated landscape using LiDAR and GIS solar radiation models. *Remote Sens. Environ.* **2014**, *154*, 387–397. [CrossRef]

7.   Mücke, W.; Hollaus, M. Modelling light conditions in forests using airborne laser scanning data. In Proceedings of the SilviLaser 2011 Conference, Hobart, Australian, 16–20 October 2011.

8.   Gonzalez-Benecke, C.; Gezan, S.A.; Samuelson, L.J.; Cropper, W.P., Jr.; Leduc, D.J.; Martin, T.A. Estimating Pinus palustris tree diameter and stem volume from tree height, crown area and stand-level parameters. *J. For. Res.* **2014**, *25*, 43–52. [CrossRef]

9.   Bortolot, Z.J.; Wynne, R.H. Estimating forest biomass using small footprint LiDAR data: An individual tree-based approach that incorporates training data. *ISPRS J. Photogramm. Remote Sens.* **2005**, *59*, 342–360. [CrossRef]

10.   Drake, J.B.; Dubayah, R.O.; Knox, R.G.; Clark, D.B.; Blair, J.B. Sensitivity of large-footprint lidar to canopy structure and biomass in a neotropical rainforest. *Remote Sens. Environ.* **2002**, *81*, 378–392. [CrossRef]

11.   Hosoi, F.; Omasa, K. Estimating vertical plant area density profile and growth parameters of a wheat canopy at different growth stages using three-dimensional portable lidar imaging. *ISPRS J. Photogramm. Remote Sens.* **2009**, *64*, 151–158. [CrossRef]

12.   Kucharik, C.J.; Norman, J.M.; Gower, S.T. Characterization of radiation regimes in nonrandom forest canopies: Theory, measurements, and a simplified modeling approach. *Tree Physiol.* **1999**, *19*, 695–706. [CrossRef] [PubMed]

13.   Mätzler, C. Microwave transmissivity of a forest canopy: Experiments made with a beech. *Remote Sens. Environ.* **1994**, *48*, 172–180. [CrossRef]

14.   Wright, W.; Wilkinson, B.; Cropper, W. Estimating Signal Loss in Pine Forests Using Hemispherical Sky Oriented Photos. *Ecol. Inf.* **2016**, *38*, 82–88. [CrossRef]

15.   Wright, W.; Wilkinson, B. Modeling GPS Signal Loss in Forests Using Terrestrial Photogrammetric Methods. In Proceedings of the 28th International Technical Meeting of The Satellite Division of the Institute of Navigation, Tampa, FL, USA, 14–18 September 2015; pp. 3094–3099.

16.   Manninen, T.; Korhonen, L.; Voipio, P.; Lahtinen, P.; Stenberg, P. Leaf area index (LAI) estimation of boreal forest using wide optics airborne winter photos. *Remote Sens.* **2009**, *1*, 1380–1394. [CrossRef]

17.   Wright, W.C. *Quantifying Global Position System Signal Attenuation as a Function of Three-Dimensional Forest Canopy Structure*; University of Florida: Gainesville, FL, USA, 2008.

18.   Frazer, G.W.; Fournier, R.A.; Trofymow, J.; Hall, R.J. A comparison of digital and film fisheye photography for analysis of forest canopy structure and gap light transmission. *Agric. For. Meteorol.* **2001**, *109*, 249–263. [CrossRef]

19.   Rich, P.M. Characterizing plant canopies with hemispherical photographs. *Remote Sens. Rev.* **1990**, *5*, 13–29. [CrossRef]

20.   Anderson, M.C. Studies of the woodland light climate: I. The photographic computation of light conditions. *J. Ecol.* **1964**, *52*, 27–41. [CrossRef]

21.   Wright, W.; Wilkinson, B.; Cropper, W. Estimating GPS Signal Loss in a Natural Deciduous Forest Using Sky Photography. *Pap. Appl. Geogr.* **2017**, *3*, 119–128. [CrossRef]

22.   Larsen, D.R.; Kershaw, J.A., Jr. Influence of canopy structure assumptions on predictions from Beer's law. A comparison of deterministic and stochastic simulations. *Agric. For. Meteorol.* **1996**, *81*, 61–77. [CrossRef]

23.   Oker-Blom, P. Photosynthetic radiation regime and canopy structure in modeled forest stands. *Acta For. Fenn.* **1986**, *197*, 1–44. [CrossRef]

24.   Neumann, H.; den Hartog, G.; Shaw, R. Leaf area measurements based on hemispheric photographs and leaf-litter collection in a deciduous forest during autumn leaf-fall. *Agric. For. Meteorol.* **1989**, *45*, 325–345. [CrossRef]

 *forests*

Article

Estimation of Forest Aboveground Biomass and Leaf Area Index Based on Digital Aerial Photograph Data in Northeast China

Dan Li [1,2], Xingfa Gu [1,*], Yong Pang [3], Bowei Chen [3] and Luxia Liu [4]

[1] Institute of Remote Sensing and Digital Earth, Chinese Academy of Sciences, Beijing 100094, China; lidan@radi.ac.cn
[2] University of Chinese Academy of Sciences, Beijing 100049, China
[3] Institute of Forest Resource Information Techniques, Chinese Academy of Forestry, Beijing 100091, China; pangy@ifrit.ac.cn (Y.P.); rs.cbw@gmail.com (B.C.)
[4] School of Forestry and Landscape Architecture, Anhui Agricultural University, Hefei 230036, China; luxia.liu@ahau.edu.cn
* Correspondence: guxingfa@radi.ac.cn; Tel.: +86-10-6485-5711

Received: 11 April 2018; Accepted: 17 May 2018; Published: 18 May 2018

Abstract: Forest aboveground biomass (AGB) and leaf area index (LAI) are two important parameters for evaluating forest growth and health. It is of great significance to estimate AGB and LAI accurately using remote sensing technology. Considering the temporal resolution and data acquisition costs, digital aerial photographs (DAPs) from a digital camera mounted on an unmanned aerial vehicle or light, small aircraft have been widely used in forest inventory. In this study, the aerial photograph data was acquired on 5 and 9 June, 2017 by a Hasselblad60 digital camera of the CAF-LiCHy system in a Y-5 aircraft in the Mengjiagang forest farm of Northeast China, and the digital orthophoto mosaic (DOM) and photogrammetric point cloud (PPC) were generated from an aerial overlap photograph. Forest red-green-blue (RGB) vegetation indices and textural factors were extracted from the DOM. Forest vertical structure features and canopy cover were extracted from normalized PPC. Regression analysis was carried out considering only DOM data, only PPC data, and a combination of both. A recursive feature elimination (RFE) method using a random forest was used for variable selection. Four different machine-learning (ML) algorithms (random forest, k-nearest neighbor, Cubist and supporting vector machine) were used to build regression models. Experimental results showed that PPC data alone could estimate AGB, and DOM data alone could estimate LAI with relatively high accuracy. The combination of features from DOM and PPC data was the most effective, in all the experiments considered, for the estimation of AGB and LAI. The results showed that the height and coverage variables of PPC, texture mean value, and the visible differential vegetation index (VDVI) of the DOM are significantly related to the estimated AGB (R^2 = 0.73, RMSE = 20 t/ha). The results also showed that the canopy cover of PPC and green red ratio index (GRRI) of DOM are the most strongly related to the estimated LAI, and the height and coverage variables of PPC, the texture mean value and visible atmospherically resistant index (VARI), and the VDVI of DOM followed (R^2 = 0.79, RMSE = 0.48).

Keywords: digital aerial photograph; aboveground biomass; leaf area index; photogrammetric point cloud; recursive feature elimination; machine-learning

---

1. Introduction

Nowadays, global climate change is becoming a major challenge for current and future generations. Forests play crucial roles in adjusting the global and regional carbon cycle and bioenergy consumption.

Forest aboveground biomass (AGB) is a key biophysical parameter. It can provide important vegetation information about growth, health, and productivity [1,2]. Also, it is often required by the implementation of effective climate policies [3,4]. Additionally, the leaf area index (LAI), another important biophysical parameter, could provide more detailed canopy structure information [5]. Therefore, the accurate estimation and prediction of these two forest parameters is of great importance.

Traditional forest investigation relies on manual ground measurements with intensive time and high costs, such as destructive individual tree sampling and non-destructive field measurements. The obvious advantage is that they can provide results with relatively high accuracy [6,7]. In an attempt to obtain a more efficient estimation of forest parameters under different scales and in various environments, remote sensing techniques have been commonly applied over the past few decades. Data obtained from different sensors, such as optical cameras [8–10], radar [11,12], and terrestrial [13–15] or airborne [16–18] light detection and ranging (LiDAR), are significantly more efficient and have a lower cost than laborious ground-based estimation methods and have become widely used to characterize forest structure [19].

It is known that AGB and LAI cannot be directly obtained using remote sensing techniques; they are usually estimated by establishing a regression relationship between parameters derived from remote sensing data. However, it is worth noting that one problem when using optical remote sensing or radar data for estimation is saturation. Therefore, under high biomass or canopy density, the estimation accuracy, especially for biomass, is often under estimation [12,20,21]. LiDAR can provide quite accurate 3-D information and a reliable estimation, but the cost to obtain data is high; therefore, it is not suitable for continuous monitoring over large areas.

In order to obtain accurate results, many researchers have tried to combine multispectral and textural information from optical remote sensing data and vertical structure information from LiDAR data to estimate forest parameters [22–26]. These findings have demonstrated that estimation accuracy could be improved by making most use of the potential by combining these two types of data [27,28].

However, due to the influence of cloud and haze, it is sometimes difficult for satellites to capture high-quality images in the short term. On the other hand, and as mentioned above, LiDAR data acquisition and processing costs are too high to perform monitoring of large forest areas [29]. Both satellite remote sensing and LiDAR data have some limitations in high phase and extensive forest inventory applications.

With the recent development of unmanned aerial vehicle (UAV) technology over the past few years, aerial photographs from a digital camera mounted on an UAV or light, small aircraft have been widely used in forest inventory [30–35], and they generally have good affordability and availability. One thing should be mentioned here is that most of the relevant research used either digital orthophoto mosaic (DOM) or photogrammetric point cloud (PPC). Dandois et al. [36] used "Structure from Motion (SfM)" computer vision algorithms to extract canopy structure and spectral attributes based on red-green-blue (RGB) aerial images. Understory digital terrain models (DTMs) and canopy height models (CHMs) were generated from leaf-on and leaf-off point clouds using procedures commonly applied to LiDAR point clouds. CHMs were strong predictors of field-measured tree heights (R^2 = 0.63–0.84). Mathews et al. [35] also used SfM computer vision algorithms to obtain high-density point clouds. Points near samples were extracted and input into a stepwise regression model to predict LAI. The final R^2 value was approximately 0.57. Ota et al. [37] investigated the capabilities of CHM derived from aerial photographs using the SfM approach to estimate AGB in a tropical forest and yielded an accurate estimate (R^2 = 0.79). These successful applications show the potential of the SfM algorithm.

In addition, recent advances in commercial software based on computer vision algorithms such as Pix4DMapper (https://pix4d.com/, Pix4D S.A. Lausanne, Switzerland) and Agisoft Photoscan (https://www.agisoft.com/, Agisoft LLC, St. Petersburg, Russia) have enabled the mass production of digital surface models (DSMs) using the SfM algorithm with a much higher level of automation and much greater ease of use [38]. Applying the SfM approach enables us to produce high spatial resolution photogrammetric point clouds with vertical structure features similar to those derived from airborne LiDAR and DOM similar to satellite optical images with horizontal features.

In terms of the regression model used for building the relationship, many researchers have applied machine-learning methods in forest inventory and have obtained better accuracy in the past few years [39,40]. It is well known that the near-infrared (NIR) domain is a good indicator of vegetation health; we imitated the structure of the NIR band to construct the vegetation indices based on only red-green-blue three spectral bands in this study. To fully explore the advantages of digital aerial photographs (only RGB bands, not including the NIR band) in forest inventory, we focus on the estimation of AGB and LAI based on only DOM, only PPC, and DOM + PPC using a machine-learning method. This study will achieve the following goals: (1) processing digital aerial photograph (DAP) data based on the SfM approach to generate PPC and DOM and (2) estimating forest AGB and LAI based on only DOM, only PPC, and a combination of both.

2. Materials and Methods

2.1. Study Area

The study area is located at the Mengjiagang forest farm (46°26′ N, 130°43′ E) of Jiamusi city, in Heilongjiang Province of China (Figure 1), which is influenced by a temperate continental climate. Annual precipitation occurs mainly in summer. The study area covers an approximate land area of 260 km^2, which varies in elevation from about 180 m to 450 m above sea level. The land is relatively flat without an extreme slope. The dominant tree species is Larch (*Larix gmelinii* (Rupr.) Rupr.), followed by Korean pine (*Pinus koraiensis* Sieb. et Zucc.), Scots pine (*Pinus sylvestris* L.var. mongolica Litv.), and Spruce (*Picea asperata* Mast).

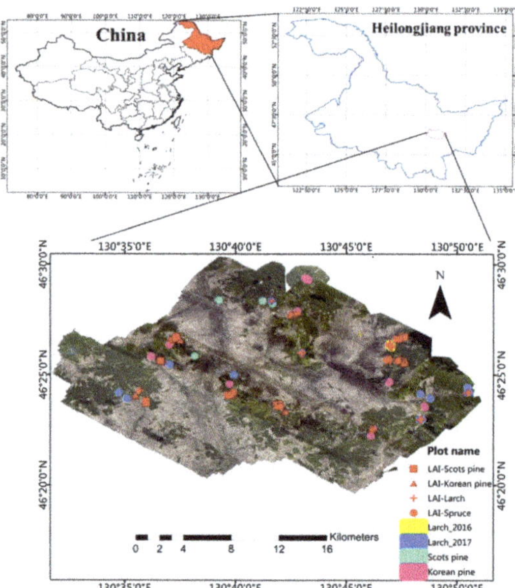

Figure 1. Location of the study area and distribution of field measurements. Red "■, ▲, +, ●" points are the sites of center coordinates of the Scots pine, Korean pine, Larch and Spruce, respectively, for LAI observations. Yellow locations are the sites of Larch plots in 2016. Dark blue locations are the sites of Larch plots in 2017. Light blue locations are the sites of Scots pine plots. Purple locations are the sites of Korean pine plots.

2.2. Data

The research data include (1) digital aerial photograph (DAP) data (year 2017), (2) 0.25 m spatial resolution DEM from LiDAR (year 2017), (3) LAI field data (year 2017), and (4) field measured data (year 2017 and 2016).

2.2.1. DAP Data Collection

DAP and LiDAR data were acquired at the same time on 5 and 9 June 2017 by the Chinese Academy of Forestry (CAF) using a Hasselblad60 digital camera and Riegl LMS-Q680i of CAF-LiCHy system in a Y-5 aircraft. The average flight altitudes were 950 m and 1250 m above ground level (AGL), and the mean ground sampling distance (GSD) of DAP images was 9 cm and 13 cm, respectively. The average flight speed was 40 m/s, and the mean forward overlap (FO) and the mean overlap between flight lines of DAP images were about 70% and 50%, respectively. In terms of aerial photograph data, the camera focal length, image size, and pixel size inside the camera are 50 mm, 8964 × 6716 pixels, and 6.0 μm respectively. The details of the flight parameters are shown in Table 1.

Table 1. Details of digital aerial photograph and LiDAR data specifications.

Flight Conditions	
Flight altitude (above-ground)	950 m, 1250 m
Flying speed	40 m/s
Acquisition time	5 and 9 June 2017
Digital Aerial Photograph: Hasselblad60	
Focal length	50 mm
Spectral bands	red, green, blue
Forward overlap (FO)	70%
Overlap between flight lines	50%
Scale	8964 × 6716 pixels
Pixel size	6 μm
Ground resolution	9 cm, 13 cm
Average density of point cloud	11–26 pts/m^2
LiDAR: Riegl LMS-Q680i	
Wavelength	1550 nm
Laser pulse length	3 ns
Waveform sampling interval	1 ns
Laser beam divergence	0.5 mrad
Vertical resolution	15 cm, 20 cm
Pulse repetition rate	300 kHZ, 200 kHZ
Scan angle	±30°
Average density of point cloud	9.6 pts/m^2, 6.3 pts/m^2

2.2.2. Field Measurements

The standard plots were set based on comprehensive field exploration. The shape of the plot was rectangular, and the area of the plot was determined according to factors such as tree age, the density of the forest, and site quality. In this study, we chose three kinds of tree ages: mature forest, middle forest, and young forest. We established three density plot types: dense young tree plots, dense middle tree plots, and sparse mature large tree plots. There should be at least 50 trees in every plot in the mature forest, at least 70 trees in the middle forest, and at least 90 trees in the young forest.

According to this principle, an original area of 400 m^2 could be set in advance, and then the numbers of individual trees in each plot should be counted to determine the area of the plot. There were six different area sizes, which were 400 m^2 (20 m × 20 m), 600 m^2 (20 m × 30 m), 900 m^2 (30 m × 30 m), 1000 m^2 (25 m × 40 m), 1200 m^2 (30 m × 40 m), 1500 m^2 (30 m × 50 m). The numbers of plots of Korean pine, Scots pine, and Larch are 16, 5, and 35 respectively. The field data of 9 Larch plots were collected in July 2016. The field data of 26 Larch plots were collected under leaf-on canopy conditions from 4 to 20 June 2017. The field data of all Korean pine and Scots pine plots were measured from 20 to 30 August 2017. In addition to these three main tree species, there are Elm (*Ulmus laciniata* (Trautv.) Mayr), Linden (*Tilia mandshurica* Rupr. et Maxim), Aspen (*Populus tomentosa* Carr), Oak (*Quercus mongolica* Fisch. ex Ledeb.), Silver birch (*Betula platyphylla* Suk.), Maple birch (*Betula costata* Trautv.), Black birch (*Betula davurica* Pall.), and Ashtree (*Fraxinus mandschurica* Rupr.) in the plots. Fifteen Korean pine plots, three Scots pine plots, and nineteen Larch plots cover 600 m^2. One Scots pine plot covers 400 m^2. Nine Larch plots and one Scots pine plot cover 900 m^2. One Scots pine plot and one Larch plot cover 1000 m^2. Two Larch pine plots cover 1200 m^2. Four Larch pine plots cover 1500 m^2.

The quadrangle boundaries of field plots were firstly measured using GPS and tape. Based on the ground base station data, the accurate locations of plots were obtained by a differential global positioning system (DGPS). The differential total accuracy was between 0.5 m and 0.8 m.

Tree species, diameter at breast height (DBH), tree height, and the crown width of all living trees with a DBH greater than 5 cm were measured in each plot using tape or meters. The lengths of the east–west and south–north directions of the tree crown were measured with the tape, and the average value of two lengths was taken as the crown width. Tree species and DBH of all dead trees were measured. The statistical results of living trees are shown in Table 2.

Table 2. Summary of the field data (living trees).

Species	Number	Tree Height (m)		DBH (cm)	
		Range	Mean	Range	Mean
Korean pine (*Pinus koraiensis*)	776	7.2–22.9	14.2	6.6–35.2	22.0
Scots pine (*Pinus sylvestris*)	215	12.9–30.6	20.3	15.3–43.6	25.0
Larch (*Larix gmelinii*)	3046	4.2–33.1	19.0	4.1–41.6	17.3
Elm (*Ulmus laciniata*)	9	8.7–15.4	11.2	7.1–11.0	8.8
Linden (*Tilia mandshurica*)	20	8.6–18.7	12.7	8.5–30.2	14.8
Aspen (*Populus tomentosa*)	6	14.3–17.3	15.9	15.0–27.0	17.9
Oak (*Quercus mongolica*)	63	7.2–19.8	11.2	6.4–24.4	9.2
Silver birch (*Betula platyphylla*)	154	9.6–21.5	14.2	6.2–26.6	11.0
Maple birch (*Betula costata*)	6	8.7–16.9	14.1	7.0–31.0	17.1
Black birch (*Betula dahurica*)	2	7.6–14.2	10.9	5.9–14.7	10.3
Ashtree (*Fraxinus mandschurica*)	10	7.5–15.5	13.8	10.0–16.3	12.1

We calculated the AGB of each measured living tree using these equations of different tree species based on tree height and DBH in Table 3.

Table 3. Aboveground biomass equations (AGB) of different tree species.

Tree Species	AGB Equation	Reference
Korean pine	$WT = 0.027847(D^2H)^{0.956544}$	
Scots pine	$WT = 0.3364D^{2.0067} + 0.2983D^{1.144} + 0.2931D^{0.8486}$	
Larch	$WT = 0.046238(D^2H)^{0.905002}$	[41]
Birch	$WT = 0.0278601(D^2H)^{0.993386}$	
Soft broad-leaf trees	$WT = 0.0495502(D^2H)^{0.952453}$	
Silver birch	$WT = 0.1193(D^2H)^{0.8372} + 0.002(D^2H)^{1.12} + 0.000015(D^2H)^{1.47}$	
Maple birch	$WT = 0.07936(D^2H)^{0.901} + 0.014167(D^2H)^{0.764} + 0.01086(D^2H)^{0.847}$	
Black birch	$WT = 0.14114(D^2H)^{0.723} + 0.00724(D^2H)^{1.0225} + 0.0079(D^2H)^{0.8085}$	
Elm	$WT = 0.03146(D^2H)^{1.032} + 0.007429D^{2.6745} + 0.002754D^{2.4965}$	[42]
Linden	$WT = 0.01275(D^2H)^{1.009} + 0.00824(D^2H)^{0.975} + 0.00024(D^2H)^{0.991}$	
Oak	$WT = 0.03141(D^2H)^{0.733} + 0.002127D^{2.9504} + 0.00321D^{2.4735}$	
Aspen	$WT = 0.2286(D^2H)^{0.6938} + 0.0247(D^2H)^{0.7378} + 0.0108(D^2H)^{0.8181}$	
Ashtree	$WT = 0.06013(D^2H)^{0.891} + 0.00652(D^2H)^{1.169} + 0.0044(D^2H)^{0.9919}$	

We only used the AGB of living trees in this study and did not consider the AGB of dead trees. The AGB of each plot was then calculated by summing up the AGB of each living tree.

We used two methods to calculate the AGB of living trees. One was on the basis of five tree species: Korean pine, Scots pine, Larch, Birch (Sliver birch, Maple birch and Black birch), and Soft broad-leaf trees (Aspen, Elm, Linden, Oak, and Ashtree). Another was on the basis of all 11 tree species. The statistical results of all plots are shown in Table 4. As we can see from Table 4, the results of the two methods are not very different. Thus, we used the AGB results of five tree species to build a model and analyze retrieval results in this study.

Table 4. Summary of field AGB (living trees).

Main Tree Species	Number of Plots	Number of Species [1]	DOM		PPC		DOM + PPC		AGB (t/ha)		
			T [2]	V [3]	T [2]	V [3]	T [2]	V [3]	Range	Mean	SD [4]
Korean pine	16	5	11	5	11	5	13	3	83.64–180.90	107.89	21.86
		11							82.79–180.90	107.71	22.07
Scots pine	5	5	4	1	4	1	4	1	94.06–182.30	147.69	34.59
		11							94.06–182.30	147.69	34.59
Larch	35	5	24	11	24	11	23	12	81.25–261.84	142.84	39.34
		11							78.14–261.84	143.62	38.61

[1] The number of tree species used to calculate AGB. [2] Training, the number of plots used to train in study result of the paper. [3] Validation, the number of plots used to validate in study result of the paper. [4] SD: standard deviation.

2.2.3. LAI Field Measurements

LAI field data were obtained from 5 to 20 June, 2017. There were 192 circular plots at a radius of 15 m. The LAI in each plot was measured using a LAI-2000 Plant Canopy Analyzer (https://www.licor.com/, LI-COR Corporate, Lincoln NE, USA). In total, 12 points in four perpendicular directions were measured in every plot, and the average value of 12 points was calculated as the field measured data. The central positions of these plots were first measured using a GPS instrument. Then, based on the ground base station data, the accurate locations of plots were obtained using DGPS. The differential total accuracy was between 0.5 m and 0.8 m. There are four species in the plots: Spruce (*Picea asperata* Mast), Korean pine (*Pinus koraiensis* Sieb. et Zucc.), Scots pine (*Pinus sylvestris* L. var. mongolica Litv.), and Larch (*Larix gmelinii* (Rupr.) Rupr.). The statistical results of the LAI field data are shown in Table 5.

Table 5. Summary of the leaf area index (LAI) field data.

Tree Species	Plot Amount	DOM		PPC		DOM + PPC		LAI Measured Value	
		T [1]	V [2]	T [1]	V [2]	T [1]	V [2]	Range	Mean
Spruce	57	43	14	40	17	38	19	3.46–6.14	4.39
Korean pine	55	36	19	39	16	42	13	2.49–5.01	3.46
Larch	42	29	13	32	10	31	11	2.13–5.42	3.51
Scots pine	38	23	15	25	13	22	16	1.48–3.05	2.22
All	192	131	61	136	56	133	59	1.48–6.14	3.54

[1] Training, the number of plots used to train in study result of the paper. [2] Validation, the number of plots used to validate in study result of the paper.

2.3. Methods

The research method consists of four parts: data preprocessing, features extraction, selection of the variables, and the model and validation (Figure 2).

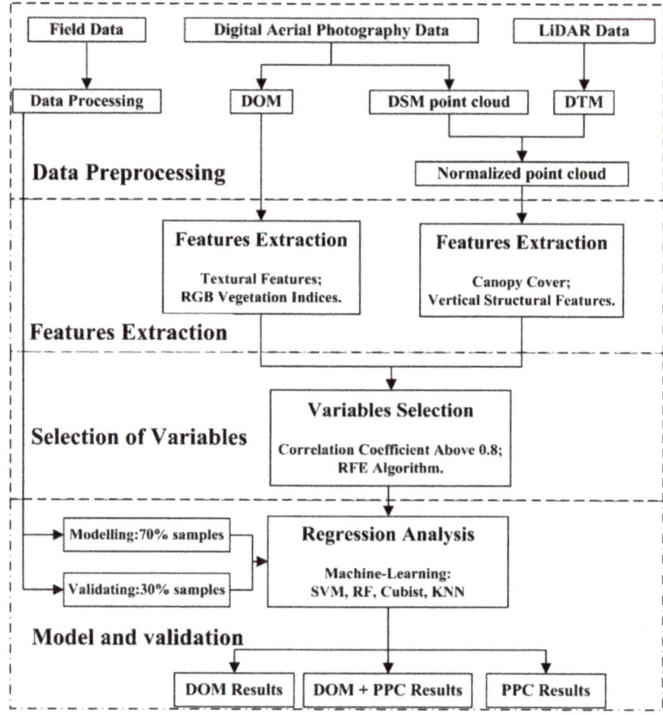

Figure 2. Technical flow chart of the process of retrieving AGB and LAI.

2.3.1. LiDAR Data Pre-Processing

LiDAR data pre-processing was carried out by the CAF. This included noise removal of the point cloud and point cloud classification, which gives ground and non-ground echoes according to the proprietary algorithm implemented in the TerraScan software (https://www.terrasolid.com, TerraSolid Ltd., Helsinki, Finland). A digital terrain model (DTM) with a 1-m spatial resolution was derived using a triangulated irregular network (TIN) interpolating method from the ground-classified points.

2.3.2. DAP Data Pre-Processing

First, we combined an onboard inertial measurement unit (IMU)/global positioning system (GPS) with camera exposure position information and ground base station data of the closest official reference points of the Heilongjing Bureau of Surveying and Mapping Geoinformation to generate the accurate position and attitude information of every photo taken using DGPS. The overall accuracy was between 10 cm and 15 cm.

We used the SfM algorithm to generate a DSM dense point cloud and DOM from an overlapping collection of digital aerial photographs in the proprietary software Pix4DMapper Professional Edition 2.1.0 (64 bit) (https://pix4d.com/, Pix4D S.A. Lausanne, Switzerland). The resolution of the DOM is 0.1 m, and the tolerance of data processing is 0.02 m. SfM is the process of estimating the 3D structure of a scene from a set of 2D images. SfM requires point correspondences between images. Corresponding points were identified either by matching features or tracking points from image 1 to image 2 [43]. The fundamental matrix describes the epipolar geometry of two images and is computed using the corresponding points of two images. The orientation and location in the specified coordinate system are returned by relative and absolute orientation calculation. The 3D locations of matched points are determined using triangulation.

We compared the DSM point cloud from the DAP data to the DSM point cloud from the LiDAR data (Figure 3). The results showed that the average range deviation between the LiDAR point cloud and the DAP point cloud results was less than 0.5 m.

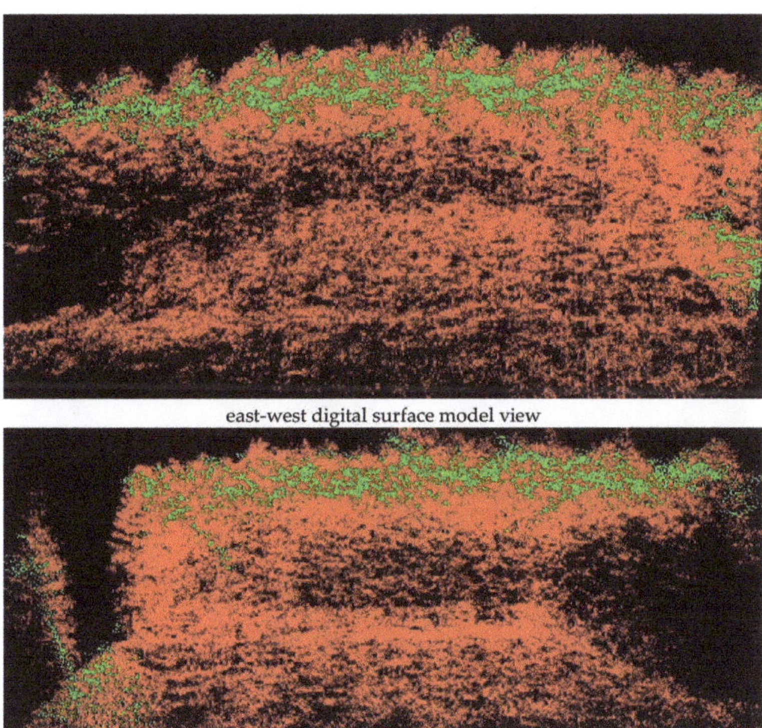

east-west digital surface model view

west-east digital surface model view

Figure 3. Matching the figures of the digital aerial photograph (DAP) point cloud and the light detection and ranging (LiDAR) point cloud. Red points are LiDAR point cloud and the green points are the DAP point cloud.

The results of DOM, DTM, and DSM in the same area are shown in Figure 4. Then, the DTM point cloud from LiDAR and the DSM point cloud from DAP were optimized in Terrascan software. The absolute heights of the point cloud were normalized by subtracting the terrain heights from DTM to obtain the relative heights.

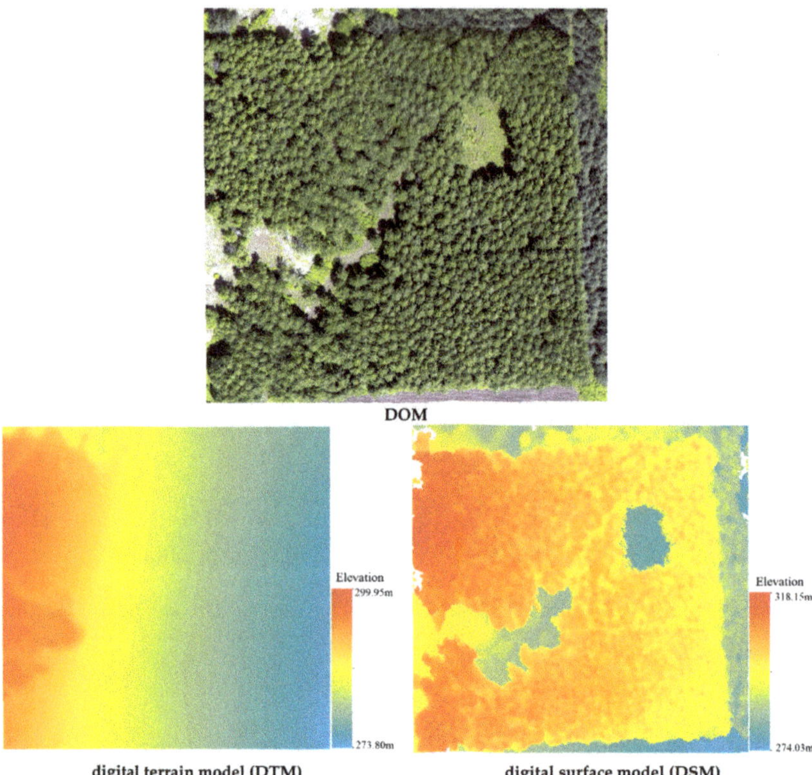

Figure 4. Data pre-processing results of digital aerial photographs and LiDAR.

2.3.3. Feature Extraction of DOM

Feature extraction of DOM can be divided into two different categories of features: (1) RGB vegetation indices and (2) textural features.

RGB vegetation indices were constructed by imitating the structure of the NIR band from 0.1-m resolution DOM data. We used the digital number (DN) value to take the place of reflectance in this study (Table 6). Six kinds of vegetation indices were created using DN values of the red, green, and blue bands.

Table 6. Six vegetation indices of red-green-blue (RGB) bands.

Vegetation Indices Name	Equations [1]	References
Visible differential vegetation index (VDVI)	$\frac{2 \times DN_G - DN_R - DN_B}{2 \times DN_G + DN_R + DN_B}$	[44]
Excess green index (EXG)	$2 \times DN_G - DN_R - DN_B$	[44]
Visible atmospherically resistant index (VARI)	$\frac{DN_G - DN_R}{DN_G + DN_R - DN_B}$	[45]
Green red ratio index (GRRI)	$\frac{DN_G}{DN_R}$	[46]
Green blue ratio index (GBRI)	$\frac{DN_G}{DN_B}$	[47]
Red blue ratio index (RBRI)	$\frac{DN_R}{DN_B}$	

[1] DN_G: digital number (DN) value of the green band. DN_R: digital number (DN) value of the red band. DN_B: digital number (DN) value of the blue band.

Four textural features (mean, homogeneity, dissimilarity, and correlation) were extracted from the first principal component of the 0.1-m DOM in this study (Table 7). Every feature includes four different window sizes of 4.5 m, 6.5 m, 10.1 m and 25.1 m.

The mean values of vegetation indices and textural features were calculated using R packages including caret [48], raster [49] and dplyr [50].

Table 7. Four textural features.

Feature Name	Equations [1]
Mean	$\sum_{i,j=0}^{N-1} iP_{i,j}$
Homogeneity	$\sum_{i,j=0}^{N-1} i\dfrac{P_{ij}}{1+(i-j)^2}$
Dissimilarity	$\sum_{i,j=0}^{N-1} iP_{i,j}\lvert i-j \rvert$
Correlation	$\sum_{i,j=0}^{N-1} iP_{i,j}\left[\dfrac{(i-mean)(j-mean)}{\sqrt{variance_i variance_j}}\right]$

[1] i is the line number of the gray level co-occurrence matrix. j is the column number of the gray level co-occurrence matrix. P_{ij} is normalized co-occurrence matrix. ME is the mean value of the gray level co-occurrence matrix. VA is the variance value of the gray level co-occurrence matrix.

2.3.4. Feature Extraction of PPC

PPC is used to provide forest vertical structure parameters. Height statistics (standard deviation (Stddev), variance, coefficient of variation (CV), skewness, kurtosis, maximum, mean, mode, average absolute deviation (AAD), l-moments, canopy relief ratio, median of absolute deviation (MADMedian), and mode of absolute deviation (MADMode)), height percentiles (IQ, P50, P75, P95, and P100), coverage statistical features (percentage above mean, percentage above mode, percentage above 2 m, 10 to 20 proportion, and 5 to 10 proportion) (Table 8), and canopy cover (CC) were extracted based on relative height values from normalized point clouds in a hierarchical way, using open source FUSION software (http://forsys.cfr.washington.edu/fusion).

CC is the index that describes the degree of canopy connection of trees, and it is the ratio of canopy projected area to woodland area. CC was generated from the normalized point cloud. We used the height threshold to distinguish between ground points and tree points.

Table 8. Forest vertical structure parameters from the normalized point cloud.

Variable Type	Variable Name	Variable Description	References
	Stddev	Standard deviation of point cloud height	
	Variance	Variance of point cloud	[51]
	CV	Coefficient of variation of point cloud height	
	Skewness	Skewness of point cloud height	[52]
	Kurtosis	Kurtosis of point cloud height	
Height statistics of point cloud	Maximum	Maximum of point cloud height	
	Mean	Mean of point cloud height	[51]
	Mode	Mode of point cloud height	
	AAD	Average absolute deviation	
	L-moments	Linear moment, including L1,L2,L3,L4	[53]
	Canopy relief ratio	$\frac{(mean-min)}{(max-min)}$	
	MADMedian	Median of absolute deviation of point cloud above median	[54]
	MADMode	Mode of absolute deviation of point cloud above median	
Height percentile of point cloud	IQ	Height 75th percentile minus 25th percentile	
	P99	Height 99th percentile	
	P95	Height 95th percentile	[51]
	P75	Height 75th percentile	
	P50	Height 50th percentile	
Coverage of point cloud	Percentage above mean	Percentage of point cloud above mean	
	Percentage above mode	Percentage of point cloud above mode	
	Percentage above 2 m	Percentage of point cloud above 2 m	[54]
	10 to 20 proportion	Proportion of point cloud from 10 and 20 m	
	5 to 10 proportion	Proportion of point cloud from 5 and 10 m	

2.3.5. Estimation of AGB and LAI

In the study, three types of data sources were used to estimate forest AGB and LAI. We merged all features from DOM, all structural variables from PPC, and a combination of both. Because there are four different texture window sizes, we need to separately merge all parameters. Because of high correlations among variables, the correlation was statistically calculated using R packages including caret [48] and corrplot [55]. We tested for multicollinearity between variables and removed variables with Pearson's r > 0.8 to reduce the redundancy of variables. In AGB retrieval, 10 variables for DOM, 6 variables for PPC, and 16 variables for a combination of both were selected after removing redundant variables. In LAI retrieval, 10 variables for DOM, 10 variables for PPC, and 20 variables for a combination of both were selected after removing redundant variables (Table 9).

Table 9. Variables from DOM and PPC after reducing redundancy.

Variable Type		Variable Name	
		LAI	AGB
PPC features	Height statistics	Stddev	
		Variance	
		CV	
		Skewness	Mean
		Kurtosis	Canopy relief ratio
	Height percentile	IQ	
			P95
	Coverage	Percentage above mean	
		10 to 20 proportion	
	Canopy cover	Mean(CC)	

Table 9. *Cont.*

Variable Type		Variable Name	
		LAI	AGB
DOM features	Texture		Mean(mean) Mean(dissimilarity) Mean(homogeneity) Mean(correlation)
	Vegetation indices		Mean(VDVI) Mean(EXG) Mean(VARI) Mean(GRRI) Mean(GBRI) Mean(RBRI)

Random forest (RF) is a natural multiclass algorithm with an internal measure of feature importance. The random forest recursive feature elimination (RF-RFE) selection method is basically a recursive process that ranks features according to some measure of their importance [56].

According to the importance of variables and referring to the model accuracy of cross validation, the influence of each variable on the cross-validation accuracy of the model was considered iteratively from the most important variable to the least important one. Following a machine-learning (ML) regression analysis, the joint RF-RFE algorithm was carried out to estimate AGB and LAI based on DOM variables, PPC variables, and a combination of both. For each of the three data sources, we used four ML model algorithms: random forest (RF), supporting vector machine (SVM), k-nearest neighbor (KNN) and Cubist to build the model. The tuning methods of the four models and RF-RFE operation were achieved using R package, including caret [48], e1071 [57], cubist [58] and randomForest [59]. In the Cubist model, the value range of number parameter committees of model trees is from 1 to 50, with a step of 1. The range of the nearest-neighbor sample neighbors is from 0 to 9, with a step of 1. The tuneLength of mtry is 7 and ntree is 1000 in the RF model. The tuneLength of k in the KNN model is 50. The values of kernel function gamma are 0.5, 1, and 2. The values of the punish coefficient cost are 0.1, 1, 10, and 100.

For DOM and PPC alone, we chose the first half of all variables. For DOM + PPC, we chose the first third of total variables based on RFE results to participate in THE ML model. Thus, five variables of DOM, three variables of PPC and six variables of DOM + PPC were used in the estimation of AGB. Five variables of DOM, five variables of PPC, and seven variables of DOM + PPC were chosen in the estimation of LAI.

In total, 70% of samples (approximately 40 plots for AGB and approximately 135 plots for LAI) were randomly selected for training, and 30% of samples (approximately 16 plots for AGB and approximately 57 plots for LAI) were selected for validation in this study. For regression analysis, we ensured the distribution of different densities and different tree species plots in training and validation samples.

2.4. Model Accuracy Evaluation

After the regression model was established, the coefficient of determination, R-Square (R^2), and root mean square error (RMSE) were used to assess the goodness and accuracy of the established models. The larger the R^2 value, and the stronger correlation. The smaller the RMSE value, the higher the predicted accuracy. R^2 and RMSE were calculated using Equations (1) and (2).

$$R^2 = 1 - \frac{\sum_{i=1}^{n}\left(y_i - y_i'\right)^2}{\sum_{i=1}^{n}\left(y_i - \overline{y_i}\right)^2} \tag{1}$$

$$RMSE = \sqrt{\frac{\sum_{i=1}^{n}\left(y_i - y_i'\right)^2}{n}} \tag{2}$$

where n is the number of plots, y_i is the ground field measurement reference value of AGB or LAI for plot i; $\overline{y_i}$ is the average value of y_i; y_i' is the model estimate value of AGB or LAI.

3. Results

About 70% of plot measurements were used to build the regression analysis model, and 30% were retained to validate the model. By training and comparing AGB and LAI retrieval results based on textural features of four window sizes using same method, we chose features of 6.5 m window size to retrieve AGB and LAI. The tuning results of four machine-learning models for the estimation of AGB and LAI from the three data sources are shown in Table 10.

Table 10. The best combination of parameters for the four models for the estimation of AGB and LAI.

Models	AGB			LAI		
	DOM	PPC	DOM + PPC	DOM	PPC	DOM + PPC
Cubist [1]	c = 2, n = 2	c = 45, n = 9	c = 2, n = 2	c = 8, n = 2	c = 6, n = 9	c = 11, n = 2
KNN	k = 5	k = 11	k = 5	k = 7	k = 11	k = 5
RF	mtry = 4	mtry = 3	mtry = 2	mtry = 2	mtry = 3	mtry = 3
SVM [2]	g = 2, c = 100	g = 0.5, c = 100	g = 0.5, c = 1	g = 1, c = 1	g = 1, c = 1	g = 0.5, c = 1

[1] c: committees, n: neighbors. [2] g: gamma, c: cost.

The variables used for final modeling are shown in Table 11, ranked according to the importance.

Table 11. The variables used in modeling.

Importance	AGB		
	DOM	PPC	DOM + PPC
1	VDVI	P95	P95
2	RBRI	Mean	P-Mean [1]
3	EXG	10 to 20 proportion	10 to 20 proportion
4	Mean		D-Mean [2]
5	GBRI		VDVI
6			CC

Importance	LAI		
	DOM	PPC	DOM + PPC
1	GRRI	CC	CC
2	Mean	P95	GRRI
3	VARI	10 to 20 proportion	P95
4	VDVI	CV	10 to 20 proportion
5	EXG	Skewness	D-Mean [2]
6			VARI
7			VDVI

[1] P-Mean: mean of point cloud height. [2] D-Mean: the textural feature from DOM.

As we can see from Table 11, vegetation indices are the most significant variables for the estimation of AGB in models obtained using DOM data. In addition, the texture mean variable is important. In those models comprising only PPC variables, height percentile and statistic are the most significant variables. Concerning DOM + PPC variables, height and coverage variables are the most significant variables; texture and vegetation index variables followed.

For estimation of LAI, the green red ratio index (GRRI) is the most significant variable in DOM models; the texture mean variable followed. In PPC models, coverage and height percentile are the most significant variables. In the combined DOM + PPC variables, the coverage variable is the most significant, followed by the vegetation index variable.

As we can see from Figure 5, for field data versus estimated data in the estimation of AGB, the optimal values of R^2 and RMSE in DOM alone are 0.65 and 21 t/ha, respectively. In PPC alone, the optimal values are 0.7 and 26 t/ha. In DOM + PPC, the values are 0.73 and 20 t/ha, respectively. The results obtained using a combination of DOM and PPC provide increased R^2 accuracy. DOM results are particularly poor.

It is worth noting that the differences between PPC and DOM + PPC models are less marked. In some cases, PPC variables alone provide better results. It is apparent that R^2 is much smaller for the model based only on DOM variables compared to the two others. PPC and DOM + PPC models provide very similar results.

The field data versus the estimated data in the estimation of LAI are shown in Figure 6. The values of R^2 and RMSE in DOM alone are 0.73 and 0.49, respectively. In PPC alone, the values are 0.65 and 0.57, respectively. In DOM + PPC, the values are 0.79 and 0.48, respectively. The results obtained using the combination of DOM and PPC provide increased R^2 accuracy. PPC results are particularly poor.

The differences between DOM and DOM + PPC models are less marked. DOM variables alone provide better results. It is apparent that R^2 is much smaller for the model based only on PPC variables compared to the two other models. In particular, DOM and DOM + PPC models provide very similar results.

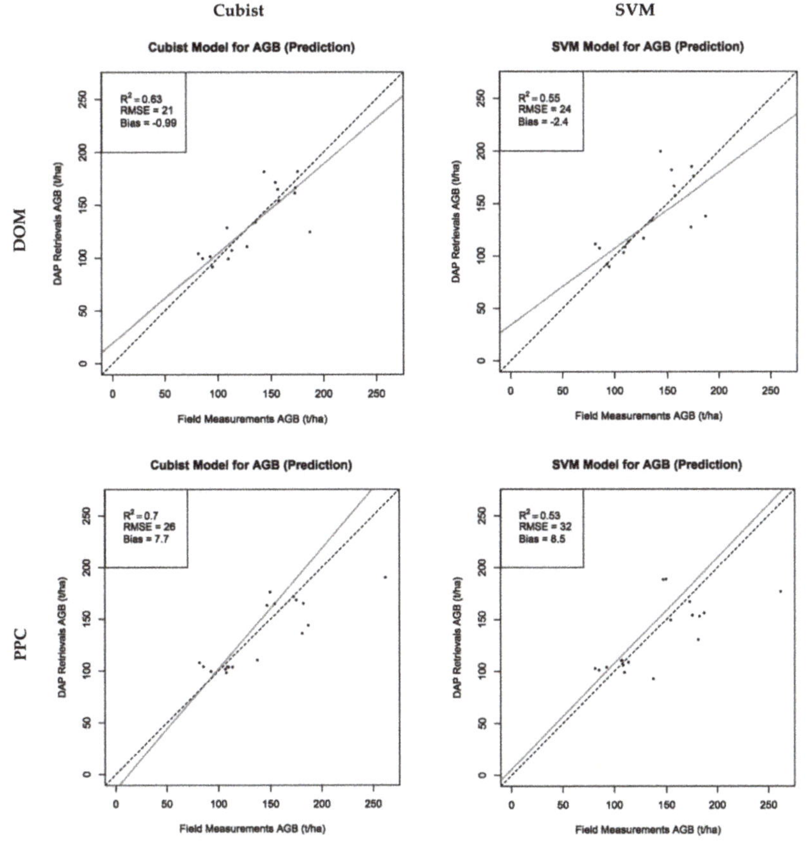

Figure 5. *Cont.*

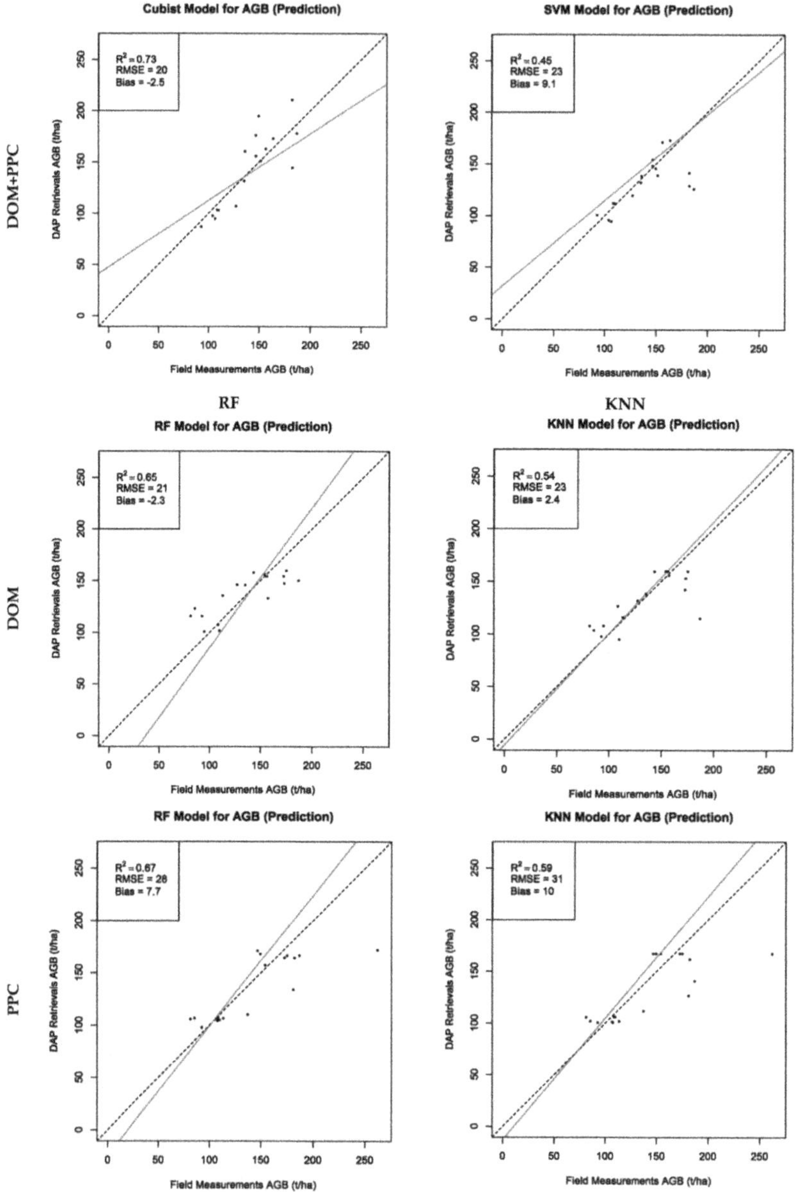

Figure 5. *Cont.*

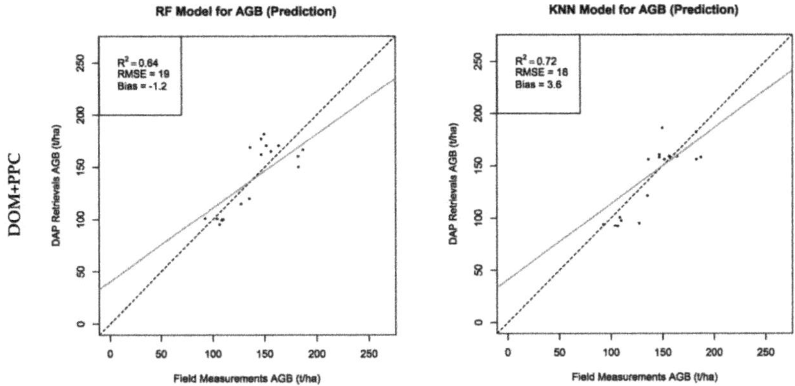

Figure 5. Prediction accuracy of the four models for AGB estimation based on the three sources of data.

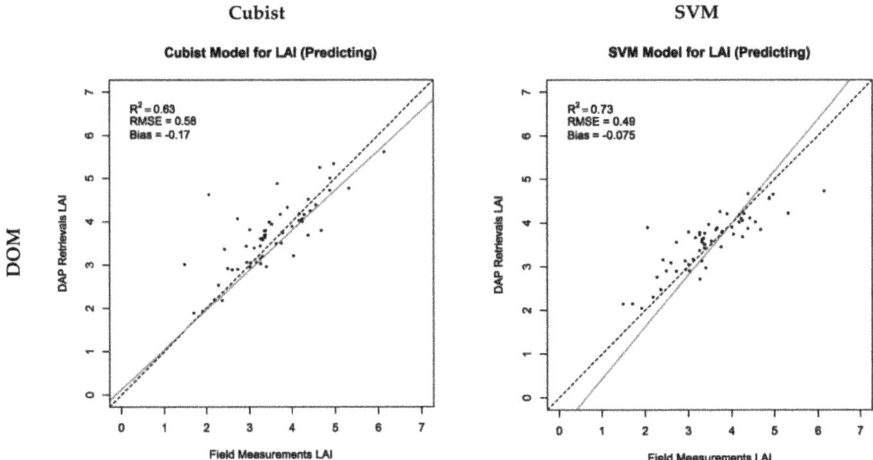

Figure 6. *Cont.*

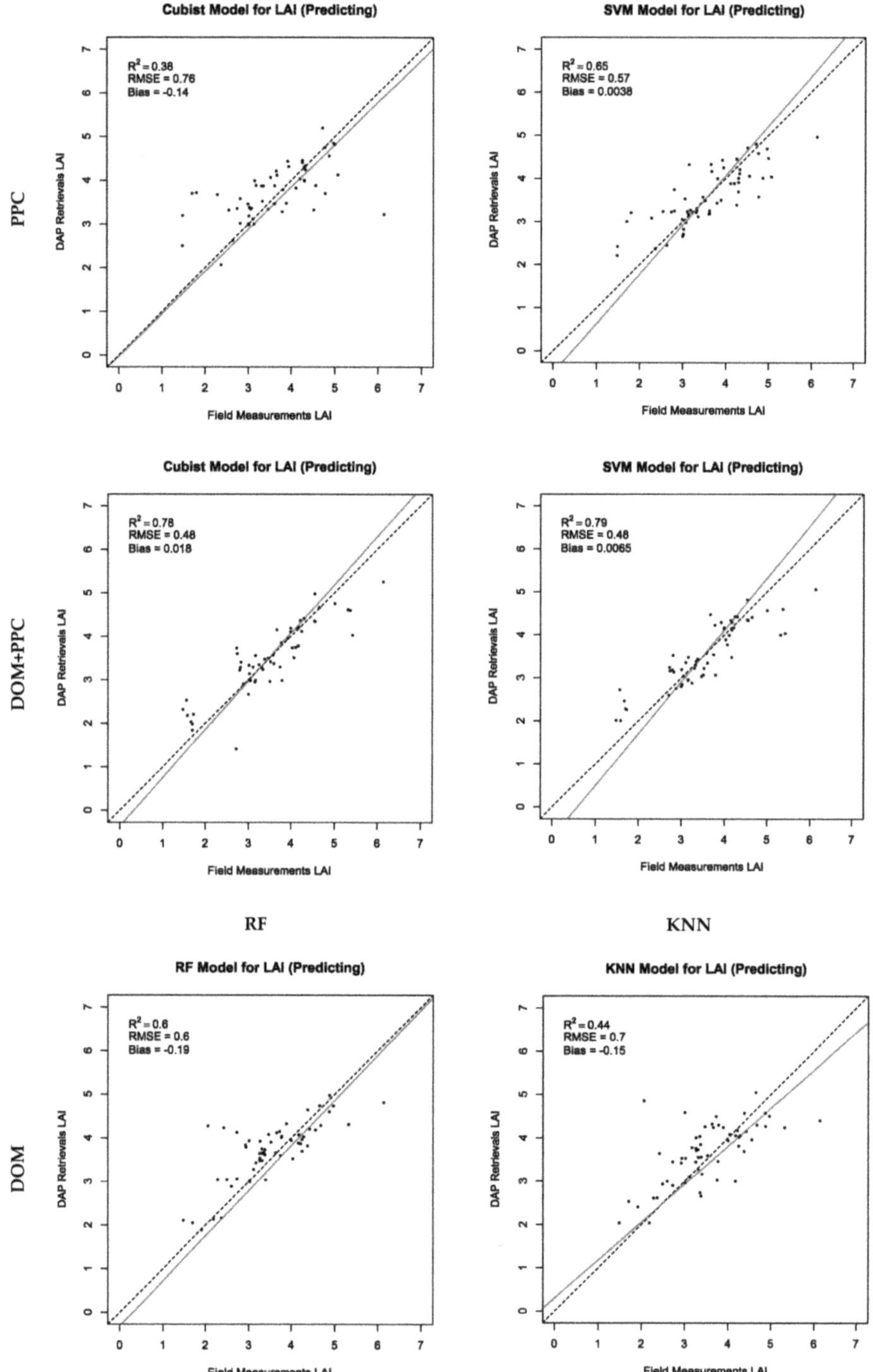

Figure 6. *Cont.*

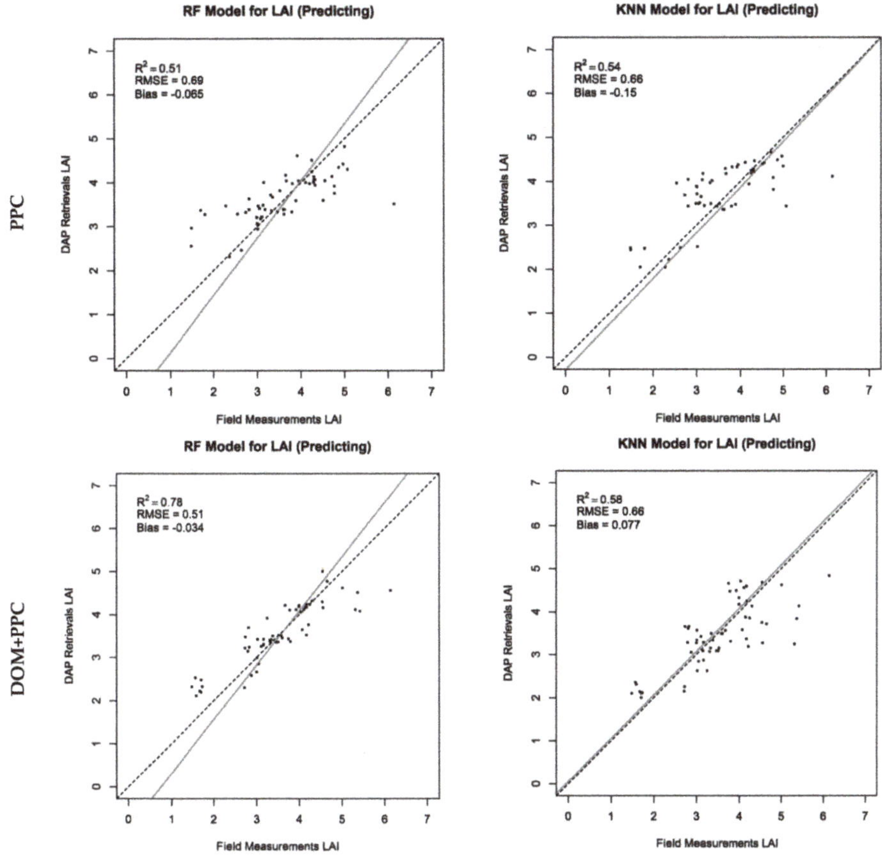

Figure 6. Prediction accuracy of the four models for LAI retrieval based on the three data sources.

4. Discussion

As important ecological and biophysical parameters, AGB and LAI were estimated using three data sources (DOM data alone, PPC data alone, and DOM + PPC data) from the DAP data in this study. Our findings showed that combining DOM and PPC data could improve estimation accuracy for AGB and LAI.

Better accuracy (R^2 = 0.73 and RMSE = 20 t/ha) for the AGB estimation based on a Cubist regression model was achieved from the combined model processing of vertical structure, vegetation indices, and textural features from the combination of DOM and PPC data. The retrieval result of AGB from height variables of PPC data was better than the textural feature and vegetation indices from DOM data. Height and coverage variables derived from PPC data were the first three selections. This shows that height is a key parameter in the estimation of AGB. Studies using airborne LiDAR or a combination of LiDAR and aerial photographs have shown how height can be used to estimate AGB [4,37,60]. Our study has confirmed that height is particularly important and a suitable index to AGB estimation when we used photogrammetric point cloud from the results of SfM approach processing. Ota et al. [37] obtained aboveground biomass using aerial photographs in a seasonal tropical forest. Canopy height models from aerial photograph DSM and LiDAR DTM yielded higher accuracy (R^2 = 0.93). Hansen et al. [61] estimated forest biomass based on empirical relationships between field-observed biomass and variables derived from LiDAR data, and a relatively lower accuracy was obtained (R^2 = 0.71, RMSE = 158 Mg/ha). Our accuracy of AGB estimation from combined DOM and PPC data is between Ota's and Hansen's results.

The R^2 and RMSE values of LAI estimation based on the SVM regression model from the combined DOM and PPC data were 0.79 and 0.48, respectively. Canopy cover (CC) from PPC data and the green red ratio index (GRRI) from DOM data contributed better results. It is worth noting that the GRRI is the second most important variable after CC. This indicates that GRRI has a strong contribution to LAI estimation. Previous studies using LiDAR, satellite images, or a combination of both showed similar results in estimating LAI. Ma et al. [62] estimated LAI based on full-waveform LiDAR data using a radiative transfer model. The R^2 and RMSE values of estimated LAI were 0.73 and 0.67, respectively. Mathews et al. [35] used a stepwise regression model to predict LAI based on a high-density point cloud using unmanned aerial vehicle (UAV) collection. The final result of the R^2 value was 0.57. Omer et al. [63] used spectral vegetation indices calculated from WorldView-2 data to predict LAI at the tree species level using support vector machines and artificial neural networks machine learning regression algorithms. They obtained better accuracy (R^2 = 0.75, RMSE = 0.05). Ma et al. [23] used the canopy height variable from LiDAR and the BRDF/Albedo variable from MODIS optical data to estimate LAI. The highest R^2 value was 0.73. Our accuracy is slightly higher than previous study results, and our study has confirmed that canopy coverage and vegetation index are particularly important and suitable indices to LAI estimation.

Compared with PPC data, DOM data had lower estimation accuracy (R^2 = 0.65, RMSE = 21 t/ha) of AGB, but the estimation accuracy (R^2 = 0.73, RMSE = 0.49) of LAI was higher. Similar findings were reported [9,63]. Summarizing these results above, we believe that the spectral and textural features had a larger contribution to LAI retrieval than vertical height features, and the forest vertical height features had greater effects on AGB retrieval. Our study is slightly different from previous studies. The differences are mainly because of the different data sources, different types of vegetation, biomass abundance, and canopy density.

As we can see in Figures 5 and 6, a combination of DOM and PPC data could improve the estimation accuracy of AGB and LAI compared with either data source alone. The combined DOM and PPC data yielded the highest estimation accuracy for AGB (R^2 = 0.73, RMSE = 20 t/ha) and LAI (R^2 = 0.79, RMSE = 0.48). These estimations were made using machine-learning with the Cubist regression model and SVM regression model, respectively. Our study has demonstrated the ability to estimate AGB and LAI using the combination of DOM and PPC from DAP data with the SfM approach.

The results of our study using only DAP data are consistent with previous studies using LiDAR and other optical remote sensing data. However, the acquisition and processing cost of DAP data is much lower than that of LiDAR, while the spatial resolution is much higher than that of optical remote sensing data. Using DAP data to estimate forest parameters is worthy of further exploration and research. Our study could provide valuable guidance for accurate AGB and LAI estimation using DOM+PPC data from DAP in boreal coniferous forests.

In this study, with the exception of DEM from LiDAR data, we used DAP data alone with only red, green, and blue visible bands to estimate AGB and LAI. In previous studies, multispectral vegetation indices were used [63]. The near-infrared band can provide much better information about vegetation health [64]. We imitated the principle of multispectral vegetation indices to construct visible light vegetation indices. The results may have been improved if we added the spectral information of the near-infrared band. This work will be performed in a future study.

For a future practical application of the results from this study in other areas, we can consider all variables (vertical features and horizontal features) in this paper as realistic approaches.

5. Conclusions

The objective of this study is to explore the ability of retrieving forest parameters only using DOM, PPC, and DOM+PPC from RGB-only DAP data. In the paper, three analyses on DOM, PPC, and combined DOM and PPC data for the estimation of AGB and LAI have been presented. The study indicates that a combination of DOM and PPC data is useful as it provides a slight increase in estimation accuracy. Height and coverage variables of PPC, texture mean value, and visible differential vegetation index (VDVI) of DOM are significantly related to the estimation of AGB (R^2 = 0.73, RMSE = 20 t/ha). The canopy cover of PPC and green red ratio index (GRRI) of DOM are the most strongly related to the estimation of LAI, followed by the height and coverage variables of PPC, texture mean value, the visible atmospherically resistant index (VARI), and the VDVI of DOM (R^2 = 0.79, RMSE = 0.48). The model derived from only DOM data provides lower accuracy than only PPC data for the estimation of AGB at the Mengjiagang forest farm. In terms of LAI estimation, the result is different. Variables from either DOM or PPC data alone can provide the majority of the explanatory contribution for LAI estimation.

The current study focuses on boreal coniferous forest areas using RGB-only DAP data, with an R2 higher than 0.7. Nevertheless, more studies should be carried on a variety of forest types to determine the validity of the defined parameters and the accuracy reported. Possible future developments of this work are (1) to consider other target variables (such as tree height and individual tree segment), (2) to use the same method in temperate broadleaved forest areas to determine the universality of the method, and (3) to add NIR spectral information.

Author Contributions: D.L. is the principal author of this manuscript, having written the majority of the manuscript and contributed at all phases of data processing, research, and field data analysis. X.G. and Y.P. contributed to the interpretation of the methods and provided suggestions regarding the structure. B.C. and L.L. contributed to the investigation and R program correction.

Acknowledgments: The study was funded by the China National Key Research and Development Program (2017YFD0600404); the China State Administration of Science, Technology and Industry for National Defense Program, "Major Special Project—the China High-Resolution Earth Observation System, Project of civil space pre-research of the 13th five-year plan" (Y7K00100KJ) and the Chinese Academy of Forestry Foundation (CAFYBB2016ZD004).

References

1. Dubayah, R.O.; Drake, J.B. Lidar remote sensing for forestry. *J. For.* **2000**, *98*, 44–46.
2. Ediriweera, S.; Pathirana, S.; Danaher, T.; Nichols, D. Estimating above-ground biomass by fusion of LiDAR and multispectral data in subtropical woody plant communities in topographically complex terrain in North-Eastern Australia. *J. For. Res.* **2014**, *25*, 761–771. [CrossRef]
3. Næsset, E.; Gobakken, T.; Bollandsås, O.M.; Gregoire, T.G.; Nelson, R.; Ståhl, G. Comparison of precision of biomass estimates in regional field sample surveys and airborne LiDAR-assisted surveys in Hedmark County, Norway. *Remote Sens. Environ.* **2013**, *130*, 108–120. [CrossRef]
4. Tuominen, S.; Haapanen, R. Estimation of forest biomass by means of genetic algorithm-based optimization of airborne laser scanning and digital aerial photograph features. *Silva Fennica.* **2013**, *47*, 20. [CrossRef]
5. Chen, J.M.; Cihlar, J. Retrieving leaf area index of boreal conifer forests using Landsat TM images. *Remote Sens. Environ.* **1996**, *55*, 179–193. [CrossRef]
6. Homolová, L.; Malenovský, Z.; Clevers, J.G.; García-Santos, G.; Schaepman, M.E. Review of optical-based remote sensing for plant trait mapping. *Ecol. Complex.* **2013**, *15*, 1–16. [CrossRef]
7. Seidel, D.; Fleck, S.; Leuschner, C.; Hammett, T. Review of ground-based methods to measure the distribution of biomass in forest canopies. *Ann. For. Sci.* **2011**, *68*, 225–244. [CrossRef]
8. Huang, J.L.; Ju, W.M.; Zheng, G.; Kang, T.T. Estimation of forest aboveground biomass using high spatial resolution remote sensing imagery. *Acta Ecol. Sin.* **2013**, *33*, 6497–6508. [CrossRef]
9. Nichol, J.E.; Sarker, M.L.R. Improved biomass estimation using the texture parameters of two high-resolution optical sensors. *IEEE Trans. Geosci. Remote Sens.* **2011**, *49*, 930–948. [CrossRef]
10. Gallaun, H.; Zanchi, G.; Nabuurs, G.J.; Hengeveld, G.; Schardt, M.; Verkerk, P.J. EU-wide maps of growing stock and above-ground biomass in forests based on remote sensing and field measurements. *For. Ecol. Manag.* **2010**, *260*, 252–261. [CrossRef]
11. Gao, S.; Niu, Z.; Huang, N.; Hou, X. Estimating the Leaf Area Index, height and biomass of maize using HJ-1 and RADARSAT-2. *Int. J. Appl. Earth Observ. Geoinf.* **2013**, *24*, 1–8. [CrossRef]

12. Mitchard, E.T.A.; Saatchi, S.S.; Woodhouse, I.H.; Nangendo, G.; Ribeiro, N.S.; Williams, M.; Ryan, C.M.; Lewis, S.L.; Feldpausch, T.R.; Meir, P. Using satellite radar backscatter to predict above-ground woody biomass: A consistent relationship across four different African landscapes. *Geophys. Res. Lett.* **2009**, *36*, L23401. [CrossRef]

13. Greaves, H.E.; Vierling, L.A.; Eitel, J.U.H. Estimating aboveground biomass and leaf area of low-stature Arctic shrubs with terrestrial LiDAR. *Remote Sens. Environ.* **2015**, *164*, 26–35. [CrossRef]

14. Moskal, M.L.; Zheng, G. Retrieving forest inventory variables with terrestrial laser scanning (TLS) in urban heterogeneous forest. *Remote Sens.* **2011**, *4*, 1–20. [CrossRef]

15. Lin, Y.; Herold, M. Tree species classification based on explicit tree structure feature parameters derived from static terrestrial laser scanning data. *Agric. For. Meteorol.* **2015**, *201*, 710–711. [CrossRef]

16. Tang, H.; Brolly, M.; Zhao, F.; Strahler, A.H.; Schaaf, C.L. Deriving and validating leaf area index (LAI) at multiple spatial scales through lidar remote sensing: A case study in Sierra National Forest, CA. *Remote Sens. Environ.* **2014**, *143*, 131–141. [CrossRef]

17. Xing, Y.Q.; Huo, D.; You, H.T.; Tian, X.; Jiao, Y.T. Estimation of birch forest LAI based on single laser penetration index of airborne LiDAR data. *J. Appl. Ecol.* **2016**, *27*, 3469–3478. (In Chinese)

18. Kankare, V.; Raty, M.; Yu, X.; Holopainen, M.; Vastaranta, M. Single tree biomass modeling using airborne laser scanning. *ISPRS J. Photogramm. Remote Sens.* **2013**, *85*, 66–73. [CrossRef]

19. Lefsky, M.A.; Cohen, W.B.; Parker, G.G.; Harding, D.J. LiDAR remote sensing for ecosystem studies. *BioScience* **2002**, *52*, 19–30. [CrossRef]

20. Lu, D. The potential and challenge of remote sensing-based biomass estimation. *Int. J. Remote Sens.* **2006**, *27*, 1297–1328. [CrossRef]

21. Ranson, K.J.; Sun, G.; Lang, R.H.; Chauhan, N.S.; Cacciola, R.J.; Kilic, O. Mapping of boreal forest biomass from spaceborne synthetic aperture radar. *J. Geophys. Res.* **1997**, *102*, 29599–29610. [CrossRef]

22. Liu, L.X.; Coops, N.C.; Aven, N.W.; Pang, Y. Mapping urban tree species using integrated airborne hyper-spectral and LiDAR remote sensing data. *Remote Sens. Environ.* **2017**, *200*, 170–182. [CrossRef]

23. Ma, H.; Song, J.L.; Wang, J.D.; Fu, D. Improvement of spatially continuous forest LAI retrieval by integration of discrete airborne LiDAR and remote sensing multi-angle optical data. *Agric. For. Meteorol.* **2014**, *189*, 60–70. [CrossRef]

24. Zolkos, S.G.; Goetz, S.J.; Dubayah, R. A meta-analysis of terrestrial aboveground biomass estimation using LiDAR remote sensing. *Remote Sens. Environ.* **2013**, *128*, 289–298. [CrossRef]

25. Luo, S.Z.; Wang, C.; Xi, X.H.; Pan, F.F.; Peng, D.L.; Zou, J.; Nie, S.; Qin, H.M. Fusion of airborne LiDAR data and hyperspectral imagery for aboveground and belowground forest biomass estimation. *Ecol. Indic.* **2017**, *73*, 378–387. [CrossRef]

26. Tonolli, S.; Dalponte, M.; Neteler, M.; Rodeghiero, M.; Vescovo, L.; Gianelle, D. Fusion of airborne LiDAR and satellite multispectral data for the estimation of timber volume in the southern Alps. *Remote Sens. Environ.* **2011**, *115*, 2486–2498. [CrossRef]

27. Tsui, O.W.; Coops, N.C.; Wulder, M.A.; Marshall, P.L.; McCardle, A. Using multi-frequency radar and discrete-return LiDAR measurements to estimate above-ground biomass and biomass components in a coastal temperate forest. *ISPRS J. Photogramm. Remote Sens.* **2012**, *69*, 121–133. [CrossRef]

28. Vaglio, L.G.; Chen, Q.; Lindsell, J.A.; Coomes, D.A.; Frate, F.D.; Guerriero, L.; Pirotti, F.; Valentini, R. Above ground biomass estimation in an African tropical forest with LiDAR and hyperspectral data. *ISPRS J. Photogramm. Remote Sens.* **2014**, *89*, 49–58. [CrossRef]

29. Pflugmacher, D.; Cohen, W.B.; Kennedy, R.E.; Yang, Z. Using Landsat-derived disturbance and recovery history and Lidar to map forest biomass dynamics. *Remote Sens. Environ.* **2014**, *151*, 124–137. [CrossRef]

30. Tuominen, S.; Pekkarinen, A. Performance of different spectral and textural aerial photograph features in multisource forest inventory. *Remote Sens. Environ.* **2005**, *94*, 256–268. [CrossRef]

31. Puliti, S.; Ørka, H.O.; Gobakken, T.; Næsset, E. Inventory of small forest areas using an unmanned aerial system. *Remote Sens.* **2015**, *7*, 9632–9654. [CrossRef]

32. Zahawi, R.A.; Dandois, J.P.; Holl, K.D.; Nadwodny, D.; Reid, J.L.; Ellis, E.C. Using lightweight unmanned aerial vehicles to monitor tropical forest recovery. *Biol. Conserv.* **2015**, *186*, 287–295. [CrossRef]

33. Puliti, S.; Gobakken, T.; Ørka, H.O.; Næsset, E. Assessing 3D point clouds from aerial photographs for species-specific forest inventories. *Scand. J. For. Res.* **2017**, *32*, 68–79. [CrossRef]

34. Nurminen, K.; Karjalainen, M.; Yu, X.W.; Hyyppä, J.; Honkavaara, E. Performance of dense digital surface models based on image matching in the estimation of plot-level forest variables. *ISPRS J. Photogram. Remote Sens.* **2013**, *83*, 104–115. [CrossRef]

35. Mathews, A.J.; Jensen, J.L.R. Visualizing and quantifying vineyard canopy LAI using an unmanned aerial vehicle (UAV) collected high density structure from motion point cloud. *Remote Sens.* **2013**, *5*, 2164–2183. [CrossRef]

36. Dandois, J.P.; Ellise, E.C. High spatial resolution three-dimensional mapping of vegetation spectral dynamics using computer vision. *Remote Sens. Environ.* **2013**, *136*, 259–276. [CrossRef]

37. Ota, T.; Ogawa, M.; Shimizu, K.; Kajisa, T.; Mizoue, N. Aboveground biomass estimation using structure from motion approach with aerial photographs in a seasonal tropical forest. *Forests* **2015**, *6*, 3883–3898. [CrossRef]

38. Fonstad, M.A.; Dietrich, J.T.; Courville, B.C; Jensen, J.L; Carbonneau, P.E. Topographic structure from motion: A new development in photogrammetric measurement. *Earth Surf. Process. Landf.* **2013**, *38*, 421–430. [CrossRef]

39. Asner, G.P.; Sinan, S.; Knapp, D.E.; Selmants, P.C.; Martin, R.E.; Hughes, R.F.; Giardina, C.P. Rapid forest carbon assessments of oceanic islands: A case study of the Hawaiian archipelago. *Carbon Balance Manag.* **2016**, *11*, 1–13. [CrossRef] [PubMed]

40. Tamm, T.; Remm, K. Estimating the parameters of forest inventory using machine learning and the reduction of remote sensing features. *Int. J. Appl. Earth Obs.* **2009**, *11*, 290–297. [CrossRef]

41. Li, H.K.; Lei, Y.C. *China Forest Vegetation Biomass and Carbon Stock Assessment*; China Forestry Publishing House: Beijing, China, 2010. (In Chinese)

42. Chen, C.G.; Zhu, J.F. *Northeast Main Forest Biomass Handbook*; China Forestry Publishing House: Beijing, China, 1989. (In Chinese)

43. MathWorks. Structure from Motion. Available online: http://www.mathworks.com/help/vision/ug/structure-from-motion.html (accessed on 4 May 2018).

44. Torres-Sánchez, J.; Peña, J.M.; Castro, A.I.; López-Granados, F. Multi-temporal mapping of the vegetation fraction in early-season wheat fields using images from UAV. *Comput. Electron. Agric.* **2014**, *103*, 104–113. [CrossRef]

45. Gitelson, A.A.; Vina, A.; Arkebauer, T.J.; Rundquist, D.C.; Keydan, G.P.; Leavitt, B.; Keydan, G. Remote estimation of leaf area index and green leaf biomass in maize canopies. *Geophys. Res. Lett.* **2003**, *30*, 335–343. [CrossRef]

46. Verrelst, J.; Schaepman, M.E.; Koetz, B.; Kneubuhler, M. Angular sensitivity analysis of vegetation indices derived from CHRIS/PROBA data. *Remote Sens. Environ.* **2008**, *112*, 2341–2353. [CrossRef]

47. Sellaro, R.; Crepy, M.; Trupkin, S.A.; Karayekov, E.; Buchovsky, A.S.; Rossi, C.; Casal, J.J. Cryptochrome as a sensor of the blue/green ratio of natural radiation in Arabidopsis. *Plant Physiol.* **2010**, *154*, 401–409. [CrossRef] [PubMed]

48. Kuhn, M.; Weston, S.; Williams, A.; Keefer, C.; Engelhardt, A.; Cooper, T.; Mayer, Z.; Kenkel, B.; the R Core Team; Benesty, M.; et al. Caret: Classification and Regression Training. R Package Version 6.0-78. Available online: http://CRAN.R-project.org/package=caret (accessed on 1 August 2017).

49. Hijmans, R.J. Raster: Geographic Data Analysis and Modeling R Package Version 2.6-7. 2017. Available online: https://CRAN.R-project.org/package=raster (accessed on 1 August 2017).

50. Wickham, H.; Francois, R.; Henry, L.; Müller, K. Dplyr: A Grammar of Data Manipulation. R Package Version 0.7.4. 2017. Available online: https://CRAN.R-project.org/package=dplyr (accessed on 1 August 2017).

51. Xycoon. Statistics-Econometrics-Forecasting. Office for Research Development and Education. Available online: http://www.xycoon.com (accessed on 2 April 2018).

52. Engineering Statistics Handbook. Available online: http://www.itl.nist.gov/div898/handbook/eda/section3/eda35b.htm (accessed on 2 April 2018).

53. Wang, O.J. Direct sample estimators of L moments. *Water Resour. Res.* **1996**, *32*, 3617–3619. [CrossRef]

54. McGaughey, R.J. *FUSION/LDV: Software for LiDAR Data Analysis and Visualization*; FUSION Version 3.60+; United States Department of Agriculture: Washington, DC, USA, 2016.

55. Wei, T.Y.; Simko, V. R package "corrplot": Visualization of a Correlation Matrix (Version 0.84). 2017. Available online: https://github.com/taiyun/corrplot (accessed on 3 August 2017).

56. Granitto, P.M.; Furlanello, C.; Biasioli, F.; Gasperi, F. Recursive feature elimination with random forest for PTR-MS analysis of agroindustrial products. *Chemometr. Intell. Lab.* **2006**, *83*, 83–90. [CrossRef]

57. Meyer, D.; Dimitriadou, E.; Hornik, K.; Weingessel, A.; Friedrich Leisch, F. e1071: Misc Functions of the Department of Statistics, Probability Theory Group (Formerly: E1071), TU Wien. R package version 1.6-8. 2017. Available online: https://CRAN.R-project.org/package=e1071 (accessed on 3 August 2017).

58. Kuhn, M.; Quinlan, R. Cubist: Rule- And Instance-Based Regression Modeling. R Package Version 0.2.1. 2017. Available online: https://CRAN.R-project.org/package=Cubist (accessed on 3 August 2017).

59. Liaw, A.; Wiener, M. Classification and Regression by randomForest. *R News* **2002**, *23*, 18–22.

60. Manuri, S.; Andersen, H.E.; McGaughery, R.J.; Brack, C. Assessing the influence of return density on estimation of lidar-based aboveground biomass in tropical peat swamp forests of Kalimantan, Indonesia. *Int. J. Appl. Earth Obs.* **2017**, *56*, 24–35. [CrossRef]

61. Hansen, E.; Gobakken, T.; Bollandsås, O.; Zahabu, E.; Næsset, E. Modeling aboveground biomass in dense tropical submontane rainforest using airborne laser scanner data. *Remote Sens.* **2015**, *7*, 788–807. [CrossRef]

62. Ma, H.; Song, J.L.; Wang, J.D. Forest canopy LAI and vertical FAVD profile inversion from airborne full-waveform LiDAR data based on a radiative transfer model. *Remote Sens.* **2015**, *7*, 1897–1914. [CrossRef]

63. Omer, G.; Mutanga, O.; Abdel-Rahman, E.M.; Adam, E. Empirical prediction of leaf area index (LAI) of endangered tree species in intact and fragmented indigenous forests ecosystems using WorldView-2 data and two robust machine learning algorithm. *Remote Sens.* **2016**, *8*, 1–26. [CrossRef]

64. Pinty, B.; Lavergne, T.; Widlowski, J.L.; Gobron, N.; Verstraete, M.M. On the need to observe vegetation canopies in the near-infrared to estimate visible light absorption. *Remote Sens. Environ.* **2009**, *113*, 10–23. [CrossRef]

Article

The Potential of High Resolution (5 m) RapidEye Optical Data to Estimate Above Ground Biomass at the National Level over Tanzania

Lorena Hojas Gascón [1,2], Guido Ceccherini [2], Francisco Javier García Haro [1], Valerio Avitabile [2] and Hugh Eva [2,*]

1   Facultat de Física, Universitat de València, Burjassot, 46100 València, Spain; lorenahojas@hotmail.com (L.H.G.); J.Garcia.Haro@uv.es (F.J.G.H.)
2   Sustainable Resources Directorate, Joint Research Centre (JRC), European Commission, Via E. Fermi, 2749, TP 261, VA 21027 Ispra, Italy; guido.ceccherini@ec.europa.eu (G.C.); valerio.avitabile@ec.europa.eu (V.A.)
*   Correspondence: hugh.eva@ec.europa.eu; Tel.: +39-0332-785016

Received: 3 December 2018; Accepted: 23 January 2019; Published: 29 January 2019

Abstract: In this paper, we review the potential of high resolution optical satellite data to reduce the significant investment in resources required for a national field survey for producing estimates of above ground biomass (AGB). We use 5 m resolution RapidEye optical data to support a country wide biomass inventory with the objective of bringing to the attention of the traditional forestry sector the advantages of integrating remote sensing data in the planning and execution of field data acquisition. We analysed the relationship between AGB estimates from a subset of the national survey field plot data collected by the Tanzania Forest Service, with a set of remote sensing biophysical parameters extracted from a sample of fine spatial (5 m) resolution RapidEye images using a regression estimator. We processed RapidEye data using image segmentation for 76 sample sites each of 20 km by 20 km (covering 2.3% of the land area of the country) to image objects of 1 ha. We extracted reflectance and texture information from those objects which overlapped with the field plot data and tested correlations between the two using four different models: Two models from inferential statistics and two models from machine learning. The best results were found using the random forests algorithm (R^2 = 0.69). The most important explicative factor extracted from the remote sensing data was the shadow index, measuring the absorption of light in the visible bands. The model was then applied to all image objects on the RapidEye images to obtain AGB for each of the 76 sample sites, which were then interpolated to estimate the AGB stock at the national scale. Using the relative efficiency measure, we assessed the improvement that the introduction of remote sensing data brings to obtain an AGB estimate at the national level, with the same precision as the full survey. The improvement in the precision of the estimate (by reducing its variance) resulted in a relative efficiency of 3.2. This demonstrates that the introduction of remote sensing data at this fine resolution can substantially reduce the number of field plots required, in this case threefold.

Keywords: forests biomass; remote sensing; REDD+; random forest; Tanzania; RapidEye

---

1. Introduction

1.1. Background to the Study—The REDD+ Initiative

Deforestation and forest degradation account for up to 12% of the global carbon emissions [1]. Most of these emissions (>80%) [2] originate from the tropics, where over 90% of humid forests are located. The United Nations Framework Convention on Climate Change (UNFCCC) aims to reduce such emissions via its REDD+ mechanism (Reducing Emissions from Deforestation and forest

Degradation), by financially rewarding developing countries for reducing greenhouse gas (GHG) emissions from deforestation and degradation. Countries may claim financial incentives by submitting reports on carbon stock changes based on the area changes in forest cover (activity data) and density of forest carbon stock (emission factors) [2].

To estimate GHG emissions, countries should use the Intergovernmental Panel on Climate Change (IPCC) guidance and guidelines as adopted (i.e., [3]) or encouraged [4]. According to such guidelines, there are three approaches for assessing forest area changes (activity data) and three tiers for assessing emissions factors, each with increasing accuracy and precision.

The activity data must be assessed with satellite data at a minimum mapping unit of up to a maximum of 1.0 ha. Area changes of forest to non-forest (i.e., deforestation) consider that a forest has a minimum tree-crown cover of 10–30%, with trees that have reached, or could reach, a minimum height of 2–5 m at maturity in the same location. Changes within the same forest class are also accounted for when there is a reduction of the carbon stock (i.e., degradation). Challenges remain for consistently mapping activity data [5], notably forest degradation [6].

For emissions factors, three tiers are proposed: Tier 1 uses default above-ground biomass (AGB) values per ecological zone and continent. Tier 2 and Tier 3 are more elaborate, based on country-specific remote sensing or permanent sample-plot data. Although the Tier 1 approach has large uncertainties, for the time being, many developing countries have to use this representative carbon stock as they lack the financial and technical capacities to implement forest carbon stock inventories to derive Tier 2 and Tier 3 emission factors [7].

While plot data are often regarded as 'ground truth', they also have many uncertainties—location, measurements [8,9] size of plots [10], and extrapolation models (i.e., allometric equations) [11–14].

A key challenge, therefore, is to enable participating countries to obtain estimates of carbon stocks (AGB) in a cost effective way.

It is in this context that we place the current work. We wish to demonstrate that by combining optical RS data with a reduced number of field measurements, substantial savings can be made in making AGB estimates, both in time and resources. The work is undertaken not as a test case, where all data are available for a chosen test site, but as an 'operational' national study, where we work with the data available—both remote sensing and field measurements. In addressing a country wide survey, we also must deal with different ecosystems manifesting different vegetation types that exist across Tanzania. In addition, the availability of satellite images and their utility varies across the country, restricted by available cloud free imagery, and vegetation conditions due to phenology. This paper attempts to address these problems.

1.2. Use of Remote Sensing Data to Map Above Ground Biomass

Mapping and monitoring of forest carbon stocks in the tropics has traditionally relied on field surveys that are often limited to particular ecosystems and locations [15]. Given the restrictions of the scope and of the costs of recurring surveys, many researchers have used available satellite and aerial remote sensing instead [16–18].

Earth Observation offers the unique opportunity to assess the state and changes of vegetation dynamics, providing data over large (e.g., continental scale) areas and long (e.g., a decade or more) periods, at spatial and temporal sampling frequencies that are potentially suitable for detecting key forest carbon stock distribution and changes [19,20].

Remote sensing-based datasets that are widely used are: (1) Light Detection and Ranging (LiDAR); (2) optical data at various spatial resolutions; and (3) Synthetic Aperture Radar (SAR). LiDAR data are able to map forest structure and height in the three dimensions [21–24]. However, acquisition costs, processing, and difficulties in replicating in both space and time often render the use of LiDAR difficult. Optical data have been extensively used for AGB modelling with very high (less than 10 m) or coarse spatial resolution (*circa* 250m) data. Generally, AGB estimation from optical remotely sensed data is carried out through regression models, based on relationships between AGB and reflectance [25].

Direct biomass estimations based on optical satellite data have been successfully carried out with very high resolution images (VHR) [26–30]. Despite good results, such approaches are only currently applicable to small area sites, due to their costs and availability of VHR data.

At the regional level, coarse spatial resolution satellite imagery have been used in combination with field plot measurements and space-borne LiDAR data to derive wall-to-wall pan-tropical biomass maps at 1 km resolution for the year, 2000 [31], and 500 m resolution for 2007 [32]. Both studies use a similar approach, correlating tree height from Geoscience Laser Altimeter System (GLAS) data points to AGB values of (spatially overlapping) field inventory plots. These are then interpolated to deduce the AGB values for all non-overlapping GLAS data points.

Langner [33] reviewed these two pantropical biomass maps and a combination of both (combined dataset) in comparison with the IPCC Tier 1 values, which led to an overestimation of AGB as the values often refer to intact forest sites [31]. The mean AGB show a good consistency between the two datasets values per pan-tropical ecological zone with a correlation coefficient (R^2) of 0.87. When restricting the regression to intact forest areas (IFA [34]), R^2 is even higher: 0.97. For non-IFA, R^2 is lower: 0.80. A comparison between these two pan-tropical maps generally shows higher AGB values from the 500 m map [32] than the 1 km map [31]. However, a detailed look at the data shows significant differences (both in spatial distribution and magnitude of AGB) between the maps, notably for tropical dry Africa. Nevertheless, Mitchard et al. [35] concluded that no single map was generally superior to the other despite the substantial differences.

The main limitations of optical satellite data are: (i) Cloud cover can significantly affect AGB estimates by limiting the number of satellite orbits acquired over a cloudy area; (ii) optical data become saturated with dense canopy and high biomass.

SAR (Synthetic Aperture Radar) data are considered as reliable indicators of AGB [36–39]. Radar backscatter is highly sensitive to vegetation structural properties that are related to biomass [40], with C, L, and P bands giving higher backscatters to leaves, branches, and trunks respectively [41–43]. The ability of SAR to penetrate cloud cover makes it particularly valuable in cloudy areas, such as the tropics, however, limitations occur in areas of high topography. L-band is the most widely used for forest AGB estimation. The physical properties of vegetation and woody components (comparable in size to the wavelengths used) influences the AGB-backscatter relationship. However, the L-band is only efficient up to around 100 t ha^{-1}, above which sensitivity to AGB is lost [44].

To overcome this limitation, data fusion of SAR and optical images has been proposed [45–47]. The combination of ALOS PALSAR 2 (Phased Array type L-band Synthetic Aperture Radar) and WorldView-2 data has been shown to be a promising approach to improve biomass estimation over a larger area in China [48]. Scientific consensus on the accuracy of SAR and optical data fusion is still not complete [49], and most of these tests are of limited area coverage, unlike this study where a country wide estimate is reviewed.

1.3. The Current Study

We use field plot data collected by Tanzania's NAFORMA project (National Forest Monitoring and Assessment) in conjunction with RapidEye, 5 m resolution optical satellite data. The choice of RapdiEye data was seen as the opportunity to have high spatial resolution data over a large area. Previous studies on linking optical data at this (national) scale rely on low (*circa* 250 m) to medium (*circa* 30 m) resolution data [50].

Nationwide field surveys for measuring biomass, such as that carried out by NAFORMA, are expensive and time consuming and contain errors, in some cases up to 30% [9]. We assess the possibility of combining a limited sample of the field survey data (*circa* 1000 field plots out of 32,000) to train and validate estimates of AGB using remotely sensed images through a regression estimator. We use a readily available fine spatial (5 m) resolution RapidEye optical data systematically distributed across Tanzania in 76 sample sites, in conjunction with a corresponding subset of field plots extracted from the national survey carried out by NAFORMA. Different parametric and non-parametric approaches were applied to develop linear and nonlinear regressions between input variables and AGB, using random forest (RF) and support vector machine (SVM) regression methods.

If by combining these two data sets (through modelling and extrapolation), we can demonstrate the potential for providing estimates of AGB at the national level with less field data, significant financial savings could be achieved. This would allow more countries to rapidly provide data on AGB for REDD+ reporting [51]. The measure of the improvement of the combined approach compared to the field plots alone is provided by the relative efficiency [10,52], which gives the improvement in precision (i.e., reduction in variance). A relative efficiency of two would mean that a ground survey of X plots with remote sensing correction would give the same precision as a ground survey of 2X plots without remote sensing data.

The novelties of this study are in the combination of: (i) The use of 5m optical data to support a country wide biomass inventory across all ecosystems; (ii) the use of object segmentation at this scale along with texture measures [26,27,53] and (iii) the provision of traditional forestry sector agencies with guidelines for integrating remote sensing data in the planning of field data acquisition. If the remote sensing parameters and field data results on AGB can be shown to be correlated, one can obtain significant cost savings by reducing field data acquisition.

The work therefore addresses two related themes: (1) Specific results and quantification of increased precision from use of RapidEye optical data, and (2) a variety of practical, operational considerations and potential applications of RapidEye data.

2. Materials and Methods

The methodology was applied in the following steps: We determined the optimal minimum mapping unit (MMU) to employ when processing the remote sensing data and applied it to the RapidEye data after cloud and cloud shadow masking. Reflectance, texture, and indices were extracted from those image objects which corresponded spatially to the field data (geometric location), and ambiguous or erroneous field data were screened and removed. We then tested models so as to relate field data on AGB with the spectral and textural parameters extracted from satellite data. The best model was then applied to all the image objects in the full Remote Sensing data set, so as to obtain AGB estimates for each of the remote sensing sample sites. These results were then interpolated by direct expansion to the national level and compared to the results from the full field survey.

2.1. Study Area

The study area covers the full land surface of mainland Tanzania, which is located on the east coast of Africa between parallels 1° S and 12° S and meridians 30° E and 40° E. Tanzania has an area of 945,090 km^2. Following the White classification scheme [54], the country is covered by five ecosystems (percentage area of the country is given in brackets); the Zambezian ecoregion (55.7%), the Zanzibar-Inhambane (8.5%) the Somali Masai (26.5%), and the Afromontane region (4.8%). The Victoria region refers to the area around Lake Victoria (4.5%) The predominant vegetation covers are respectively: Miombo woodland, heavily disturbed coastal forests, arid shrub-lands, and dense forests (Figure 1) [55].

Almost all of the forests are naturally regenerated, with only 1% of the forest cover being considered as 'primary'. The national survey does not give AGB levels by ecological zone, however, estimates are given by vegetation type (Table 1) from which we see that the country average will be close to that of woodlands, which are pre-dominant.

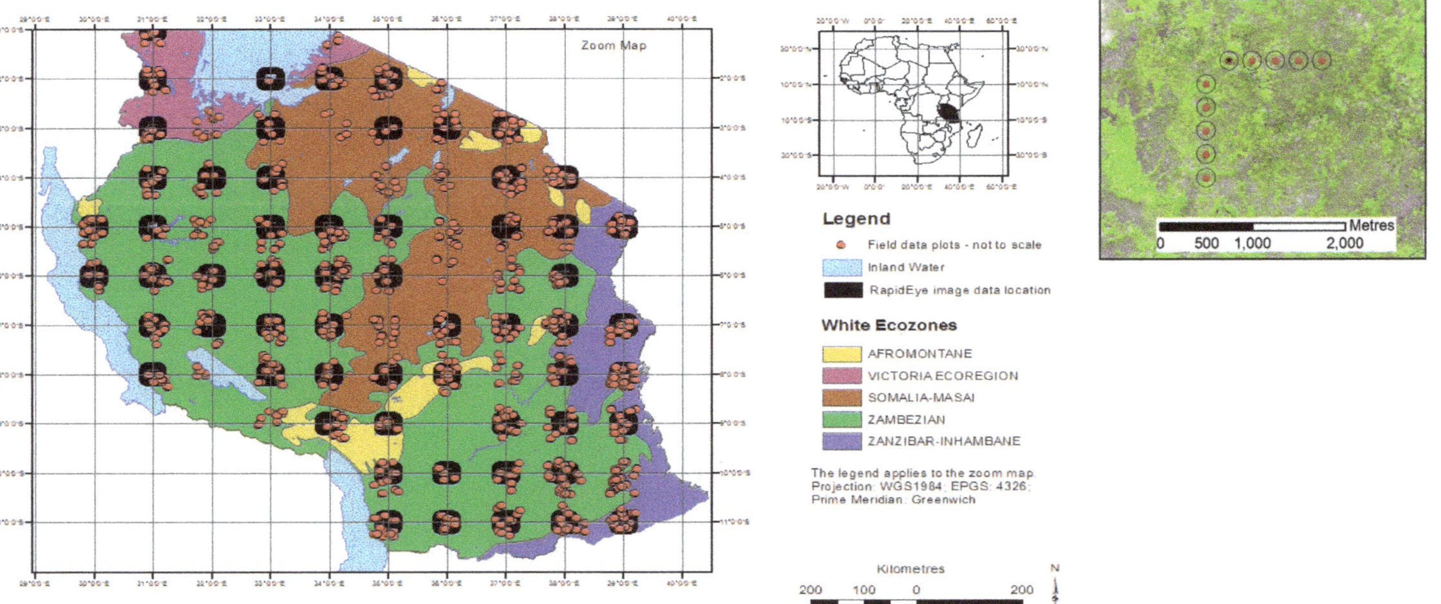

Figure 1. The ecological zone map of Tanzania [54] with the location of the field plots (not to scale) to coincide with the satelite data. An excert of the points for one cluster overlaid on a RapidEye image is shown to the right.

Table 1. Above ground biomass (AGB) by vegetation type: National Forestry Resources monitoring and Assessment (NAFORMA) Main Report [55].

Vegetation	Area (ha)	AGB (t/ha)	Share of Carbon Stock (%)
Forest	3,364,457	59.5	11.5
Woodland	44,726,246	27.7	73.5
Bushland	6,445,471	11	4.4
Grassland	8,242,245	2.9	2.3
Cultivated land	22,248,092	5.9	8
Water	1,162,552		0.3

2.2. NAFORMA Field Data

The National Forestry Resources monitoring and Assessment (NAFORMA) project was set up by the Ministry of Natural Resources of Tanzania (MNRT) and the United Nations Programme on Reducing Emissions from Deforestation and Forest Degradation (UN-REDD), with the technical support of FAO (the Food and Agriculture Organization) and Finland [56,57] for assessing and monitoring forest carbon pools compatible with REDD+ requirements. The national field data set collected by NAFORMA was designed to provide a national estimate of forest area, wood volume, and growing stock [58], along with a set of social indicators and a soil database. The survey design was stratified using potential biomass and accessibility to facilitate the field visits [59,60]. Some 32,000 plots were visited over a period of 3 years, collecting the biophysical and social variables. The final results were released in 2015 [55].

The sampling design produced clusters each of 10 plots, five plots running west-east and five north-south, with a distance of 250 m between plots (see inset in Figure 1). The size of the plots (15 m radius circle) was devised for efficiency in the field, however, this may not be optimum for linking with remote sensing data [10]. The field work started in May 2010 and was completed in June 2013. Data for over half the field plots were collected during the period of RapidEye satellite acquisition. The remaining images were acquired up to a maximum of 14 months after the field data (see Supplementary Materials). For each plot, data were collected on canopy cover, tree and shrub height, trunk diameter (DBH), species, dead wood, and soil. The tree measurements (DBH $\geq$ 1 cm) were conducted in concentric circular plots. Species, health, and diameter were measured for all trees, and height, stump diameter, and bole height measured for tally trees (i.e., every 5th tree) [58,61]. We used the data from individual plots, rather than aggregating the data at the cluster scale, due to the observed heterogeneity between adjacent plots.

The biophysical data on basal area and tree height were then used by NAFORMA to calculate bole volume in a model and hence carbon biomass for each plot, from which national and regional level estimates were made. Allometric equations were used to calculate tree volume using the DBF and estimated height for each plot. Different models were used, three for commonly used plantation tree species and two generic models for the remaining natural species. To obtain AGB, the tree volume was multiplied by the wood specific density, which was given as 0.58 for forest species and 0.50 for woodlands [58]. While we use these data as our reference, we also need to be aware that the models used to calculate AGB from field data also suffer from bias [62].

Through a collaborative agreement with the Tanzania Forest Service, our institution, the European Commission's Joint Research Centre (JRC), was given access to a subset of the NAFORMA field data. This subset consisted of those plots falling within 20 km by 20 km boxes centered on the 76 latitude-longitude confluence points (i.e., the crossing of the integral longitude and latitude) in Tanzania, a total of around 1000 plots. These sites were chosen as they corresponded to the satellite image sample sites used for the Global Forest Resources Assessment Remote Sensing Survey 2010 carried out by the JRC and United Nations Food and Agriculture Organization (FAO) [63]. Their geographical distribution means that the remote sensing data cover all types of different ecosystems in the country.

2.3. RapidEye Satellite Data

To support the FAO FRA remote sensing survey, the European Union's GMES (Global Monitoring for Environment and Security) provide a RapidEye dataset [64] over the 76 confluence points in Tanzania. For providing continental estimates of forest change between 2000 and 2010, the FRA project, in conjunction with the JRC, used this systematic sample of satellite images across the globe with samples based on the latitude-longitude confluence points (i.e., the crossing for the integral longitude and latitude) [65]. These data were the only freely available 5 m resolution data set which covered Tanzania. Over 80% of the images were acquired in the second half of 2010, with the remaining images coming from the first quarter of 2011. On average, the image acquisition precedes the field data collection by 10 months (see Supplementary Materials Figure S1 for image acquisitions and differences between field data collection).

RapidEye data come from a constellation of five radiometrically inter-calibrated satellites [66], each with a swath width of approximately 500 km, acquiring data in 5 spectral bands (Table 2), from blue to near infra-red (including a 'red edge' band). The data revisit period is a nominal of 1 day on request, however, the effective acquisition is limited due to recording and receiving capacities, relatively high cloud cover over the target area, and priority demands for commercial acquisitions. The latter is especially true for Africa, where lower data availability can occur due to the demand for images over the European agricultural zone. The 'best' (cloud and artefact free) images from RapidEye satellites 3, 4, and 5 were selected and mosaicked using all bands to cover a 20 km by 20 km box around each confluence point (Figure 2).

Table 2. RapidEye Spectral Channels.

Channel	Spectral Band Name	Spectral Range (nm)
1	Blue	440–510
2	Green	520–590
3	Red	630–685
4	Red Edge	690–730
5	Near infra-red	760–850

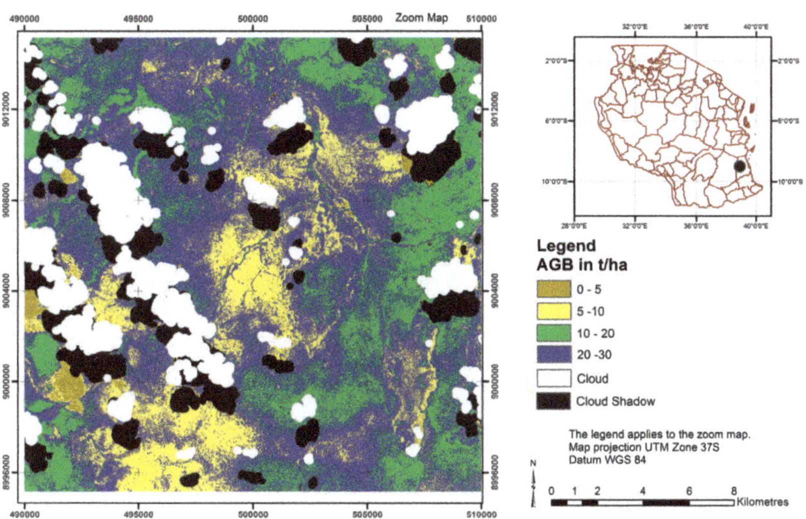

(a)

Figure 2. *Cont.*

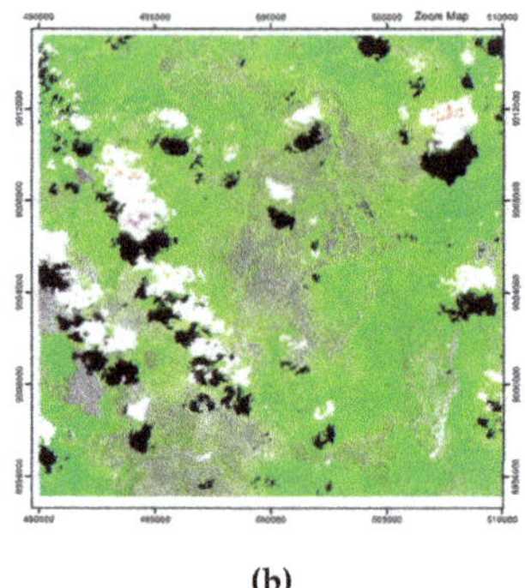

(b)

Figure 2. The RapidEye ABG map (**a**) for a sample centered on the 9° S 38° E confluence point and (**b**) the satellite image false colour composite (Shortwave Infrared (SWIR), Near-Infrared (NIR), Red). Cloud and cloud shadow are masked out in the first stages of the pre-processing, then image segments are assigned AGB values using the model developed using random forest.

2.4. RapidEye Preprosessing—Radiometric Calibration

RapidEye level 3A data are provided as 5 band layer stacked Geo-Tiff files, with a nominal 5 m pixel size stored as 16 bit data. The data were converted to at-sensor radiance (W m^{-2} sr^{-1} µm^{-1}), and then the top of atmosphere reflectance was calculated using the local solar zenith angle and sun irradiance supplied for each band in the metadata and published calibration coefficients [64]. To obtain at sensor radiance in watts per steradian per square meter (W m^{-2} sr^{-1} µm^{-1}), a scale factor is applied as follows:

$$L_\lambda = DN_\lambda \times ScaleFactor(\lambda) \tag{1}$$

where ScaleFactor(λ) = 0.01.

The Top of Atmosphere reflectance is calculated by:

$$\varrho_\lambda = \pi \times L_\lambda \times d^2/ESUN_\lambda \times \cos \theta_{SZ} \tag{2}$$

where:

ϱ_λ = TOA reflectance for band λ

L_λ = Radiance for band λ

θ_{SZ} = Local solar zenith angle

d = (1 − 0.01672 × cos (0.01745 × (0.9856 × (Julian Day Image − 4)))

The mean exoatmospheric solar irradiance, ESUN λ, in W/m^2/µm) for each channel is respectively: ESUN $\lambda_{1\text{-}5}$ = (B1 = 1997.8 B2 = 1863.5 B3 = 1560.4 B4 = 1395.0 B5 = 1124.4).

The reflectance values, ranging between 0 and 1, are rescaled to 16 bit Unsigned Integer (0–10,000) with a linear multiplication factor of 10,000. The required formulas and parameters are taken from the literature [64]. We performed an 'evergreen forest normalization' [67,68]), based on the theory of dark object subtraction [69], which reduces the variance in reflectance between images acquired at different dates and locations.

2.5. RapidEye Preprosessing—Image Segmentation to Obtain The MMU

The UNFCC, at its 7th Convention of the Parties (COP), proposed that countries should choose an MMU no greater than 1 ha [70]. For its submission to the UNFCCC, Tanzania used a 0.5 ha MMU [71].

Image segmentation was chosen to map landscape units at a cartographic MMU corresponding as closely as possible to that set by the national authority responsible for forests (the Tanzania Forest Service) and international guidelines. Almost all national forest definitions are based on tree height and tree cover with a minimum mapping unit. Remote sensing data are imaged at the pixel level and so segmentation allows us to cluster pixels of a similar reflectance and texture into objects (vector polygons) that correspond to landscape units (e.g., 'a forest') [72]. The segmentation of the image into the objects is achieved through the software package, eCognition [73].

To select the segmentation parameters which determine the aggregation of pixels to objects (or polygons), we effectuated a series of tests changing the shape factor, the compactness, and reflectance weights of the segmentation process executed in the eCognition software package. We found that a high compact value (0.9) provided results that were more in line with traditional photo interpretation results. Results with lower compactness gave us highly fragmented polygons.

The shape factor is the parameter (integer > 1) that controls the size of the objects [74]. The greater the shape factor, the larger the objects. However, the results of the segmentation (i.e., object size) are dependent not only on the shape factor, but on the image itself, i.e., the image contrast, variance and landscape fragmentation. Therefore, a different value of the shape factor parameter may be required to obtain objects of the same size on different images. In conjunction with the FAO, the JRC developed an iterative routine applied to each image to create initial, small, landscape units, and then in a stepwise process, increased these units until the mean object area of each of the image corresponded to the MMU (Figure 3). Objects smaller than the MMU were then dissolved and added to the adjacent object with the closest spectral signature [72]. As the mean object area must be calculated only on land cover features, objects corresponding to clouds and cloud shadows had to be detected, masked out, and omitted from the routine. An example of the increasing shape factor iteration is given in the Supplementary Materials Figure S3 with the final result in Figure S4.

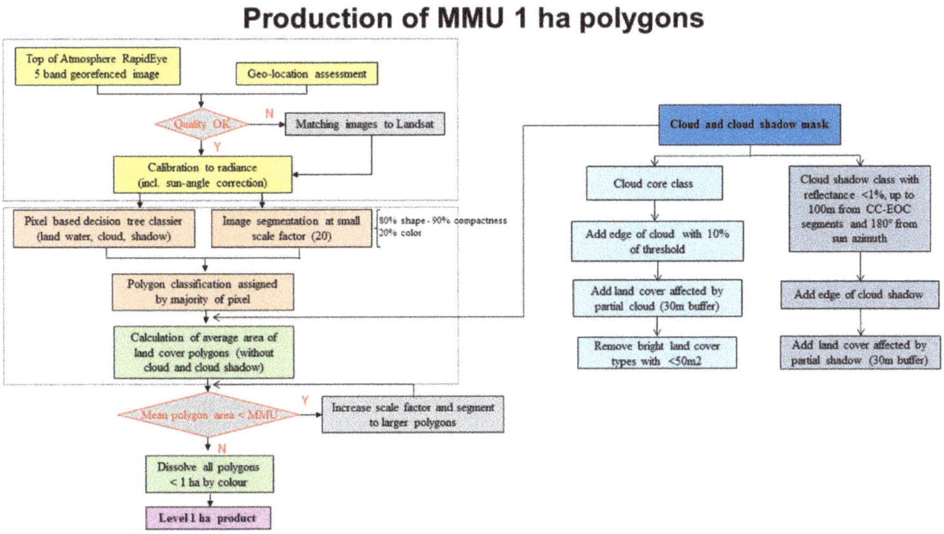

Production of MMU 1 ha polygons

Figure 3. Processing scheme to obtain 1 ha MMUs.

The texture of the image's reflectance can be an effective parameter for measuring AGB [26,27,75]. To have a consistent measure of texture, the scale of analysis (kernel) needs to be large enough to cover the variation in the target landcover with respect to the resolution of the satellite data. At fine spatial scales (e.g., 1 m), the texture of tree crowns shows high variance. This variation remains high until the unit of observation, i.e., the kernel, is larger than a single tree. We carried out a series of tests extracting texture measures from the RapidEye data using a set of incrementally increasing circular units in homogeneous landscapes from 0.25 ha to 6.25 ha to find a MMU that gave stable (i.e., invariant) results (Supplementary Materials Figure S2). Some texture measurements were highly sensitive to the size of the image object, but stabilized at 1 ha. This was then used as the MMU for our objects. One hectare is larger than the final national specification (0.5 ha), but is in line with international guidelines.

2.6. Cloud and Cloud Shadow Masking

The RapidEye data exhibited two artefacts that needed to be addressed: (i) Cloud and cloud shadow displayed a locational shift of several pixels between image bands, (ii) the sun azimuth angle, which is used in classification software for detecting cloud shadow, was missing in the metadata.

The locational shift (i) between bands is as a result of the push-broom design of the RapidEye sensor, which means that different spectral channels are acquired at slightly different times. This effect is combined with a parallax effect, which depends on the height of the cloud in the individual images [76]. We therefore used an incremental approach whereby an initial (high) reflectance threshold of 50% was applied to detect core cloud areas in each band [77], creating 'core cloud objects'. These objects were then merged together and then a lower (10%) threshold was applied only to those areas adjacent to the core areas (i.e., edge of cloud). This has the effect of expanding the area classified as cloud.

To address (ii) that of the missing the sun azimuth angle, an interface was developed to visually estimate the sun azimuth angle. An operator used the position of clouds and their respective shadows to provide an angle for the missing value to replace the missing metadata.

2.7. Extraction of Remote Sensing Parameters for Developing Models for AGB

After cloud and cloud shadow masking, the segmentation iteration loop, as described in Section 2.5, was run to obtain image objects at the MMU (1 ha) for all images. A set remote sensing parameters were then extracted from the level 1 ha objects corresponding (spatially) to the field data plots. These parameters were then used as the basis for generating a model for relating AGB to remotely sensed data (Figure 4).

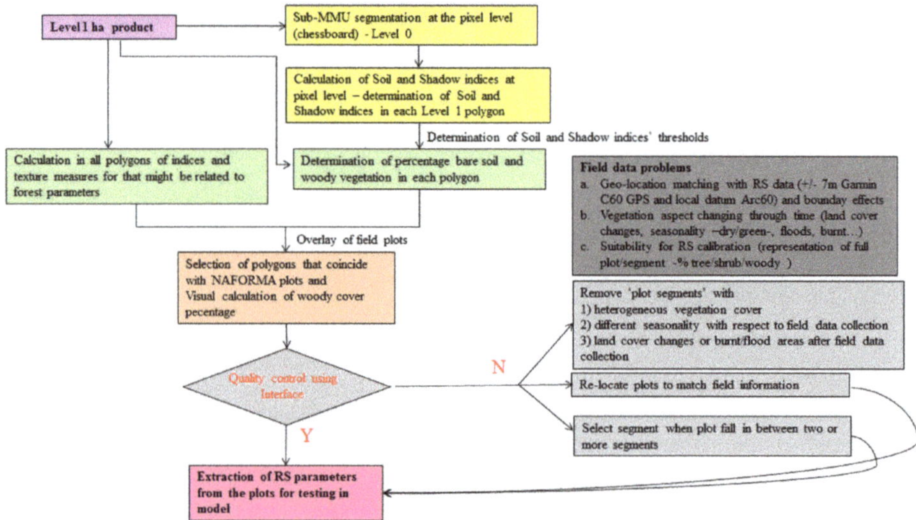

Figure 4. Data extraction scheme.

The parameters fall into five categories: Simple reflectance means and standard deviations for the objects; derived vegetation indices designed to highlight vegetation characteristics, advanced texture measures based on Grey Level Co-occurrence Matrices (GLCM) [78], and a categorical class giving the percentages of the polygon belonging either to bare soil, grasslands, or woody vegetation.

Spectral indices have been used for detecting vegetation and more specifically forest parameters in a number of studies [79,80]. The shortwave infrared bands (1.6 and 2.7 μm) are found to be highly correlated with forest parameters and canopy cover [52]. Unfortunately, these bands are absent from the RapidEye sensor. Indices, such as the shadow index, have also been used for forest applications [81,82]. Improvements in assessing AGB have been achieved using temporal series to monitor phenological changes [83], however, in our study, we were limited to single data imagery.

As satellite data of finer spatial resolution have become more readily available, the use of the texture measures has become more common (e.g., [84]). The number of texture indices available is extensive (200 are available in the image processing software, eCognition). Each measure (column 1 of Table 2) can be assessed in 5 different directions for each individual band or for all bands combined. Since these indices are computationally time consuming, we used pairwise correlation tests on the parameters to identify those indices that were highly inter-correlated (see correlation matrix in the Supplementary Materials Figure S5). As a result, we reduced the number of potential texture indices to 25.

For the percentages of bare soil, woody vegetation, and photosynthetically active non-woody vegetation, we used the Shadow Index (SI), the Bare Soil Index (BSI) [85], and the Modified Chlorophyll Absorption Ratio Index (MCARI) [86]. To restrict our final models to 'woody' vegetation, we produced a classification of each of the RapidEye images based on a decision tree approach [77], which assigns each object to either forest, shrub, or non-forest. The data extraction processing chain (Figure 4) was implemented in eCognition (the rule set is available both as text and an eCognition dcp file in the Supplementary Materials). The list of reflectance and texture measures extracted from the RapidEye objects is shown in Table 3.

Table 3. List of reflectance and texture measures extracted from the RapidEye objects.

Single Band Reflectance of Objects	Equation	Acronym
Mean bands 1 to 5	BLUE, GREEN, RED, REdge, NIR	R1–R5
Standard deviation bands 1 to 5		SD1–SD5
Simple Ratios		
Ratio of mean to standard deviation for all bands	[Mean band #]/[SD band #]	e.g., R1SD1
Standard deviation. ratios	[SD band #]/[SD band #]	e.g., SD1SD2
Red Vegetation Index	NIR/RED	RVI
Green Vegetation Index	NIR/GREEN	GVI
Green Red Vegetation Index	GREEN/RED	GRVI
Other Indices		
Normalised Difference Vegetation Index [87]	(NIR − RED)/(NIR + RED)	NDVI
Enhanced Vegetation Index [88]	$2.5 \times ((NIR - RED))/((NIR + 6 \times RED - 7.5 \times BLUE + 1))$	EVI
Soil Adjusted Vegetation Index [89]	$(NIR - RED)/((NIR + RED + L)) \times (1 + L)$	SAVI
Shadow Index [85]	$((1 - BLUE) \times (1 - GREEN) \times (1 - RED))^{0.33}$	Shadow_Index
Modified Transverse Vegetation Index [86]	$(1.5 \times (1.2 \times ([R5] - [R2]) - 2.5 \times ([R3] - [R2])))/((2 \times [R5] + 1)^2 - (6 \times [R5] - 5 \times ([R3])^{0.5}) - 0.5)^{0.5}$	MTVI
Modified Chlorophyll Absorption Reflectance Index [90]	$((([NIR] - [RED]) - 0.2 \times ([NIR] - [GREEN])) \times ([NIR]/[RED])$	MCARI
Bare Soil Index	$(((([NIR] + [RED]) - ([REdge] + [BLUE]))/((([NIR] + [RED]) + ([REdge] + [BLUE]))) + 1$	Bare_soil
Shadow to Soil Ratio	ShadowIndex/Bare_Soil	Shadsoil
GLCM Texture Features	(1) In all directions, 0°, 45°, 90°, 135°	
	(2) For combined and for individual bands	
Homogeneity	See Haralick [91]	GCLM_Hom
Contrast		GCLM_Con
Dissimilarity		GCLM_Diss
Angular 2nd moment		GCLM_Ang2
Entropy		GCLM_Entro
Mean		GCLM_Mean
Variance		GCLM_StdDev
Correlation		GCLM_Corr
Object Proportions		
Bare Soil	Percentage classed as bare soil	Rel_soil
Shadow	Percentage classed as woody	Rel_shad

2.8. Reviewing the Field Plot Data with Respect to the Rapideye Data

To assess the compatibility between field data and image data, we developed an interface displaying the image data, the plot location, and the plot information (see Supplementary Materials Figure S6). A number of problems came into evidence during this review. These related to spatial mismatches, temporal mismatches between image acquisition and field visits, and the impact of mixed land cover within the sample sites.

Specifically:

- The field data (15 m circle plots—*circa* 700 m^2) cover 27 RapidEye pixels. However, the precision of the plot geolocation taken in the field (Garmin C60) had limited accuracy ($+/-7$ m). There are also known problems in changing between the Arc60 datum of topographic maps of East Africa used in the field survey, and that of the satellite reference datum, UTM-WGS84 [92];
- there were a number of discrepancies in the geolocation of various RapidEye scenes, which after the review, were addressed by shifting images to a reference data base of Landsat scenes. Over 50 sample site images were shifted by up to 12 pixels, 30 in the X direction and 41 in the Y direction;
- temporal differences; the field data collection took place over three years; satellite data were only available for one of these years. On average, the difference between field data collection and satellite image acquisition was 10 months (see Supplementary Materials Figure S1), with most of the field data collected after the image acquisition;
- data collected in the field gave no systematic estimation of the respective cover of trees, shrubs, or other land cover classes, despite being foreseen in the original protocol. On a number of occasions, when given, the canopy cover did not correspond to that seen from the satellite image—perhaps due to problems of geolocation between the data sets, or differences in the time between the field visit and the image acquisition. Also, canopy density is known to be difficult to measure with accuracy from the ground [8];
- the land cover classification given to the field teams was not adapted to providing adequate field data for calibrating remote sensing data. The vast majority of plots were classified as 'woodland', without further elaboration;
- even if the land cover has not changed throughout the year, its condition does, especially in the tropics, predominantly due to seasonality. It may be in a lush green phase, drying out, exceptionally dry, burnt, or flooded. All these present different spectral signatures for the same land cover. We removed 33 plots that were burnt, 9 that were flooded, and 13 that had cloud or cloud shadow; and
- finally, we found that the field plot was not always representative of the 1 ha image object on the remote sensing data.

2.9. Models for Predicting Above Ground Biomass

We tested four predictive models to relate the remotely sensed parameters to AGB and evaluated their accuracies. Two models using inferential statistics (i.e., a generalised linear model and a generalised exponential model) and two machine learning models (i.e., a random forest model and a support vector machine (SVM) model). The predictors of the models are image bands, their textures, and spectral indices, while the response variable is the AGB. A scatterplot of each input variable (predictors in the y axis) and biomass is shown in the Supplementary Materials Figure S7.

In the generalised linear model, the dependent variables are assumed to be a linear function of several independent variables, where each of them has a weight (i.e., a regression coefficient). Generalised linear models require a linear relation between predictors and response variables.

Allometric equations relating tree height and basal area to volume, hence calculating AGB, are generally non-linear [11]. We therefore employed a logarithmic transformation to the dependent variables, which is a generalised exponential regression.

Random forest (RF) [93] is an ensemble method that builds several decision trees (weak learners) and outputs the mean prediction of the individual decision tree models. The improved RF strategy alleviates the often reported overfitting problem of simple trees. The decorrelation of the decision trees is achieved through the random selection of the input explanatory variables [94] by bootstrap methods. In this case, 63% of the data is used for training (in-bag data) and the remaining 37% (out-of-bag data) for validation. The choice of random forest was mainly motivated by: (i) RF runs efficiently on large databases (i.e., it is relatively fast to build and even faster to predict), (ii) RF is resistant to outliers and over-training, (iii) RF does not require cross-validation for model selection, (iv) RF provides further information about the most relevant variables inputted in the model, and (v) RF is computationally parallelizable. In a decision tree, an input is entered at the top and as it moves down the decision tree the data gets binned into smaller and smaller sets. The main principle behind ensemble methods is that a group of "weak learners" can come together to form a "strong learner", for details see [93]. Two parameters are important in the RF algorithm: (i) The number of decision trees used in the forest (*ntree*) and (ii) the number of random variables used in each decision tree (*mtry*). In our RF model, we have 500 decision trees (i.e., *ntree* = 500) and the default *mtry* value (i.e., *mtry* = 9). To find the correct *ntree* number, we built RF models with different *ntree* values, recorded the error, and computed the number of decision trees needed to reach the minimum error estimate. RF provides the percentage of variance explained and the most relevant input variables in the model. For reproducibility, we used a fixed random number parameter, which forces the generator to give the same random numbers in the random forest function. Since random forest performances generally increase as the size of training data increases, we trained the final model with the entire ground truth dataset.

Random forest is also able to provide information about the importance of the predictors. Specifically, the mean decrease accuracy ("%IncMSE") (Figure 5) has been used as a measure of variable importance. The %IncMSE quantifies by how much the removal of each variable reduces the accuracy of the model. The higher the %IncMSE is, the more important the variable. We ran random forest 100 times to find a robust estimate of these predictors, with the average values and the quartile distributions of the 20 most relevant predictors shown in Figure 5. The Shadow Index was found to be the most important variable. R2SD2 (that is the ratio of band 2 reflectance to its standard deviation) and MCARI are also able to influence AGB variability. Other predictors are ranked lower in their relative variable importance. The main limitation of the random forest model is the lack of capacity to predict beyond the range of the response values in the training data.

The support vector machines (SVM) are supervised learning methods used for regression tasks that originated from statistical learning theory [95]. SVM methods perform a linear regression in a high dimension using kernels. This allows it to capture nonlinear tendencies in the original input data. SVM has some parameters that need to be selected (i.e., the kernel) and tuned (i.e., cost and gamma) to obtain a better performance. Among all the kernels available in the literature, we chose the radial basis function (RBF) kernel, due to theoretical and computational convenience. In order to optimise the SVM, we have tested several different cost and gamma values, and returned the one which minimizes the mean squared error for a 10-fold cross validation. Specifically, the optimal values were found as 10 and 0.5 for the cost and gamma parameters, respectively.

The ground truth dataset is the national field data set collected by NAFORMA that consists of 512 points where we have AGB information (*circa* 500 other points were excluded due to very low or zero biomass, detected on the basis of the decision tree classifier—Section 2.7). The ground truth dataset was split, 85% into a training dataset (435 points) and 15% into a validation dataset (77 points). Random forest already separates partial data for an internal validation (i.e., 63% of the data is used for training and 37% for validation as discussed in Section 2.9), but to compare results with other models, it has been trained and validated with the same validation dataset as for the other models. This scheme ensures: (i) That identical datasets are used for training and (ii) an independent validation with a dataset not included in the training phase.

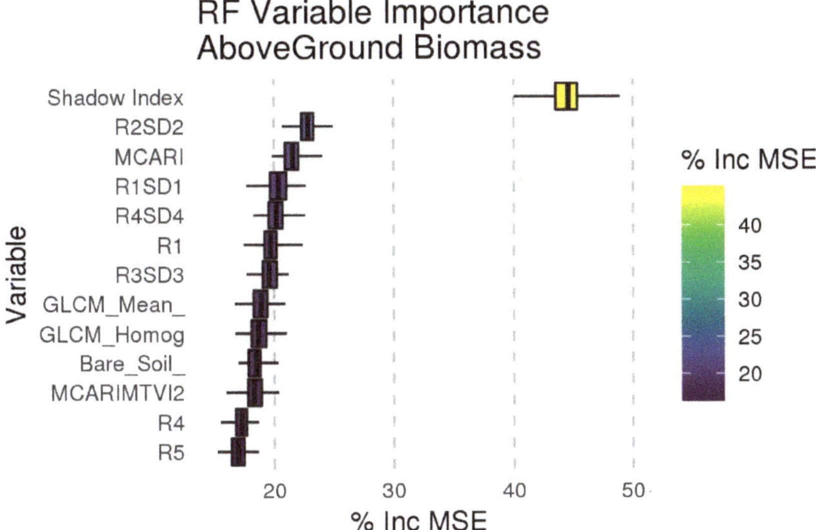

Figure 5. Box-plot of the importance of variables for AGB in the final Random forest model, showing the increase in mean standard error (MSE) when removing a variable (*y*-axis), and their importance on the x-axis. For the sake of simplicity, only the first 12 variables are shown. *RXSDX* denotes the ratio of the band X reflectance to its standard deviation.

Once we obtained the best model for AGB, we created the final model using all of the ground truth dataset (for further details see next section). The final model was then applied to all the image objects pre-classified as 'woody' in the full Remote Sensing data set, so as to obtain AGB estimates for each of the remote sensing sample sites.

Four statistical indicators were used to evaluate the performances of the models: (i) Root mean square error (RMSE); (ii) mean absolute error (MAE); (iii) relative root mean square error (relRMSE); and (iv) the coefficient of determination (R^2). Note that results of models refer only to the validation dataset, however, the performance of the model with the training dataset is shown in the Supplementary Materials Figure S8.

2.10. Interpolating the Results to the National Level

To obtain the country level estimates, we interpolate by direct expansion to using the Direct Expansion Estimator. The application of the direct expansion method has been used in various studies [52].

$$AGB = D \times \bar{y} \tag{3}$$

where D is the total study area and $\bar{y}$ is the average AGB ha^{-1} of the sample areas.

2.11. The Relative Efficiency to Measure the Improvement in Precision Brought by the Remote Sensing Data

We calculated REff, the relative efficiency (i.e., the ratio of the variance of the estimates for the AGB calculated from the RapdiEye data combined with field data (VAR$_{REFD}$) to the variance of the AGB from the field plots alone (VAR$_{FD}$). This measures the improvement in precision (as opposed to accuracy) that the introduction of the remote sensing data brings [96]. In theory, a relative efficiency of X would mean that the improvement brought about by the introduction of the remotely sensed data could be achieved by using X times as many field plots [10].

$$REff = VAR_{REFD}/VAR_{FD} \tag{4}$$

An estimate [52] can be made with:

$$\text{REff} = \frac{1}{1 - R^2} \tag{5}$$

where R^2 is the square of the Pearson coefficient of correlation between the remote sensing data and the field data.

3. Results

3.1. Model Results

Table 4 summarizes these main statistical indicators, with scatterplots of the difference between the observed and modelled AGB shown in Figure 6. All performance measures (i.e., RMSE, MAE, relRMSE, R^2, and observed vs. modelled scatterplots) were assessed using only the validation dataset, with random forest exhibiting the best correlation (R^2 = 0.69).

Table 4. Statistical indicators of modelled data versus observed AGB.

Model	RMSE	MAE	R^2	relRMSE
Generalized Exponential Regression	44.55	36.44	0.32	21%
Generalized Linear Regression	41.81	35.07	0.39	20%
Random Forest	30.00	21.64	0.69	14%
SVM	39.08	29.76	0.42	19%

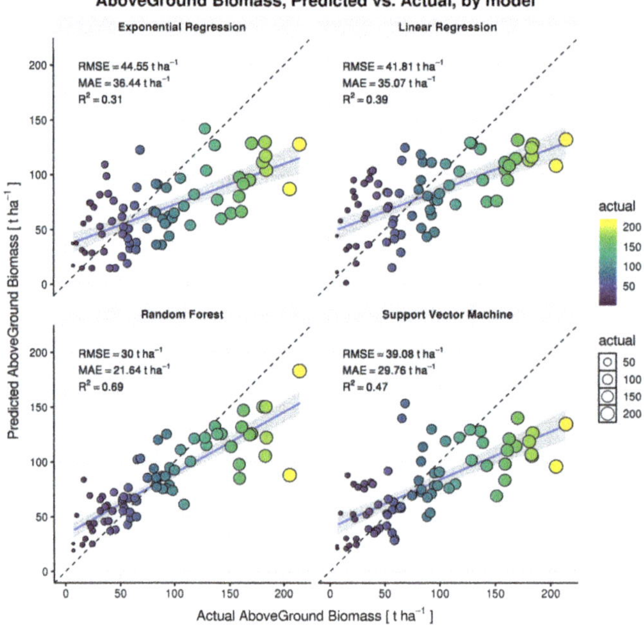

Figure 6. Scatterplot of predicted vs. modelled AGB in tha^{-1} for the four models using the independent validation dataset. Both circle size and colour refer to the actual AGB. Model performance indicators are also shown in Table 4. The blue line indicates the linear regression between the actual and modelled data and the grey area is the 95% confidence level interval.

The RMSE of the random forest model is ~30, with an MAE of ~20 and an relRMSE of 15%. The performance of SVM is less satisfactory: RMSE is ~40, MAE is ~30, and relative RMSE is ~40%. As expected, the four methods presented a gain in performance on the training dataset, showing the RF with very high accuracies (i.e., RMSE~15, $R^2 = 0.93$), while the exponential and linear regression methods still presented poor (biased) estimates of AGB (see Supplementary Materials Figure S8). However, all four methods produce biased results, overestimating low AGB and underestimating AGB, as indicated by the slopes' deviation from the 1:1 line.

3.2. Analysis at the National Level and by Ecoregion

For our sample sites, covering 2.3% of the country, we find an average of 22.1 tons per ha of above ground biomass (AGB). This is only for those landscape units that are classed as being 'woody biomass'. Using the areas provided by NAFORMA for forests and woodlands (54,534,500 ha), this results in a total of 602.4 MtC for the country and compared to the estimate from the field plots of 624.9 MtC, a difference of 4%. The standard deviation is high—25%, however, this is not unusual, as land cover is not generally distributed in a regular manner, especially in Tanzania, which exhibits land covers of highly different levels of biomass.

Using the location of the individual sample sites, we calculated the average AGB for the respective ecoregions (Table 5). Due to the limited number of samples in the other ecoregions, the informative results are for the Somalia-Masai and Zambezian regions.

Table 5. Estimates of average AGB (t ha^{-1}) by White ecoregion. *, Refers to the number of RapidEye sites in each ecoregion.

	Afromontane	Zanzibar-Inhambane	Somalia-Masai	Zambezian	Lake Victoria
Area (km^2)	43,500	77,100	238,965	502,052	40,300
Count *	5	6	19	41	4
Minimum	17.3	15.1	9.8	11.6	17.8
Maximum	27.7	30.4	22.4	34.3	25.9
Mean	20.9	21.5	16.7	25.2	22.3
SD	4.9	5.6	4.1	4.6	3.3

3.3. Relative Efficiency

The relative efficiency (REff) is calculated using Equation (5), which gives a REff of 3.2. The introduction of the remote sensing data can therefore reduce the field data collection by three, obtaining the same precision.

This result helps in answering one of the main questions of the study: "How much RapidEye can improve the precision of field-based inventory estimates of above-ground biomass in forests and woodlands of Tanzania?"

4. Discussion

A key point that this paper wishes to address is how to improve the link between remote sensing parameters and data collected in the field.

4.1. Practical and Operational Consderations to Imprvve AGB Estimates from RapidEye Data

In preparing data for testing models to relate remote sensing parameters to field data (AGB), we found that a major effort was required in 'cleaning' the field data. By 'cleaning' we mean removing those aspects of the field data that reduce their potential to calibrate remote sensing data for biomass estimation. The country wide RS data acquired under the GMES initiative was restricted to 2010–2011, whereas the field data were collected over a period of three years. Hence a temporal difference often occurs between a specific image and the corresponding field measurement. However, few plots would have undergone a change during this time lapse. To avoid cloud contamination, most images come from the dry season. At this time, the vegetation is senescent, with trees and shrubs losing their leaves. In this situation, the differences in reflectance between grasslands, shrub and tree savannah, and open woodlands is reduced.

4.1.1. Co-Location of the Data Sets

The site visits for field data plots were organized without using remote sensing imagery. Field teams, using the national topographic maps (in Arc60 datum), navigated to the pre-selected sites. Such field visits should be supported by cartographic and digital extracts of the field locations. Both high cost (high precision GPS with inbuilt satellite and/or map visualisation) and low cost (smart phone/pad and PDF maps) exist to facilitate both navigation and on site location. When local datum are used, so as relate to the national mapping datum, dual GPS locations with WGD 1984 should be employed.

4.1.2. Data Collected

The data collected for the national assessment were suited for the statistical assessment of biomass and for other parameters. However, we found that they were not specific enough to easily correlate them to the satellite data. Optical satellite data record reflectance values within the plot that can be related to the vegetation canopy cover, distribution condition (dry/green/burnt/flooded), and structure. The field data collected need to correspond to these vegetation parameters, which should be recorded and mapped on site during the survey with the aid of an orthomap produced from the remote sensing data to be used in calibration.

4.1.3. Data Cleaning

The inclusion of non-woody, low biomass sites was found to introduce high variance in the spectral signatures for similar values of AGB. These sites have a high variation in land cover and land cover condition. They include barren surfaces, agriculture exploitations, grasslands, and park savannah, each of which manifested different states—burnt, flooded, green flush, senescence, bare soil. We therefore removed them from the data set, reducing the available points from *circa* 1000 to *circa* 500.

4.1.4. Spectral Signatures

The reflectance of forest canopy depends on a number of factors, structure, or row orientation (for young artificial forests), optical properties of the background (soil and understory), and canopy geometry [97]. The Shadow Index was found to be the most relevant parameter related to AGB, followed by texture measures. The value of the Shadow Index increases as the forest density increases, hence it is appropriate for relating to woody biomass.

Mature, natural forest stands tend to have more heterogeneous surfaces than non-forest and young regrowth, creating more shadows. The shortwave (1.6 μm and 2.7 μm) infrared bands (SWIR) have been shown to be highly successful in mapping forests [97]. This is mainly because at these wavelengths, there is very low diffuse light, hence shadows are more contrasted. In the absence of SWIR bands on the RapidEye sensor, we used the shadow index as defined by Rikimaru et al. (2002) [58], which uses the visible bands only. Similarly, texture measures, at finer spatial resolutions, differentiate between heterogeneous and homogenous surfaces (Table 3).

4.1.5. Ancillary Data

The parameters entered into the model came exclusively from the single date RapidEye data. While these (RapidEye) are superior in terms of resolution, they are limited for temporal analysis. With the arrival of the Sentinel 2a & 2b sensors, 10 m resolution data are now available at a 5 day frequency. Simulation of this scenario with the SPOT4take5 program [98] has already shown that differences in forest canopy cover (and as a result AGB) can be determined [99].

4.2. Reviewing the Results

4.2.1. Estimation by Ecozone

The Somalia-Masai shows a significantly lower biomass than the Zambezian region (Table 5) as expected, the former is dominated by shrub formations, the latter by woodlands. The other zones are too small to consider using this methodology due to the small number of RapidEye images available in each zone. The Zanzibar-Inhambane, home of the Coastal forests, should have a high average AGB, however, it is these forests that have been depleted most by the impact of exploitation by residents of Dares Salaam and Morogoro [100].

4.2.2. Relative Efficiency

It is difficult to quantify the potential financial savings the inclusion of remote sensing data would bring to the project without knowing the direct costs of the field survey. In the NAFORMA project, the total spent on salaries for the field survey component alone was *circa* $5 million [63]. However, this includes household surveys ($n = 3500$) and soil data collections ($n = 4000$). The costs of overheads, vehicles, and fuel would need to be added. In any case, a relative efficiency of over three indicates that when combined with remote sensing data, the field component could be more than halved to obtain the same precision. The cost of RapidEye data at the time was around $1 per square kilometre, giving a commercial cost of just over $30,000 for the satellite data used in this study. Full country coverage by RapidEye would be around $1 million. Such a coverage would also provide a wall-to-wall map of biomass, which could allow for the estimation of biomass in small ecozones described above.

4.2.3. Model Results

The final model has to cover the full variation in vegetation types and conditions across Tanzania. Clearly, studies on small areas [45], single ecosystems [39], or a mono-species plantation [46] will provide better results, however, these are not comparable to a country wide assessment.

It should be noted that two field plots may have the same AGB, but with totally different characteristics, in terms of vegetation type, cover, and condition. We can imagine a plot with one tree surrounded by grass having the same AGB as a plot fully covered by shrub. This is where the strength of random forest lies, applying a decision tree to associate different input parameters (in this case RS variables) to the same class.

Compared with the other methods available, random forest considerably reduces the dispersion of residuals (e.g., RMSE) and increases the model fit (R^2). Despite this improvement in accuracy and precision, the AGB estimate presents a reduced range (40–150 t ha^{-1} in the case of random forest), which is much lower than the response values in the training data (in the range of 10–200 t ha^{-1}). Specifically, the statistics by the biomass range show that RF tends to overestimate biomass in the range of 0–40 t ha^{-1} and underestimates it at higher values (i.e., above 150 t ha^{-1}).

Note that this behavior is present for all the four models and is minimized for RF. This behaviour has been observed in many other AGB products [101]. A solution to this would be to compute AGB error statistics for biomass ranges (i.e., 0–50 t ha^{-1}, 50–100 t ha^{-1}). It also has to be remembered that the biomass estimates themselves are interpolated from field measures using allometric equations, which in themselves may have bias for certain ranges of AGB [62].

The AGB overestimation at low biomass and underestimation at high biomass is still an open issue, which has been widely reported in the literature using either optical or radar data (e.g., [102]). Although the presence of a bias can be regarded as a main drawback of such maps, this caveat does not mean that the model result is without merit, as all models are affected by various limitations, and their identification allows a proper use of the outputs. For example, if the map accuracy is not sufficient for certain applications (e.g., local level management), other applications (e.g., stratification, regional analysis) can benefit from knowing the spatial distribution of biomass in relative terms. In any case, quantile analysis on the results of RF regression (Supplementary Materials Figure S9) shows that RF is able to discriminate four main levels of AGB with a relatively good accuracy (e.g., 70% and almost 80% for low and high values, respectively) and the bias in the extremes is not so important.

Underestimation of high biomass values may be due to the following reasons:

(i)   The limited sensitivity of satellite sensors to variations in canopy height and tree diameter in dense forests. Specifically, optical sensors' radiometers (such as RapidEye) tend to saturate at high biomass in dense forest where there is a weak reflectance-biomass relationship, e.g., [103]. For this reason, the combined use optical data in combination with SAR and LiDAR data would improve results as shown in [104,105];

(ii)   we were limited to the use of single date imagery with most of the images acquired in the dry season when many seasonal forests have lost their leaves and are spectrally similar to low biomass grasslands; and

(iii)   RF is trained on a dataset where high biomass values represent the tail of the frequency distribution and therefore its performance decreases as the biomass increases.

With regards to low biomass values, there could be two explanations: (i) Confusion between understory vegetation and forest canopy where shrubs might be confused with the canopy thus overestimating AGB; and (ii) tree savanna, isolated trees may tend to have a larger canopy, which is relatively similar (as seen from satellite) to that of a denser forest.

The use of diverse data sources may overcome this problem: Combining space borne SAR (e.g., ALOS PALSAR 2, Envisat ASAR, Sentinel-1), optical (e.g., Landsat 7 & 8, Sentinel-2), LiDAR (ICESAT), and auxiliary datasets with multiple estimation procedures as executed by Avitabile et al. [106].

Our work demonstrated that forest services face major hurdles when trying to integrate remote sensing data into a national inventory. These hurdles include obtaining cloud free data for all reference (field) measurements at dates as close to the data collection campaign as possible; ensuring a good geolocation of the two data sets; and having an adequate representation of the distribution of the biomass levels across the full country. We provide potential users with techniques to cope with these problems and show that optical data of finer spatial resolution, in this case 5 m resolution, can be used to obtain biomass estimates at a national level, albeit, with some bias. The arrival of the freely available Sentinel-2 satellite data (although at 10 m resolution) provides an opportunity for more frequent wall-to-wall national coverage. Since 2016, wide area coverage has been available from the commercial 3 m spatial resolution PlanetScope set of satellites [107].

In our study, we were limited by the sampling scheme and the acquisition dates of the available RapidEye data. The scheme, a regular systematic grid, was devised for a continental scale forest change estimates. A more appropriate approach would be to use a stratified random scheme with more (smaller) samples (e.g., [50]) so as to better capture the variance of the biomass across the different ecosystems. While the sampling approach is an efficient method for obtaining biomass estimates, is does not provide a wall-to-wall cartographic product, which would be a benefit for forest management planning and provide orthomaps to support the field survey and stratified estimation [108].

5. Conclusions

We found that remote sensing parameters extracted from 5 m resolution satellite data combined with modelling using the random forest algorithm were an effective method for improving the precision of AGB estimates.

The Shadow Index was found to the most relevant parameter related to AGB, followed by texture measures. The use of both vegetation indices and texture data confirms our hypothesis that a combination of different information extracted from the RS data enhances the precision of AGB estimation, but also can increase bias. This is because the relationship between reflectance and AGB over dense forests is affected by the saturation phenomenon. In contrast, textures and non-woody vegetation cover information (i.e., shadow index) might be more sensitive to canopy architecture and structure. As expected, the high performances of random forest are more evident when the response variables are a result of non-linear and complex interactions between multiple predictors, as has been demonstrated in numerous previous studies. The RF was shown to be quite robust for including a large number of heterogeneous input variables. Also, many predictors of AGB are highly inter-correlated and RF handles both these and redundant variables well.

In terms of supporting REDD+ initiatives, this work demonstrates that nationwide AGB inventories can be carried out in a more rapid and cost effective way by combining 5 m resolution optical satellite data with field plots.

The codes and data used for this study have been made permanently and publicly available in the Supplementary Materials so that results are entirely reproducible.

Supplementary Materials: The following are available online at http://www.mdpi.com/1999-4907/10/2/107/s1, Figure S1: Acquisition dates of the RapidEye images (top) and the differences in months between field plots and associated image (bottom). Figure S2: The impact on texture indices resulting from changing the mapping unit (top) by analysing increasing circles over the RapidEye data (bottom) where the red circle shows the different areas used. Figure S3: The application of successive shape factors to RapidEye data; From left to right SF = 20, 60, 100, 124; The resulting number of objects falls from 59198 to 8059 to 3190 to 2217 objects as the MMU of 1 ha is approached through the iteration. Figure S4: In the final step all polygons less than 1 ha are dissolved and we arrive at 1395 objects Figure S5: Cross correlation results carried out so as to reduce the number of texture measures by removing 'duplicate' measures—i.e., those that are highly correlated—blue and red. Due to text sizes the names of the indices are reduced. Figure S6: The interface developed to review the plot data. The plots are seen on the middle panel overlaid on the RapidEye data. On the left panel high resolution images from GoogleEarth. The plot characteristics as collected by NAFORMA are displayed on the right. Figure S7: Scatterplot of each input variable versus Above Ground Biomass. Regression Line is in blue, while the 95% confidence interval is in grey. Figure S8: Scatterplot of Predicted vs. Modelled AGB in tha^{-1} for the four models using the training dataset. Both circle size and colour refer to the actual AGB. The blue line indicates the linear regression between actual and modelled data and the grey area is the 95% confidence level interval. Figure S9: Random Forest model's quality. Figure shows the result of arranging all scores or predicted values in sorted quantiles, from worst to best, and how the classification goes compared to the test set. Specifically we have used 4 AGB quantiles using the validation subset. The x axis represents the 4 quantiles of AGB from NAFORMA plots. The y axis shows how much the random forest derived AGB falls in each quantiles. File 1: Rule set used to segment the RapidEye images and extract the spectral and textural information for correlation with the field data on AGB: RapidEye2018RuleSet.dcp. File 2: A text version of the above: RapidEye2018RuleSetDocu.txt. File 3: The extracted remote sensing parameter with the field data on AGB by plot: DataBiomassRS.csv. File 4: The R statistical package code used to set up and test the four models: Code_AGB_RS.R.

Author Contributions: L.H.G., F.J.G.H. and H.E. conceived the paper in the framework of L.H.G.'s Ph.D. L.H.G. carried out the remote sensing processing. L.H.G. and G.C. produced the statistical analysis supported by F.J.G.H. V.A. reviewed the draft text, providing a number of new incites to the direction of the paper. Field survey was designed and carried out by H.E. and L.G.H.

Acknowledgments: This work is funded by the European Union and supported by CIFOR—the Center for International Forest Research. The support of the Tanzania Forest Service, particularly Messers. N. Chamuya, J. Otieno and S. Dalsgard was particularly important. We wish to thank colleagues at the UN FAO, in particular A. Pekkarinen. The authors also thank those responsible for the efforts on providing the free tools (R) used in this work. The R packages were obtained from the Comprehensive R Archive Network (http://cran.r-project.org/) or R-forge (https://r-forge.r-project.org/). The AGB dataset and the materials necessary to completely reproduce the analysis and figures included in this article are available as "Supplementary Files". Source materials for reproducing this manuscript. This is a compressed .zip directory containing the main source file in R format, as well as the additional files necessary to completely recreate the original manuscript submitted to Remote Sensing. For development versions of the AGB modelling script please see https://github.com/guidolavespa/Forest. For the latest archived version of the library, use DOI https://doi.org/10.5281/zenodo.1324052 [109].

Conflicts of Interest: The authors declare no conflict of interest.

References

1. Van der Werf, G.R.; Morton, D.C.; DeFries, R.S.; Olivier, J.G.J.; Kasibhatla, P.S.; Jackson, R.B.; Collatz, G.J.; Randerson, J.T. CO2 emissions from forest loss. *Nat. Geosci.* **2009**, *2*, 737–738. [CrossRef]
2. UNFCCCC. *Report of the Conference of the Parties on Its Nineteenth Session, Held in Warsaw from 11 to 23 November 2013. Part One: Proceedings*; United Nations Framework Convention on Climate Change: Bonn, Germany, 2014.
3. IPCC. *Revised 1996 IPCC Guidelines for National Greenhouse Gas Inventories*; IPCC: Geneva, Switzerland, 1996.
4. IPCC. *Good Practice Guidance for Land Use, Land-Use Change and Forestry (GPG-LULUCF)*; IPCC: Geneva, Switzerland, 2003.
5. GOFC-GOLD; Achard, F.; Boschetti, L.; Brown, S.; Brady, M.; DeFries, R.; Grassi, G.; Herold, M.; Mollicone, D.; Pandey, D.; et al. A Sourcebook of Methods and Procedures for Monitoring and Reporting Anthropogenic Greenhouse Gas Emissions and Removals Associated with Deforestation, Gains and Losses of Carbon Stocks in Forests Remaining Forests, and Forestation. Available online: https://www.researchgate.net/publication/283417201_A_sourcebook_of_methods_and_procedures_for_monitoring_and_reporting_anthropogenic_greenhouse_gas_emissions_and_removals_associated_with_deforestation_gains_and_losses_of_carbon_stocks_in_forests_remai (accessed on 28 January 2019).
6. Herold, M.; Román-Cuesta, R.M.; Mollicone, D.; Hirata, Y.; Van Laake, P.; Asner, G.P.; Souza, C.; Skutsch, M.; Avitabile, V.; MacDicken, K. Options for monitoring and estimating historical carbon emissions from forest degradation in the context of REDD+. *Carbon Balance Manag.* **2011**, *6*, 1–7. [CrossRef] [PubMed]
7. Maniatis, D.; Mollicone, D. Options for sampling and stratification for national forest inventories to implement REDD+ under the UNFCCC. *Carbon Balance Manag.* **2010**, *5*, 9. [CrossRef] [PubMed]
8. Jennings, S.; Brown, N.; Sheil, D. Assessing canopies and understorey illumination: Canopy closure, canopy cover and other measures. *Forestry* **1999**, *72*, 60–73. [CrossRef]
9. Gonçalves, F.; Treuhaft, R.; Law, B.; Almeida, A.; Walker, W.; Baccini, A.; dos Santos, J.; Graça, P. Estimating Aboveground Biomass in Tropical Forests: Field Methods and Error Analysis for the Calibration of Remote Sensing Observations. *Remote Sens.* **2017**, *9*, 47. [CrossRef]
10. Hansen, E.; Gobakken, T.; Solberg, S.; Kangas, A.; Ene, L.; Mauya, E.; Næsset, E. Relative Efficiency of ALS and InSAR for Biomass Estimation in a Tanzanian Rainforest. *Remote Sens.* **2015**, *7*, 9865–9885. [CrossRef]
11. Chave, J.; Andalo, C.; Brown, S.; Cairns, M.A.; Chambers, J.Q.; Eamus, D.; Fölster, H.; Fromard, F.; Higuchi, N.; Kira, T.; et al. Tree allometry and improved estimation of carbon stocks and balance in tropical forests. *Oecologia* **2005**, *145*, 87–99. [CrossRef]
12. Henry, M.; Picard, N.; Trotta, C.; Manlay, R.J.; Valentini, R.; Bernoux, M.; Saint-André, L. Estimating tree biomass of sub-Saharan African forests: a review of available allometric equations. *Silva Fennica* **2011**, *45*, 477–569. [CrossRef]
13. Lewis, S.L.; Sonke, B.; Sunderland, T.; Begne, S.K.; Lopez-Gonzalez, G.; van der Heijden, G.M.F.; Phillips, O.L.; Affum-Baffoe, K.; Baker, T.R.; Banin, L.; et al. Above-ground biomass and structure of 260 African tropical forests. *Philos. Trans. R. Soc. B Biol. Sci.* **2013**, *368*, 20120295. [CrossRef]
14. Feldpausch, T.R.; Banin, L.; Phillips, O.L.; Baker, T.R.; Lewis, S.L.; Quesada, C.A.; Affum-Baffoe, K.; Arets, E.J.M.M.; Berry, N.J.; Bird, M.; et al. Height-diameter allometry of tropical forest trees. *Biogeosciences* **2011**, *8*, 1081–1106. [CrossRef]
15. Goetz, S.; Dubayah, R. Advances in remote sensing technology and implications for measuring and monitoring forest carbon stocks and change. *Carbon Manag.* **2011**, *2*, 231–244. [CrossRef]
16. Timothy, D.; Onisimo, M.; Riyad, I. Quantifying aboveground biomass in African environments: A review of the trade-offs between sensor estimation accuracy and costs. *Trop. Ecol.* **2016**, *57*, 393–405.
17. Goetz, S.J.; Hansen, M.; Houghton, R.A.; Walker, W.; Laporte, N.; Busch, J. Measurement and monitoring needs, capabilities and potential for addressing reduced emissions from deforestation and forest degradation under REDD+. *Environ. Res. Lett.* **2015**, *10*, 123001. [CrossRef]
18. Avitabile, V.; Herold, M.; Henry, M.; Schmullius, C. Mapping biomass with remote sensing: A comparison of methods for the case study of Uganda. *Carbon Balance Manag.* **2011**, *6*, 7. [CrossRef]
19. Baccini, A.; Laporte, N.; Goetz, S.J.; Sun, M.; Dong, H. A first map of tropical Africa's above-ground biomass derived from satellite imagery. *Environ. Res. Lett.* **2008**, *3*, 045011. [CrossRef]

20. Hansen, M.C.; Potapov, P.V.; Moore, R.; Hancher, M.; Turubanova, S.A.; Tyukavina, A.; Thau, D.; Stehman, S.V.; Goetz, S.J.; Loveland, T.R.; et al. High-Resolution Global Maps of 21st-Century Forest Cover Change. *Science* **2013**, *342*, 850–853. [CrossRef]

21. Asner, G.P.; Mascaro, J.; Muller-Landau, H.C.; Vieilledent, G.; Vaudry, R.; Rasamoelina, M.; Hall, J.S.; van Breugel, M. A universal airborne LiDAR approach for tropical forest carbon mapping. *Oecologia* **2012**, *168*, 1147–1160. [CrossRef] [PubMed]

22. Asner, G.P.; Mascaro, J. Mapping tropical forest carbon: Calibrating plot estimates to a simple LiDAR metric. *Remote Sens. Environ.* **2014**, *140*, 614–624. [CrossRef]

23. Hajj, M.E.; Baghdadi, N.; Fayad, I.; Vieilledent, G.; Bailly, J.-S.; Minh, D.H.T. Interest of Integrating Spaceborne LiDAR Data to Improve the Estimation of Biomass in High Biomass Forested Areas. *Remote Sens.* **2017**, *9*, 213. [CrossRef]

24. Nelson, R.; Margolis, H.; Montesano, P.; Sun, G.; Cook, B.; Corp, L.; Andersen, H.-E.; deJong, B.; Pellat, F.P.; Fickel, T.; et al. Lidar-based estimates of aboveground biomass in the continental US and Mexico using ground, airborne, and satellite observations. *Remote Sens. Environ.* **2017**, *188*, 127–140. [CrossRef]

25. Foody, G.M.; Boyd, D.S.; Cutler, M.E.J. Predictive relations of tropical forest biomass from Landsat TM data and their transferability between regions. *Remote Sens. Environ.* **2003**, *85*, 463–474. [CrossRef]

26. Ploton, P.; Pélissier, R.; Barbier, N.; Proisy, C.; Ramesh, B.R.; Couteron, P. Canopy Texture Analysis for Large-Scale Assessments of Tropical Forest Stand Structure and Biomass. In *Treetops at Risk*; Lowman, M., Devy, S., Ganesh, T., Eds.; Springer: New York, NY, USA, 2013; pp. 237–245. ISBN 978-1-4614-7160-8.

27. Bastin, J.-F.; Barbier, N.; Couteron, P.; Adams, B.; Shapiro, A.; Bogaert, J.; De Cannière, C. Aboveground biomass mapping of African forest mosaics using canopy texture analysis: Toward a regional approach. *Ecol. Appl.* **2014**, *24*, 1984–2001. [CrossRef] [PubMed]

28. Ploton, P. Assessing aboveground tropical forest biomass using Google Earth canopy images. *Ecol. Appl.* **2012**, *22*, 993–1003. [CrossRef]

29. Baynes, J. Assessing forest canopy density in a highly variable landscape using Landsat data and FCD Mapper software. *Aust. For.* **2004**, *67*, 247–253. [CrossRef]

30. Gonzalez, P.; Asner, G.P.; Battles, J.J.; Lefsky, M.A.; Waring, K.M.; Palace, M. Forest carbon densities and uncertainties from Lidar, QuickBird, and field measurements in California. *Remote Sens. Environ.* **2010**, *114*, 1561–1575. [CrossRef]

31. Saatchi, S.S.; Harris, N.L.; Brown, S.; Lefsky, M.; Mitchard, E.T.; Salas, W.; Zutta, B.R.; Buermann, W.; Lewis, S.L.; Hagen, S.; et al. Benchmark map of forest carbon stocks in tropical regions across three continents. *Proc. Natl. Acad. Sci. USA* **2011**, *108*, 9899–9904. [CrossRef] [PubMed]

32. Baccini, A.; Goetz, S.J.; Walker, W.S.; Laporte, N.T.; Sun, M.; Sulla-Menashe, D.; Hackler, J.; Beck, P.S.A.; Dubayah, R.; Friedl, M.A.; et al. Estimated carbon dioxide emissions from tropical deforestation improved by carbon-density maps. *Nat. Clim. Chang.* **2012**, *2*, 182–185. [CrossRef]

33. Langner, A.; Achard, F.; Grassi, G. Can recent pan-tropical biomass maps be used to derive alternative Tier 1 values for reporting REDD+ activities under UNFCCC? *Environ. Res. Lett.* **2014**, *9*, 124008. [CrossRef]

34. Potapov, P.; Yaroshenko, A.; Turubanova, S.; Dubinin, M.; Laestadius, L.; Thies, C.; Aksenov, D.; Egorov, A.; Yesipova, Y.; Glushkov, I.; et al. Mapping the World's Intact Forest Landscapes by Remote Sensing. *Ecol. Soc.* **2008**, *13*, 51. [CrossRef]

35. Mitchard, E.T.; Saatchi, S.S.; Baccini, A.; Asner, G.P.; Goetz, S.J.; Harris, N.L.; Brown, S. Uncertainty in the spatial distribution of tropical forest biomass: A comparison of pan-tropical maps. *Carbon Balance Manag.* **2013**, *8*, 10. [CrossRef]

36. Peregon, A.; Yamagata, Y. The use of ALOS/PALSAR backscatter to estimate above-ground forest biomass: A case study in Western Siberia. *Remote Sens. Environ.* **2013**, *137*, 139–146. [CrossRef]

37. Soja, M.J.; Sandberg, G.; Ulander, L.M.H. Regression-Based Retrieval of Boreal Forest Biomass in Sloping Terrain Using P-Band SAR Backscatter Intensity Data. *IEEE Trans. Geosci. Remote Sens.* **2013**, *51*, 2646–2665. [CrossRef]

38. Cartus, O.; Kellndorfer, J.; Walker, W.; Franco, C.; Bishop, J.; Santos, L.; Fuentes, J.M.M. A National, Detailed Map of Forest Aboveground Carbon Stocks in Mexico. *Remote Sens.* **2014**, *6*, 5559–5588. [CrossRef]

39. Pham, T.D.; Yoshino, K.; Bui, D.T. Biomass estimation of Sonneratia caseolaris (L.) Engler at a coastal area of Hai Phong city (Vietnam) using ALOS-2 PALSAR imagery and GIS-based multi-layer perceptron neural networks. *GISci. Remote Sens.* **2017**, *54*, 329–353. [CrossRef]

40. Le Toan, T.; Quegan, S.; Davidson, M.W.J.; Balzter, H.; Paillou, P.; Papathanassiou, K.; Plummer, S.; Rocca, F.; Saatchi, S.; Shugart, H.; et al. The BIOMASS mission: Mapping global forest biomass to better understand the terrestrial carbon cycle. *Remote Sens. Environ.* **2011**, *115*, 2850–2860. [CrossRef]

41. Englhart, S.; Keuck, V.; Siegert, F. Aboveground biomass retrieval in tropical forests—The potential of combined X- and L-band SAR data use. *Remote Sens. Environ.* **2011**, *115*, 1260–1271. [CrossRef]

42. Yu, Y.; Saatchi, S. Sensitivity of L-Band SAR Backscatter to Aboveground Biomass of Global Forests. *Remote Sens.* **2016**, *8*, 522. [CrossRef]

43. Schlund, M.; Davidson, M.W.J. Aboveground Forest Biomass Estimation Combining L- and P-Band SAR Acquisitions. *Remote Sens.* **2018**, *10*, 1151. [CrossRef]

44. Joshi, N.; Mitchard, E.T.A.; Brolly, M.; Schumacher, J.; Fernández-Landa, A.; Johannsen, V.K.; Marchamalo, M.; Fensholt, R. Understanding 'saturation' of radar signals over forests. *Sci. Rep.* **2017**, *7*, 3505. [CrossRef] [PubMed]

45. Vafaei, S.; Soosani, J.; Adeli, K.; Fadaei, H.; Naghavi, H.; Pham, T.D.; Tien Bui, D. Improving Accuracy Estimation of Forest Aboveground Biomass Based on Incorporation of ALOS-2 PALSAR-2 and Sentinel-2A Imagery and Machine Learning: A Case Study of the Hyrcanian Forest Area (Iran). *Remote Sens.* **2018**, *10*, 172. [CrossRef]

46. Pham, T.D.; Yoshino, K.; Le, N.N.; Bui, D.T. Estimating aboveground biomass of a mangrove plantation on the Northern coast of Vietnam using machine learning techniques with an integration of ALOS-2 PALSAR-2 and Sentinel-2A data. *Int. J. Remote Sens.* **2018**, *39*, 7761–7788. [CrossRef]

47. Ho Tong Minh, D.; Ndikumana, E.; Vieilledent, G.; McKey, D.; Baghdadi, N. Potential value of combining ALOS PALSAR and Landsat-derived tree cover data for forest biomass retrieval in Madagascar. *Remote Sens. Environ.* **2018**, *213*, 206–214. [CrossRef]

48. Deng, S.; Katoh, M.; Guan, Q.; Yin, N.; Li, M. Estimating Forest Aboveground Biomass by Combining ALOS PALSAR and WorldView-2 Data: A Case Study at Purple Mountain National Park, Nanjing, China. *Remote Sens.* **2014**, *6*, 7878–7910. [CrossRef]

49. Hame, T.; Rauste, Y.; Antropov, O.; Ahola, H.A.; Kilpi, J. Improved Mapping of Tropical Forests with Optical and SAR Imagery, Part II: Above Ground Biomass Estimation. *IEEE J. Sel. Top. Appl. Earth Obs. Remote Sens.* **2013**, *6*, 92–101. [CrossRef]

50. Potapov, P.; Hansen, M.C.; Stehman, S.V.; Loveland, T.R.; Pittman, K. Combining MODIS and Landsat imagery to estimate and map boreal forest cover loss. *Remote Sens. Environ.* **2008**, *112*, 3708–3719. [CrossRef]

51. Köhl, M.; Lister, A.; Scott, C.T.; Baldauf, T.; Plugge, D. Implications of sampling design and sample size for national carbon accounting systems. *Carbon Balance Manag.* **2011**, *6*, 10. [CrossRef] [PubMed]

52. Gallego, F.J. Remote sensing and land cover area estimation. *Int. J. Remote Sens.* **2004**, *25*, 3019–3047. [CrossRef]

53. Gong, P.; Marceau, D.J.; Howarth, P.J. A comparison of spatial feature extraction algorithms for land-use classification with SPOT HRV data. *Remote Sens. Environ.* **1992**, *40*, 137–151. [CrossRef]

54. White, F. *The Vegetation of Africa: A Descriptive Memoir to Accompany the UNESCO/AETFAT/UNSO Vegetation Map of Africa*; UNESCO: Paris, France, 1983; ISBN 92-3-101955-4.

55. MNRT. *NAFORMA Main Results*; MNRT: Dodoma, Tanzania, 2015.

56. UN-REDD. *Draft Action Plan for Implementation of National Strategy for REDD+*; UN-REDD: Geneva, Switzerland, 2012.

57. UN-REDD. *Estimating the Cost Elements of REDD+ in Tanzania*; UN-REDD: Geneva, Switzerland, 2012.

58. NAFORMA. *Field Manual, Biophysical Survey*; Ministry of Natural Resources and Tourism, Forestry and Beekeeping Division: Dar es Salaam, Tanzania, 2010.

59. Tomppo, E.; Katila, M.; Peräsaari, J.; Malimbwi, R.; Chamuya, N.; Otieno, J.; Dalsgaard, S.; Leppänen, M. *A Report to the Food and Agriculture Organization of the United Nations (FAO) in Support of Sampling Study for National Forestry Resources Monitoring and Assessment (NAFORMA) in Tanzania*; Sokoine University of Agriculture: Morogoro, Tanzania, 2010.

60. Tomppo, E.; Malimbwi, R.; Katila, M.; Mäkisara, K.; Henttonen, H.M.; Chamuya, N.; Zahabu, E.; Otieno, J. A sampling design for a large area forest inventory: case Tanzania. *Can. J. For. Res.* **2014**, *44*, 931–948. [CrossRef]

61. Hojas Gascón, L.; Eva, H. *Field Guide for Forest Mapping with High Resolution Satellite Data*; Publications Office: Luxembourg, 2014; ISBN 978-92-79-44012-0.

62. Mascaro, J.; Litton, C.M.; Hughes, R.F.; Uowolo, A.; Schnitzer, S.A. Minimizing Bias in Biomass Allometry: Model Selection and Log-Transformation of Data. *Biotropica* **2011**, *43*, 649–653. [CrossRef]

63. FAO; JRC; SDSU; UCL. *The 2010 Global Forest Resources Assessment Remote Sensing Survey: FRA Working Paper 155*; FAO: Rome, Italy, 2009.

64. Tyc, G.; Tulip, J.; Schulten, D.; Krischke, M.; Oxfort, M. The RapidEye mission design. *Acta Astronaut.* **2005**, *56*, 213–219. [CrossRef]

65. Beuchle, R.; Grecchi, R.C.; Shimabukuro, Y.E.; Seliger, R.; Eva, H.D.; Sano, E.; Achard, F. Land cover changes in the Brazilian Cerrado and Caatinga biomes from 1990 to 2010 based on a systematic remote sensing sampling approach. *Appl. Geogr.* **2015**, *58*, 116–127. [CrossRef]

66. RapidEye. *Planet Labs San Francisco Satellite Imagery Product Specifications*, version 6.1; RapidEye: Berlin, Germany, 2016.

67. Beuchle, R.; Eva, H.D.; Stibig, H.-J.; Bodart, C.; Brink, A.; Mayaux, P.; Johansson, D.; Achard, F.; Belward, A. A satellite data set for tropical forest area change assessment. *Int. J. Remote Sens.* **2011**, *32*, 7009–7031. [CrossRef]

68. Hojas Gascón, L.; Eva, H.; Laporte, N.; Simonetti, D.; Fritz, S. The Application of Medium-Resolution MERIS Satellite Data for Continental Land-Cover Mapping over South America: Results and Caveats. In *Remote Sensing of Land Use and Land Cover: Principles and Applications*; CRC Press/Taylor & Francis: Boca Raton, FL, USA, 2012.

69. Hansen, M.C.; Roy, D.P.; Lindquist, E.; Adusei, B.; Justice, C.O.; Altstatt, A. A method for integrating MODIS and Landsat data for systematic monitoring of forest cover and change in the Congo Basin. *Remote Sens. Environ.* **2008**, *112*, 2495–2513. [CrossRef]

70. UNFCCC. *Report of the Conference of the Parties on Its Seventh Session, Held at Marrakesh from 29 October to 10 November 2001*; United Nations Framework Convention on Climate Change: Bonn, Germany, 2001.

71. The United Republic of Tanzania. Tanzania's Forest Reference Emission Level Submission to the UNFCCC 2016. Available online: https://redd.unfccc.int/files/frel__for__tanzania_december2016_27122016.pdf (accessed on 28 January 2019).

72. Raši, R.; Bodart, C.; Stibig, H.-J.; Eva, H.; Beuchle, R.; Carboni, S.; Simonetti, D.; Achard, F. An automated approach for segmenting and classifying a large sample of multi-date Landsat imagery for pan-tropical forest monitoring. *Remote Sens. Environ.* **2011**, *115*, 3659–3669. [CrossRef]

73. Flanders, D.; Hall-Beyer, M.; Pereverzoff, J. Preliminary evaluation of eCognition object-based software for cut block delineation and feature extraction. *Can. J. Remote Sens.* **2003**, *29*, 441–452. [CrossRef]

74. Blaschke, T. Object based image analysis for remote sensing. *ISPRS J. Photogramm. Remote Sens.* **2010**, *65*, 2–16. [CrossRef]

75. Dube, T.; Mutanga, O. Investigating the robustness of the new Landsat-8 Operational Land Imager derived texture metrics in estimating plantation forest aboveground biomass in resource constrained areas. *ISPRS J. Photogramm. Remote Sens.* **2015**, *108*, 12–32. [CrossRef]

76. Konstanski, H. *Apparent Cloud Shift in RapidEye Image Data*; RapidEye: Berlin, Germany, 2012.

77. Simonetti, D.; Simonetti, E.; Szantoi, Z.; Lupi, A.; Eva, H.D. First results from the phenology-based synthesis classifier using Landsat 8 imagery. *IEEE Geosci. Remote Sens. Lett.* **2015**, *12*, 1496–1500. [CrossRef]

78. Haralick, R.M.; Shanmugam, K.; Dinstein, I. Textural features for image classification. *IEEE Trans. Syst. Man Cybern.* **1973**, *SMC-3*, 610–621. [CrossRef]

79. Haboudane, D. Hyperspectral vegetation indices and novel algorithms for predicting green LAI of crop canopies: Modeling and validation in the context of precision agriculture. *Remote Sens. Environ.* **2004**, *90*, 337–352. [CrossRef]

80. Xue, J.; Su, B. Significant Remote Sensing Vegetation Indices: A Review of Developments and Applications. *J. Sens.* **2017**, *2017*, 1353691. [CrossRef]

81. Lu, D.; Mausel, P.; Brondízio, E.; Moran, E. Relationships between forest stand parameters and Landsat TM spectral responses in the Brazilian Amazon Basin. *For. Ecol. Manag.* **2004**, *198*, 149–167. [CrossRef]

82. Lu, D.; Chen, Q.; Wang, G.; Liu, L.; Li, G.; Moran, E. A survey of remote sensing-based aboveground biomass estimation methods in forest ecosystems. *Int. J. Digit. Earth* **2016**, *9*, 63–105. [CrossRef]

83. Gwenzi, D.; Helmer, E.; Zhu, X.; Lefsky, M.; Marcano-Vega, H. Predictions of Tropical Forest Biomass and Biomass Growth Based on Stand Height or Canopy Area Are Improved by Landsat-Scale Phenology across Puerto Rico and the U.S. Virgin Islands. *Remote Sens.* **2017**, *9*, 123. [CrossRef]

84. Eckert, S. Improved Forest Biomass and Carbon Estimations Using Texture Measures from WorldView-2 Satellite Data. *Remote Sens.* **2012**, *4*, 810–829. [CrossRef]

85. Rikimaru, A.; Roy, P.S.; Miyatake, S. Tropical forest cover density mapping. *Trop. Ecol.* **2002**, *43*, 39–47.

86. Eitel, J.U.H.; Long, D.S.; Gessler, P.E.; Smith, A.M.S. Using in-situ measurements to evaluate the new RapidEye satellite series for prediction of wheat nitrogen status. *Int. J. Remote Sens.* **2007**, *28*, 4183–4190. [CrossRef]

87. Tucker, C.J.; Pinzon, J.E.; Brown, M.E.; Slayback, D.A.; Pak, E.W.; Mahoney, R.; Vermote, E.F.; El Saleous, N. An extended AVHRR 8-km NDVI dataset compatible with MODIS and SPOT vegetation NDVI data. *Int. J. Remote Sens.* **2005**, *26*, 4485–4498. [CrossRef]

88. Huete, A.; Didan, K.; Miura, T.; Rodriguez, E.P.; Gao, X.; Ferreira, L.G. Overview of the radiometric and biophysical performance of the MODIS vegetation indices. *Remote Sens. Environ.* **2002**, *83*, 195–213. [CrossRef]

89. Huete, A. A soil-adjusted vegetation index (SAVI). *Remote Sens. Environ.* **1988**, *25*, 295–309. [CrossRef]

90. Haboudane, D.; Miller, J.R.; Tremblay, N.; Zarco-Tejada, P.J.; Dextraze, L. Integrated narrow-band vegetation indices for prediction of crop chlorophyll content for application to precision agriculture. *Remote Sens. Environ.* **2002**, *81*, 416–426. [CrossRef]

91. Haralick, R.M. Statistical and structural approaches to texture. *Proc. IEEE* **1979**, *67*, 786–804. [CrossRef]

92. Moses, M.; Stevens, T.S.; Bax, G. GIS Data Ineroperability in Uganda. *Int. J. Spat. Data Infrastruct. Res.* **2012**, *7*, 488–507.

93. Breiman, L. Random Forests. *Mach. Learn.* **2001**, *45*, 5–32. [CrossRef]

94. Hastie, T.; Tibshirani, R.; Friedman, J. *The Elements of Statistical Learning—Data Mining, Inference, and Prediction*; Springer Series in Statistics; Springer: New York, NY, USA, 2001; Volume 1.

95. Vapnik, V.N. *The Nature of Statistical Learning Theory*; Springer, Inc.: New York, NY, USA, 1995; ISBN 978-0-387-94559-0.

96. Payandeh, B. Relative Efficiency of Two-Dimensional Systematic Sampling. *For. Sci.* **1970**, *16*, 271–276.

97. Guyot, G.; Guyon, D.; Riom, J. Factors affecting the spectral response of forest canopies: A review. *Geocarto Int.* **1989**, *4*, 3–18. [CrossRef]

98. Hagolle, O.; Huc, M.; Dedieu, G.; Sylvander, S. SPOT4 (Take 5) Times series over 45 sites to prepare Sentinel-2 applications and methods. In Proceedings of the ESA's Living Planet Symposium, Edinburgh, UK, 9–13 September 2013; Volume 11.

99. Hojas-Gascon, L.; Belward, A.; Eva, H.; Ceccherini, G.; Hagolle, O.; Garcia, J.; Cerutti, P. Potential improvement for forest cover and forest degradation mapping with the forthcoming Sentinel-2 program. *ISPRS Int. Arch. Photogramm. Remote Sens. Spat. Inf. Sci.* **2015**, *XL-7/W3*, 417–423. [CrossRef]

100. Hojas-Gascon, L.; Eva, H.D.; Ehrlich, D.; Pesaresi, M.; Achard, F.; Garcia, J. *Urbanization and Forest Degradation in East Africa—A Case Study around Dar es Salaam, Tanzania*; IEEE: Piscataway, NJ, USA, 2016; pp. 7293–7295.

101. Avitabile, V.; Camia, A. An assessment of forest biomass maps in Europe using harmonized national statistics and inventory plots. *For. Ecol. Manag.* **2018**, *409*, 489–498. [CrossRef]

102. Gao, Y.; Lu, D.; Li, G.; Wang, G.; Chen, Q.; Liu, L.; Li, D. Comparative Analysis of Modeling Algorithms for Forest Aboveground Biomass Estimation in a Subtropical Region. *Remote Sens.* **2018**, *10*, 627. [CrossRef]

103. Mutanga, O.; Adam, E.; Cho, M.A. High density biomass estimation for wetland vegetation using WorldView-2 imagery and random forest regression algorithm. *Int. J. Appl. Earth Obs. Geoinf.* **2012**, *18*, 399–406. [CrossRef]

104. Shao, Z.; Zhang, L. Estimating Forest Aboveground Biomass by Combining Optical and SAR Data: A Case Study in Genhe, Inner Mongolia, China. *Sensors* **2016**, *16*, 834. [CrossRef] [PubMed]

105. Attarchi, S.; Gloaguen, R. Improving the Estimation of Above Ground Biomass Using Dual Polarimetric PALSAR and ETM+ Data in the Hyrcanian Mountain Forest (Iran). *Remote Sens.* **2014**, *6*, 3693–3715. [CrossRef]

106. Avitabile, V.; Herold, M.; Heuvelink, G.B.M.; Lewis, S.L.; Phillips, O.L.; Asner, G.P.; Armston, J.; Ashton, P.S.; Banin, L.; Bayol, N.; et al. An integrated pan-tropical biomass map using multiple reference datasets. *Glob. Chang. Biol.* **2016**, *22*, 1406–1420. [CrossRef] [PubMed]

107. Hoa, N.H. Comparison of various spectral indices for estimating mangrove covers using PlanetScope data: A case study in Xuan Thuy national park, Nam Dinh province. *J. For. Sci. Technol.* **2017**, *5*, 74–83.

108. McRoberts, R.E.; Holden, G.R.; Nelson, M.D.; Liknes, G.C.; Gormanson, D.D. Using satellite imagery as ancillary data for increasing the precision of estimates for the Forest Inventory and Analysis program of the USDA Forest Service. *Can. J. For. Res.* **2005**, *35*, 2968–2980. [CrossRef]
109. Guidolavespa. *Guidolavespa/Forest v1.0*; Zenodo: Geneva, Switzerland, 2018.

 *forests*

Article

The Potential of Multisource Remote Sensing for Mapping the Biomass of a Degraded Amazonian Forest

Clément Bourgoin [1,2,*], Lilian Blanc [1,2], Jean-Stéphane Bailly [3,4], Guillaume Cornu [1,2], Erika Berenguer [5,6], Johan Oszwald [7], Isabelle Tritsch [8], François Laurent [9], Ali F. Hasan [9], Plinio Sist [1,2] and Valéry Gond [1,2]

[1] CIRAD, Forêts et Sociétés, F-34398 Montpellier, France; lilian.blanc@cirad.fr (L.B.); guillaume.cornu@cirad.fr (G.C.); plinio.sist@cirad.fr (P.S.); valery.gond@cirad.fr (V.G.)
[2] Forêts et Sociétés, University Montpellier, CIRAD, 34398 Montpellier, France
[3] LISAH, University Montpellier, INRA, IRD, Montpellier SupAgro, 34398 Montpellier, France; bailly@agroparistech.fr
[4] Department SIAFEE AgroParisTech, 75231 Paris, France
[5] Environmental Change Institute, University of Oxford, Oxford OX1 3QY, UK; erikaberenguer@gmail.com
[6] Lancaster Environment Centre, Lancaster University, Lancaster LA1 4YQ, UK
[7] UMR CNRS LETG 6554, Laboratory of Geography and Remote Sensing COSTEL, Université de Rennes 2, 35043 Rennes, France; johan.oszwald@univ-rennes2.fr
[8] Centre de Recherche et de Documentations sur les Amériques (CREDA), UMR 7227, Université Sorbonne Nouvelle, Paris 3, 75006 Paris, France; isabelle.tritsch@gmail.com
[9] UMR CNRS ESO (Espaces et Sociétés), Le Mans Université, 72000 Le Mans, France; francois.laurent@univ-lemans.fr (F.L.); alihassanali78@yahoo.com (A.F.H.)
* Correspondence: clement.bourgoin@cirad.fr; Tel.: +33-467-593-787

Received: 13 April 2018; Accepted: 25 May 2018; Published: 29 May 2018

Abstract: In the agricultural frontiers of Brazil, the distinction between forested and deforested lands traditionally used to map the state of the Amazon does not reflect the reality of the forest situation. A whole gradient exists for these forests, spanning from well conserved to severely degraded. For decision makers, there is an urgent need to better characterize the status of the forest resource at the regional scale. Until now, few studies have been carried out on the potential of multisource, freely accessible remote sensing for modelling and mapping degraded forest structural parameters such as aboveground biomass (AGB). The aim of this article is to address that gap and to evaluate the potential of optical (Landsat, MODIS) and radar (ALOS-1 PALSAR, Sentinel-1) remote sensing sources in modelling and mapping forest AGB in the old pioneer front of Paragominas municipality (Para state). We derived a wide range of vegetation and textural indices and combined them with in situ collected AGB data into a random forest regression model to predict AGB at a resolution of 20 m. The model explained 28% of the variance with a root mean square error of 97.1 Mg·ha^{-1} and captured all spatial variability. We identified Landsat spectral unmixing and mid-infrared indicators to be the most robust indicators with the highest explanatory power. AGB mapping reveals that 87% of forest is degraded, with illegal logging activities, impacted forest edges and other spatial distribution of AGB that are not captured with pantropical datasets. We validated this map with a field-based forest degradation typology built on canopy height and structure observations. We conclude that the modelling framework developed here combined with high-resolution vegetation status indicators can help improve the management of degraded forests at the regional scale.

Keywords: forest degradation; multisource remote sensing; modelling aboveground biomass; random forest; Brazilian Amazon

1. Introduction

Deforestation and forest degradation are major sources of greenhouse gas emissions [1,2], contributing to forest carbon losses [3], global climate change, affecting biodiversity [4] and the entire forest ecosystem. While deforestation refers to the rapid conversion from forest to non-forest areas, degradation implies changes in the forest structure with no change in land use [5,6]. In Amazonia, over the last decades, deforestation and forest degradation have shaped the rural landscape, resulting in a complex mosaic of fragmented forests associated with agricultural lands [7]. A total area of 766,448.5 km^2 was cleared in 2015 [8], representing 20% of Amazonia [9,10].

Since 2005, deforestation in Brazil has drastically decreased thanks to coercive measures taken by the Brazilian government associated with private initiatives (soy and beef moratoria), among other factors [11]. However, these measures are not effective for reducing forest degradation [12,13]. Most of the remaining forested lands are degraded due to the accumulation over time and space of severe degradation processes mainly triggered by anthropogenic impacts through unsustainable logging practices, fire, shifting cultivation and charcoal production [14,15].

Reducing forest degradation is a major challenge given the rapid need to reduce carbon emissions to the atmosphere, conserve biodiversity, limit soil erosion and regulate the water cycle [16]. Forest monitoring based on the forest/non-forest approach used to quantify deforestation is not relevant for providing information on the forest status [17,18]. The biomass value of a forest is a relevant indicator to quantify the intensity of degradation [19]. Forest biomass mapping is therefore a critical step to reach the challenge [20].

At the pantropical scale, two maps of biomass density that present the spatial distribution of the biomass of all forest types at a moderate resolution [21,22] have been used as baselines for the tropical belt. More recently, a harmonized reference aboveground biomass (AGB) map has been released that significantly improves the estimation and local distribution of AGB using the combination of in-situ collected data, remote sensing and regional biomass maps [23].

At the local scale, most of the approaches integrate field-collected data with Light Detection and Ranging (LIDAR) to scale up forest biomass natural distribution, which normally requires spatial interpolation of in-situ biomass [24]. LIDAR can map the forest canopy in three dimensions and can retrieve accurate forest biomass through forest canopy height and structure [25–29]. It is also sensitive to the carbon density of the different types of degraded forests, from logging at a low impact to forest stands burned multiple times [30]. Satellite LIDAR has been used in validating AGB maps [31], calibrating local regressions between in situ AGB data and metrics derived from LIDAR footprints and extrapolating using different remote sensing sources [28,32]. However, most airborne and satellite LIDAR datasets are often difficult to access (acquisition cost) and to replicate in both time and space [33].

At the meso-scale (regional), many studies demonstrated the potential of optical and radar remote sensing-derived indicators to characterize degraded forests [34]. The study of degraded forests requires the analysis of vertical and horizontal disturbances within the forest structure [5,33,35–37]. Optical images can provide information on the photosynthetic activity and moisture of the forest canopy [38]. Spectral unmixing approaches are recognized to be the most effective method to assess the status of degraded forests using the percentage of active vegetation, dead vegetation and bare soil at the pixel scale [15,39,40]. Radar images are sensitive to the texture of the impacted forest canopy [41]. Canopy texture-derived indicators based on co-occurrence matrices use the variance of the signal in a given window to spatially quantify the distribution of tree crowns structure [42–44]. Estimating the biomass from the radar data generally concerns wavelengths up to the meter (band P or L) with signal saturation thresholds around 200 Mg/ha of AGB [45–47].

This review of recent remote sensing methods illustrates the trade-off that needs to be made between resolution, accuracy, area covered, cost and frequency to map forest biomass. It also highlights the fact that there is remarkably little information at the regional scale on the potential of open access optical and radar remote sensing to model and map the aboveground biomass of degraded forests.

Most of the approaches tend to capture the local variation of AGB following environmental variables, i.e., climate, topography and natural forest dynamic gradients, but do not particularly capture the distribution along anthropogenic disturbance gradients [30].

In this sense, there is a need to better understand how remotely sensed indicators perform with AGB modelling at the regional scale in order to provide relevant information to decision makers on the state of the resource of remnants forest.

In order to answer this need, this paper aims (i) to assess the potential of multisource, multi-indicator and open access remote sensing in modelling and mapping aboveground biomass of degraded forests; (ii) to quantify the spatial distribution of degraded forest biomass in comparison with the pantropical AGB map; and (iii) to evaluate the relevance of this regional forest AGB mapping at the stand scale.

This quantification is highly informative for forested land use planning and policy makers. Many South American governments and global NGOs are seeking more accurate and definitive information about the scale of degradation so they can propose policies and actions to ameliorate and reduce the level of degradation [48].

2. Materials and Methods

2.1. Study Area

The study was carried out in the municipality of Paragominas, located in the northeastern part of the State of Para, Brazil, and covering an area of 19,342 km^2 with a population size of 108,547 [49]. The municipality was founded in 1965 along the BR-010 road connecting Brasilia to Belém (Figure 1). The colonization process led to a large conversion of lands into pasture, with cattle ranching becoming the dominant land use. The municipality went through a succession of different economical models that have drastically shaped the landscape [50]. The boom in the logging industry started from the 1980s, where most of the timber was transformed in the 350 sawmills located along the main road. Deforestation and forest degradation were accentuated with the grain agribusiness boom in the 2000s (soybean and maize cultivation) and charcoal production. In 2007, the municipality was red-listed by the federal government as one of the most deforested Amazonian municipalities. The consequences were an immediate loss of access to credit and market for any commodities. Many charcoal plants and illegal sawmills were shut down. In response to this governmental ban, Paragominas became the first "green municipality" in the country in 2008 in order to end illegal deforestation, to tend to zero net deforestation by 2014, and to promote alternative production systems and reforestation. Land management was also improved through the Rural Environmental Registry [51].

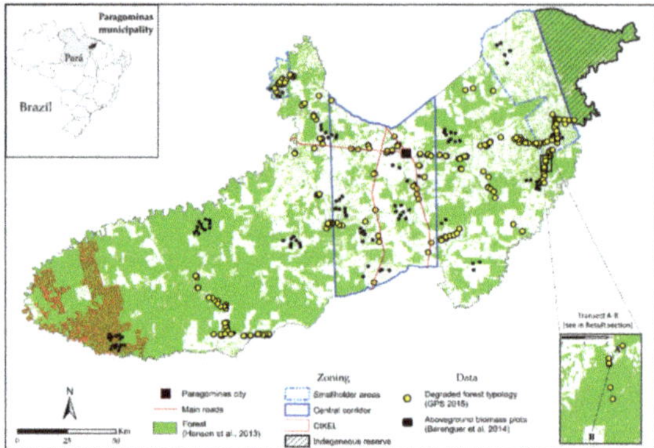

Figure 1. Map of Paragominas municipality in the northeastern part of the State of Para with the location of the biomass-collected plots, the degraded forest typology and the main zoning areas (extracted from [9]).

2.2. In Situ AGB Collection

Aboveground biomass data were collected during the dry seasons in 2009 and 2010 by teams from the Sustainable Amazon Network (Rede Sustentavel Amazonia—RAS) [19,52,53]. In total, 18 study sites were randomly selected in the municipality of Paragominas. For each site, plots of 250 m by 20 m were designed, resulting in a total of 121 plots. Each plot represents a homogeneous type of forest, classified into four types: undisturbed forest, logged forest, logged forest and burned and secondary forest. Aboveground biomass was estimated using Chave's allometric equations [54] and was based on the measurements and identification of all live trees, palms and lianas ($\geq$10 cm DBH). The results of this study showed significant differences in biomass for each type of forest (Table 1). We chose a threshold of 286 Mg·ha^{-1} to distinguish between degraded and sustainably managed or conserved forests. Mazzei et al. [55] showed that undisturbed forests store on average 409.8 Mg·ha^{-1} in the Cikel Fazenda, located in the western part of the municipality (Figure 1). The mean AGB total lost under reduced-impact logging operations is 94.5 Mg·ha^{-1}, which represents around 23% of biomass lost [55]. The threshold is thus the percentage of biomass lost applied to the undisturbed primary forest biomass identified by RAS.

The aboveground vegetation that corresponds to the largest carbon storage compartment is very sensitive to the type of disturbance, with about a 40% of difference in ABG between mature forests and secondary forests. We used the RAS AGB dataset in shapefile format to calibrate the Random Forest model presented in Section 2.4.

Table 1. AGB collection data in the municipality of Paragominas (see [19]).

Forest Type	Mean AGB (Mg·ha^{-1})	Standard Deviation (Mg·ha^{-1})	No. of Plots
Undisturbed primary forest	371.8	96.9	13
Logged primary forest	229.5	79.2	44
Logged and burnt primary forest	145.8	73.4	44
Young secondary forest	1.1	0.5	2
Intermediate secondary forest	57.6	38.3	12
Old secondary forest	92.2	58.4	5
Abandoned plantation	54.1	/	1

2.3. Remote Sensing Multisource Data: Image Acquisition, Pre-Processing and Biophysical Indicators Variables Extraction

We used 38 indicators derived from passive (MODIS, Landsat-8) and active (ALOS-1, Sentinel-1) remote sensing sources. These indicators are correlated with different vegetation parameters such as photosynthesis activity and vegetation structure. All the satellite images used in this study are freely available and span the globe.

- MODIS

To quantify forest canopy health and temporal dynamics, we used the enhanced vegetation index (EVI) from the MODIS sensor. EVI was extracted from the '16-Day L3 Global 250m product (MOD13Q1 c5)' from January 2001 to December 2014. EVI is directly related to photosynthetic activity [56]. It does not saturate quickly for high values of chlorophyll activity and provides improved sensitivity for high biomass areas such as tropical forests [38]. The 16-day composite was built by choosing within the 16 daily acquisitions the two pixels with the highest value of NDVI (Normalized Difference Vegetation Index). Of these two values, the pixel with the smallest viewing angle was chosen in order to minimize the residual angle [56]. A quality filter was then applied to the composite to remove clouds and reduce atmospheric contamination [38]. We extracted three indicators from this dataset: the mean and the standard deviation calculated for the whole time series based on annual average EVI and the pooled variance which is the weighted sum of annual variance based on the number of values available for each pixel for each year (giving more weight to the pixels that were not under cloud coverage during the 16 days × 14-year period).

$$\text{Pooled variance} = \frac{\sum_{i=1}^{14} Variance(i) \times available\ values(i)}{\sum_{i=1}^{14} available\ values(i) - 14} \tag{1}$$

These three indicators provide global information on the stability of the forest canopy photosynthetic activity.

- Landsat 8

We used three Landsat 8 images taken during the dry season of 2014, at 30 m resolution. We acquired the surface reflectance data with the pre-processing already performed with the algorithm developed by the NASA Goddard Space Flight Center (GSFC). We used 5 spectral bands (blue, green, red, near-infrared and short-wave infrared) and we derived 13 indicators using the Orfeo toolbox [57]. We used the Carnegie Landsat Analysis System-lite (Claslite) to derive the Photosynthetic Vegetation (PV), Non Photosynthetic Vegetation (NPV) and Bare Soil indexes [39].

- ALOS-1 PALSAR

We downloaded seven images ALOS-1 PALSAR (L-band) taken in 2010 from the JAXA (Japan Aerospace Agency, http://www.eorc.jaxa.jp/ALOS/en/index.htm) platform. These images have 25-m spatial resolution ("ALOS-1 mosaic 25 m" product) and are dual-polarized (HH and HV). The incidence angle varies between 35° (near range) to 42° (far range). The images were correctly geo-referenced, so we only processed the conversion from digital raw number (DN) to gamma and sigma following these two equations [58,59]:

$$\text{gamma [dB]} = 10 \times log10(DN^2) - 83 \tag{2}$$

$$\text{sigma [dB]} = 10 \times log10(DN^2) - 83 + 10 \times log10(cos\theta) \tag{3}$$

We tested these two indicators, expecting a strong correlation between backscatter coefficients and aboveground biomass [47].

- Sentinel-1

We acquired one image Sentinel-1 (C-band, dual polarization VV/VH, descending pass direction) taken in May 2015, at 10-m resolution and in Interferometric Wide (IW) swath mode. We performed the pre-processing using the free software Sentinel Toolbox, which allows the derivation of backscatter coefficients and processing of the range Doppler terrain corrections using the 3 arc-seconds SRTM Digital Elevation Model. We derived 9 indicators from the grey level co-occurrence matrix (GLCM) that are based on the statistical relationship between the values of the pixels within a 9 × 9 pixels window [60]. These indicators are relevant to quantify forest canopy texture [61].

The source, description of the remote sensing images and the derived indicators are detailed in the Appendix A, Table A1.

2.4. Random Forest Regression Model

We used a random forest regression tree to explore the performance of the different remote sensing data sources and derived indicators for AGB modelling and mapping. Regression trees are particularly efficient for remotely sensed indicators that show unknown multivariate patterns and nonlinear relationships [62].

2.4.1. Data Preparation

We resampled the indicators at 20-m resolution using GRASS libraries and stacked them all together to make sure they are georeferenced in the same system. This resolution matches with the size of the plot measurements. For each indicator, we generated an automatic process to extract the mean and standard deviation within the extent of each plot. We then compiled all the data in a file with the identification of each plot (row 1 to 121) and the estimated AGB followed by the mean and standard deviation of the 38 indicators (columns). Finally, this dataset is randomly mixed to avoid any biases related to its original structure and is split into 10 folds (Figure 2). This number of folds (k) is often suggested in order to balance bias limitation (lower value of k) and variability (higher value of k) [63].

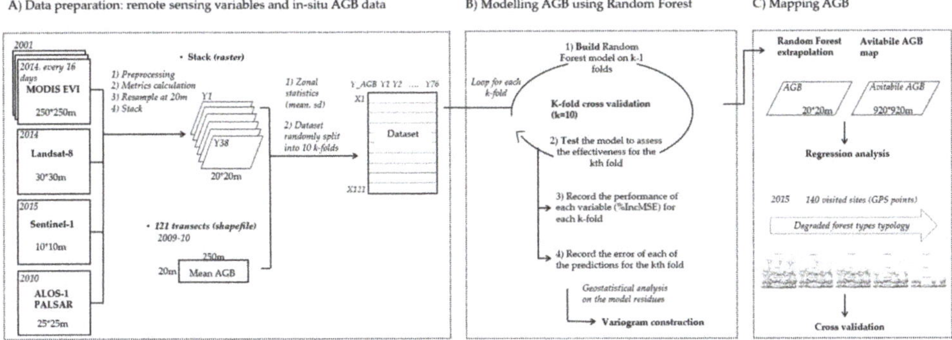

Figure 2. Workflow of the evaluation of the indicators performance in AGB modelling and mapping.

2.4.2. Modelling AGB with Random Forest

For each independent indicator, regression trees recurrently split the data into more homogeneous samples and identify the most significant indicator that gives best homogeneous sets of samples [64]. Random Forest grows multiple trees (500 trees in our study) by randomization of data subsampling in order to improve the predictive power of regression and to limit overfitting which is the most practical difficulty for decision tree models. Random forest provides for each independent indicator an increase in mean-squared error (Percentage of IncMSE), which quantifies how much MSE increases when that indicator is randomly permuted. This error measures the relative importance of each indicator, where a low IncMSE implies that the indicator does not have much weight on the model prediction and inversely [65]. Random Forest is used as a regression and mapping tool, but also as an investigation tool to assess the importance of each indicator in the creation of the model and on its global error. Random Forest is used more and more to estimate carbon reservoirs and perform biomass regression [25,66,67].

Due to the limited number of AGB plots, we used the k-fold cross validation technique to estimate regression performance [63,68]. This technique involves reserving a particular sample of the dataset on which the Random Forest model will be tested, while the rest of the dataset is used for training. For each of the 10 folds, the Random Forest model is trained on 9 folds (k − 1) and the 10th fold is used to test the model and check its effectiveness. This process goes through each of the 10 folds until each of the kfolds has served as a test set. After the end of each loop, we record the percentage of IncMSE calculated for each of the 76 indicators and the predicted value of AGB. The average RMSE will serve as the performance metric for the model.

We finally invert the model using the values of the remote sensed indicators across Paragominas municipality to predict AGB.

2.5. Geostatistical Analysis of Random Forest Model Residuals

Geostatistical analysis is used to explore the spatial autocorrelation of the model residuals and evaluate whether model residuals data are independent (in the case of absence of autocorrelation) [69]. The presence of spatial autocorrelation or spatial structure of the model residuals would refer to any patterns of gradients or cluster within the data that the AGB Random Forest model would not have adequately been able to capture. To explore this spatial structure, we estimated an omnidirectional variogram at short distances (6 km) on Random Forest residuals, and we used the usual permutation test to compute a bi-lateral confidence band of pure spatial randomness [70,71].

2.6. Comparison with Avitabile et al. [23] AGB Dataset

The available Avitabile et al. [23] forest pantropical biomass map that displays aboveground biomass density in units of $Mg \cdot ha^{-1}$ at a 920-m spatial resolution was used to compare the spatial distribution of aboveground biomass of the degraded forest with the Random Forest predicted map. In the Amazon basin, the Avitabile AGB map provides lower RMSE and bias compared to the Baccini and Saatchi pantropical maps [21–23]. However, at the regional scale (e.g., Paragominas municipality), the Avitabile map can present error patterns because the quality reference data of the highly degraded forest were lacking and could not be used to calibrate the model [23].

2.7. Comparison with Degraded Forest Typology

To conduct a field validation of the predicted aboveground biomass values, we built a degraded forest typology based on the observation of 140 forest sites in May 2015. Each observation is associated with a GPS point (Garmin 60CSx, Garmin, Olathe, KS, USA), a description of the forest site and illustrative photos. We extracted the values of the predicted AGB at the location of each forest site. We finally used one-way ANOVA with post hoc Tukey tests to evaluate differences in predicted AGB between the different forest classes of the typology.

This typology is a result of the combination of in-situ qualitative indicators of forest degradation and semi-quantitative observations of the forest structure. First, we noted the presence or absence of fire and logging marks (burned trees, strains, trunks, logging trails), of pioneer species (mainly *Cecropia* species), which may indicate a recent opening of the canopy, and of trees with a diameter at breast height (DBH) greater than 80 cm. Then, we measured canopy height and the number of vegetative strata using a laser rangefinder and estimated the forest canopy texture (roughness) and the percentage of gaps between emergent trees. These four forest structure indicators provide relevant information on the vertical and horizontal process of forest degradation.

In order to make the typology representative of the diversity of degraded forest types (conserved, legally logged, illegally logged and/or burned), we made sure that the sampling covered the different forest landscapes and main zoning that can be found in Paragominas (see Figure 1).

2.8. Computational Aspects

Except for the statistics performed with the software ArcGIS (Esri, Redlands, CA, USA), all developments were programmed under the R environment. We used the Raster, Random Forest, shapefile, rgdal, geoR and gstat packages [62,71–75].

3. Results

3.1. Model and Indicator Performance

The mean variance explained by the random forest model is 28%, with a root mean squared residual error (RMSE) of 97.1 $Mg \cdot ha^{-1}$. Depending on the AGB calibrated data and the associated remotely sensed indictors, the random forest model performs differently with an explained variance that ranges between 24% and 30% and an RMSE that ranges between 75.7 and 101.2 $Mg \cdot ha^{-1}$ (Table 2).

Table 2. Random Forest model performance (mean squared residuals and percentage of variance explained for each of the 10 k-fold random forest models).

Random Forest k-fold [1]	1	2	3	4	5	6	7	8	9	10	Average
Mean of squared residuals (Mg·ha^{-1})	97.8	95.5	100.5	91.8	99.6	100.3	93.9	75.7	97.8	101.2	97.1
Percentage of variance explained	26	24	25	30	26	22	27	30	26	25	28

[1] Number of Trees: 500, No. of Indicators Tried at Each Split: 25.

Six indicators (out of the 76) contribute the most to the 10 regression models, showing the highest and most stable IncMSE scores (Figure 3). Three are derived from MODIS: mean of annual standard deviation EVI, mean and standard deviation of annual mean EVI and three from Landsat: mean infra-red (MIR), mean and standard deviation of bare soils.

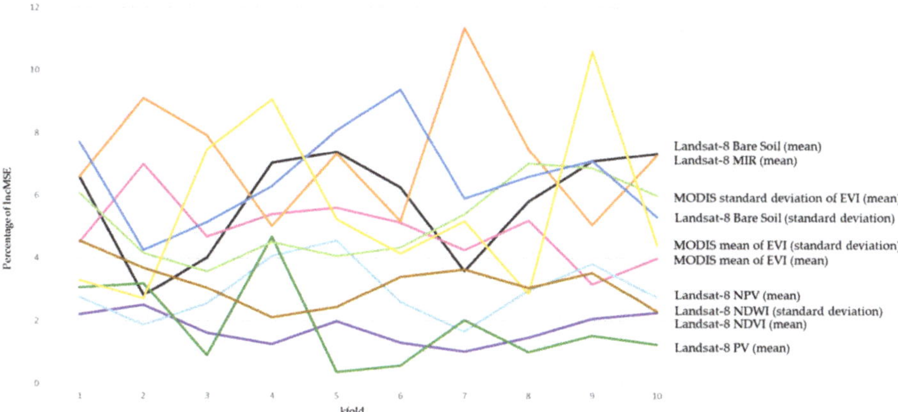

Figure 3. Explained indicator performance (Percentage of IncMSE) of the Random Forest model.

3.2. Geostatistical Analysis of Random Forest Model Residuals

The main spatial variability in the AGB data was captured by the Random Forest model, and no additive spatial variability can be explained through model residual interpolation. Figure 4 shows a flat empirical variogram contained within the confidence band which validates the absence of spatial structure within the Random Forest model residuals. We found similar results at smaller (0 to 1500 m) and larger (0–200,000 m) spatial scales. The distribution of the residuals shows an overall overestimation (high frequency of positive values, Figure 4B), which is important to consider when predicting AGB for the Paragominas municipality.

3.3. Above Ground Biomass Map

The range of AGB predicted values was large, spanning from 57 to 454 Mg·ha^{-1}. AGB was unequally spatially distributed over the municipality (Figure 5). The forests in the 80-km-wide central corridor have the lowest AGB values and are highly fragmented. The forests in the far-eastern and western part of the municipality contain the highest AGB. The percentage of degraded forest (below the threshold of 286 Mg·ha^{-1}, see Section 2.2) reached 87%.

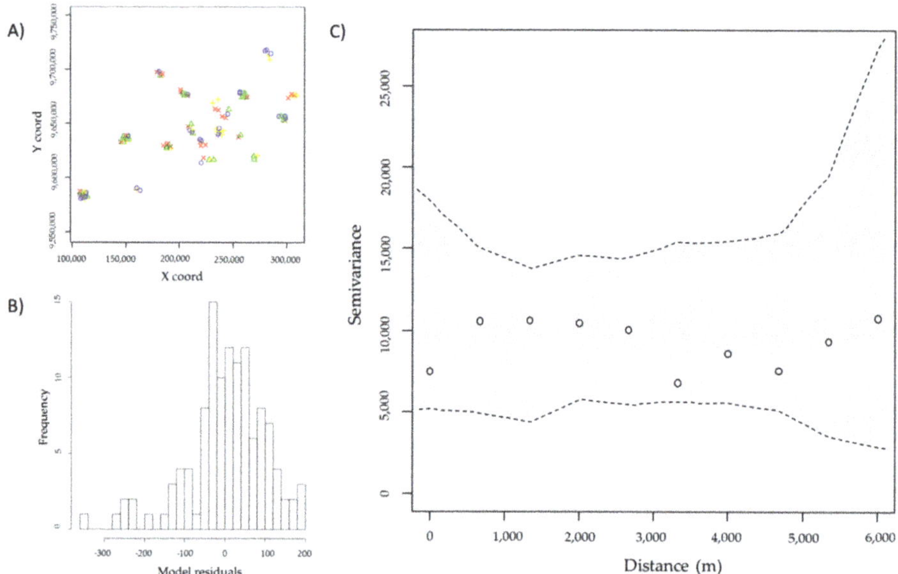

Figure 4. (**A**) Spatial distribution of the model residuals (blue, green, yellow and red colors for the respective four quartiles); (**B**) Histogram of the model residuals; (**C**) Variogram of biomass model residuals (The grey shape shows the confidence band interval expected for each distance class).

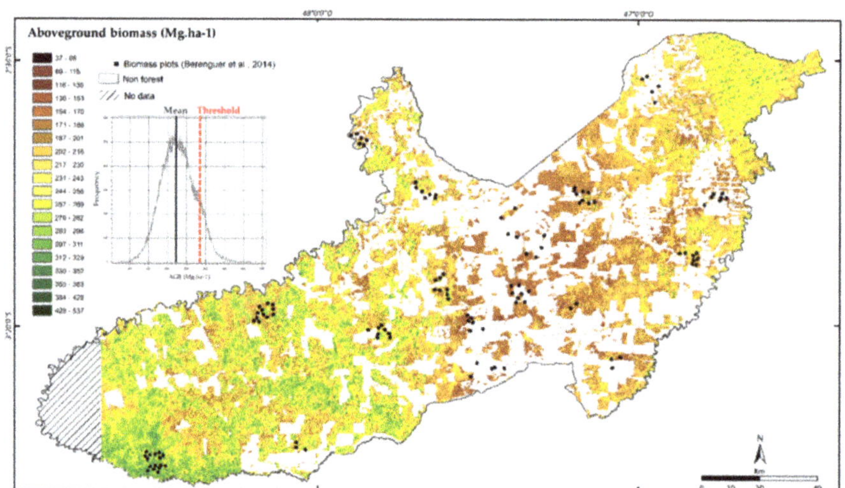

Figure 5. Random Forest predicted values of Aboveground Biomass across Paragominas municipality. In the western part of the municipality (stripped area), Sentinel-1 data was not available.

3.4. Comparison with Avitabile Pantropical Biomass Map

Figure 6 shows that our Random Forest model is more accurate than the Avitabile AGB map (R^2 higher and RMSE lower). The two models tend to overestimate for values of AGB lower than 200 Mg·ha^{-1} and underestimate for values higher than 300 Mg·ha^{-1}. Despite the fact that the Random

Forest has a higher explained variance and lower RMSE, the Avitabile dataset has a better potential to capture the large variability of AGB, particularly for extreme values where the dispersion follows the identity line. Random forest data are overall less scattered than Avitabile but display a much lower deviation between estimated and observed AGB for values from 100 to 300 Mg·ha^{-1} (Random Forest R^2 = 0.28, Avitabile R^2 = 0.14). This result is particularly interesting in the case of degraded forests where biomass values are contained within this window (Table 1).

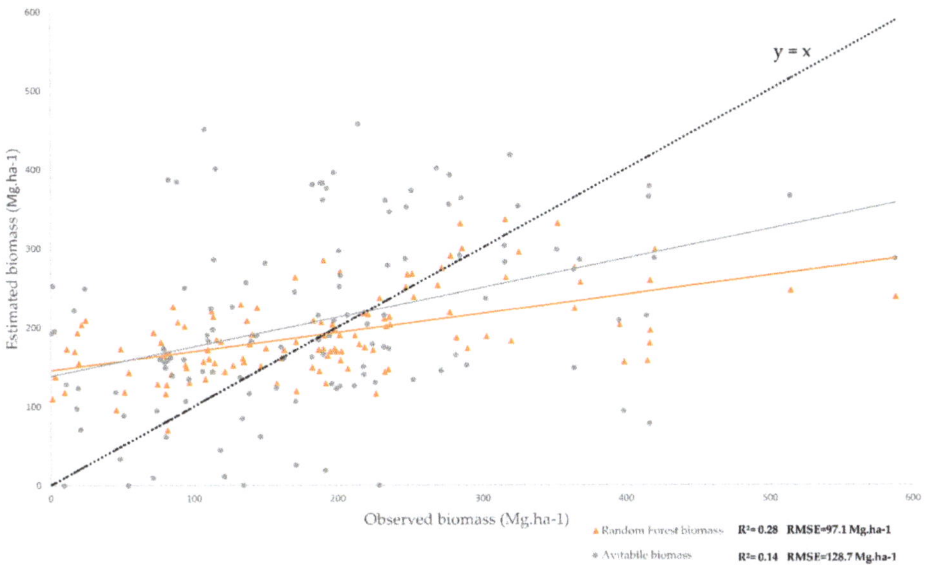

Figure 6. Observed (see Section 2.2) and estimated biomass from Avitabile [23] and the Random Forest model.

The Avitabile map is less detailed than the Random Forest biomass map. It can be explained by the difference of spatial resolution between the two datasets and also the contribution of local remotely sensed indicators. After aggregating the Random Forest biomass map at the resolution of Avitabile's (Figure 7(3)), the difference between these two datasets shows similar AGB estimation in non-degraded areas and an overestimation of AGB values in degraded (logged) forests for Avitabile's dataset (differences lower than −100 Mg·ha^{-1}). The Random Forest map captures small-scale forest disturbances such as roads, i.e., skids trails and log-landing areas and canopy gaps (pictures 1 and 2). It also displays lower biomass values around the areas impacted by selective logging. These finely detailed forest disturbances are not translated into the 920-m resolution Avitabile map. In this figure, we can see that the forest-non forest transition is much better detailed with the Random Forest map than with Avitabile maps. The transect (Figure 7(4)) and the map show that the north forest edge is more degraded (probably burned with agriculture encroachment) than the south edge where the transition between the two land uses is much clearer and sharper (Transect A–B, see Figure 1 for its location). At this local scale, Avitabile data appear to stretch the values of AGB and thus smoothen the distribution of AGB in transition areas.

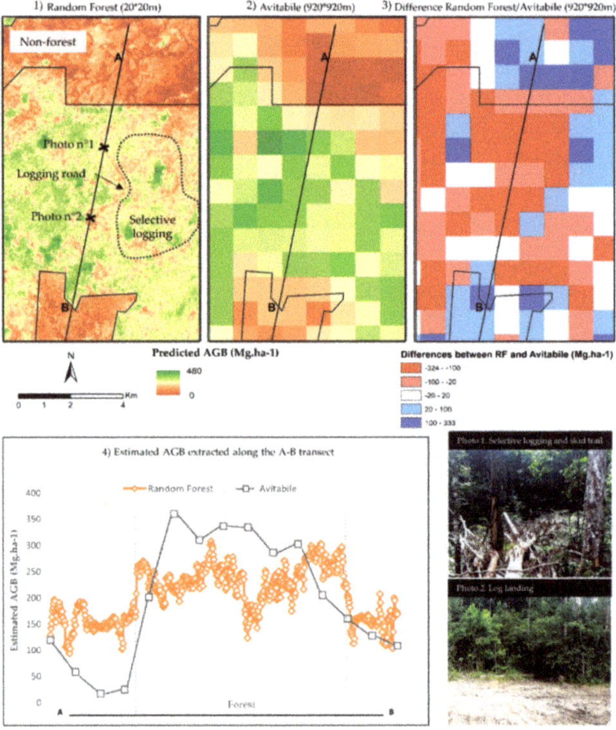

Figure 7. Comparison of AGB estimations, Random Forest, Avitabile [23] and the difference map (1–3 respectively) in a selected zone representing the transition between non-forest to forest area (see Figure 1 for precise location in Paragominas). The selective logging area and the logging road were validated during the field trip.

3.5. Comparison with Degraded Forest Typology

The less degraded forest type (type F1) presents a closed canopy above 35 m, no signs of human impact and 4 vegetative strata defined as follows: the shrub layer below 5 m, the under canopy layer located around 15 m, the canopy layer at 25 m and the emergent layer that goes beyond 35 m. The most valuable tree species harvested first are the emergent trees with a large diameter (higher than 80 cm). This selective logging can cause small degradation vertically, with an impacted and lower canopy and also horizontal damages generated by the extraction of the tree (type F2). When the selective logging becomes more intense (type F3), all the trees taller than 35 m and larger than 80 cm diameter are harvested, which causes a lowering of the canopy line at 25 m and big gaps inside the forest structure with the presence of skid tracks, broken and unrooted trees, log landing and other logging roads. Consequently, the degraded forest is much more sensitive to drought and fires, which can lead to severe disturbance, lowering the canopy at 15 m with only two remaining vegetative strata. This degraded forest type (F4) is often characterized by a high density of vegetative regrowth (pioneers species). Trees with DBH less than 80 cm are also harvested which causes an opened and destructed canopy. Finally, the intensification of fire leads to the most degraded type (F5) with only the shrub layer remaining and a few trees from the under canopy layer (Figure 8).

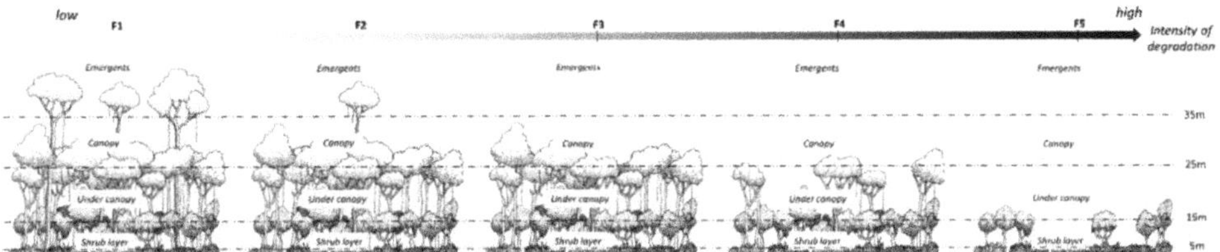

Figure 8. Five classes of degraded forest typology based on the observation of 140 forest sites in May 2015.

The typology of the degraded forest allows us to evaluate the relevance of the regional forest AGB mapping at the stand scale. The boxplots of Figure 9 show a relationship between the intensity of degradation identified in the field and the Random Forest prediction. We can assume the homogeneity of variances in the five degraded forest types (p-value = 0.09 > 0.05). Among the five types, only type F1 is significantly different from the other types F2, F3, F4, F5 (ANOVA test with p-value < 0.05). Degraded type F1 presented the highest values of AGB (average of 270 Mg·ha^{-1}). Types F2, F3, F4, F5 had statistically similar values of AGB (200 Mg·ha^{-1}), although we found a decreasing trend in the mean predicted AGB.

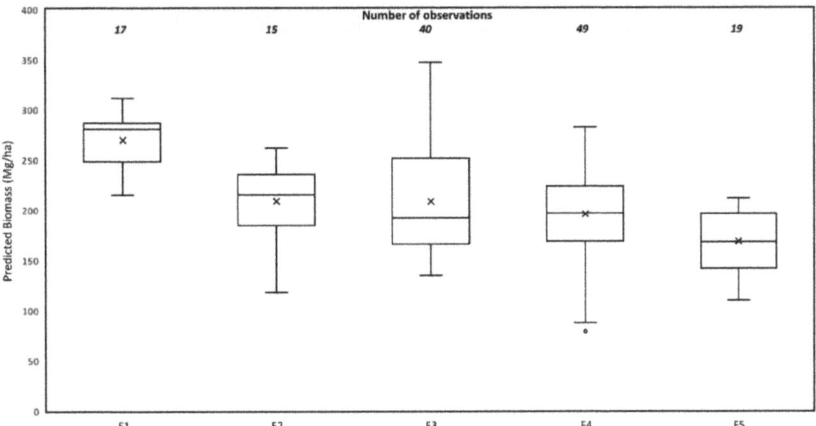

Figure 9. Boxplots of predicted values of AGB calculated for the five-class typology of degraded forest with the range (1st and 3rd quartile), the mean (line) and the median (cross).

4. Discussion and Conclusions

In this paper, we propose a novel approach to model and map AGB of degraded forests at the regional scale, which provides more detailed and accurate information than the pantropical dataset. We demonstrated that the model captures all spatial variability of the AGB data and identified the most robust remotely sensed indicators. Besides the model bias in underestimating AGB values greater than 300 Mg·ha^{-1}, this AGB map constitutes key spatial information for the future management of degraded forest.

4.1. Model Performance Analysis

- Extraction and selection of suitable indicators from remote sensing and future promising development in AGB mapping

There is a need to rank and prioritize the most suitable indicators to characterize degraded forests from multisource remote sensing [33,34,76,77]. Our results show that Landsat spectral unmixing bare soil, mid infrared and MODIS EVI standard deviation from the 2001 to 2014 time series were the most suitable indicators to model AGB and are the less data input driven. These indicators thus have the highest potential to map AGB. Unmixing approaches did perform well in our case, which confirms reports of other authors [15,40] within the Amazon forest landscape. The bare soil Landsat indicator best translates the proportion of bare soil within a 30 × 30 m Landsat pixels while mid infrared is sensitive to humidity and therefore to forest structure and functioning [33]. These results are coherent with the degradation process happening in the Paragominas forest. The emergent and most valuable trees are targeted first through legal and illegal selective logging which triggers a vertical and horizontal opening of the canopy [78]. This type of degradation directly affects the internal forest structure, which leads to forest drying and a decrease in the local forest evapotranspiration and photosynthetic activity. Another degradation type is located along forest edges, which are always very degraded and impacted by local drying effect, fire or slow encroachment due to agriculture expansion (i.e., swidden agriculture) [79]. The MODIS compiled standard deviation of EVI represents the stability over time of the annual mean EVI and thus annual photosynthetic activity. A drop in this activity in one particular year will lead to a high standard deviation value and information on potential degradation activity [79].

The other indicators have a much lower performance score, which indicates a lower sensitivity to the AGB range of values. The usual vegetation indices extracted from Landsat such as NDVI or SAVI are not sensitive enough for AGB of degraded forests. This may be related to the over saturation of the signal in the case of a tropical forest canopy and the measured photosynthesis activity, which do not allow differentiation between two types of degraded forests [56].

For radar data, one single data and related derived indicators (intensity coefficients and texture) are not sensitive enough to model AGB. To further develop this modelling approach, radar needs to be used as time series data, which would provide better information on the forest canopy status. Further investigation is also required on the preprocessing steps (e.g., filtering noise method) and on the derived texture indicators. To the extent of our knowledge, degraded forest structure and status remain understudied, with a lot of technical aspects (frequency of the time series, spatial resolution, polarizations, incident angles and frequency bands) that need to be tackled [80–82].

The integration of Sentinel-1 and Sentinel-2 time series would be an interesting way to improve the model performance combined with daily Planet images that can map at 3-m resolution the forest canopy. Very high resolution remote sensing offers a unique opportunity to characterize degraded forest structure using canopy texture mapping [60,83]. These analysis provide perspective for future space-borne LIDAR and RADAR data satellites (US GEDI mission and ESA Biomass) which will enable us to provide data sets on forest structure dynamics and forest biomass around the pantropical belt [84].

- Identification of proper algorithms to develop biomass estimation models and their related uncertainty analysis

Although the model is robust because it has been trained and tested 10 times on independent and randomly selected datasets, we identified three potential error sources. First, the limitations can be linked to the large diversity of degraded forest types that are modelled and the limited number of ground truth data. More in-situ AGB data would help to train the model on large portion of the data set and better capture its underlying trend [85]. Then, a certain time gap between the collection of AGB and the remote sensing data needs to be noted and accounted. This gap is due to remote sensing data quality and availability and can be up to 5 years in the case of Sentinel-1 images. In the case of MODIS time series, there is also a temporal gap as we modelled single date forest status (i.e., AGB) with temporal indicators that are sensitive to photosynthetic activity dynamics. These temporal mismatches could introduce some uncertainty in the model. However, in light of the minimal increases in AGB within burned and illegally logged forests (representing 88 out of 121 field plots) over a five-year period, we do not consider this uncertainty to be meaningful [86]. Finally, the resampling of the remote sensing indicators to a higher resolution (20 m resolution) may introduce bias referring to the "ecological fallacy" problems [87]. These problems are recurrent in the aggregation/disaggregation of remote sensing data and could be minimized by adapting the field collection protocol to the remote sensing data.

Despite these limitations and the low variance explained, we demonstrated that all spatial variability of the AGB data has been captured by the model through its input indicators. This gives interesting insights into how to improve local models to characterize degraded forests using biomass and how to quantify the influence of input indicators on the final modelling result.

- Comparison with existing AGB maps

When comparing with pantropical datasets such as the Avitabile AGB map, we noted differences in the statistical accuracy with local AGB datasets and in the levels of details which are particularly important for degraded forest biomass mapping. Avitabile data do not capture all the details of local AGB distribution in small-scale degraded and complex mosaic forest due mainly to their coarse spatial resolution. Transition zones from forest/non-forest areas are better and more precisely described using local trained models, where logging roads, degraded forest edges and other distribution are mapped.

4.2. AGB Spatial Distribution in the Municipality

In the map produced, the spatial distribution of AGB varies depending on land-uses and landscape organization [88]. Degraded forests are dominant in the Paragominas forest landscape. The central corridor is a mosaic of agricultural lands (agribusiness) dominated by large-scale soybean cultivation and pasturelands and with fragmented patches of degraded forests. Forest biomass values range between 100 to 150 Mg·ha^{-1} with the lowest values within the first 200 m of forest edge which is consistent with previous studies [79]. Forest fragmentation is even more important in smallholder areas (see Figure 1) where population density is high. These areas were major charcoal production hotspots (prohibited since 2008), which caused severe impacts on the forest resource. The transition between these areas and the indigenous protected reserve is sharp within the forest AGB values (with AGB higher than 250 Mg·ha^{-1} in the reserve).

4.3. Characterization of Degraded Forests

From a field point of view, degraded forests are a gradient of forest types marked by a certain canopy height and forest structure that vary depending on the intensity of past degradation trajectories. The five-class typology does not represent all the complexity of degraded forests but remains relevant in terms of modelled forest biomass. Besides the fact that only type F1 was significantly different from the four others, we found an interesting trend in the modelled biomass that requires further investigation. Type F3 presented the highest variability, which could be linked with an over-evaluation of the level of degradation in the field and the disproportionate size sample of this class. Type 5, the most degraded stage, presents the lowest predicted biomass (around 150 Mg·ha). These findings accord with those identified by the RAS team during the collection of in-situ AGB data (Table 1). This typology based on the observation of structural parameters (canopy height, number of vegetative strata, canopy rugosity, presence of emergent trees) constitutes a first step in the characterization of degraded forests which could help in the calibration and validation of other forest biomass and carbon stock assessment or the monitoring of forest degradation [76,89]. It also summarizes a one-shot time visualization of the status of degraded forests. From this, the next priority is to monitor the dynamics of degraded forest status over time using time series remote sensing [90].

4.4. How Can These Data Be Useful for Forest Management at the Regional Scale?

Besides the limitations of the proposed modelling approach and the evaluated bias, we showed the importance of regional AGB mapping in Paragominas human-modified forest landscapes, in particular with the capacity to identify patterns of different forest status values linked with degradation and agricultural activities at the regional scale. This agricultural frontier is characterized by fragmented forest impacted by the accumulation of small-scale human disturbances that cannot be captured by pantropical databases, by forest field inventories or by high detailed/low extent coverage remote sensing sources (LIDAR).

In the context of zero deforestation commitment, this forest AGB modelling and mapping is particularly important in order to provide to decision makers with spatially detailed information on the status of the 50% remnant forest [91,92]. The agricultural expansion over forest areas is now severely restricted [50]. Hence, the sustainable management of degraded forests is becoming a priority in order to enhance forest resources at different levels of disturbance [93] as much as the urgent obligation to prevent further degradation. The quantification of the forest status and the understanding of the drivers of degradation are necessary in order to improve the management of forest landscape.

Author Contributions: Study conception and design: C.B., L.B., V.G., J.O., I.T., P.S., F.L., A.F.H. Acquisition of data: C.B., L.B., I.T., E.B., V.G., F.L., A.F.H. Data Analysis and model building: C.B., J.-S.B., G.C. Drafting of the manuscript: C.B. Critical revisions: C.B., L.B., J.O., V.G., E.B., J.-S.B.

Funding: This study is part of a PhD project funded by the Cirad and the CIAT. The authors are grateful to the ECOTERA project funded by the French National Agency for Research (ANR) (grant agreement ANR-13-AGRO-0003). The authors thank the Forest Tree and Agrofrestry program of CGIAR for their financial support for open access fees.

Acknowledgments: This is publication number 67 of the Sustainable Amazon Network series. The authors are grateful to the Rede Amazônia Sustentável (RAS) team for sharing the AGB data and the Platform in partnership Amazonia (https://www.dp-amazonie.org/en). The authors thank Joice Ferreira, Solen Le Clec'h and Nicolas Baghdadi for valuable discussions.

Conflicts of Interest: The authors declare no conflict of interest.

Appendix A

Table A1. Remote sensing metadata and derived indicators.

Satellite Sources	Links	Date	Resolution	ID	Preprocessing	Spectral Bands	Indicators
MODIS	http://modis.gsfc.nasa.gov/	2001–2014 every 16 days	250 m	MOD13Q1:h13v09	Georeferenced, removed cloud covered pixels, atmospheric corrections	Red (620–670 nm), Near infrared (841–876 nm)	EVI mean, EVI standard deviation, EVI variance
Landsat 8	http://earthexplorer.usgs.gov/	27 October 2014 16 September 2014 12 June 2014	30 m	LC8220622014300LGN00 LC8220622014259LGN00 LC8220622014163LGN00	Georeferenced and reflectance product	Blue (450–515 nm), Green (525–600 nm), Red (630–680 nm), Near Infrared (845–885 nm), Short Wave Infrared (1560–1660 nm)	PV, NPV, Bare Soil *(Claslite unmixing indicators)* B, G, R, NIR, SWIR NDVI, RVI, RI, SAVI, NDWI, MSAVI, GEMI, WDVI, NDTI, TSAVI, NDPI, IPVI, TNDVI *(Orfeo Toolbox derived indicators)*
ALOS-1 PALSAR	http://global.jaxa.jp/	26 May 2010 18 July 2010 23 July 2010 28 July 2010 7 August 2010 9 September 2010	25 m	N00W050	Georeferenced and calibrated: (1) gamma [dB] (2) sigma [dB]	L-band (1.27 GHz), dual-polarisation (HH, HV)	Gamma [dB], Sigma [dB]
Sentinel-1	https://scihub.esa.int/	4 May 2015	10 m	L1 (Ground Range Detected)	Calibration and georeferenced (Sentinel-Toolbox Software)	C-band dual-polarisation (VV, VH)	Sigma [dB] Grey Level Co-occurence matrix (9 × 9 pixels): Contrast, Dissimilarity, Energy, Entropy, Correlation, Mean, Variance, Homogeneity, Maximum *(Sentinel Toolbox derived indicators)*

References

1. Houghton, R.A. The emissions of carbon from deforestation and degradation in the tropics: Past trends and future potential. *Carbon Manag.* **2013**, *4*, 539–546. [CrossRef]
2. Simula, M.; Mansur, E. A global challenge needing local response. *Unasylva* **2011**, *62*, 238.
3. Baccini, A.; Walker, W.; Carvalho, L.; Farina, M.; Sulla-Menashe, D.; Houghton, R.A. Tropical forests are a net carbon source based on aboveground measurements of gain and loss. *Science* **2017**, *358*, 230–234. [CrossRef] [PubMed]
4. Barlow, J.; Lennox, G.D.; Ferreira, J.; Berenguer, E.; Lees, A.C.; Nally, R.M.; Thomson, J.R.; Ferraz, S.F.D.B.; Louzada, J.; Oliveira, V.H.F.; et al. Anthropogenic disturbance in tropical forests can double biodiversity loss from deforestation. *Nature* **2016**, *535*, 144–147. [CrossRef] [PubMed]
5. Thompson, I.D.; Guariguata, M.R.; Okabe, K.; Bahamondez, C.; Nasi, R.; Heymell, V.; Sabogal, C. An operational framework for defining and monitoring forest degradation. *Ecol. Soc.* **2013**, *18*, 20. [CrossRef]
6. Putz, F.E.; Redford, K.H. The importance of defining 'forest': Tropical forest degradation, deforestation, long-term phase shifts, and further transitions. *Biotropica* **2010**, *42*, 10–20. [CrossRef]
7. Lamb, D.; Erskine, P.D.; Parrotta, J.A. Restoration of degraded tropical forest landscapes. *Science* **2005**, *310*, 1628–1632. [CrossRef] [PubMed]
8. PRODES—Coordenação-Geral de Observação da Terra. Available online: http://www.obt.inpe.br/OBT/assuntos/programas/amazonia/prodes (accessed on 13 April 2018).
9. Hansen, M.C.; Stehman, S.V.; Potapov, P.V. Quantification of global gross forest cover loss. *Proc. Natl. Acad. Sci. USA* **2010**, *107*, 8650–8655. [CrossRef] [PubMed]
10. Fearnside, P.M. Deforestation in Brazilian Amazonia: History, rates, and consequences. *Conserv. Biol.* **2005**, *19*, 680–688. [CrossRef]
11. Nepstad, D.; Soares-Filho, B.S.; Merry, F.; Lima, A.; Moutinho, P.; Carter, J.; Bowman, M.; Cattaneo, A.; Rodrigues, H.; Schwartzman, S.; et al. The end of deforestation in the Brazilian Amazon. *Science* **2009**, *326*, 1350–1351. [CrossRef] [PubMed]
12. DEGRAD—Coordenação-Geral de Observação da Terra. Available online: http://www.obt.inpe.br/OBT/assuntos/programas/amazonia/degrad (accessed on 10 April 2018).
13. Florestas Do Brasil em Resumo 2013. Available online: http://www.florestal.gov.br/publicacoes/572-florestas-do-brasil-em-resumo-2013 (accessed on 10 April 2018).
14. Asner, G.P.; Knapp, D.E.; Broadbent, E.N.; Oliveira, P.J.; Keller, M.; Silva, J.N. Selective logging in the Brazilian Amazon. *Science* **2005**, *310*, 480–482. [CrossRef] [PubMed]
15. Souza, C., Jr.; Siqueira, J.V.; Sales, M.H.; Fonseca, A.V.; Ribeiro, J.G.; Numata, I.; Cochrane, M.A.; Barber, C.P.; Roberts, D.A.; Barlow, J.; et al. Ten-Year landsat classification of deforestation and forest degradation in the Brazilian Amazon. *Remote Sens.* **2013**, *5*, 5493–5513. [CrossRef]
16. Thompson, I.; Mackey, B.; McNulty, S.; Mosseler, A. Forest Resilience, Biodiversity, and Climate Change. A synthesis of the biodiversity/resilience/stability relationship in forest ecosystems. *Secr. Conv. Biol. Divers. Montr.* **2009**, *43*, 1–67.
17. Potapov, P.; Hansen, M.C.; Laestadius, L.; Turubanova, S.; Yaroshenko, A.; Thies, C.; Smith, W.; Zhuravleva, I.; Komarova, A.; Minnemeyer, S.; et al. The last frontiers of wilderness: Tracking loss of intact forest landscapes from 2000 to 2013. *Sci. Adv.* **2017**, *3*, e1600821. [CrossRef] [PubMed]
18. Bernier, P.Y.; Paré, D.; Stinson, G.; Bridge, S.R.J.; Kishchuk, B.E.; Lemprière, T.C.; Thiffault, E.; Titus, B.D.; Vasbinder, W. Moving beyond the concept of 'primary forest' as a metric of forest environment quality. *Ecol. Appl.* **2016**, *27*, 349–354. [CrossRef] [PubMed]
19. Berenguer, E.; Ferreira, J.; Gardner, T.A.; Aragão, L.E.O.C.; De Camargo, P.B.; Cerri, C.E.; Durigan, M.; De Oliveira Junior, R.C.; Vieira, I.C.G.; Barlow, J. A large-scale field assessment of carbon stocks in human-modified tropical forests. *Glob. Chang. Biol.* **2014**, *20*, 3713–3726. [CrossRef] [PubMed]
20. Bustamante, M.M.C.; Roitman, I.; Aide, T.M.; Alencar, A.; Anderson, L.O.; Aragão, L.; Asner, G.P.; Barlow, J.; Berenguer, E.; Chambers, J.; et al. Toward an integrated monitoring framework to assess the effects of tropical forest degradation and recovery on carbon stocks and biodiversity. *Glob. Chang. Biol.* **2016**, *22*, 92–109. [CrossRef] [PubMed]

21. Saatchi, S.S.; Harris, N.L.; Brown, S.; Lefsky, M.; Mitchard, E.T.A.; Salas, W.; Zutta, B.R.; Buermann, W.; Lewis, S.L.; Hagen, S.; et al. Benchmark map of forest carbon stocks in tropical regions across three continents. *Proc. Natl. Acad. Sci. USA* **2011**, *108*, 9899–9904. [CrossRef] [PubMed]

22. Baccini, A.; Goetz, S.J.; Walker, W.S.; Laporte, N.T.; Sun, M.; Sulla-Menashe, D.; Hackler, J.; Beck, P.S.A.; Dubayah, R.; Friedl, M.A.; et al. Estimated carbon dioxide emissions from tropical deforestation improved by carbon-density maps. *Nat. Clim. Chang.* **2012**, *2*, 182–185. [CrossRef]

23. Avitabile, V.; Herold, M.; Heuvelink, G.B.M.; Lewis, S.L.; Phillips, O.L.; Asner, G.P.; Armston, J.; Ashton, P.S.; Banin, L.; Bayol, N.; et al. An integrated pan-tropical biomass map using multiple reference datasets. *Glob. Chang. Biol.* **2016**, *22*, 1406–1420. [CrossRef] [PubMed]

24. Malhi, Y.; Wood, D.; Baker, T.R.; Wright, J.; Phillips, O.L.; Cochrane, T.; Meir, P.; Chave, J.; Almeida, S.; Arroyo, L.; et al. The regional variation of aboveground live biomass in old-growth Amazonian forests. *Glob. Chang. Biol.* **2006**, *12*, 1107–1138. [CrossRef]

25. Asner, G.P.; Joseph Mascaro, J.; Muller-Landau, H.C.; Vieilledent, G.; Vaudry, R.; Rasamoelina, M.; Hall, J.S.; van Breugel, M. A universal airborne LiDAR approach for tropical forest carbon mapping. *Oecologia* **2012**, *168*, 1147–1160. [CrossRef] [PubMed]

26. Asner, G.P. Tropical forest carbon assessment: Integrating satellite and airborne mapping approaches. *Environ. Res. Lett.* **2009**, *4*, 034009. [CrossRef]

27. Asner, G.P.; Mascaro, J. Mapping tropical forest carbon: Calibrating plot estimates to a simple LiDAR metric. *Remote Sens. Environ.* **2014**, *140*, 614–624. [CrossRef]

28. Fayad, M.; Baghdadi, N.; Fayad, I.; Vieilledent, G.; Bailly, J.-S.; Minh, D. Interest of integrating spaceborne LiDAR data to improve the estimation of biomass in high biomass forested areas. *Remote Sens.* **2017**, *9*, 213. [CrossRef]

29. Mascaro, J.; Detto, M.; Asner, G.P.; Muller-Landau, H.C. Evaluating uncertainty in mapping forest carbon with airborne LiDAR. *Remote Sens. Environ.* **2011**, *115*, 3770–3774. [CrossRef]

30. Longo, M.; Keller, M.; dos-Santos, M.N.; Leitold, V.; Pinagé, E.R.; Baccini, A.; Saatchi, S.; Nogueira, E.M.; Batistella, M.; Morton, D.C. Aboveground biomass variability across intact and degraded forests in the Brazilian Amazon. *Glob. Biogeochem. Cycles* **2016**, *30*, 1639–1660. [CrossRef]

31. Baccini, A.; Laporte, N.; Goetz, S.J.; Sun, M.; Dong, H. A first map of tropical Africa's above-ground biomass derived from satellite imagery. *Environ. Res. Lett.* **2008**, *3*, 045011. [CrossRef]

32. Fayad, I.; Baghdadi, N.; Bailly, J.-S.; Barbier, N.; Gond, V.; Hajj, M.E.; Fabre, F.; Bourgine, B. Canopy height estimation in French Guiana with LiDAR ICESat/GLAS data using principal component analysis and random forest regressions. *Remote Sens.* **2014**, *6*, 11883–11914. [CrossRef]

33. Hirschmugl, M.; Gallaun, H.; Dees, M.; Datta, P.; Deutscher, J.; Koutsias, N.; Schardt, M. Methods for mapping forest disturbance and degradation from optical earth observation data: A review. *Curr. For. Rep.* **2017**, *3*, 32–45. [CrossRef]

34. Herold, M.; Román-Cuesta, R.M.; Mollicone, D.; Hirata, Y.; Van Laake, P.; Asner, G.P.; Souza, C.; Skutsch, M.; Avitabile, V.; Macdicken, K.; et al. Options for monitoring and estimating historical carbon emissions from forest degradation in the context of REDD+. *Carbon Balance Manag.* **2011**, *6*, 13. [CrossRef] [PubMed]

35. Rappaport, D.; Morton, D.C.; Longo, M.; Keller, M.; Dubayah, R.; dos-Santos, M.N. Quantifying long-term changes in carbon stocks and forest structure from Amazon forest degradation. *Environ. Res. Lett.* **2018**. [CrossRef]

36. Achard, F.; Boschetti, L.; Brown, S.; Brady, M.; DeFries, R.; Grassi, G.; Herold, M.; Mollicone, D.; Mora, B.; Pandey, D.; et al. *A Sourcebook of Methods and Procedures for Monitoring and Reporting Anthropogenic Greenhouse Gas Emissions and Removals Associated with Deforestation, Gains and Losses of Carbon Stocks in Forests Remaining Forests, and Forestation*; GOFC-GOLD: Wageningen, The Netherlands, 2014.

37. Lambin, E.F. Monitoring forest degradation in tropical regions by remote sensing: some methodological issues. *Glob. Ecol. Biogeogr.* **1999**, *8*, 191–198. [CrossRef]

38. Gond, V.; Fayolle, A.; Pennec, A.; Cornu, G.; Mayaux, P.; Camberlin, P.; Doumenge, C.; Fauvet, N.; Gourlet-Fleury, S. Vegetation structure and greenness in Central Africa from Modis multi-temporal data. *Philos. Trans. R. Soc. B Biol. Sci.* **2013**, *368*, 20120309. [CrossRef] [PubMed]

39. Asner, G.P.; Knapp, D.E.; Balaji, A.; Páez-Acosta, G. Automated mapping of tropical deforestation and forest degradation: CLASlite. *J. Appl. Remote Sens.* **2009**, *3*, 033543. [CrossRef]

40. Tritsch, I.; Sist, P.; Narvaes, I.D.S.; Mazzei, L.; Blanc, L.; Bourgoin, C.; Cornu, G.; Gond, V. Multiple patterns of forest disturbance and logging shape forest landscapes in Paragominas. *Braz. For.* **2016**, *7*, 315. [CrossRef]
41. Joshi, N.; Mitchard, E.T.; Woo, N.; Torres, J.; Moll-Rocek, J.; Ehammer, A.; Collins, M.; Jepsen, M.R.; Fensholt, R. Mapping dynamics of deforestation and forest degradation in tropical forests using radar satellite data. *Environ. Res. Lett.* **2015**, *10*, 034014. [CrossRef]
42. Kuplich, T.M.; Curran, P.J.; Atkinson, P.M. Relating SAR image texture to the biomass of regenerating tropical forests. *Int. J. Remote Sens.* **2005**, *26*, 4829–4854. [CrossRef]
43. Luckman, A.J.; Frery, A.C.; Yanasse, C.C.F.; Groom, G.B. Texture in airborne SAR imagery of tropical forest and its relationship to forest regeneration stage. *Int. J. Remote Sens.* **1997**, *18*, 1333–1349. [CrossRef]
44. Haralick, R.M.; Shanmugam, K.; Dinstein, I. Textural features for image classification. *IEEE Trans. Syst. Man Cybern.* **1973**, *6*, 610–621. [CrossRef]
45. Morel, A.C.; Saatchi, S.S.; Malhi, Y.; Berry, N.J.; Banin, L.; Burslem, D.; Nilus, R.; Ong, R.C. Estimating aboveground biomass in forest and oil palm plantation in Sabah, Malaysian Borneo using ALOS PALSAR data. *For. Ecol. Manag.* **2011**, *262*, 1786–1798. [CrossRef]
46. Mitchard, E.T.A.; Saatchi, S.S.; Lewis, S.L.; Feldpausch, T.R.; Woodhouse, I.H.; Sonké, B.; Rowland, C.; Meir, P. Measuring biomass changes due to woody encroachment and deforestation/degradation in a forest–savanna boundary region of central Africa using multi-temporal L-band radar backscatter. *Remote Sens. Environ.* **2011**, *115*, 2861–2873. [CrossRef]
47. Englhart, S.; Keuck, V.; Siegert, F. Aboveground biomass retrieval in tropical forests—the potential of combined X- and L-band SAR data use. *Remote Sens. Environ.* **2011**, *115*, 1260–1271. [CrossRef]
48. Imazon—Instituto Do Homem e Meio Ambiente da Amazônia. Available online: http://imazon.org.br/en/ (accessed on 18 May 2018).
49. IBGE, Paragominas. Available online: https://cidades.ibge.gov.br/brasil/pa/paragominas/panorama (accessed on 17 May 2018).
50. Piketty, M.-G.; Poccard-Chapuis, R.; Drigo, I.; Coudel, E.; Plassin, S.; Laurent, F.; Marcello, T. Multi-level governance of land use changes in the Brazilian Amazon: Lessons from Paragominas, State of Pará. *Forests* **2015**, *6*, 1516–1536. [CrossRef]
51. Viana, C.; Coudel, E.; Barlow, J.; Ferreira, J.; Gardner, T.; Parry, L. How does hybrid governance emerge? Role of the elite in building a Green Municipality in the Eastern Brazilian Amazon: Role of the elite in building a green municipality. *Environ. Policy Gov.* **2016**, *26*, 337–350. [CrossRef]
52. Gardner, T.A.; Burgess, N.D.; Aguilar-Amuchastegui, N.; Barlow, J.; Berenguer, E.; Clements, T.; Danielsen, F.; Ferreira, J.; Foden, W.; Kapos, V.; et al. A framework for integrating biodiversity concerns into national REDD+ programmes. *Biol. Conserv.* **2012**, *154*, 61–71. [CrossRef]
53. Gardner, T.A.; Ferreira, J.; Barlow, J.; Lees, A.C.; Parry, L.; Vieira, I.C.G.; Berenguer, E.; Abramovay, R.; Aleixo, A.; Andretti, C.; et al. A social and ecological assessment of tropical land uses at multiple scales: The Sustainable Amazon Network. *Philos. Trans. R. Soc. B Biol. Sci.* **2013**, *368*, 20120166. [CrossRef] [PubMed]
54. Chave, J.; Andalo, C.; Brown, S.; Cairns, M.A.; Chambers, J.Q.; Eamus, D.; Fölster, H.; Fromard, F.; Higuchi, N.; Kira, T.; et al. Tree allometry and improved estimation of carbon stocks and balance in tropical forests. *Oecologia* **2005**, *145*, 87–99. [CrossRef] [PubMed]
55. Mazzei, L.; Sist, P.; Ruschel, A.; Putz, F.E.; Marco, P.; Pena, W.; Ferreira, J.E.R. Above-ground biomass dynamics after reduced-impact logging in the Eastern Amazon. *For. Ecol. Manag.* **2010**, *259*, 367–373. [CrossRef]
56. Huete, A.; Didan, K.; Miura, T.; Rodriguez, E.P.; Gao, X.; Ferreira, L.G. Overview of the radiometric and biophysical performance of the MODIS vegetation indices. *Remote Sens. Environ.* **2002**, *83*, 195–213. [CrossRef]
57. Orfeo ToolBox—Orfeo ToolBox Is Not a Black Box. Available online: https://www.orfeo-toolbox.org/ (accessed on 13 April 2018).
58. Mermoz, S.; le Toan, T.; Villard, L.; Réjou-Méchain, M.; Seifert-Granzin, J. Biomass assessment in the Cameroon savanna using ALOS PALSAR data. *Remote Sens. Environ.* **2014**, *155*, 109–119. [CrossRef]
59. Shimada, M.; Itoh, T.; Motooka, T.; Watanabe, M.; Shiraishi, T.; Thapa, R.; Lucas, R. New global forest/non-forest maps from ALOS PALSAR data (2007–2010). *Remote Sens. Environ.* **2014**, *155*, 13–31. [CrossRef]

60. Ploton, P.; Barbier, N.; Couteron, P.; Antin, C.M.; Ayyappan, N.; Balachandran, N.; Barathan, N.; Bastin, J.F.; Chuyong, G.; Dauby, G.; et al. Toward a general tropical forest biomass prediction model from very high resolution optical satellite images. *Remote Sens. Environ.* **2017**, *200*, 140–153. [CrossRef]

61. Champion, I.; Germain, C.; da Costa, J.P.; Alborini, A.; Dubois-Fernandez, P. Retrieval of forest stand age from SAR image texture for varying distance and orientation values of the gray level co-occurrence matrix. *IEEE Geosci. Remote Sens. Lett.* **2014**, *11*, 5–9. [CrossRef]

62. Breiman, L. Random forests. *Mach. Learn.* **2001**, *45*, 5–32. [CrossRef]

63. Kohavi, R. A study of cross-validation and bootstrap for accuracy estimation and model selection. *IJCAI* **1995**, *14*, 1137–1145.

64. Strobl, C.; Malley, J.; Tutz, G. An introduction to recursive partitioning: Rationale, application, and characteristics of classification and regression trees, bagging, and random forests. *Psychol. Methods* **2009**, *14*, 323–348. [CrossRef] [PubMed]

65. Mascaro, J.; Asner, G.P.; Knapp, D.E.; Kennedy-Bowdoin, T.; Martin, R.E.; Anderson, C.; Higgins, M.; Chadwick, K.D. A tale of two 'Forests': Random forest machine learning aids tropical forest carbon mapping. *PLoS ONE* **2014**, *9*, e85993. [CrossRef] [PubMed]

66. Goetz, S.J.; Baccini, A.; Laporte, N.T.; Johns, T.; Walker, W.; Kellndorfer, J.; Houghton, R.A.; Sun, M. Mapping and monitoring carbon stocks with satellite observations: A comparison of methods. *Carbon Balance Manag.* **2009**, *4*, 2. [CrossRef] [PubMed]

67. Mutanga, O.; Adam, E.; Cho, M.A. High density biomass estimation for wetland vegetation using WorldView-2 imagery and random forest regression algorithm. *Int. J. Appl. Earth Obs. Geoinf.* **2012**, *18*, 399–406. [CrossRef]

68. Borra, S.; di Ciaccio, A. Measuring the prediction error. A comparison of cross-validation, bootstrap and covariance penalty methods. *Comput. Stat. Data Anal.* **2010**, *54*, 2976–2989. [CrossRef]

69. Rossi, R.E.; Mulla, D.J.; Journel, A.G.; Franz, E.H. Geostatistical tools for modeling and interpreting ecological spatial dependence. *Ecol. Monogr.* **1992**, *62*, 277–314. [CrossRef]

70. Gräler, B.; Pebesma, E.; Heuvelink, G. Spatio-temporal interpolation using GSTAT. *RFID J.* **2016**, *8*, 204–218.

71. Ribeiro, J.; Diggle, P.J. geoR: A package for geostatistical analysis. *R News* **2001**, *1*, 14–18. Available online: http://www.leg.ufpr.br/geoR/ (accessed on 25 May 2018).

72. CRAN—Package Gstat. Available online: https://cran.r-project.org/web/packages/gstat/index.html (accessed on 13 April 2018).

73. CRAN—Package Raster. Available online: https://cran.r-project.org/web/packages/raster/index.html (accessed on 13 April 2018).

74. Stabler, B. CRAN—Package Shapefiles. Available online: https://cran.r-project.org/web/packages/shapefiles/ (accessed on 23 May 2018).

75. Bivand, R.; Keitt, T.; Rowlingson, B.; Pebesma, E.; Sumner, M.; Hijmans, R.; Rouault, E.; Ooms, J. CRAN—Package Rgdal. Available online: https://cran.r-project.org/web/packages/rgdal/ (accessed on 23 May 2018).

76. Mitchell, A.L.; Rosenqvist, A.; Mora, B. Current remote sensing approaches to monitoring forest degradation in support of countries measurement, reporting and verification (MRV) systems for REDD+. *Carbon Balance Manag.* **2017**, *12*, 9. [CrossRef] [PubMed]

77. De Sy, V.; Herold, M.; Achard, F.; Asner, G.P.; Held, A.; Kellndorfer, J.; Verbesselt, J. Synergies of multiple remote sensing data sources for REDD+ monitoring. *Curr. Opin. Environ. Sustain.* **2012**, *4*, 696–706. [CrossRef]

78. Ferreira, J.; Blanc, L.; Kanashiro, M.; Lees, A.C.; Bourgoin, C.; Freitas, J.V.D.; Gama, M.B.; Laurent, F.; Martins, M.B.; Moura, N.; et al. *Degradação Florestal na Amazônia: Como Ultrapassar os Limites Conceituais, Científicos e Técnicos Para mUdar Esse Cenário*; Embrapa Amazônia Oriental: Belém, Brazil, 2015.

79. Briant, G.; Gond, V.; Laurance, S.G.W. Habitat fragmentation and the desiccation of forest canopies: A case study from Eastern Amazonia. *Biol. Conserv.* **2010**, *143*, 2763–2769. [CrossRef]

80. Trisasongko, B.H. The use of polarimetric SAR data for forest disturbance monitoring. *Sens. Imaging Int. J.* **2010**, *11*, 1–13. [CrossRef]

81. Deutscher, J.; Perko, R.; Gutjahr, K.; Hirschmugl, M.; Schardt, M. Mapping tropical rainforest canopy disturbances in 3D by COSMO-SkyMed Spotlight InSAR-Stereo data to detect areas of forest degradation. *Remote Sens.* **2013**, *5*, 648–663. [CrossRef]

82. Solberg, S.; Astrup, R.; Breidenbach, J.; Nilsen, B.; Weydahl, D. Monitoring spruce volume and biomass with InSAR data from TanDEM-X. *Remote Sens. Environ.* **2013**, *139*, 60–67. [CrossRef]

83. Barbier, N.; Couteron, P.; Proisy, C.; Malhi, Y.; Gastellu-Etchegorry, J.-P. The variation of apparent crown size and canopy heterogeneity across lowland Amazonian forests: Amazon forest canopy properties. *Glob. Ecol. Biogeogr.* **2010**, *19*, 72–84. [CrossRef]

84. Le Toan, T.; Quegan, S.; Davidson, M.W.J.; Balzter, H.; Paillou, P.; Papathanassiou, K.; Plummer, S.; Papathanassiou, K.; Rocca, F.; Saatchi, S.; et al. The BIOMASS mission: Mapping global forest biomass to better understand the terrestrial carbon cycle. *Remote Sens. Environ.* **2011**, *115*, 2850–2860. [CrossRef]

85. Guitet, S.; Hérault, B.; Molto, Q.; Brunaux, O.; Couteron, P. Spatial structure of above-ground biomass limits accuracy of carbon mapping in rainforest but large scale forest inventories can help to overcome. *PLoS ONE* **2015**, *10*, e0138456. [CrossRef] [PubMed]

86. Blanc, L.; Echard, M.; Herault, B.; Bonal, D.; Marcon, E.; Chave, J.; Baraloto, C. Dynamics of aboveground carbon stocks in a selectively logged tropical forest. *Ecol. Appl.* **2009**, *19*, 1397–1404. [CrossRef] [PubMed]

87. Robinson, W. Ecological correlations and the behavior of individuals. *Int. J. Epidemiol.* **2009**, *38*, 337–341. [CrossRef] [PubMed]

88. Laurent, F.; Arvor, D.; Daugeard, M.; Osis, R.; Tritsch, I.; Coudel, E.; Piketty, M.-G.; Piraux, M.; Viana, C.; Dubreuil, V.; et al. Le tournant environnemental en Amazonie: Ampleur et limites du découplage entre production et déforestation. *EchoGéo* **2017**, *41*, 36–57. [CrossRef]

89. DeVries, B.; Verbesselt, J.; Kooistra, L.; Herold, M. Robust monitoring of small-scale forest disturbances in a tropical montane forest using Landsat time series. *Remote Sens. Environ.* **2015**, *161*, 107–121. [CrossRef]

90. DeVries, B.; Decuyper, M.; Verbesselt, J.; Zeileis, A.; Herold, M.; Joseph, S. Tracking disturbance-regrowth dynamics in tropical forests using structural change detection and Landsat time series. *Remote Sens. Environ.* **2015**. [CrossRef]

91. Climate-KIC | The EU's Main Climate Innovation Initiative. Available online: http://www.climate-kic.org/ (accessed on 18 May 2018).

92. Observatory of the Dynamics of Interactions between Societies and Environnement in the Amazon, ODYSSEA. Available online: https://odyssea-amazonia.org/ (accessed on 18 May 2018).

93. Goldstein, J.E. The afterlives of degraded tropical forests: New value for conservation and development. In *Environment and Society: Advances in Research*; West, P., Brockington, D., Eds.; Berghahn Books: New York, NY, USA, 2014; Volume 5, pp. 124–140.

 *forests*

Article

Forest Above-Ground Biomass Estimation Using Single-Baseline Polarization Coherence Tomography with P-Band PolInSAR Data

Haibo Zhang, Changcheng Wang *, Jianjun Zhu, Haiqiang Fu, Qinghua Xie and Peng Shen

School of Geosciences and Info-Physics, Central South University, Changsha 410083, China;
haibozhang@csu.edu.cn (H.Z.); zjj@csu.edu.cn (J.Z.); haiqiangfu@csu.edu.cn (H.F.); csuxqh@csu.edu.cn (Q.X.);
shenpengcsu@163.com (P.S.)
* Correspondence: wangchangcheng@csu.edu.cn; Tel.: +86-731-8883-6931

Received: 8 February 2018; Accepted: 21 March 2018; Published: 23 March 2018

Abstract: Forest above ground biomass (AGB) extraction using Synthetic Aperture Radar (SAR) images has been widely used in global carbon cycle research. Classical AGB inversion methods using SAR images are mainly based on backscattering coefficients. The polarization coherence tomography (PCT) technology which can generate vertical profiles of forest relative reflectivity, has the potential to improve the accuracy of biomass inversion. The relationship between vertical profiles and forest AGB is modeled by some parameters defined based on geometric characteristics of the relative reflectivity distribution curve. But these parameters are defined without physical characteristics. Among these parameters, tomographic height (*TomoH*) is considered as the most important one. However, *TomoH* only corresponds to the highest volume relative reflectivity, which is lower than the actual forest height, affecting the accuracy of forest height and AGB inversion. In this paper, we introduce a new parameter, the canopy height (H_{ac}), for AGB inversion by analyzing the vertical backscatter power loss. Then, we construct an inversion model based on the combination of the new parameter (H_{ac}) and other parameters from the tomographic profile. The P-band polarimetric SAR datasets of the European Space Agency (ESA) BioSAR 2008 campaign acquired over Krycklan Catchment are selected for the verification experiment at two different flight directions. The results show that H_{ac} performs better in estimating forest height and AGB than *TomoH* does. The inversion root mean square error (RMSE) of the proposed method is 18.325 t ha^{-1}, and the result of using *TomoH* is 21.126 t ha^{-1}.

Keywords: forest above ground biomass (AGB); polarization coherence tomography (PCT); P-band PolInSAR; tomographic profiles

---

1. Introduction

Forest ecosystems cover around 30% of the land surface, accounting for 75% of terrestrial gross primary production and about 80% of the global plant biomass [1,2]. So, they play an important role in the global carbon balance and climate change [3]. An important parameter reflecting the forest carbon cycle change is above-ground biomass (AGB). Many different techniques have been used to estimate AGB and AGB changes [4–6]. Among them, remote sensing techniques perform better in large-scale forest AGB mapping [7,8] than traditional forest inventory techniques.

Over the last two decades, airborne and spaceborne sensors have been used to estimate forest AGB [9–11]. Optical remote sensing datasets (e.g., Moderate Resolution Imaging Spectroradiometer, MODIS and Landsat Thematic Mapper, TM) have been successfully used for the estimation of forest parameters and assessment of woody biomass with different quality results, mainly by revealing the correlation between vegetation indices (e.g., normalized difference vegetation index, NDVI) or spectral responses and ground inventory data [12]. However, the retrieved AGB values using optical remote sensing data are usually troubled with saturation effects, especially in the high carbon stock forests [13,14]. Due to the limitation of the penetration in vegetated areas, the spectral responses recorded in optical images are mainly related to the interaction between the solar radiance and forest stand canopies [15], which mainly contains vegetation information in the horizontal direction. The saturation points for optical remote sensing range from 15 t ha^{-1} to 70 t ha^{-1} [16].

Compared with optical remote sensing, the synthetic aperture radar (SAR) has the capability to penetrate cloud and vegetation canopies. Therefore, SAR systems can observe the ground surface in all weather conditions and with continuous temporal coverage. This technique has been widely used in earthquake [17,18], landslide [19,20], glacier [21,22], agriculture [23], and forestry monitoring [9]. In particular, the long-wavelength SAR data are more sensitive to forest AGB [24–26] at HV [27] and HH polarizations [28–30]. Generally, the most frequently used methods in forest parameter estimation with SAR can be classified into several types. The 2D method based on backscattering coefficients from Polarimetric SAR (PolSAR) data can provide an estimation of forest AGB [31]. Most studies used the logarithm of biomass [32,33], square root [34] or cube root [35] of the biomass and the backscattering coefficient (in dB) for biomass prediction. However, the 2D method also has a saturation problem, which depends upon different wavelengths, polarizations, and incidence angles [36–38]. Interferometric SAR (InSAR) data or Polarimetric SAR interferometry (PolInSAR) data have proven to be effective for forest AGB estimation, as the ground elevation and tree height can be obtained from interferometric phase and coherence [39–42], and then, this tree height can be converted into forest AGB by allometric equations and other models [43,44]. This approach involving interferometry has the potential to overcome the saturation problem to some extent, and allows the estimation of forest AGB over a wider range of values at least for homogeneous forests [45]. However, the forest biomass is not only related to forest height, but also tree species, canopy density, and vertical structure. So, it is imperfect to use only forest height to estimate AGB, especially in forests with high heterogeneity in their three-dimensional structure [46].

Recently, some studies have indicated that the vertical profile of relative reflectivity, which is a structure parameter describing the variation of backscatter signal along the vertical direction, is a good indicator for estimating AGB. The tomography profile can be obtained using the multi-baseline InSAR/PolInSAR [47–49], SAR tomography [50–52], or polarization coherence tomography (PCT) [53]. However, either the multi-baseline InSAR or SAR tomography requires multiple SAR data. As PCT can overcome those limitations, it uses a priori information of volume height and topographic phase to reconstruct vertical profiles. Single baseline PCT is one of the simplest ways to invert the vertical distribution of relative reflectivity using single baseline PolInSAR data [54]. Cloude [53] pointed out that the PCT technology had potential application value in forest biomass estimation. Luo et al. [55] defined nine parameters (P1–P9) to characterize the average vertical profile of relative reflectivity with the L-band SAR data. Li et al. [56] improved this approach and replaced the parameter P8 with the tenth parameter P10 for AGB inversion. In addition, they defined the mean of the canopy profile's Gaussian fitting as the tomographic height (*TomoH*), which is more sensitive to the forest AGB than forest heights. However, the *TomoH* only represents the height of the maximum relative reflectance in the canopy, which is usually lower than the real forest height, especially for long-wavelength (i.e., P-band) SAR data, due to the strong penetration. Meanwhile, these parameters are defined based on geometric characteristics [56], without considering the backscattering signal attenuation in the forest canopy.

In this study, the average tomographic profiles are produced using the PCT method with single baseline P-band PolInSAR data. We introduce a new parameter, the canopy height (H_{ac}), by analyzing the variation of the backscattering power of forest for AGB inversion. Then, the performance of this parameter is assessed by a comparative analysis with the *TomoH*. The paper is organized as follows. Section 2 provides information on the study area and datasets. Section 3 describes the methods used for vertical profile reconstruction, forest parameters retrieval, model construction, and validation.

Results and discussion are presented in Sections 4 and 5, respectively. Finally, the major conclusions of this work are given in Section 6.

2. Study Area and Data Sets

2.1. Study Area

The test site is located in the Krycklan river catchment (64°16′ N, 19°46′ E) in northern Sweden (Figure 1), which is about 50 km northwest of Umea and covers approximately 9390 ha. It has become the test site for field-based forest research at the Faculty of Forest Sciences, Swedish University of Agricultural Sciences. The dominating forest type is mixed coniferous, including Norway spruce, Scots pine, and Birch [57]. In addition, there are some other small deciduous trees such as Aspen and Rowan. The dominating soil is moraine, with variations in thickness. This area is hilly, with elevation variations from 100 to 400 m throughout the whole scene, and surface slopes of up to 20° [44].

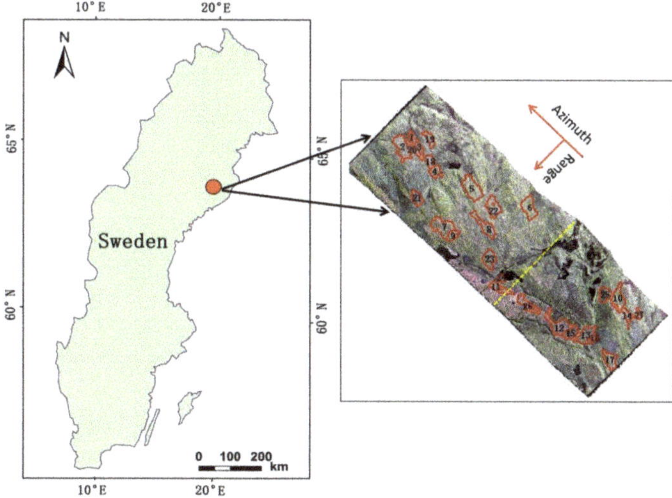

Figure 1. The test site: P-band synthetic aperture radar (SAR) image in the Pauli basis. The red polygons indicate the forest stands.

2.2. Field Data

The field survey data were collected and processed as a part of the BioSAR campaign in 2008 [58]. Twenty-seven forest stands with sizes ranging between 3.07 and 24.34 ha (boundaries are showed in Figure 1) were selected for the validation experiment. Within each stand, eight to 13 circular sample plots were laid out with a systematic spacing of 50 to 160 m, depending on the size of the area. The spacing in each stand was confirmed with the aim of obtaining 10 plots, having a radius of 10 m. The total number of plots was 310. In these plots, all trees with a diameter at breast height (DBH) greater than 4 cm were calipered and recorded. On the basis of probability proportional to their basal area, 1.5 sample trees were randomly selected from each plot to measure the height and age. Other parameters were also collected, such as the vegetation type and soil type. The biomass of different tree species, including the stem, bark, branches, and needles, but excluding the stump and roots, was calculated based on Petersson's biomass functions [59]. In addition, the biomass of each tree species was divided into three components, namely trunks, branches, and leaves. Due to seasonal reason, the leaf biomasses of the deciduous forests were not calculated. Table 1 describes the main features of the 27 forest stands.

Table 1. Main features of the in-situ data.

ID	Mean DBH (cm)	Mean Height (m)	Mean Age (Year)	Biomass (t ha^{-1})
1	17.6	13.07	77.45	72.39
2	20.8	14.08	61.22	69.22
3	17.11	13.12	57.05	58.56
4	23.76	16.94	107.46	103.68
5	23.1	17.69	11.28	134.51
6	28.35	21.41	12.24	182.54
7	26.54	18.06	131.92	119.13
8	16.95	15.41	57.38	106.24
9	28.31	19.59	143.72	139.76
10	28.84	20.15	140.52	167.11
11	26.37	17.36	157.15	116.46
12	21.41	15.39	63.09	74.28
13	22.67	17.13	127.8	86.03
14	25.64	18.81	114.4	158.23
15	17.34	12.31	52.99	35.08
16	18.28	13.23	83.72	85.6
17	21.32	15.77	124.83	89.31
18	21.36	15.73	124.87	108.99
19	18.76	14.26	90.67	119.8
20	19.17	13.87	86.88	84.45
21	12.02	9.56	34.85	42.71
22	21.01	16.58	76.19	95.75
23	27.53	17.37	144.12	111.96
24	20	13.93	117.7	110.78
25	23.6	15.99	93.27	73.8
26	8.65	7.51	30.62	27.46
27	22.01	15.77	101.06	112.82

DBH: diameter at breast height.

2.3. Polarimetric SAR Data

The P-band fully polarimetric SAR data over this study area were acquired in the framework of the European Space Agency (ESA) BioSAR 2008 campaign in the repeat-pass mode. The SAR system used in this campaign was the German Aerospace Center's (DLR) E-SAR airborne system [58]. The platform height was about 4091 m above ground and the pixel spacing was 1.6 m and 2.12 m in the azimuth direction and slant range, respectively. Four fully polarimetric SAR images were selected for the interferometric process. There were two master images and two slave images with two different flight tracks (314° and 134° from north). In this study area, there were 12 P-band fully polarimetric SAR data in two different flight tracks. These data could constitute five pairs of PolInSAR data in each direction and their baselines were 8 m, 8 m, 16 m, 24 m, and 32 m, respectively. Kugler et al. evaluated the multifaceted effect of the effective spatial baseline by analyzing the vertical wavenumber (Kz). He concluded that a good choice to get reliable forest height estimates with sufficient accuracy for single baseline acquisitions is to select Kz ranging from 0.05 to 0.15 rad/m [60]. According to our statistics of five pairs of PolInSAR datasets in the two flight directions, we found that the baselines of 32 m in both two flight directions can better fulfill the above condition of Kz. Detailed information on these two baselines of PolInSAR datasets is listed in Table 2. The basic data processing including terrain-correction and image-registration were done by the DLR. In addition, an airborne LiDAR measurement was also conducted in the study area to serve as a reference for the parameters estimation. The LiDAR measurement as a part of the BIOSAR 2008 campaign was performed on August 2008 with the TopEye system S/N 425 mounted on a helicopter. Details of the LiDAR data can be found in [58].

Table 2. Parameters of the SAR data.

Sensor	Band	Polarization	Flight Tracks	Purpose	Acquisition Time	Temporal Baseline	K_z
E-SAR	P	Full	314°	Master	2008-10-14 11:44:14	70 min	0.05–0.24
E-SAR	P	Full	314°	Slave	2008-10-14 12:55:21		
E-SAR	P	Full	134°	Master	2008-10-14 11:52:35	71 min	0.02–0.26
E-SAR	P	Full	134°	Slave	2008-10-14 13:03:38		

3. Methodology

In order to obtain a new parameter from the tomographic profile and achieve a more accurate forest AGB estimation, we firstly acquire two scattering mechanisms by the Phase Diversity (PD) coherence optimal algorithm and calculate the forest height and ground phase. Secondly, the vertical profile of a single pixel is obtained based on the PCT technique with the volume scattering mechanism. Then, we calculate the average vertical distribution of the relative reflectance, according to the polygonal boundary of the forest stand. Thirdly, a new parameter for biomass estimation is proposed by establishing the backscatter power loss area. Finally, the estimation model of the forest AGB is constructed by combining the new parameter and other parameters from the tomographic profile described in [56]. The flowchart is shown in Figure 2.

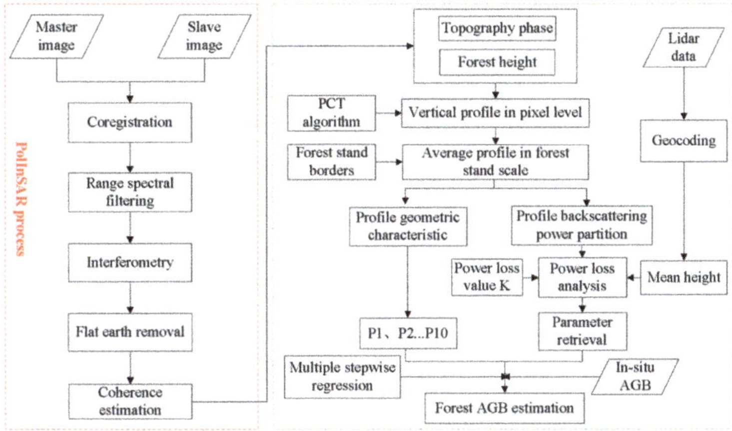

Figure 2. Flowchart of forest AGB estimation. AGB: above ground biomass; PCT: polarization coherence tomography; PolInSAR: Polarimetric SAR interferometry; P1–P10: parameters.

3.1. Polarization Coherence Tomography

The vertical profile function in penetrable volume scattering is reconstructed by the single-baseline PCT technique. The main observable in PCT is the volume scattering complex interferometric coherence, which depends on vertical structure variations [53]. Therefore, this dependence relationship can be used to build 3-D imaging and extract physical parameters. The complex coherence is shown as follows [53]:

$$\tilde{\gamma}(w) = e^{ik_z Z_0} \frac{\int_0^{h_v} f(Z)e^{ik_z Z}dZ}{\int_0^{h_v} f(Z)dZ} = e^{i\Phi_0} \frac{\frac{h_v}{2}e^{i\frac{k_z h_v}{2}}\int_{-1}^{1}(1+f(Z'))e^{i\frac{k_z h_v}{2}Z'}dZ'}{\frac{h_v}{2}\int_{-1}^{1}(1+f(Z'))dZ'} \tag{1}$$

where

$$0 \le |\widetilde{\gamma}| \le 1; \ Z' = \frac{2Z}{h_v} - 1$$

The complex interferometric coherence $\widetilde{\gamma}$ is related to the polarization state w. The vertical wavenumber k_Z is related to the interferometric baseline, Z_0 is the position of the bottom of the vegetation layer, hv is the vegetation height, and ϕ_0 is the topographic phase. $f(Z)$ is the vertical structure function, which physically represents the vertical variations of the scattering signal at a point in the SAR image [61]. It is bounded from the underlying surface to the top of the vegetation layer and can be developed efficiently in a Fourier-Legendre series as shown in (2):

$$f(Z') = \sum_n a_n P_n(Z'), \ a_n = \frac{2n+1}{2} \int_{-1}^{1} f(Z') P_n(Z') dZ' \tag{2}$$

where a_n denotes the Legendre coefficient, and $P_n(Z')$ represents the Legendre polynomials with vertical variable Z'. Then, the complex coherence for $f(Z)$ can be rewritten as [59]:

$$\widetilde{\gamma}(w) e^{-i(\phi_0 + k_v)} = \widetilde{\gamma}_k = f_0 + a_{10} f_1 + a_{20} f_2 \ldots + a_{i0} f_i \tag{3}$$

where $f_i (i = 1, 2, \ldots, n)$ represents the Legendre polynomial parameter at order i, which is a function of the single parameter $k_v = k_Z h_v / 2$, and a_{i0} is the Legendre coefficient. Equation (3) shows that the coherence can be seen as an algebraic sum of a series of structural functions.

Although multi-baseline PCT provides the potential to improve the resolution of the tomographic profile [53], it increases the number of unknown parameters, thereby increasing the computational complexity. However, single-baseline PCT only requires the second order of the Legendre series to describe the variations of the vertical profile, which is the simplest way to be implemented. In this case, obtaining the second order vertical structure function $f(Z')$ requires estimating two unknown coefficients (a_{10}, a_{20}) in Equation (3). We use a matrix inversion of Equation (4) to achieve this purpose.

$$\begin{bmatrix} 1 & 0 & 0 \\ 0 & -if_1 & 0 \\ 0 & 0 & f_2 \end{bmatrix} \begin{bmatrix} a_{00} \\ a_{10}(w) \\ a_{20}(w) \end{bmatrix} = \begin{bmatrix} 1 \\ \mathrm{Im}(\widetilde{\gamma}_k) \\ \mathrm{Re}(\widetilde{\gamma}_k - f_0) \end{bmatrix}$$

$$\Rightarrow [F]a(w) = b \Rightarrow \hat{a}(w) = [F]^{-1}\hat{b} \tag{4}$$

$$\hat{f}_{L2}(w,z) = \frac{1}{\hat{h}_v}\left\{ 1 - \hat{a}_{10}(w) + \hat{a}_{20}(w) + \frac{2z}{\hat{h}_v}(\hat{a}_{10}(w) - 3\hat{a}_{20}(w)) + \hat{a}_{20}(w)\frac{6z^2}{\hat{h}_{v^2}} \right\} \tag{5}$$

where $\hat{f}_{L2}(w,z)$ is the vertical structure function, which varies with the polarization state, $L2$ represents the second order expansion of Fourier-Legendre polynomials, and z is the vertical position from $(0, h_v)$.

Before implementing the PCT algorithm, we need to estimate the priori information of vegetation height h_v and topographic phase ϕ_0. We exploit the widely used random volume over ground (RVOG) model and three-stage inversion method to obtain forest height and topographic phase [41,62]. In addition, we also note that the PCT algorithm involves two polarization modes: the dominant volume scattering and the dominant ground surface scattering. Cloude [53] proposed two polarization channels, which are HV and HH-VV, representing the volume and ground surface scattering, respectively. However, the HV channel also has some ground surface scattering contributions, which leads to pure volume coherence error to vertical profile reconstruction. Therefore, we use the phase diversity (PD) coherence optimization algorithm to find the optimal polarization modes (i.e., $\mathrm{PD}_{\mathrm{high}}$ and $\mathrm{PD}_{\mathrm{low}}$) in the polarized space, thereby separating the two-phase centers maximally. The higher phase $\mathrm{PD}_{\mathrm{high}}$ corresponds to the "pure" volume scattering phase and the lower phase $\mathrm{PD}_{\mathrm{low}}$ corresponds to the "pure" ground surface scattering phase [63].

3.2. The Retrieval Method of Canopy Height (H_{ac})

According to the PCT algorithm, the tomographic profile at the pixel scale can be obtained with a 0.2 m interval in vegetation heights [54,64]. The profile values represent relative reflectivity values. However, the relative reflectivity of a single pixel is unrelated to forest biomass, because it is randomly distributed and unable to represent the vertical structure [56]. Therefore, the average of the stand scale is used to obtain change rules of the vertical structure. As shown in Figure 3, we use stands No. 6 and No. 13 of the study area to analyze the vertical distribution of average relative reflectivity. In Figure 3, the vertical distribution curve of the average relative reflectivity is divided into upper and lower parts by *h1*, and the envelope between *h1* and *h3* is the distribution of the relative reflectivity values of forest canopy. *h1* is the height of the forest corresponding to the inflection point of the upper and lower parts of the curve, and *h3* represents the height of the upper half of the curve with the relative reflectivity value closest to 0.001. These data between *h1* and *h3* constitute the first envelope. *h2* is the height position where the relative reflectivity value is the maximum in the first envelope, and the data between 0 and *h2* form the second envelope.

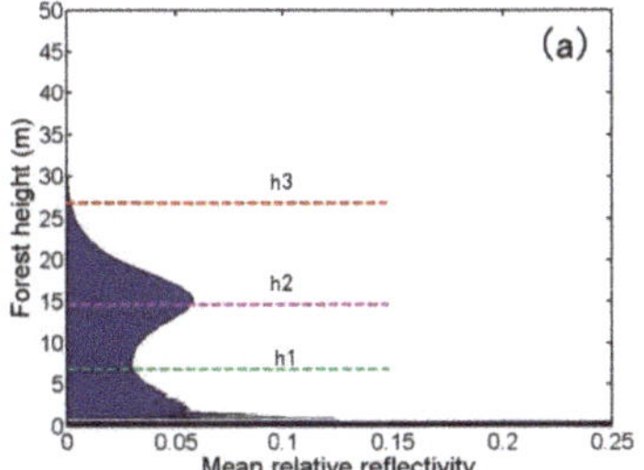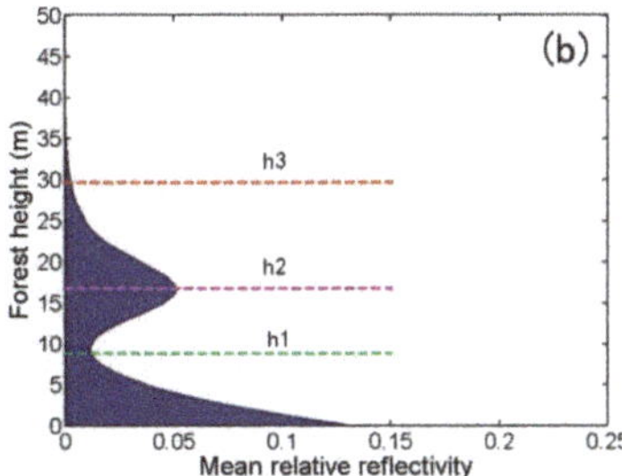

Figure 3. Tomographic profiles of (**a**) stand No. 6 and (**b**) stand No. 13.

After obtaining the vertical distribution of average relative reflectivity, the relationship between the curve and the forest biomass needs to be established. The parameters (Table 3) were obtained by parameterizing the average relative reflectivity [56]. P1 is the ratio of the peak value to the first envelope span. P2 is the integral of the relative reflectivity multiplied by the height in the first envelope. P3, P4 (*TomoH*), and P5 are the maximum probability, the mean, and the standard deviation of the fitted Gaussian function of the first envelope, respectively. P6 and P7 are the reciprocal of the relative reflectivity summation of the first and second envelopes, respectively. P8 represents the ratio of P6 to P7. P9 is the relative reflectivity summation between *h1* and *h2* multiplied by the relative reflectivity summation between *h2* and *h3*. P10 represents the integral of the relative reflectivity multiplied by the corresponding height from 0 to *h3*. These parameters were defined without physical characteristics [56]. Among these parameters, P4 (the tomographic height (*TomoH*)) is considered as the most important parameter for forest AGB estimation. However, *TomoH* only corresponds to the highest volume relative reflectivity, which is lower than the actual forest height.

In order to find a parameter closer to the true canopy height of the stand, we characterize the average relative reflectivity as the backscattered power distribution, and then divide the average relative reflectivity distribution of the forest canopy into three parts, taking the No. 13 stand as an example (Figure 4). The first part (between *h1* and *h3*) corresponds to the canopy phase zone, where most of the backscatter is concentrated. Its maximum value corresponds to the stand canopy phase center position, i.e., *h2*. The second part (between *h2* and *h3*) is the backscatter power loss zone, where the backscatter undergoes a loss along the vertical direction from *h2* to *h1*. The third part (above *h3*) is the noise zone, where the backscatter power is mostly contributed by noise, unlikely to be associated with physically relevant components. A parameter related to the average height of the stand can be extracted by analyzing the power loss value in the first envelope [50–52], which is:

$$H(i) = arg\,min\{\,|P(z,i) - P(\overline{H}_{PC},i) - K|\,\}$$ (6)

where $\overline{H}_{PC}$ is the height of the average phase center of each stand, i is the stand serial number, $P(\overline{H}_{PC},i)$ is the backscattering power at location $\overline{H}_{PC}$ of stand i, and z is the value ranging from $\overline{H}_{PC}$ to $h3$. The power loss value K ranges from 0 to 1 with the step of 0.1. The optimal K is determined by the forest height that is closest to the true value obtained by LiDAR [51]. We name the corresponding forest height as the average canopy height H_{ac}.

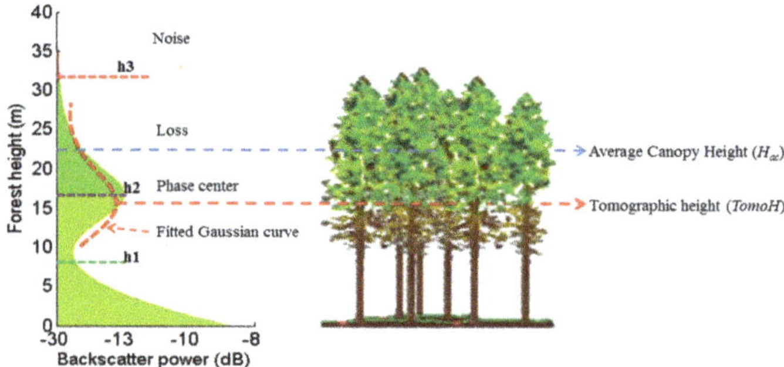

Figure 4. The schematic view of the stand vertical backscatter distribution.

Table 3. Parameters of parameterizing average relative reflectivity.

Parameter	Description
P1	the ratio of the peak value to the first envelope span
P2	the integral of the relative reflectivity multiplied with the height in the first envelope
P3	the maximum probability of the fitted Gaussian function of the first envelope
P4	the mean of the fitted Gaussian function of the first envelope
P5	the standard deviation of the fitted Gaussian function of the first envelope
P6	the reciprocal of the relative reflectivity summation of the first envelopes
P7	the reciprocal of the relative reflectivity summation of the second envelopes
P8	the ratio of P6 to P7
P9	the relative reflectivity summation between $h1$ and $h2$ multiplied by the relative reflectivity summation between $h2$ and $h3$
P10	the integral of the relative reflectivity multiplied by the corresponding height from 0 to $h3$

3.3. Model Construction and Validation

Since the power function can be used to effectively express the relationship between SAR parameters and biomass [56,65], we use a stepwise regression method to construct a multiple linear biomass estimation model with the natural logarithm of the SAR parameters.

$$ln(B) = a_0 + a_1ln(X_1) + a_2ln(X_2) + \ldots + a_iln(X_n)$$ (7)

where B is the biomass (t ha^{-1}), and a_i is the coefficient corresponding to the extracted parameters Xn from tomographic profiles.

The stepwise regression method is employed to find the best combination of independent variables to predict the dependent variable. The method introduces the independent variables to the regression model one by one (from small to large), and only those satisfying the test value F remain in the model. Usually, the remaining variables are significant and have large contributions to dependent variable estimation. When no more variables are eligible for inclusion or removal, the process is terminated. After obtaining the estimation model, we must also consider the problem of collinearity between variables. Generally, this problem can be indicated by the values of tolerance and variance inflation factor (VIF) [66].

In this paper, the three independent variables H_{ac}, P6, and P7 are considered more sensitive to forest biomass than other parameters. The canopy height H_{ac} is a new parameter proposed by analyzing the variation of the backscattering power of forest, and P6 and P7 are the reciprocal of the relative reflectivity summation of the first ($h1$–$h3$) and second (0–$h1$) envelopes, respectively. In this study, we use the m-fold (13-fold) cross-validation experiments to test the accuracy of the multiple linear biomass estimation models. We divide the forest stands into 10 groups equally, one of which is used for validation and the others are applied to regression models.

4. Experimental Results and Analysis

4.1. Tomographic Profiles

The vertical structure functions are reconstructed at PD_{high}, PD_{low}, HV, and HH-VV polarimetric channels. The tomographic profiles in the 2500th line (yellow line in Figure 1) along the range direction are demonstrated in Figure 5, where the horizontal axis is the pixel position, the vertical axis is the forest height, and the different colors represent the intensity of relative reflectivity.

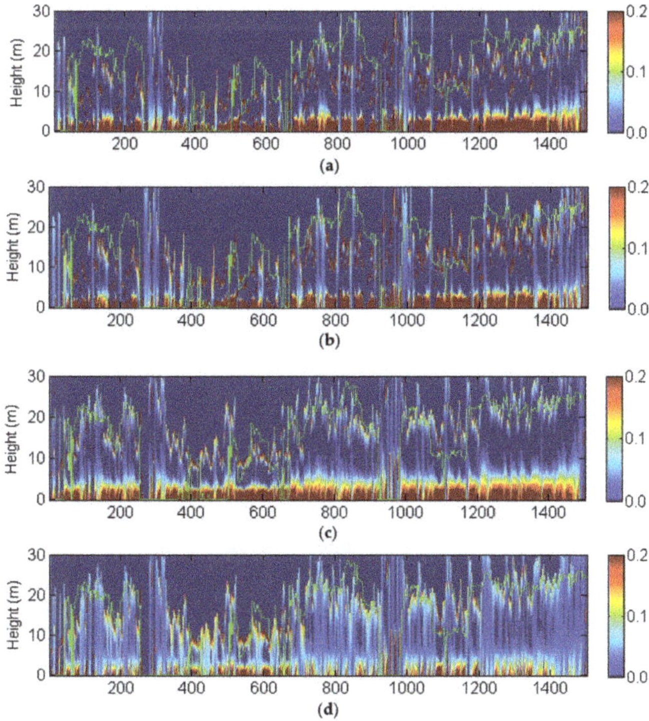

Figure 5. Tomographic profiles for different polarimetric channels along the range direction in the line 2500, see the yellow dashed line in Figure 1. (**a**) HH-VV channel; (**b**) HV channel; (**c**) PD_{low} channel; (**d**) PD_{high} channel. The green lines denote the LiDAR height. PD: phase diversity.

From Figure 5, it can be observed that the effect of different polarimetric channels on the tomographic profile reconstruction is significant. In the HH-VV polarimetric channel, where the surface scattering is dominant, the relative reflectivity values are mainly distributed in the ground surface layer. In the HV polarimetric channel, volume scattering is the dominant scattering mechanism and its scattering center of the interferometric phase is closer to the vegetation canopy top than that of the HH-VV channel. However, the HV channel still retains an amount of ground surface scattering contributions, which leads to pure volume coherence error affecting tomographic profile reconstruction. It means the relative reflectivity values still have a high distribution in the ground surface. From the visual perspective of Figure 5a,b, the difference of relative reflectivity vertical distribution is relatively small.

The relative reflectivity distributions are obviously different in PD_{low} (Figure 5c) and PD_{high} (Figure 5d) channels. We note that the ground surface has high reflectivity values distribution in the PD_{high} channel, which may be caused by an underlying shrub layer there. In addition, compared with the other three channels, the PD_{high} channel contains more abundant vegetation information, which is conducive to extract vegetation parameters. Hence, in this paper, we use the tomographic profile of the PD_{high} channel for analysis.

4.2. Parameter Retrieval

According to the method described in Section 3.2, we retrieve the new parameter of average canopy height H_{ac} from the average tomographic profile of each stand (Figure 4). Using the top-of-canopy height LiDAR model [51], we can get the parameter's location corresponding to a power loss value in each stand, with respect to the stand canopy phase center, ranging from -10 dB to 0 dB. Figures 6 and 7 present the average bias of the stand height and the root mean square error (RMSE) with respect to the LiDAR measurements in different flight tracks, respectively. In the power loss interval of the 314 deg. flight direction, the average bias of 27 forest stands reaches the maximum at the stand canopy phase center, which is 3.162 m. The minimum value is -6.409 m, and the overall trend increases with the power loss from the phase center, which is close to 0 m at -1.549 dB. The RMSE also reaches the maximum at the stand canopy phase center, which is 8.255 m, while the minimum RMSE is found at -1.549 dB. In the power loss interval of the 134 deg. flight direction, the average bias also reaches the maximum at the stand canopy phase center, which is 7.126 m. The minimum value is -2.765 m, and the overall trend is the same as the 314 deg. flight direction increasing with the power loss from the phase center, which is close to 0 m at -6.0206 dB (Figure 7a). The RMSE also reaches the maximum at the stand canopy phase center, which is 7.862 m, while the minimum RMSE is 2.954 at -6.0206 dB (Figure 7b). These results show that -1.549 dB and -6.0206 dB are the optimal power loss values for retrieving the new parameter H_{ac} in the two different flight directions, respectively.

The forest heights corresponding to the power loss values of -1.549 dB and -6.0206 dB are extracted for each forest stand in different flight directions, respectively. We use *TomoH* [56] to evaluate the performance of parameter H_{ac} in the estimation of AGB. In order to assess the sensitivity of the parameters H_{ac} and *TomoH* to the forest biomass, respectively, the simple linear model is used to estimate the biomass of 27 forest stands. The estimated biomasses are tested with in situ AGB for each forest stand. As shown in Figures 8 and 9, in the 314 deg. flight direction, the value of the correlation coefficient R^2 using the parameter H_{ac} is 0.630, which is much higher than that of *TomoH* (0.548). Also, the RMSEs obtained by H_{ac} and *TomoH* are 28.063 t ha^{-1} and 33.317 t ha^{-1}, respectively. In the 134 deg. flight direction, the RMSEs obtained by H_{ac} and *TomoH* are 27.061 t ha^{-1} and 31.311 t ha^{-1}, respectively. Obviously, the sensitivity and the accuracy of H_{ac} inversion are higher than the values of the *TomoH* inversion in the two different flight directions, respectively, which means that H_{ac} is more helpful for the estimation of forest biomass with P-band ESAR data in the test site.

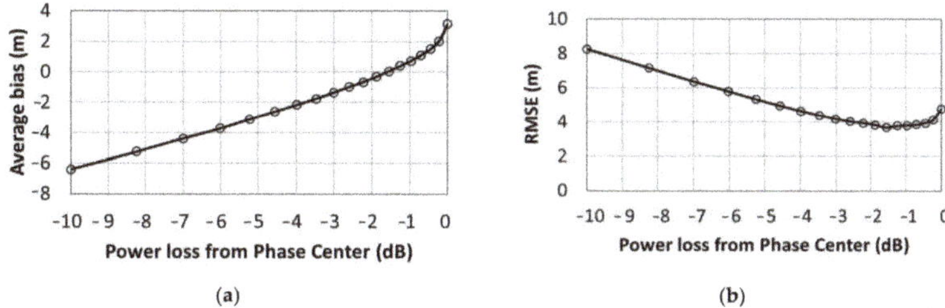

Figure 6. Average bias and root mean square error (RMSE) in 314 deg. flight direction (**a**) Stand height average bias and (**b**) RMSE versus power loss with respect to stand average canopy phase center elevation.

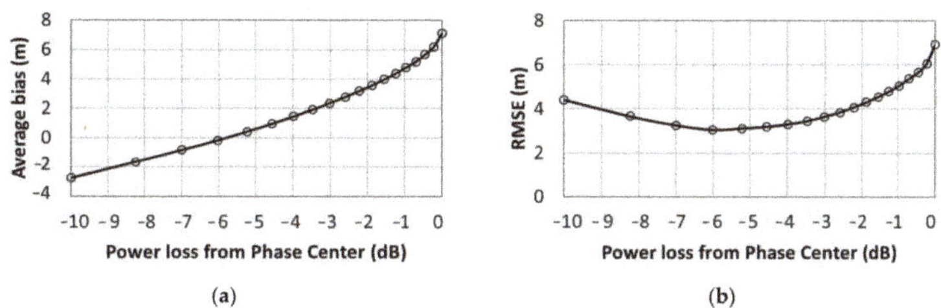

Figure 7. Average bias and RMSE in 134 deg. flight direction (**a**) Stand height average bias and (**b**) RMSE versus power loss with respect to stand average canopy phase center elevation.

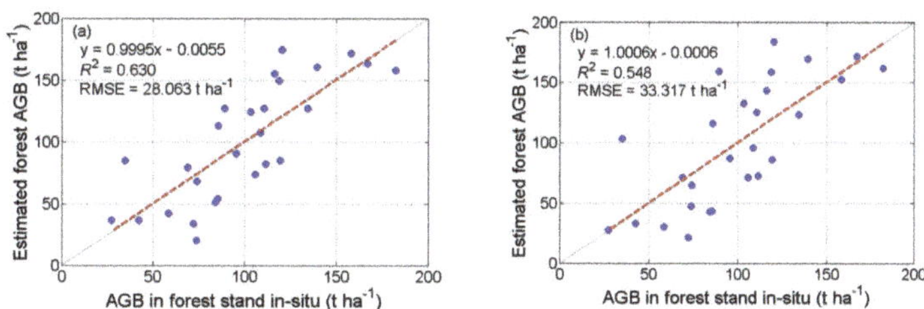

Figure 8. Forest AGB estimation results using (**a**) H_{ac} and (**b**) *TomoH* in 314 deg. flight direction.

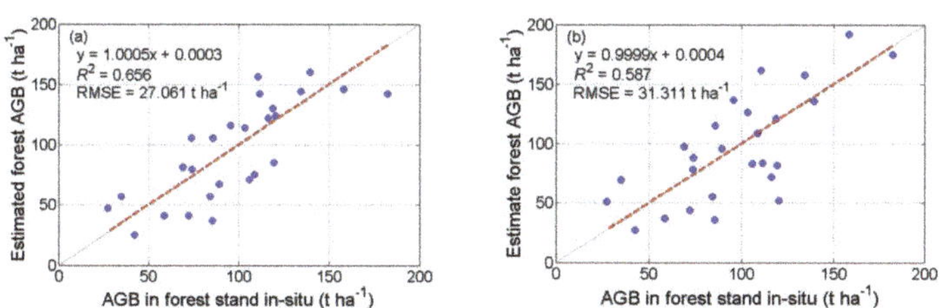

Figure 9. Forest AGB estimation results using (**a**) H_{ac} and (**b**) *TomoH* in 134 deg. flight direction.

4.3. Forest AGB Estimation

In this section, we construct a reasonable forest AGB estimation model in the test site using parameter H_{ac} and the nine parameters (P1–P3 and P5–P10) described in [56]. Table 4 shows the results of estimation models obtained through the multiple linear stepwise regression method with the F test. $H_{ac}1$ to $H_{ac}10$ are the ten stepwise regression models with parameter H_{ac} and the nine parameters (P1–P3 and P5–P10) in the 314 deg. flight direction. Based on different parameters, these models can be divided into two categories: one category includes parameters H_{ac} and P6 ($H_{ac}1$, $H_{ac}3$, $H_{ac}5$, $H_{ac}7$, $H_{ac}9$, $H_{ac}11$, $H_{ac}13$); and the other includes H_{ac}, P6, and P7 ($H_{ac}2$, $H_{ac}4$, $H_{ac}6$, $H_{ac}8$, $H_{ac}10$, $H_{ac}12$, $H_{ac}14$), which indicates that H_{ac}, P6, and P7 are more sensitive to forest biomass than other parameters. According to these models, we can choose one of the best models for biomass estimation in the two categories, respectively. When using the same method to obtain estimation models with Li's ten parameters (P1–P10), including *TomoH*, ten models are also obtained in the 314 deg. flight direction. However, based on the same rules, we only choose the best two models for comparison, namely *TomoH*1 and *TomoH*2. In addition, we also list the best models using different parameters in the 134 deg. flight direction, named $H_{ac}15$, $H_{ac}16$, *TomoH*3, and *TomoH*4 (Table 4).

According to Table 4, from models $H_{ac}1$ to $H_{ac}14$, considering only collinearity between variables cannot select the best models, because these parameters do not appear to have serious collinearity problems according to the tolerance and variance inflation factor. However, when we consider the correlation coefficient in this table, $H_{ac}3$ and $H_{ac}8$ will perform relatively better than other models, whose R^2 values are 0.734 and 0.798, respectively. When considering the validated error, tolerance, and variance inflation factor together, we still find that $H_{ac}3$ and $H_{ac}8$ are more suitable for forest AGB estimation than other models in the two categories. Then, we choose these four models for analysis in the two different flight directions, respectively. $H_{ac}3$, $H_{ac}8$, *TomoH*1, and *TomoH*2 belong to the 314 deg. flight direction, whereas $H_{ac}15$, $H_{ac}16$, *TomoH*3, and *TomoH*4 belong to the 134 deg. flight direction.

The comparison results with field data are showed in Figures 10 and 11. In the 314 deg. flight direction (Figure 10), the correlation coefficient value between field data and AGB derived by $H_{ac}8$ is 0.761, which is higher than the values obtained by $H_{ac}3$, *TomoH*1, and *TomoH*2 (0.728, 0.573, and 0.687, respectively). Additionally, the RMSEs of $H_{ac}3$ and $H_{ac}8$ are 19.941 t ha^{-1} and 18.824 t ha^{-1}, respectively, which are lower than those of *TomoH*1 (24.834 t ha^{-1}) and *TomoH*2 (21.219 t ha^{-1}). Obviously, the performance of $H_{ac}8$-based inversion is better than that of other models. In the 134 deg. Flight direction (Figure 11), the performance of $H_{ac}16$-based inversion is better than that of other models, whose correlation coefficient value is 0.776 and RMSE is 18.325 t ha^{-1}.

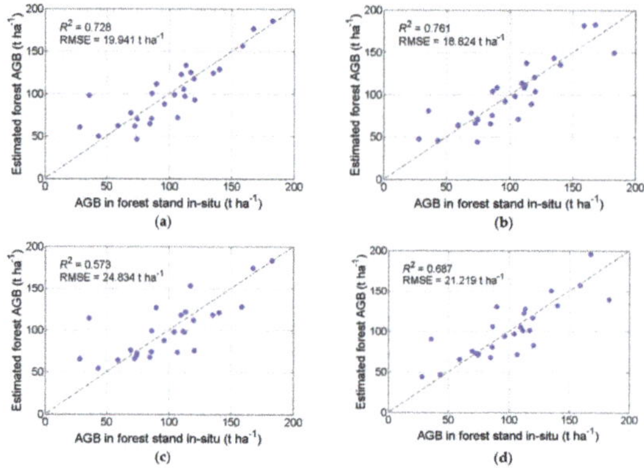

Figure 10. Comparison between in-situ AGB and estimated forest AGB derived from model in 314 deg. flight direction (**a**) $H_{ac}3$, (**b**) $H_{ac}8$, (**c**) *TomoH*1, and (**d**) *TomoH*2.

Table 4. Models for forest AGB estimation.

Models	Computing Formula	R^2	Tolerance	VIF	Validated Error
$H_{ac}1$	$\ln(B) = -3.797 + 1.866\ln(H_{ac}) - 2.977\ln(P6)$	0.704	0.919, 0.919	1.088, 1.088	0.392, 0.138
$H_{ac}2$	$\ln(B) = -9.363 + 2.414\ln(H_{ac}) - 6.276\ln(P6) - 1.563\ln(P7)$	0.769	0.668, 0.341, 0.363	1.497, 2.930, 2.757	0.397, 0.091
$H_{ac}3$	$\ln(B) = -3.829 + 2.019\ln(H_{ac}) - 2.563\ln(P6)$	0.734	0.904, 0.904	1.106, 1.106	0.163, 0.276
$H_{ac}4$	$\ln(B) = -10.007 + 2.401\ln(H_{ac}) - 6.288\ln(P6) - 1.671\ln(P7)$	0.791	0.655, 0.299, 0.327	1.527, 3.342, 3.056	0.025, 0.524
$H_{ac}5$	$\ln(B) = -3.899 + 2.034\ln(H_{ac}) - 2.586\ln(P6)$	0.720	0.894, 0.894	1.118, 1.118	0.167, 0.485
$H_{ac}6$	$\ln(B) = -10.072 + 2.414\ln(H_{ac}) - 6.319\ln(P6) - 1.667\ln(P7)$	0.789	0.656, 0.300, 0.333	1.526, 3.336, 3.007	0.061, 0.529
$H_{ac}7$	$\ln(B) = -4.010 + 2.020\ln(H_{ac}) - 2.751\ln(P6)$	0.720	0.898, 0.898	1.113, 1.113	-1.025, 0.466
$H_{ac}8$	$\ln(B) = -10.334 + 2.421\ln(H_{ac}) - 6.508\ln(P6) - 1.746\ln(P7)$	0.798	0.658, 0.311, 0.343	1.519, 3.214, 2.918	0.043, 0.515
$H_{ac}9$	$\ln(B) = -3.307 + 1.887\ln(H_{ac}) - 2.422\ln(P6)$	0.704	0.907, 0.907	1.102, 1.102	0.267, 0.257
$H_{ac}10$	$\ln(B) = -8.508 + 2.195\ln(H_{ac}) - 5.567\ln(P6) - 1.447\ln(P7)$	0.756	0.667, 0.308, 0.336	1.498, 3.246, 2.979	0.171, 0.236
$H_{ac}11$	$\ln(B) = -3.784 + 1.883\ln(H_{ac}) - 2.918\ln(P6)$	0.708	0.917, 0.917	1.091, 1.091	0.055, 0.388
$H_{ac}12$	$\ln(B) = -9.415 + 2.230\ln(H_{ac}) - 6.270\ln(P6) - 1.580\ln(P7)$	0.773	0.670, 0.344, 0.367	1.492, 2.908, 2.726	0.032, 0.395
$H_{ac}13$	$\ln(B) = -4.081 + 2.033\ln(H_{ac}) - 2.777\ln(P6)$	0.715	0.888, 0.888	1.127, 1.127	0.069, 0.478
$H_{ac}14$	$\ln(B) = -10.166 + 2.418\ln(H_{ac}) - 6.387\ln(P6) - 1.686\ln(P7)$	0.790	0.661, 0.316, 0.353	1.513, 3.167, 2.831	0.061, 0.528
$TomoH1$	$\ln(B) = -3.544 + 1.789\ln(TomoH) - 3.277\ln(P6)$	0.615	0.840, 0.840	0.493, 0.233, 0.274	0.136, 0.055
$TomoH2$	$\ln(B) = -13.199 + 2.437\ln(TomoH) - 9.115\ln(P6) - 2.539\ln(P7)$	0.723	0.502, 0.230, 0.266	1.992, 4.352, 3.758	-0.087, 0.042
$H_{ac}15$	$\ln(B) = 2.172 + 1.588\ln(H_{ac}) + 2.419\ln(P6)$	0.744	0.954, 0.954	1.049, 1.049	0.165, 0.132
$H_{ac}16$	$\ln(B) = 3.876 + 1.468\ln(H_{ac}) + 3.505\ln(P6) + 0.417\ln(P7)$	0.785	0.868, 0.592, 0.539	1.152, 1.689, 1.854	0.201, 0.023
$TomoH3$	$\ln(B) = 3.333 + 0.691\ln(TomoH) + 2.951\ln(P6)$	0.706	0.905, 0.905	1.105, 1.105	-0.307, 0.216
$TomoH4$	$\ln(B) = 5.189 + 0.646\ln(TomoH) + 4.736\ln(P6) - 0.514\ln(P7)$	0.730	0.880, 0.456, 0.426	1.136, 2.191, 2.347	-0.353, 0.069

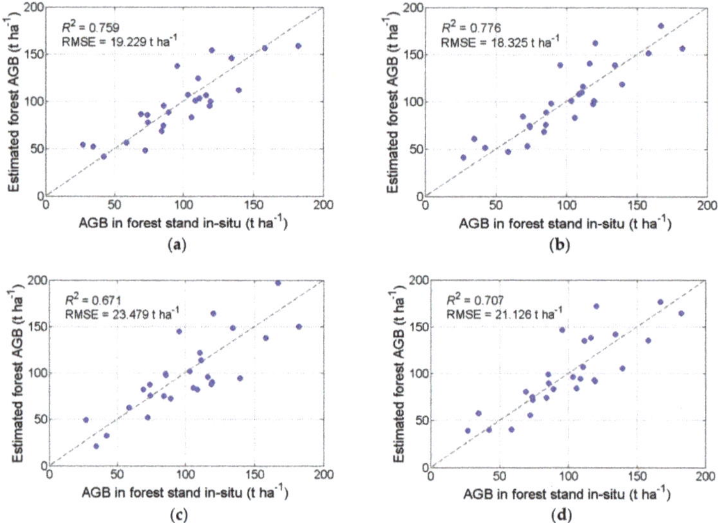

Figure 11. Comparison between in-situ AGB and estimated forest AGB derived from model in 134 deg. flight direction (**a**) $H_{ac}15$, (**b**) $H_{ac}16$, (**c**) *TomoH3*, and (**d**) *TomoH4*.

5. Discussion

Traditional studies on the PCT technology usually focus on forest vertical profile generation [59,61], with rare consideration of the relationship between the vertical profile and the forest parameters. In this study, all the parameters for AGB inversion were extracted by analyzing the geometric characteristics of the average vertical profiles within the forest stand [55,56]. These parameters are mathematically defined without considering physical scattering mechanisms of forest. In references [55,56], *TomoH* (i.e., P4) is the most important parameter for the estimation of forest AGB. Nevertheless, *TomoH* only corresponds to the highest volume relative reflectivity, which is lower than the actual forest height, especially for long wavelengths (such as P-band) SAR data [55]. In this study, we use the PCT technique and P-band airborne PolInSAR data to generate the average vertical profile. A new parameter (i.e., H_{ac}) for estimating forest AGB is retrieved through analyzing the variation of the vertical backscatter power, which considers the physical scattering attenuation when radar echo penetrates into the forest. The experiment results (Figures 8 and 9) show that better accuracies are achieved for AGB inversion, which demonstrates that H_{ac} performs relatively better than *TomoH*. Additionally, we combine H_{ac} with Li's nine parameters to construct the forest AGB estimation model by a multiple linear stepwise regression method. The results in Table 4, as well as Figures 10 and 11, show that the combination has a better performance than only using Li's ten parameters, which also indirectly verify the validity and reliability of the new parameter in estimating forest AGB.

Forest biomass estimation is a timely topic, and many scholars have used different data and methods to estimate biomass in different regions. In the test site of this paper (krycklan area), Ulander et al. [67] used a multiple linear regression model based on P-band multi-polarization backscatter to estimate biomass, and the RMSE varied in the range 29–42 t ha^{-1}. Neumann et al. [44] used parametric (linear regression) and non-parametric (random forest, support vector machine) methods to assess biomass estimation performance with polarimetric interferometric synthetic aperture radar (PolInSAR) data at L- and P-band in this test area, and the cross-validated biomass RMSE was reduced to 23 t ha^{-1} in the best case at P-band. Compared with these methods, PCT technology can extract forest vertical structures from the vertical distribution of relative reflectivity and has the potential to improve biomass estimation. In addition, the potential of P-band TomoSAR to characterize forest structure was assessed in a number of studies relating forest vertical structure to forest biomass [50–52]. Compared with TomoSAR technology, the single baseline PCT reduces the amount of data, but it has the problem of inverting the instability of the vertical profile of relative reflectivity in some places (Figure 5) and multi-baseline PCT technology may improve the phenomenon. Moreover, we note that the optimal power loss value for forest parameter extraction in this test site is different at two different flight tracks, and the value also should be different for other test sites, since power loss is always related to the vertical resolution, forest types, density, etc. [58].

6. Conclusions

In this paper, we propose a method based on the polarimetric coherence tomography for forest AGB inversion. A new parameter of average canopy height (H_{ac}) considering scattering attenuation is introduced for AGB inversion. Two pairs of P-band E-SAR full polarimetric SAR data covering the Krycklan river catchment in northern Sweden are selected for experiments. The results show that the tomographic profiles are greatly influenced by the polarimetric channels, and the PD algorithm can provide two more suitable polarimetric channels for the PCT technology. Referenced to the LiDAR forest height, in the 314 deg. flight direction, the power loss of -1.549 dB relative to the canopy phase center's power is confirmed to be the optimal value and is used to retrieve the new parameter of average canopy height H_{ac} with average bias and RMSE. In contrast with the tomographic height (*TomoH*), the new parameter H_{ac} is shown to be closer to the top of the stands, and has more advantages for forest AGB inversion. Following this, a high performance and precision inversion model is constructed by combining parameter H_{ac} with other parameters (Li's parameters). The RMSE is 18.824 t ha^{-1} in the test area. The same conclusion is reached with the data we use in the other flight direction. The RMSE is 18.325 t ha^{-1}. In this study, we try to introduce a parameter considering scattering characteristics, rather than just from geometric characteristics, for improving the accuracy of forest AGB inversion.

Acknowledgments: The work was supported by the National Natural Science Foundation of China (No. 41531068, 41671356 and 41371335), the Natural Science Foundation of Hunan Province, China (No. 2016JJ2141), and Innovation Foundation for Postgraduate of Central South University, China (No. 2017zzts179). Additionally, the experimental datasets are supported by PA-SB ESA EO Project Campaign (ID. 14655, 14751).

Author Contributions: Haibo Zhang conceived the idea, performed the experiments, and wrote and revised the paper; Changcheng Wang contributed some ideas, analyzed the experimental results, and revised the paper; Jianjun Zhu supervised the work and contributed some ideas; Haiqiang Fu, Qinghua Xie, and Peng Shen contributed to the discussion of the results.

Conflicts of Interest: The authors declare no conflict of interest.

References

1. Bonan, G.B. Forests and climate change: Forcings, Feedbacks, and the Climate Benefits of Forests. *Science* **2008**, *320*, 1444–1449. [CrossRef] [PubMed]
2. Kindermann, G.E.; McCallum, I.; Fritz, S.; Obersteiner, M. A global forest growing stock, biomass and carbon map based on FAO statistics. *Silva Fenn.* **2008**, *42*, 387–396. [CrossRef]
3. Pan, Y.; Birdsey, R.A.; Houghton, R.; Kauppi, P.E.; Kurz, W.A.; Phillips, O.L.; Shvidenko, A.; Lewis, S.L.; Canadell, J.G.; Ciais, P.; et al. A Large and Persistent Carbon Sink in the World's Forests. *Science* **2011**, *333*, 988–993. [CrossRef] [PubMed]
4. Askne, J.I.H.; Fransson, J.E.S.; Santoro, M.; Soja, M.J.; Ulander, L.M.H. Model-based biomass estimation of a Hemi-Boreal forest from multitemporal TanDEM-X acquisitions. *Remote Sens.* **2013**, *5*, 5725–5756. [CrossRef]
5. Houghton, R.A. Aboveground forest biomass and the global carbon balance. *Glob. Chang. Biol.* **2005**, *11*, 945–958. [CrossRef]
6. Houghton, R.A.; Hall, F.; Goetz, S.J. Importance of biomass in the global carbon cycle. *J. Geophys. Res. Biogeosci.* **2009**, *114*. [CrossRef]
7. Lu, D. The potential and challenge of remote sensing-based biomass estimation. *Int. J. Remote Sens.* **2006**, *27*, 1297–1328. [CrossRef]

8. Powell, S.L.; Cohen, W.B.; Healey, S.P.; Kennedy, R.E.; Moisen, G.G.; Pierce, K.B.; Ohmann, J.L. Quantification of live aboveground forest biomass dynamics with Landsat time-series and field inventory data: A comparison of empirical modeling approaches. *Remote Sens. Environ.* **2010**, *114*, 1053–1068. [CrossRef]

9. Gibbs, H.K.; Brown, S.; Niles, J.O.; Foley, J.A. Monitoring and estimating tropical forest carbon stocks: Making REDD a reality. *Environ. Res. Lett.* **2007**, *2*, 045023. [CrossRef]

10. Uddin, K.; Gilani, H.; Murthy, M.S.R.; Kotru, R.; Qamer, F.M. Forest Condition monitoring using very-high-resolution satellite imagery in a remote mountain watershed in Nepal. *Mt. Res. Dev.* **2015**, *35*, 264–277. [CrossRef]

11. Yu, Y.; Saatchi, S. Sensitivity of L-band SAR backscatter to aboveground biomass of global forests. *Remote Sens.* **2016**, *8*, 522. [CrossRef]

12. Anderson, G.L.; Hanson, J.D.; Haas, R.H. Evaluating Landsat thematic mapper derived vegetation indices for estimating above-ground biomass on semiarid rangelands. *Remote Sens. Environ.* **1993**, *45*, 165–175. [CrossRef]

13. Steininger, M.K. Satellite estimation of Tropical Sencondary forest above-ground biomass: Data from Brazil and Bolivia. *Int. J. Remote Sens.* **2000**, *21*, 1139–1157. [CrossRef]

14. Baccini, A.; Laporte, N.; Goetz, S.J.; Sun, M.; Dong, H. A first map of tropical Africa's above-ground biomass derived from satellite imagery. *Environ. Res. Lett.* **2008**, *3*, 45011–45019. [CrossRef]

15. Sinha, S.; Jeganathan, C.; Sharma, L.K.; Nathawat, M.S. A review of radar remote sensing for biomass estimation. *Int. J. Environ. Sci. Technol.* **2015**, *12*, 1779–1792. [CrossRef]

16. Su, Y.J.; Guo, Q.H.; Xue, B.L.; Hu, T.Y.; Alvarez, O.; Tao, S.L.; Fang, J.Y. Spatial distribution of forest aboveground biomass in China: Estimation through combination of spaceborne lidar, optical imagery, and forest inventory data. *Remote Sens. Environ.* **2016**, *173*, 197–199. [CrossRef]

17. Funning, G.J.; Parsons, B.; Wright, T.J. Fault slip in the 1997 Manyi, Tibet earthquake from linear elastic modelling of InSAR displacements. *Geophys. J. Int.* **2007**, *169*, 988–1008. [CrossRef]

18. Furuya, M.; Satyabala, S.P. Slow earthquake in Afghanistan detected by InSAR. *Geophys. Res. Lett.* **2008**, *35*, 160–162. [CrossRef]

19. Schulz, W.H.; Kean, J.W.; Wang, G. Landslide movement in southwest Colorado triggered by atmospheric tides. *Nat. Geosci.* **2009**, *2*, 863–866. [CrossRef]

20. Cai, J.; Wang, C.; Mao, X.; Wang, Q. An adaptive offset tracking method with SAR images for landslide displacement monitoring. *Remote Sens.* **2017**, *9*, 830–845. [CrossRef]

21. Dowdeswell, J.A.; Unwin, B.; Nuttall, A.; Wingham, D.J. Velocity structure, flow instability and mass flux on a large Arctic ice cap from satellite radar interferometry. *Earth Planet. Sci. Lett.* **1999**, *167*, 131–140. [CrossRef]

22. Goldstein, R.M.; Engelhardt, H.; Kamb, B.; Frolich, R.M. Satellite radar interferometry for monitoring ice sheet motion: Application to an antarctic ice stream. *Science* **1993**, *262*, 1525–1530. [CrossRef] [PubMed]

23. Lopez-Sanchez, J.M.; Vicente-Guijalba, F.; Erten, E.; Campos-Taberner, M.; Garcia-Haro, F.J. Retrieval of vegetation height in rice fields using polarimetric sar interferometry with tandem-x data. *Remote Sens. Environ.* **2017**, *192*, 30–44. [CrossRef]

24. Le Toan, T.; Beaudoin, A.; Riom, J.; Guyon, D. Relating forest biomass to SAR data. *IEEE Trans. Geosci. Remote Sens.* **1992**, *30*, 403–411. [CrossRef]

25. Le Toan, T.; Quegan, S.; Woodward, I.; Lomas, M.; Delbart, N.; Picard, G. Relating radar remote sensing of biomass to modelling of forest carbon budgets. *Clim. Chang.* **2004**, *67*, 379–402. [CrossRef]

26. Mermoz, S.; Réjou-Méchain, M.; Villard, L.; Le Toan, T.; Rossi, V.; Gourlet-Fleury, S. Decrease of L-band SAR backscatter with biomass of dense forests. *Remote Sens. Environ.* **2015**, *159*, 307–317. [CrossRef]

27. Santos, J.; Lacruz, M.; Araujo, L.; Keil, M. Savanna and tropical rainforest biomass estimation and spatialization using JERS-1 data. *Int. J. Remote Sens.* **2002**, *23*, 1217–1229. [CrossRef]

28. Saatchi, S.; Marlier, M.; Chazdon, R.; Clark, D.; Russell, A. Impact of spatial variability of tropical forest structure on radar estimation of aboveground biomass. *Remote Sens. Environ.* **2011**, *115*, 2836–2849. [CrossRef]

29. Carreiras, J.; Melo, J.B.; Vasconcelos, M.J. Estimating the above-ground biomass in Miombo Savanna woodlands (Mozambique, East Africa) using L-band synthetic aperture radar data. *Remote Sens.* **2013**, *5*, 1524–1548. [CrossRef]

30. Mermoz, S.; Le Toan, T.; Villard, L.; Réjou-Méchain, M.; Seifert-Granzin, J. Biomass assessment in the Cameroon savanna using ALOS PALSAR data. *Remote Sens. Environ.* **2014**, *155*, 109–119. [CrossRef]

31. Sandberg, G.; Ulander, L.M.H.; Fransson, J.E.S.; Holmgren, J.; Le Toan, T. L- and P-band backscatter intensity of biomass retrieval in hemiboreal forest. *Remote Sens. Environ.* **2011**, *115*, 2874–2886. [CrossRef]

32. Ranson, K.J.; Sun, G. Mapping biomass of a northern forest using multi-frequency SAR data. *IEEE Trans. Geosci. Remote Sens.* **1994**, *32*, 388–396. [CrossRef]

33. Saatchi, S.; Halligan, K.; Despain, D.G.; Crabtree, R.L. Estimation of forest fuel load from radar remote sensing. *IEEE Trans. Geosci. Remote Sens.* **2007**, *45*, 1726–1740. [CrossRef]

34. Robinson, C.; Saatchi, S.; Neumann, M.; Gillespie, T. Impacts of spatial variability on aboveground biomass estimation from L-band radar in a temperate forest. *Remote Sens.* **2013**, *5*, 1001–1023. [CrossRef]

35. Ranson, K.J.; Sun, G. Effects of environmental conditions on boreal forest classification and biomass estimates with SAR. *IEEE Trans. Geosci. Remote Sens.* **2000**, *38*, 1242–1252. [CrossRef]

36. Santos, J.R.; Freitas, C.C.; Araujo, L.S.; Dutra, L.V.; Mura, J.C.; Gama, F.F.; Soler, L.S.; SantAnna, S.J.S. Airborne P-band SAR applied to the aboveground biomass studies in the Brazilian tropical rainforest. *Remote Sens. Environ.* **2003**, *87*, 482–493. [CrossRef]

37. Cartus, O.; Santoro, M.; Kellndorfer, J. Mapping forest aboveground biomass in the Northeastern United States with ALOS PALSAR dual-polarization L-band. *Remote Sens. Environ.* **2012**, *124*, 466–478. [CrossRef]

38. Pulliainen, J.T.; Kurvonen, L.; Hallikainen, M.T. Multitemporal behavior of L- and C-band SAR observations of boreal forests. *IEEE Trans. Geosci. Remote Sens.* **1999**, *37*, 927–937. [CrossRef]

39. Fu, H.; Wang, C.; Zhu, J.; Xie, Q.; Zhao, R. Inversion of vegetation height from PolInSAR using complex least squares adjustment method. *Sci. China Earth Sci.* **2015**, *58*, 1018–1031. [CrossRef]

40. Wang, C.; Wang, L.; Fu, H.; Xie, Q.; Zhu, J. The impact of forest density on forest height inversion modeling from Polarimetric InSAR data. *Remote Sens.* **2016**, *8*, 291. [CrossRef]

41. Fu, H.; Wang, C.; Zhu, J.; Xie, Q.; Zhang, B. Estimation of pine forest height and underlying dem using multi-baseline P-band PolInSAR data. *Remote Sens.* **2017**, *9*, 363. [CrossRef]

42. Xie, Q.; Zhu, J.; Wang, C.; Fu, H.; Lopez-Sanchez, J.M.; Ballester-Berman, J.D. A modified Dual-Baseline PolInSAR method for forest height estimation. *Remote Sens.* **2017**, *9*, 819. [CrossRef]

43. Le Toan, T.; Quegan, S.; Davidson, M.; Balzter, H.; Paillou, P.; Papathanassiou, K.; Plummer, S.; Rocca, F.; Saatchi, S.; Shugart, H.; et al. The BIOMASS mission: Mapping global forest biomass to better understand the terrestrial carbon cycle. *Remote Sens. Environ.* **2011**, *115*, 2850–2860. [CrossRef]

44. Neumann, M.; Saatchi, S.S. Assessing performance of L- and P-band polarimetric interferometric SAR data in estimating boreal forest above-ground biomass. *IEEE Trans. Geosci. Remote Sens.* **2012**, *50*, 714–726. [CrossRef]

45. Hansen, E.H.; Gobakken, T.; Solberg, S.; Kangas, A.; Ene, L.; Mauya, E.; Næssse, E. Relative efficiency of ALS and InSAR for biomass estimation in a Tanzanian Rainforest. *Remote Sens.* **2015**, *7*, 9865–9885. [CrossRef]

46. Pardini, M.; Kugler, F.; Lee, S.K.; Sauer, S.; Torano-Caicoya, A.; Papathanassiou, K. Biomass estimation from forest vertical structure: Potentials and Challenges for multi-baselin Pol-InSAR techniques. In Proceedings of the European Conference on PolInSAR, Rome, Italy, 24–28 January 2011; ESA: Paris, France, 2011.

47. Treuhaft, R.N.; Chapman, B.D.; Dos Santos, J.R.; Goncalves, F.G.; Dutra, L.V.; Graca, P.; Drake, J.B. Vegetation profiles in Tropical forests from multibaseline Interferometric synthetic aperture radar, field and lidar measurements. *J. Geophys. Res.* **2009**, *114*, D23110. [CrossRef]

48. Treuhaft, R.N.; Goncalves, F.G.; Drake, J.B.; Chapman, B.D.; Dos Santos, J.R.; Dutra, L.V.; Graca, P.M.L.A.; Purcell, G.H. Biomass estimation in a Tropical Wet forest using Fourier transforms of profiles from lidar or interferometric SAR. *Geophys. Res. Lett.* **2010**, *37*, L23403. [CrossRef]

49. Tebaldini, S. Algebraic synthesis of forest scenarios from multibaseline PolInSAR data. *IEEE Trans. Geosci. Remote Sens.* **2009**, *47*, 4132–4144. [CrossRef]

50. Tebaldini, S.; Rocca, F. Multibaseline polarimetric SAR tomography of a Boreal forest at P- and L-bands. *IEEE Trans. Geosci. Remote Sens.* **2012**, *50*, 232–246. [CrossRef]

51. Minh, D.H.T.; Tebaldini, S.; Rocca, F.; Le Toan, T.; Villard, L.; Dubois-Femandez, P.C. Capabilities of BIOMASS Tomography for Investigating Tropical Forests. *IEEE Trans. Geosci. Remote Sens.* **2015**, *53*, 965–975. [CrossRef]

52. Minh, D.H.T.; Le Toan, T.; Rocca, F.; Tebaldini, S.; Villard, L.; Réjou-Méchain, M.; Phillips, O.L.; Feldpausch, T.R.; Dubois-Fernandez, P.; Scipal, K.; et al. SAR tomography for the retrieval of forest biomass and height: Cross-validation at two tropical forest sites in French Guiana. *Remote Sens. Environ.* **2016**, *175*, 138–147. [CrossRef]

53. Cloude, S.R. Polarization coherence tomography. *Radio Sci.* **2006**, *41*, RS4017. [CrossRef]

54. Cloude, S.R. *Polarisation: Applications in Remote Sensing*; Oxford University Press: New York, NY, USA, 2010.

55. Luo, H.M.; Chen, E.X.; Li, Z.Y.; Cao, C. Forest above ground biomass estimation methodology based on polarization coherence tomography. *J. Remote Sens.* **2011**, *15*, 1138–1155. [CrossRef]
56. Li, W.M.; Chen, E.X.; Li, Z.Y.; Ke, Y.H.; Zhan, W.F. Forest aboveground biomass estimation using polarization coherence tomography and PolSAR segmentation. *Int. J. Remote Sens.* **2015**, *36*, 530–550. [CrossRef]
57. Askne, J.I.H.; Soja, M.J.; Ulander, L.M.H. Biomass estimation in a boreal forest from TanDEM-X data, lidar DTM, and the interferometric water cloud model. *Remote Sens. Environ.* **2017**, *196*, 266–278. [CrossRef]
58. DLR Microwaves and Radar Institute; Swedish Defense Research Agency; Politecnico di Milano POLIMI. BIOSAR 2008: Data Acquisition and Processing Report. 2008. Available online: https://earth.esa.int/c/document_library/get_file?folderId=21020&name=DLFE-903.pdf (accessed on 17 February 2017).
59. Petersson, H. *Biomassafunktioner för Trädfraktioner av Tall, Gran och Björk i Sverige (in Swedish with English Summary)*; Swedish University of Agrivlutural Sciences: Umeå, Sweden, 1999.
60. Kugler, F.; Lee, S.K.; Hajnsek, I. Forest height estimation by means of Pol-InSAR data inversion: The role of the vertical wavenumber. *IEEE Trans. Geosci. Remote Sens.* **2015**, *53*, 5294–5311. [CrossRef]
61. Luo, H.M.; Li, X.W.; Chen, E.X.; Cheng, J.; Cao, C.X. Analysis of forest backscattering characteristics based on polarization coherence tomography. *Sci. China Technol. Sci.* **2010**, *53*, 166–175. [CrossRef]
62. Cloude, S.R.; Papathanassiou, K.P. Three-stage inversion process for polarimetric SAR interferometry. *IEEE Proc. Radar Sonar Navig.* **2003**, *150*, 125–134. [CrossRef]
63. Tabb, M.; Orrey, J.; Flynn, T.; Carande, R. Phase diversity: A decomposition for vegetation parameter estimation using polarimetric SAR interferometry. In Proceedings of the 4th European Synthetic Aperture Radar Conference (EUSAR), Cologne, Germany, 4–6 June 2002; pp. 721–724.
64. Albinet, C.; Borderies, P.; Hamadi, A.; Dubois-Fernandez, P.; Koleck, T.; Angelliaume, S. High-resolution vertical polarimetric imaging of pine forests. *Radio Sci.* **2014**, *49*, 231–241. [CrossRef]
65. Luckman, A.; Baker, J.; Honzák, M.; Lucas, R. Tropical forest biomass density estimation using JERS-1 SAR: Seasonal variation, confidence limits, and application to image mosaics. *Remote Sens. Environ.* **1998**, *63*, 126–139. [CrossRef]
66. O'Brien, R.M. A caution regarding rules of thumb for variance inflation factors. *Qual. Quant.* **2007**, *41*, 673–690. [CrossRef]
67. Ulander, L.M.H.; Sandberg, G.; Soja, M.J. Biomass retrieval algorithm based on P-band biosar experiments of boreal forest. Presented at the IGARSS, Vancouver, BC, Canada, 24–29 July 2011.

Article

Forest Structure Estimation from a UAV-Based Photogrammetric Point Cloud in Managed Temperate Coniferous Forests

Tetsuji Ota [1,*], Miyuki Ogawa [2], Nobuya Mizoue [3], Keiko Fukumoto [2] and Shigejiro Yoshida [3]

[1] Institute of Decision Science for a Sustainable Society, Kyushu University, 6-10-1 Hakozaki, Fukuoka 812-8581, Japan

[2] Graduate School of Bioresource and Bioenvironmental Sciences, Kyushu University, 6-10-1 Hakozaki, Higashi-ku, Fukuoka 812-8581, Japan; myyyy91@yahoo.co.jp (M.O.); aobozuholic@gmail.com (K.F.)

[3] Faculty of Agriculture, Kyushu University, 6-10-1 Hakozaki, Fukuoka 812-8581, Japan; mizoue@agr.kyushu-u.ac.jp (N.M.); syoshida@agr.kyushu-u.ac.jp (S.Y.)

* Correspondence: chochoji1983@gmail.com; Tel.: +81-92-642-2868

Received: 18 August 2017; Accepted: 6 September 2017; Published: 13 September 2017

Abstract: Here, we investigated the capabilities of a lightweight unmanned aerial vehicle (UAV) photogrammetric point cloud for estimating forest biophysical properties in managed temperate coniferous forests in Japan, and the importance of spectral information for the estimation. We estimated four biophysical properties: stand volume (V), Lorey's mean height (H_L), mean height (H_A), and max height (H_M). We developed three independent variable sets, which included a height variable, a spectral variable, and a combined height and spectral variable. The addition of a dominant tree type to the above data sets was also tested. The model including a height variable and dominant tree type was the best for all biophysical property estimations. The root-mean-square errors (RMSEs) for the best model for V, H_L, H_A, and H_M, were 118.30, 1.13, 1.24, and 1.24, respectively. The model including a height variable alone yielded the second highest accuracy. The respective RMSEs were 131.74, 1.21, 1.31, and 1.32. The model including a spectral variable alone yielded much lower estimation accuracy than that including a height variable. Thus, a lightweight UAV photogrammetric point cloud could accurately estimate forest biophysical properties, and a spectral variable was not necessarily required for the estimation. The dominant tree type improved estimation accuracy.

Keywords: managed temperate coniferous forests; point cloud; spectral information; structure from motion (SfM); unmanned aerial vehicle (UAV)

---

1. Introduction

Up-to-date and spatially detailed information on forest biophysical properties is fundamental to allowing managers to ensure sustainable forest management [1]. Thus, a methodology that captures the spatial and periodic information of forest biophysical properties is required. Remote sensing is an important option for capturing the spatio-temporal information of forest biophysical properties. Measuring three-dimensional (3D) forest structure as a point cloud is an established way to capture the information.

Airborne light detection and ranging (Lidar) is an active remote sensing system that directly measures 3D structures by emitting laser pulses from an aircraft-borne sensor. Because the emitted laser pulses can reach the ground by penetrating a dense forest canopy, airborne Lidar can provide terrain height, as well as a point cloud. The relative height between the point cloud and the local terrain height is well suited for measuring stand-level forest biophysical properties, including stand volume [2,3], and tree height [2,4]. Currently, airborne Lidar is the most accurate remote sensing system for obtaining specific stand-level forest biophysical properties [1,5]. However, because of cost limitations, repeat surveys using airborne Lidar data are often difficult [6]. Therefore, we need alternative approaches to obtain forest biophysical properties.

Digital aerial photographs are an alternative option to generate a point cloud. Recent advances in computer science make it possible to generate a point cloud semi-automatically using the Structure from Motion (SfM) approach [7,8]. Unlike airborne Lidar, digital aerial photographs cannot provide terrain height information, but they can provide the point cloud of the upper canopy surface [9], especially under dense forest canopy conditions. However, the relative height between the point cloud derived from digital aerial photographs and the terrain height provided by another data source, such as airborne Lidar, can be used to estimate forest biophysical properties [10–13]. Thus, periodic acquisitions of digital aerial photographs may be a practical option in areas where accurate terrain height information is available (e.g., a digital terrain model (DTM) derived from an existing Airborne Lidar dataset).

Lightweight unmanned aerial vehicles (UAVs) may be a suitable platform for acquiring digital aerial photographs for small areas at low cost [14,15]. Because the material and operational cost of lightweight UAVs is low [16], they can acquire digital photographs at a lower cost than a manned aerial vehicle with increased spatial and temporal resolution. If we can estimate forest biophysical properties from the lightweight UAV photogrammetric point cloud as accurately as a point cloud derived from digital aerial photographs acquired by a manned aerial vehicle, the former may become an alternative option to measure forest biophysical properties. At present, the digital images derived from lightweight UAVs may differ from those derived from manned airborne vehicles. In particular, lightweight UAVs are sometimes equipped with consumer-grade digital cameras and inexpensive global navigation satellite system (GNSS), which may lead to distortions or positioning errors, respectively [17,18].

Several studies have assessed the value of UAV photogrammetric point clouds in predicting biophysical properties [19–24]. However, these focused mainly on dominant tree heights (e.g., [25]) or individual tree height (e.g., [19–21]). Relatively few studies have evaluated the accuracy of predicting other stand-level forest biophysical properties, such as stand volume. In a limited study, Puliti et al. [17] demonstrated that the use of a UAV can provide accurate forest characteristics in boreal coniferous forests. Thus, more experiments should be conducted in a variety of forest types to determine the validity of the defined parameters and the accuracy of the reported data [17].

Compared with airborne Lidar, digital aerial photography has the advantage that multispectral information is automatically captured, but digital aerial photographs have a disadvantage that terrain height cannot be acquired under dense forest canopy conditions. The spectral information data provide detailed information on the 3D structural change of the forest canopy [22], and previous studies using airborne Lidar suggested that adding multispectral or hyperspectral data improved the estimation of biophysical properties [26–28]. Thus, spectral information can improve the accuracy of biophysical property estimations when both point cloud and spectral information derived from a UAV are used. However, Puliti et al. [17] showed that adding spectral information into a biophysical property estimation using a UAV photogrammetric point cloud resulted in a limited improvement. One reason for this limited improvement was that the UAV photographs in the study were acquired in late fall, which may result in a low spectral response from vegetation [17]. Thus, further research is needed to evaluate the importance of spectral information when using a lightweight UAV equipped with a consumer-grade digital camera.

In this study, we investigated the capabilities of the lightweight UAV photogrammetric point cloud using the SfM approach to estimate forest biophysical properties in managed temperate coniferous forests. We estimated four biophysical properties—stand volume (V), Lorey's mean height (H_L), mean height (H_A), and max height (H_M)—using variables derived from a UAV photogrammetric point cloud. For the estimation, six independent variable sets, which included height variables alone, spectral variables alone, and a combination of height and spectral variables, were compared.

Finally, we investigated the capability of the UAV photogrammetric point cloud and the importance of spectral information for estimating biophysical properties.

2. Materials and Methods

2.1. Study Area

The study was conducted in a temperate forest area (131°32′3″ E, 33°6′51″ N) in Oita prefecture located in southwestern Japan (Figure 1). The area is dominated by plantations of evergreen conifers trees, including Sugi (*Cryptomeria japonica*) and Hinoki (*Chamaecyparis obtusa*). The stand age of the coniferous forests ranges from eight to 62 years. The elevation ranges from 520 to 775 m above sea level.

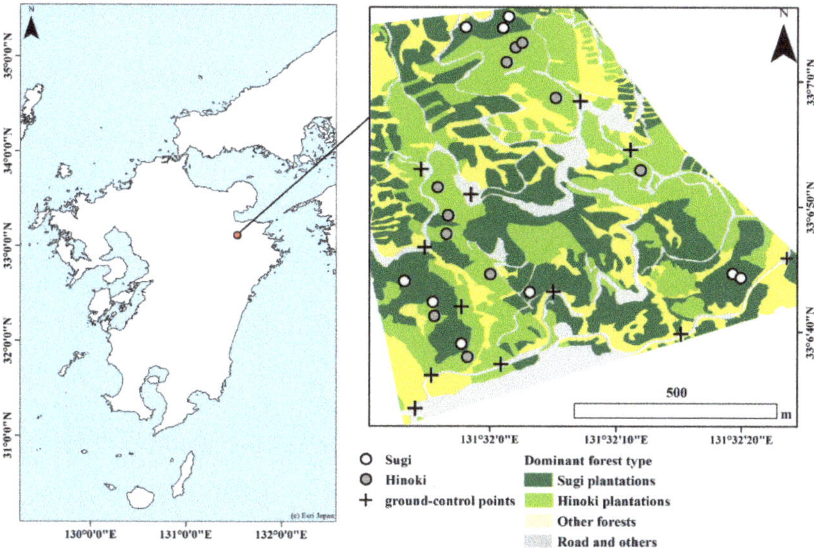

Figure 1. Location of the study area. The municipal boundaries provided by ESRI Japan were used as the country border. The forest cover map was created by visual interpretation based on aerial photographs acquired by a lightweight unmanned aerial vehicle (UAV).

2.2. Field Measurements

Field measurements in Sugi-dominated stands and Hinoki-dominated stands were conducted at previously established permanent plots as a part of different ongoing field studies. Since we used already established plots, there were two sizes of rectangular permanent plots: 400 m^2 (20 m × 20 m) and 225 m^2 (15 m × 15 m). The two Hinoki plots were 225 m^2, and others were 400 m^2. There were 9 and 11 plots in the Sugi and Hinoki stands, respectively. Field data were collected between September 2016 and October 2016. Within each plot, diameter at breast height (DBH) and tree height for all trees with DBH >5 cm were measured. In total, 1197 trees were measured. The plot locations were recorded with a GNSS (MobileMapper 120, Spectra Precision, Westminster, CO, USA). It should be noted that we could not measure the heights of 12 trees because they were inclined from the perpendicular. For trees without a height measurement, we estimated their heights using the Näslund equation [29].

For the UAV photograph survey, 11 SfM ground-control points (GCPs) were distributed across the survey area prior to the acquisition of UAV photographs (Figure 1). Each GCP was approximately 55 × 65 cm with a red cross marked on a white sheet. The GCP locations were also acquired by MobileMapper 120, which is a differential GNSS providing sub-meter accuracy after post-processing.

We calculated the stem volume for each measured tree using tree volume equations developed by the Forest Agency, Japan [30]. Then, V for each plot was calculated by summing the volume of each tree and dividing by the plot size. We also derived H_L, H_A, and H_M for each plot from field data (Table 1).

The plot and GCP location data were post-processed using the nearest GNSS-based control stations constructed by the Geospatial Information Authority of Japan (GSI) with MobileMapper Office 4.6 (Spectra Precision, Westminster, CO, USA).

Table 1. Summary of the biophysical properties, which were stand volume (V), Lorey's mean height (H_L), mean height (H_A), and max height (H_M), calculated from field measurements (SD indicates the standard deviation).

Dominant Tree Type	The Number of Plots	V (m^3/ha)		H_L (m)		H_A (m)		H_M (m)	
		Mean	SD	Mean	SD	Mean	SD	Mean	SD
Sugi	9	712.37	142.20	18.67	2.06	18.17	2.07	21.91	1.98
Hinoki	11	491.75	249.36	15.67	5.25	15.21	5.13	18.50	5.92

2.3. Remote Sensing Data

UAV photographs were acquired under leaf-on canopy conditions between September and October 2016, which corresponded to late summer and early fall. A Phantom 4 UAV was used to acquire the photographs. The UAV has an integrated camera, with a 1/2.3 CMOS sensor that can capture red–green–blue (RGB) spectral information. The UAV was operated manually with visual confirmation in accordance with Japanese laws. The flying altitude was between 70 and 110 m above ground level, and the flying speed was approximately 2.5 m/s. The average ground sampling distance was 4.3 cm. The aerial photographs were captured at 5-s intervals to achieve an overlap of more than 80%. It should be noted that an additional UAV photograph acquisition flight was conducted in January 2017, which corresponded to winter conditions, to add the rightmost GCP in Figure 1. Since there were overlaps between the first and additional acquisition, we concurrently created a photogrammetric point cloud. While the acquisition was conducted during the winter season, the acquired photographs did not include any plots. Thus, the photographs were not used for the biophysical parameters estimation but used just for adding a GCP.

2.4. Remote Sensing Data

2.4.1. Processing of the UAV Photographs

The UAV photographs were processed using PhotoScan Professional version 1.2.6 ([31]; Petersburg, Russia) to generate a photogrammetric point cloud that includes both height and spectral information (i.e., red, green, and blue). PhotoScan Professional is a commercial software that uses the SfM approach to generate a 3D reconstruction from a collection of overlapping photographs, which can provide a dense and accurate 3D point cloud [17]. Briefly, the workflow composed of the "Align photos" stage and the "Build a dense point cloud" stage. "Align photos" is the stage at which camera location and orientation, and internal parameters are determined [32]. We selected "High accuracy" and "Reference" as settings for "Accuracy" and "Pair preselection", respectively. Eleven GCPs were used to improve the accuracy of the "Align photos" stage. We imported the post-processed GCP location data into PhotoScan Professional and manually identified each GCP within the UAV photographs. The overall RMSE of GCP matching process was 2.94 m (Table 2). The "Build a dense point cloud" stage generates dense 3D, point cloud data based on the camera location and orientation, and internal parameters determined by the "Align photos" stage. We selected "Medium quality" and "Mild" as settings for "Quality" and "Depth filtering", respectively, following previous studies [17].

We constructed a 1-m resolution grid of a digital surface model (DSM) and for the spectral data from the generated 3D point cloud data. The DSM was created by assigning the highest value of the 3D point cloud for each grid cell. The 1-m resolution grid of the spectral data of each band (i.e., red, green, and blue) was created by assigning the average value of the 3D point cloud. Then, the three band values of each grid were normalized radiometrically by dividing each band value in the grid by the sum of the values of all bands in the corresponding grid, as described in previous studies [17,33,34].

Table 2. The errors of ground-control points.

Ground-Control Points	X Error (m)	Y Error (m)	Z Error (m)	Total Error (m)
1	1.15	5.19	−1.42	5.50
2	0.48	0.26	−0.07	0.55
3	−3.26	−1.08	−3.87	5.17
4	0.92	−0.15	0.44	1.03
5	2.19	−0.54	−0.26	2.27
6	−1.01	−0.81	0.15	1.30
7	0.81	−1.89	−0.85	2.23
8	−0.48	−0.63	0.90	1.20
9	1.16	−0.30	−0.01	1.20
10	−1.53	1.68	1.87	2.94
11	−0.49	−1.77	3.14	3.64
RMSE	1.47	1.89	1.71	2.94

2.4.2. Calculation of a Canopy Height Model (CHM) and Variable Extractions

A CHM was calculated as the relative height by subtracting the terrain height from each grid value of the DSM. We used a digital terrain model (DTM) developed by the Geospatial Information Authority of Japan as the terrain height. The DTM consists of 5-m spatial resolution data derived from airborne Lidar. The vertical accuracy is expressed as the standard deviation within 2 m (GSI, 2014). Because the spatial resolution of the DTM was 5 m, we interpolated the 5-m resolution data into 1-m data using the inverse distance weighted interpolation method.

Then, 13 variables derived from the CHM were calculated within each field plot, including mean canopy height (h_{mean}, the average value of the relative height), deciles of the height percentiles (h_{10}, h_{20}, ..., h_{100}), and standard deviation (h_{sd}). In addition to the height variables, we calculated the means ($RGB_{\mathrm{R, mean}}$, $RGB_{\mathrm{G, mean}}$, and $RGB_{\mathrm{R, mean}}$) and standard deviations ($RGB_{\mathrm{R, sd}}$, $RGB_{\mathrm{G, sd}}$, and $RGB_{\mathrm{R, sd}}$) of the spectral values.

2.4.3. Statistical Analysis

A regression model approach was applied to develop the biophysical properties estimation model following previous studies [10,35]:

$$F = \beta_0 h^{\beta_1} RGB^{\beta_2} e^{bz} \tag{1}$$

where F is the biophysical property calculated from the field data; β_0, β_1, and β_2 are the regression coefficients; h is the height variable (i.e., H_{mean}, h_{10}, ..., h_{100}, h_{sd}); RGB is the spectral variable; b is the regression coefficient of a dummy variable; and z is the dummy variable. The dummy variable was used to assess the influence of dominant tree type (d_{type}) on the regression. The dummy variable (i.e., z) has a value of 1 for Hinoki-dominated stands. For the Sugi-dominated stands, the dummy variable has a value of 0. To avoid problems of collinearity, we did not include more than one variable for each height and spectral variable, while we calculated 13 variables (i.e., H_{mean}, h_{10}, ..., h_{100}, h_{sd}) and six variables (i.e., $RGB_{\mathrm{R, mean}}$, $RGB_{\mathrm{B, mean}}$, ..., $RGB_{\mathrm{R, sd}}$) for the height and spectral properties, respectively. Equation (1) was log transformed to solve the equation as a linear regression using Equation (2):

$$\log F = log\beta_0 + \beta_1 \log(h) + \beta_2 \log(RGB) + bz \tag{2}$$

To evaluate the importance of each height and spectral variable for estimating forest biophysical properties, we initially regressed the biophysical properties against a height variable alone and a spectral variable alone. Then, we regressed the biophysical properties against a combination of a height variable and a spectral variable. Finally, we evaluated the importance of dominant tree type by adding dominant tree type as an independent variable, resulting in a combination of a height variable and dominant tree type, a spectral variable and dominant tree type, and a height variable, a spectral variable, and dominant tree type.

The accuracies of the estimates of the best model were validated using a coefficient of determination (R^2), the root-mean-square error (RMSE), a Bayesian information criterion (BIC), and the relative RMSE expressed as a percentage. The relative RMSE was defined as the RMSE divided by the mean value of the field data. Because the number of field plots was limited (i.e., 20), R^2 and RMSE were calculated using a leave-one-out cross-validation. R^2 was calculated using linear regression of the observed versus estimated values. The number of independent variables was the same within each dataset. Thus, for each independent variable dataset, we selected the best model using R^2. For comparisons between different independent variable datasets, we used the BIC. The statistical analysis was conducted using R ver. 3.12 [36].

3. Results

For each biophysical property, the variables derived from the UAV photogrammetric point cloud were regressed using six independent variable sets, and the selected models were summarized (Table 3). When the biophysical properties were regressed against a height variable alone, h_{90} was selected for the models. The R^2 values of the selected models for V, H_L, H_A, and H_M were 0.71, 0.93, 0.91 and 0.93, respectively. The respective RMSEs were 131.74, 1.21, 1.31, and 1.32, respectively. The model that included a spectral variable alone yielded lower estimation accuracy than the one including a height variable, and the R^2 of the selected models for V, H_L, H_A, and H_M, were 0.26, 0.23, 0.21, and 0.26, respectively. The selected model including both a height variable and a spectral variable yielded almost the same R^2 as the model including a height variable. When the biophysical properties were regressed with the addition of the dominant tree type information, in terms of R^2, adjR^2, RMSE, and relative RMSE, the accuracies of the estimation were comparable to those obtained using the respective independent variable sets without the dominant tree type information.

In terms of the BIC, the model that included a height variable with the addition of dominant tree type information was the best model for all the biophysical property estimations within the six independent variable sets. The selected models used h_{90} and dominant tree type for all the biophysical property estimations. The RMSEs for V, H_L, H_A, and H_M were 118.30, 1.13, 1.24, and 1.24, respectively. The relative RMSEs of the selected models for H_L, H_A, and H_M were between 6.17 and 7.50. The relative RMSE of V, which was 20.02%, was larger than that of the other biophysical properties. Figure 2 shows a scatterplot of the observed versus predicted values for each biophysical property using the best models. The regression lines for H_L, H_A, and H_M were almost in line with the 1:1 line.

Table 3. Summary of the selected estimation models for four forest biophysical properties: stand volume (V), Lorey's mean height (H_L), mean height (H_A), and max height (H_M) (bold indicates the best model for each biophysical property).

Dependent Variables	Independent Variable	Selected Variables	R^2	$AdjR^2$	RMSE	Relative RMSE	BIC
	h	h_{90}	0.71	0.70	131.74	22.29	7.14
	RGB	$RGB_{G,\,sd}$	0.26	0.21	291.80	49.37	58.99
V	$h + RGB$	$h_{90},\ RGB_{B,\,sd}$	0.68	0.64	143.15	24.22	9.95
	$h + d_{type}$	**$h_{90},\ d_{type}$**	**0.78**	**0.75**	**118.30**	**20.02**	**2.99**
	$RGB + d_{type}$	$RGB_{B,\,sd},\ d_{type}$	0.20	0.11	303.16	51.29	53.01
	$h + RGB + d_{type}$	$h_{90},\ RGB_{R,\,sd},\ d_{type}$	0.80	0.76	112.97	19.11	3.83

Table 3. *Cont.*

Dependent Variables	Independent Variable	Selected Variables	R^2	$AdjR^2$	RMSE	Relative RMSE	BIC
H_L	h	h_{90}	0.93	0.92	1.21	7.08	−41.14
	RGB	$RGB_{B,\,sd}$	0.23	0.19	4.31	25.29	19.73
	$h + RGB$	$h_{90},\, RGB_{B,\,mean}$	0.94	0.92	1.19	7.00	−39.41
	$\mathbf{h + d_{type}}$	$\mathbf{h_{90},\, d_{type}}$	0.92	0.93	1.13	6.65	−42.56
	$RGB + d_{type}$	$RGB_{B,\,sd},\, d_{type}$	0.90	0.10	4.69	27.57	22.73
	$h + RGB + d_{type}$	$h_{90},\, RGB_{G,\,mean},\, d_{type}$	0.93	0.92	1.15	6.7	−39.71
H_A	h	h_{90}	0.91	0.91	1.31	7.92	−35.96
	RGB	$RGB_{B,\,sd}$	0.21	0.16	4.30	25.97	21.12
	$h + RGB$	$h_{90},\, RGB_{B,\,mean}$	0.92	0.91	1.25	7.53	−35.26
	$\mathbf{h + d_{type}}$	$\mathbf{h_{70},\, d_{type}}$	**0.90**	**0.91**	**1.24**	**7.50**	**−34.12**
	$RGB + d_{type}$	$RGB_{B,\,sd},\, d_{type}$	0.89	0.08	4.62	27.96	24.11
	$h + RGB + d_{type}$	$h_{90},\, RGB_{B,\,mean},\, d_{type}$	0.92	0.91	1.24	7.51	−34.29
H_M	h	h_{90}	0.93	0.92	1.32	6.61	−42.74
	RGB	$RGB_{B,\,sd}$	0.26	0.22	4.62	23.05	14.92
	$h + RGB$	$h_{90},\, RGB_{R,\,mean}$	0.94	0.92	1.32	6.57	−40.37
	$\mathbf{h + d_{type}}$	$\mathbf{h_{90},\, d_{type}}$	**0.92**	**0.93**	**1.24**	**6.17**	**−44.89**
	$RGB + d_{type}$	$RGB_{B,\,sd},\, d_{type}$	0.90	0.13	5.04	25.17	17.91
	$h + RGB + d_{type}$	$h_{90},\, RGB_{R,\,mean},\, d_{type}$	0.93	0.94	1.13	5.63	−44.30

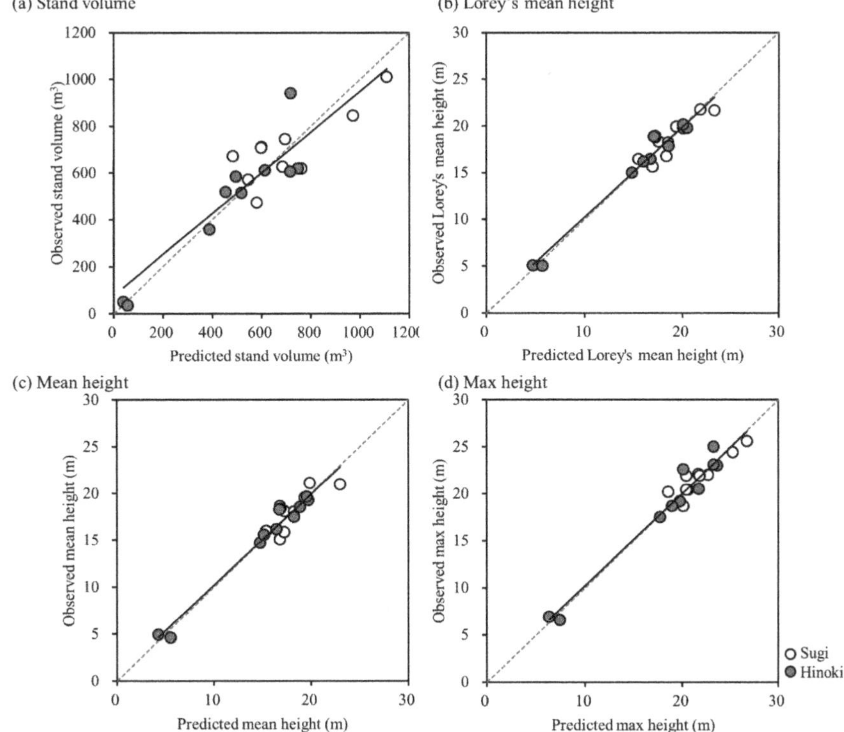

Figure 2. Observed biophysical properties versus predicted biophysical properties from the best model of each of the canopy height models. A diagonal dotted line and a solid line indicate the 1:1 line and regression line between the predicted biophysical properties and observed biophysical properties, respectively.

4. Discussion

In this study, we assessed the accuracy of forest biophysical property estimation using a lightweight UAV photogrammetric point cloud in managed temperate coniferous forests. Previous studies assessed the accuracy of the estimation in temperate broadleaved forests and conifer-dominated boreal forests. These studies focused especially on tree heights such as dominant tree height and Lorey's mean height (e.g., [17,25]). They showed relative RMSEs of 8.4% in temperate broadleaved forests [25] and 13.3% in conifer-dominated boreal forests [17]. Studies focusing on stand volume or aboveground biomass estimation are rare. Puliti et al. [17] fitted variables derived from a UAV photogrammetric point cloud for V and H_L in conifer-dominated boreal forests in Norway. The study showed a relative RMSE of 15.0%. Kachamba et al. [37] similarly evaluated the aboveground biomass estimation accuracy using a UAV photogrammetric point cloud. They showed a relative RMSE of 46.7%. In the present study, R^2 values were between 0.72 and 0.93, and the relative RMSEs were between 6.92% and 22.84% in managed temperate coniferous forests. Thus, we conclude that the estimation results in managed temperate coniferous forests are comparable to those of conifer-dominated boreal forests and superior to those of dry tropical forests. One reason why our results are superior to those of dry tropical forests is that Kachamba et al. [37] did not use the terrain height data derived from airborne Lidar, but data generated by filtering a UAV photogrammetric point cloud.

The models including a spectral variable alone yielded a lower model accuracy than the models including a height variable alone for all the biophysical properties. The accuracy of the models including both a height variable and a spectral variable also yielded lower accuracy than that of the model including a height variable alone. Thus, we conclude that the spectral information did not improve the forest biophysical property estimations. Spectral information is often used for tree species recognition in forest inventory applications using 3D point clouds, since height information alone typically does not identify different tree species [38]. However, previous studies, which used a point cloud derived from airborne Lidar or digital aerial photographs, suggested that spectral information helped to refine the biophysical property estimation [17,26–28]. In contrast, the present study showed that adding spectral information did not improve forest biophysical property estimations using a lightweight UAV photogrammetric point cloud. One possible reason is that we used the consumer-grade digital cameras and, therefore, spectral variables from only the visible range were used. It is important to note that, in our study, a spectral variable can estimate forest biophysical properties, but the R^2 was approximately 0.2. One advantage of using a spectral variable is that terrain height information is not required for forest biophysical property estimations. Because we need accurate terrain height information to use height variables for forest biophysical property estimations, using a height variable may be difficult in areas, such as tropical forests, where accurate terrain height information cannot be obtained [37]. Thus, spectral information may become an important estimator. Further research is required to explore the importance of spectral information in these regions.

The model that includes a height variable and dominant tree type yielded the lowest BIC for all the biophysical property estimations. Thus, we conclude that adding dominant tree type improved the accuracy of forest biophysical property estimations. However, it should be noted that the improvement was limited in terms of R^2 and RMSE, compared with the model that included a height variable alone. Thus, we also conclude that the dominant tree type is not necessarily required because of the limited improvement, although this information did improve model accuracy. The importance of adding forest or dominant tree type information has been examined previously, especially for estimating aboveground biomass using airborne Lidar. These studies showed that adding forest or dominant tree type information has a limited effect on improving the estimation accuracy in boreal coniferous [39] and tropical forests [35]. While studies that examined the importance of adding forest or dominant tree type information using a photogrammetric point cloud are limited, Ota et al. [10] also obtained similar results using a manned aircraft-derived, photogrammetric point cloud in tropical forests. We confirmed that the UAV photogrammetric point cloud can also estimate forest biophysical properties independently of dominant tree type in managed coniferous forests. This becomes a strong advantage when using point cloud-based forest biophysical property estimations in areas where several forest types occur in mixed communities because of the simplified calibration process and forest inventory [10].

5. Conclusions

The main conclusion of this study is that a UAV photogrammetric point cloud can accurately estimate forest biophysical properties, including V, H_L, H_A, and H_M, in managed temperate coniferous forests. For the estimations, a height variable alone is required, while a spectral variable is not necessarily required. The dominant tree type, if it is available, improves the estimation accuracy. While a spectral variable yielded lower estimation accuracy, it may be informative in areas where accurate terrain height is not available. Thus, further study is required to explore the importance of spectral information in these areas.

Acknowledgments: This study was supported by JSPS KAKENHI (grant number 16K18721) and grants from the Project of the NARO Bio-oriented Technology Research Advancement Institution (the special scheme project on regional developing strategy). We thank Leonie Seabrook, from Edanz Group (www.edanzediting.com/ac) for editing the revised draft of this manuscript.

Author Contributions: All authors contributed extensively to the work. Tetsuji Ota and Miyuki Ogawa performed the processing and wrote the manuscript. Nobuya Mizoue, Keiko Fukumoto, and Shigejiro Yoshida supervised the research work, reviewed the manuscript, and provided comments and suggestions to improve the manuscript.

Conflicts of Interest: The authors declare no conflict of interest.

References

1. White, J.C.; Coops, N.C.; Wulder, M.A.; Vastaranta, M.; Hilker, T.; Tompalski, P. Remote Sensing Technologies for Enhancing Forest Inventories: A Review. *Can. J. Remote Sens.* **2016**, *42*, 619–641. [CrossRef]
2. Næsset, E. Airborne laser scanning as a method in operational forest inventory: Status of accuracy assessments accomplished in Scandinavia. *Scand. J. For. Res.* **2007**, *22*, 433–442. [CrossRef]
3. Ioki, K.; Imanishi, J.; Sasaki, T.; Morimoto, Y.; Kitada, K. Estimating stand volume in broad-leaved forest using discrete-return LiDAR: Plot-based approach. *Landsc. Ecol. Eng.* **2010**, *6*, 29–36. [CrossRef]
4. Næsset, E.; Økland, T. Estimating tree height and tree crown properties using airborne scanning laser in a boreal nature reserve. *Remote Sens. Environ.* **2002**, *79*, 105–115. [CrossRef]
5. Wulder, M.A.; Bater, C.W.; Coops, N.C.; Hilker, T.; White, J.C. The role of LiDAR in sustainable forest management. *For. Chron.* **2008**, *84*. [CrossRef]
6. Hird, N.J.; Montaghi, A.; McDermid, J.G.; Kariyeva, J.; Moorman, J.B.; Nielsen, E.S.; McIntosh, C.A. Use of Unmanned Aerial Vehicles for Monitoring Recovery of Forest Vegetation on Petroleum Well Sites. *Remote Sens.* **2017**, *9*. [CrossRef]
7. Fonstad, M.A.; Dietrich, J.T.; Courville, B.C.; Jensen, J.L.; Carbonneau, P.E. Topographic structure from motion: A new development in photogrammetric measurement. *Earth Surf. Process. Landf.* **2013**, *38*, 421–430. [CrossRef]
8. Snavely, N.; Seitz, S.M.; Szeliski, R. Modeling the world from Internet photo collections. *Int. J. Comput. Vis.* **2008**, *80*, 189–210. [CrossRef]
9. White, J.C.; Wulder, M.A.; Vastaranta, M.; Coops, N.C.; Pitt, D.; Woods, M. The utility of image-based point clouds for forest inventory: A comparison with airborne laser scanning. *Forests* **2013**, *4*, 518–536. [CrossRef]
10. Ota, T.; Ogawa, M.; Shimizu, K.; Kajisa, T.; Mizoue, N.; Yoshida, S.; Takao, G.; Hirata, Y.; Furuya, N.; Sano, T.; et al. Aboveground Biomass Estimation Using Structure from Motion Approach with Aerial Photographs in a Seasonal Tropical Forest. *Forests* **2015**, *6*, 3882–3898. [CrossRef]
11. Bohlin, J.; Wallerman, J.; Fransson, J.E.S. Forest variable estimation using photogrammetric matching of digital aerial images in combination with a high-resolution DEM. *Scand. J. For. Res.* **2012**, *27*, 692–699. [CrossRef]
12. Balenović, I.; Simic Milas, A.; Marjanović, H. A Comparison of Stand-Level Volume Estimates from Image-Based Canopy Height Models of Different Spatial Resolutions. *Remote Sens.* **2017**, *9*, 205. [CrossRef]

13. Vastaranta, M.; Wulder, M.A.; White, J.C.; Pekkarinen, A.; Tuominen, S.; Ginzler, C.; Kankare, V.; Holopainen, M.; Hyyppä, J.; Hyyppä, H. Airborne laser scanning and digital stereo imagery measures of forest structure: Comparative results and implications to forest mapping and inventory update. *Can. J. Remote Sens.* **2013**, *39*, 382–395. [CrossRef]

14. Matese, A.; Toscano, P.; Di Gennaro, S.F.; Genesio, L.; Vaccari, F.P.; Primicerio, J.; Belli, C.; Zaldei, A.; Bianconi, R.; Gioli, B. Intercomparison of UAV, Aircraft and Satellite Remote Sensing Platforms for Precision Viticulture. *Remote Sens.* **2015**, *7*, 2971–2990. [CrossRef]

15. Salamí, E.; Barrado, C.; Pastor, E. UAV Flight Experiments Applied to the Remote Sensing of Vegetated Areas. *Remote Sens.* **2014**, *6*, 11051–11081. [CrossRef]

16. Tang, L.; Shao, G. Drone remote sensing for forestry research and practices. *J. For. Res.* **2015**, *26*, 791–797. [CrossRef]

17. Puliti, S.; Ørka, H.O.; Gobakken, T.; Næsset, E. Inventory of small forest areas using an unmanned aerial system. *Remote Sens.* **2015**, *7*, 9632–9654. [CrossRef]

18. Torresan, C.; Berton, A.; Carotenuto, F.; Di Gennaro, S.F.; Gioli, B.; Matese, A.; Miglietta, F.; Vagnoli, C.; Zaldei, A.; Wallace, L. Forestry applications of UAVs in Europe: A review. *Int. J. Remote Sens.* **2016**, *38*, 2427–2447.

19. Goodbody, T.R.H.; Coops, N.C.; Marshall, P.L.; Tompalski, P.; Crawford, P. Unmanned aerial systems for precision forest inventory purposes: A review and case study. *For. Chron.* **2017**, *93*, 71–81. [CrossRef]

20. Wallace, L.; Lucieer, A.; Malenovský, Z.; Turner, D.; Vopěnka, P. Assessment of Forest Structure Using Two UAV Techniques: A Comparison of Airborne Laser Scanning and Structure from Motion (SfM) Point Clouds. *Forests* **2016**, *7*, 62. [CrossRef]

21. Birdal, A.C.; Avdan, U.; Türk, T. Estimating tree heights with images from an unmanned aerial vehicle. *Geomat. Nat. Hazards Risk* **2017**, 1–13. [CrossRef]

22. Dandois, J.P.; Ellis, E.C. High spatial resolution three-dimensional mapping of vegetation spectral dynamics using computer vision. *Remote Sens. Environ.* **2013**, *136*, 259–276. [CrossRef]

23. Zarco-Tejada, P.J.; Diaz-Varela, R.; Angileri, V.; Loudjani, P. Tree height quantification using very high resolution imagery acquired from an unmanned aerial vehicle (UAV) and automatic 3D photo-reconstruction methods. *Eur. J. Agron.* **2014**, *55*, 89–99. [CrossRef]

24. Dandois, J.; Olano, M.; Ellis, E. Optimal Altitude, Overlap, and Weather Conditions for Computer Vision UAV Estimates of Forest Structure. *Remote Sens.* **2015**, *7*, 13895–13920. [CrossRef]

25. Lisein, J.; Pierrot-Deseilligny, M.; Bonnet, S.; Lejeune, P. A photogrammetric workflow for the creation of a forest canopy height model from small unmanned aerial system imagery. *Forests* **2013**, *4*, 922–944. [CrossRef]

26. Tonolli, S.; Dalponte, M.; Neteler, M.; Rodeghiero, M.; Vescovo, L.; Gianelle, D. Fusion of airborne LiDAR and satellite multispectral data for the estimation of timber volume in the Southern Alps. *Remote Sens. Environ.* **2011**, *115*, 2486–2498. [CrossRef]

27. Popescu, S.C.; Wynne, R.H.; Scrivani, J.A. Fusion of small-footprint lidar and multispectral data to estimate plot-level volume and biomass in deciduous and pine forests in Virginia, USA. *For. Sci.* **2004**, *50*, 551–565.

28. Luo, S.; Wang, C.; Xi, X.; Pan, F.; Peng, D.; Zou, J.; Nie, S.; Qin, H. Fusion of airborne LiDAR data and hyperspectral imagery for aboveground and belowground forest biomass estimation. *Ecol. Indic.* **2017**, *73*, 378–387. [CrossRef]

29. Näslund, M. Skogsforsöksastaltens gallringsforsök i tallskog. *Medd. Statens Skogsforsöksanstalt* **1936**, *29*, 1–169.

30. Forest Agency of Japan Timber volume table (Western Japan). Forestry Investigation Committee: Tokyo, Japan, 1970.

31. Agisoft PhotoScan. Available online: http://www.agisoft.com (accessed on 22 May 2017).

32. Agisoft, L.L.C. *Agisoft Photoscan User Manual*; Professional Edition, Version 1.2; Agisoft: St. Petersburg, Russia, 2016; Available online: http://www.agisoft.com/pdf/photoscan-pro_1_2_en.pdf (accessed on 6 July 2017).

33. Dalponte, M.; Ørka, H.O.; Ene, L.T.; Gobakken, T.; Næsset, E. Tree crown delineation and tree species classification in boreal forests using hyperspectral and ALS data. *Remote Sens. Environ.* **2014**, *140*, 306–317. [CrossRef]

34. Näsi, R.; Honkavaara, E.; Lyytikäinen-Saarenmaa, P.; Blomqvist, M.; Litkey, P.; Hakala, T.; Viljanen, N.; Kantola, T.; Tanhuanpää, T.; Holopainen, M. Using UAV-based photogrammetry and hyperspectral imaging for mapping bark beetle damage at tree-level. *Remote Sens.* **2015**, *7*, 15467–15493. [CrossRef]

35. Ota, T.; Kajisa, T.; Mizoue, N.; Yoshida, S.; Takao, G.; Hirata, Y.; Furuya, N.; Sano, T.; Ponce-Hernandez, R.; Ahmed, O.S.; et al. Estimating aboveground carbon using airborne LiDAR in Cambodian tropical seasonal forests for REDD+ implementation. *J. For. Res.* **2015**, *20*, 484–492. [CrossRef]

36. R Core Team R: A Language and Environment for Statistical Computing. Available online: https://www.r-project.org/ (accessed on 26 May 2017).

37. Kachamba, J.D.; Ørka, O.H.; Gobakken, T.; Eid, T.; Mwase, W. Biomass Estimation Using 3D Data from Unmanned Aerial Vehicle Imagery in a Tropical Woodland. *Remote Sens.* **2016**, *8*. [CrossRef]

38. Dalponte, M.; Bruzzone, L.; Gianelle, D. Tree species classification in the Southern Alps based on the fusion of very high geometrical resolution multispectral/hyperspectral images and LiDAR data. *Remote Sens. Environ.* **2012**, *123*, 258–270. [CrossRef]

39. Næsset, E. Estimation of above-and below-ground biomass in boreal forest ecosystems. In *International Society of Photogrammetry and Remote Sensing. International Archives of Photogrammetry, Remote Sensing and Spatial Information Sciences*; Thies, M., Kock, B., Spiecker, H., Weinacker, H., Eds.; International Society of Photogrammetry and Remote Sensing (ISPRS): Freiburg, Germany, 2004; pp. 145–148.

 *forests*

Article

Individual Tree Detection from Unmanned Aerial Vehicle (UAV) Derived Canopy Height Model in an Open Canopy Mixed Conifer Forest

Midhun Mohan [1,2,*], Carlos Alberto Silva [3], Carine Klauberg [4], Prahlad Jat [5], Glenn Catts [1], Adrián Cardil [6], Andrew Thomas Hudak [4] and Mahendra Dia [7]

1   Department of Forestry and Environmental Resources, North Carolina State University, 2800 Faucette Drive, Raleigh, NC 27695, USA; glenn_catts@ncsu.edu

2   Department of Operations Research, North Carolina State University, 2310 Stinson Drive, Raleigh, NC 27695, USA

3   Department of Natural Resources and Society, University of Idaho, 708 South Deakin Street, Moscow, ID 83844, USA; carlos_engflorestal@outlook.com

4   USDA Forest Service Rocky Mountain Research Station, Forestry Sciences Laboratory, 1221 South Main Street, Moscow, ID 83843, USA; carine_klauberg@hotmail.com (C.K.); ahudak@fs.fed.us (A.T.H.)

5   Department of Environmental Sciences and Engineering, UNC-Chapel Hill, 135 Dauer Drive, Chapel Hill, NC 27599, USA; jat@email.unc.edu

6   Tecnosylva, Parque Tecnológico de León, 24009 León, Spain; adriancardil@gmail.com

7   Department of Horticultural Sciences, North Carolina State University, 2721 Founders Drive, Raleigh, NC 27695, USA; mdia@ncsu.edu

*   Correspondence: mmohan2@ncsu.edu; Tel.: +1-919-771-9348

Received: 28 July 2017; Accepted: 8 September 2017; Published: 11 September 2017

Abstract: Advances in Unmanned Aerial Vehicle (UAV) technology and data processing capabilities have made it feasible to obtain high-resolution imagery and three dimensional (3D) data which can be used for forest monitoring and assessing tree attributes. This study evaluates the applicability of low consumer grade cameras attached to UAVs and structure-from-motion (SfM) algorithm for automatic individual tree detection (ITD) using a local-maxima based algorithm on UAV-derived Canopy Height Models (CHMs). This study was conducted in a private forest at Cache Creek located east of Jackson city, Wyoming. Based on the UAV-imagery, we allocated 30 field plots of 20 m × 20 m. For each plot, the number of trees was counted manually using the UAV-derived orthomosaic for reference. A total of 367 reference trees were counted as part of this study and the algorithm detected 312 trees resulting in an accuracy higher than 85% (*F*-score of 0.86). Overall, the algorithm missed 55 trees (omission errors), and falsely detected 46 trees (commission errors) resulting in a total count of 358 trees. We further determined the impact of fixed tree window sizes (FWS) and fixed smoothing window sizes (SWS) on the ITD accuracy, and detected an inverse relationship between tree density and FWS. From our results, it can be concluded that ITD can be performed with an acceptable accuracy (*F* > 0.80) from UAV-derived CHMs in an open canopy forest, and has the potential to supplement future research directed towards estimation of above ground biomass and stem volume from UAV-imagery.

Keywords: structure from motion (SfM); 3D point cloud; remote sensing; local maxima; fixed tree window size

---

1. Introduction

Sustainable forest management requires an understanding of how macroscopic patterns of forests emerge, in a timely and accurate manner, in order to make informed decisions [1,2]. Detailed information on tree-level attributes, such as tree counts, tree heights, crown base heights and diameter at breast height (DBH) are essential for monitoring forest regeneration, quantitative analysis of forest structure and dynamics, and evaluating forest damage [3–6]. However, as several forest study areas are vast and not easily accessible, with a plethora of tree species with varying shapes and sizes, a cost-effective and accurate method to acquire forest attributes such as tree density (tree/ha), and tree characteristics such as height (Ht), basal area (BA), and stem volume (V) are essential to management and conservation activities [7]. Although traditional field surveys can be used to gather detailed information regarding these forest characteristics, they can become uneconomical, time consuming and exhausting, and hence are not ideal for studies dealing with periodic data collection [8,9].

Over the years, remote sensing techniques have been increasingly used for assessing forest resources, both directly and indirectly [10–13]. Aerial photography, light detection and ranging (LiDAR) and airborne multispectral, and hyperspectral images had been perceived as potential tools for observing forest areas and for performing broad-scale analysis of forest systems. These methods have the ability to quantify the composition and structure of the forest at different temporal and geographical scales with the support of various statistical methods, and therefore can supplement forest inventory related expeditions [14–21].

Advances in the fields of the Unmanned Aerial Vehicle (UAV) technology and data processing have broadened the horizons of remote sensing of forestry, and made the acquisition of high-resolution imagery and 3D data more easily available and affordable [22–28]. In fact, UAVs can be obtained at reasonable costs and can be perceived as a forester's eye in the sky, capable of performing forest inventory and analysis on a periodic basis [22,29]. These light-weight machines can be remotely operated from the ground and can fly below cloud cover. With the availability of a wide range of sensors, these UAVs allow the end users to define the spatial resolutions, thereby opening new opportunities to forest managers [30]. In the past decade, studies have focused on exploring the usage of UAV-derived Canopy Height Models (CHMs), suggesting its potential in detection of tree tops, delineation of tree crowns, and subsequently estimation of parameters of crown morphology such as height, diameter, and surface curvature [9,22,23,25,30–32].

Previously, individual trees from UAV imagery were detected using mainly image segmentation using textural features, but photogrammetric 3D point clouds supplanted them with the progression of dense image matching methods and computing power [4,33–39]. For processing large amounts of imaging data, researchers today use so-called structure-from-motion (SfM) and multi-view stereopsis (MVS) techniques, which do not require the information on the 3D position of the camera or the 3D location of multiple control points, unlike traditional digital photogrammetric methods [40–44]. Here, 3D point clouds are generated through the matching of features in multiple images and ground control points (GCPs) are used for geo-referencing and scaling of these point clouds [45]. Nonetheless, very few studies have investigated possibilities of detecting individual tree counts from the CHMs that can be derived from these point clouds [41,46–48].

In the past two decades, LiDAR has become the dominant remote sensing technology for ITD, mainly because it can quickly provide highly accurate and spatially detailed information about forest attributes across an entire forested landscape [7]. There are a variety of approaches used to detect and delineate individual trees from LiDAR-derived CHMs: local maxima detection, valley following (VF), template matching (TM), scale-space (SS) theory, Markov random fields (MRFs), and marked point processes (MPP) [7,35,49–54]. These algorithms, when applied on CHMs derived from aforementioned methodologies, promise efficient ITD performance. For instance, the application of LM algorithm on LiDAR-derived CHMs is a well-known established framework, which has been incorporated in several recent studies and the researchers reported adequate ITD accuracy [7,55]. Nevertheless, these methods were primarily designed for measuring large spaces or objects, and require expensive sensors, well-trained personnel, and precise computational technology to obtain accurate results. Therefore, how to collect high resolution data for individual tree attributes estimation in the case of smaller study areas over time, considering the cost associated with it, is a key challenge [22,23].

As UAV remote sensing techniques are undergoing rapid improvement—along with the availability of high spatial resolution remotely sensed imagery—there is potential for conducting and automating high accuracy forest inventory and analysis in a cost-effective manner [56]. We hypothesize that it is possible to automatically detect individual trees from the UAV-derived CHM with satisfactory accuracy using the LM algorithm, primarily designed for ITD from LiDAR data. In this paper, we aim to address two criteria: (i) Evaluate the applicability of UAV-derived Canopy Height Models for ITD; (ii) Determine the impacts of fixed treetop window size (FWS) and fixed smoothing window size (SWS) on the performance of the local maxima algorithm for ITD.

2. Methods

2.1. Study Area Description

The study area (Figure 1) was comprised of 32 ha of private forest at Cache Creek located east of Jackson city, in the state of Wyoming, with an elevation ranging from 1950 m to 2100 m. The open canopy forest covered about 80% of the total study site and was comprised of a variety of four tree species which included Lodgepole pine (*Pinus contorta* Douglas), Engelmann spruce (*Picea engelmannii* Parry), Subalpine fir (*Abies lasiocarpa* Nuttall) and Douglas fir (*Pseudotsuga menziesii* Franco). The climate of the region was characterized as humid continental climate, with warm to hot summers and cold winters. Annual average precipitation was 33 mm; average temperature ranged from a minimum of—15 °C in the coolest month (January) to a 27.7 °C in the hottest month (July).

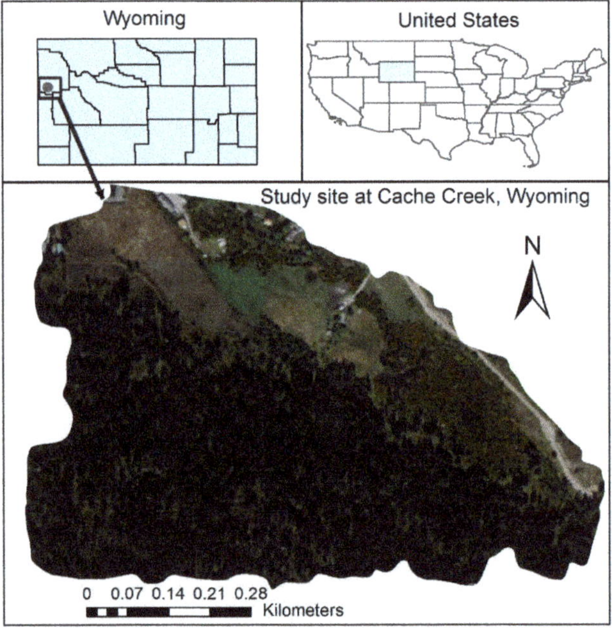

Figure 1. Study Area at Cache Creek, Wyoming, United States of America.

2.2. Data Acquisition and Processing

For this study, the aerial survey was conducted on 13 July 2015 using a DJI phantom 3 quadcopter. It had a compact RGB digital camera—PowerShot S100 (Canon, Tokyo, Japan; zoom lens 5.2 mm; 12.1 Megapixel CMOS sensor; 4000 × 3000 resolution)—attached to it; the physical dimension of the sensor (7.5 mm × 5.6 mm, 1.87 μm pixel size) was 71.5 × 56.6 degrees. The RGB camera automatically triggered (1 image/s) during the flight, capturing approximately an area of 14.42×10.77 m^2 with a pixel resolution of 3.5 mm flying at 10 m over the target and was installed on a stabilized gimbal (Photohigher AV200, CARVEC Systems, Hull, UK) along with the spectrometer to reduce the impact of the mechanical vibrations on the scientific instruments.

A total of 383 images from the flights conducted over the study area were used to generate point clouds and CHMs using the Agisoft Photoscan Professional v1.0.0 (www.agisoft.com) software (Agisoft LLC, St. Petersburg, Russia) [57]. Such software implements modern SfM algorithms on RGB photographs and thereby produces 3D reconstruction models based on the location of images with respect to each other as well as to the objects viewed within them [58–61]. In this study, the images had an overlap of 80% which made the stitching process more efficient. The processing steps included automatic aerial triangulation (AAT), bundle block adjustment (BBA), noise filtering, Digital Elevation Model (DEM) and ortho-mosaic creation. All the processes were performed under the default settings and in a fully automated way. The imagery was synchronized using the GPS position, and the ortho-mosaic was generated using absolute GPS coordinates; four random Ground Control Points (GCP) were used for enhancing accuracy. Some specific information of the camera locations and image overlaps related to UAV-image processing conducted in PhotoScan is presented in Table 1.

Table 1. Summary of UAV-image processing using PhotoScan.

Attribute	Value
Number of images	383
Flying altitude	115.29 m
Ground resolution	0.03 m·pix^{-1}
Coverage area	0.42 km^{-2}
Camera stations	351
Tie-points	87,635
Error	0.76 pix

The UAV-derived point cloud was used to compute a digital terrain model (DTM) and a CHM. First, ground points were classified using a Progressive Triangulated Irregular Network (TIN) densification algorithm implemented in lasground (settings: step is 10 m, bulge is 0.5 m, spike is 1 m, offset is 0.05 m), LAStools [62], and a 1 m DTM was created using the GridSurfaceCreate functions in FUSION/LDV 3.42 [63]. Afterwards, the UAV-derived point cloud was normalized to height above ground by subtraction of the DTM elevation from the Z coordinate of each point projected on the ground using the ClipData tool. CanopyModel function, also in FUSION/LDV, was used to compute the CHM with 0.5 m of spatial resolution for the study site.

2.3. Individual Tree Identification (ITD)

In this study, we used the local maximum (LM) algorithm [56,64,65], implemented in the rLiDAR package in R [56,64] for ITD on the UAV-derived CHM. FindTreesCHM function was used for automatic detection of tree tops. This function is based on the LM algorithm and it offers an option to search for tree tops in the CHM via a moving window with a fixed tree window size (FWS). To achieve optimal tree detection, we tested four FWS ($3 \times 3, 5 \times 5, 7 \times 7$ and 9×9 pixels) first on an unsmoothed CHM, and then on smoothed CHM by a mean smooth filter with fixed smoothing window sizes (SWS) of $3 \times 3, 5 \times 5$ and 7×7 pixels.

For validation of the ITD, we selected 30 random plots of 400 m^2 (20 m $\times$ 20 m) area each and performed a comparison between UAV based automatic ITD and an ITD done via independent visual assessment of on the UAV-derived orthomosaic, 3d point cloud and CHM. As the number of plots was considerably small, we randomly allocated the sample plots to enhance the robustness of the workflow. Further, we chose the parameters of FWS and SWS which had the best results in ITD with respect to the reference data and performed accuracy assessment for gaining a better understanding of the statistical factors and for identifying possible drawbacks. In particular, we evaluated the accuracy in terms of true positive (TP, correct detection), false negative (FN, omission error) and false positive (FP, commission error), as well as with respect to recall (r), precision (p) and F-score (F) as explained in Li et al. [66], using the following equations [67,68]:

$$r = TP/TP + FN \tag{1}$$

$$p = TP/TP + FP \tag{2}$$

$$F = 2 \times r \times p/r + p \tag{3}$$

Here, recall can be viewed as a measure of trees-detected, as it is inversely related to omission error; precision represents a measure of trees that were correctly detected, as it is inversely related to commission error, and F-score represents the harmonic mean of recall and precision, and hence higher p and r values result in higher F-scores. Figure 2 provides an overview of the study methodology.

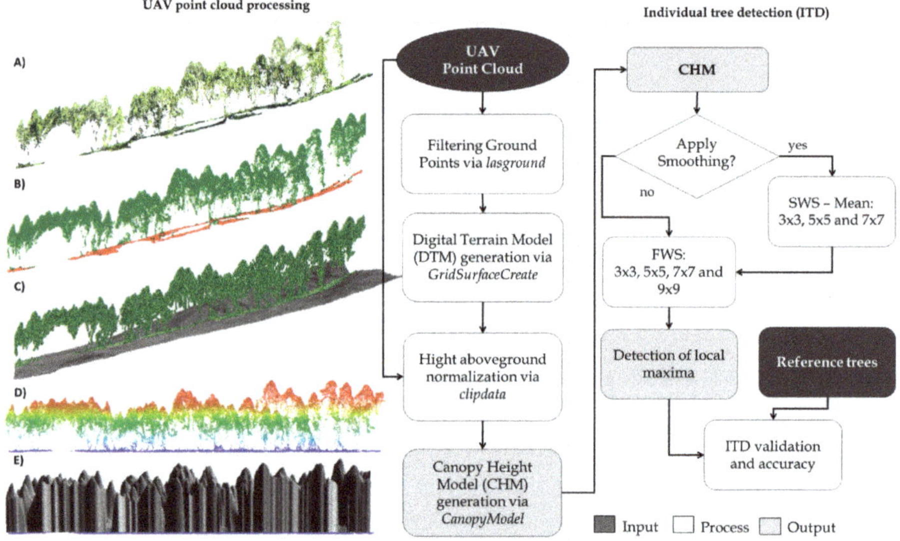

Figure 2. Workflow of the UAV data processing and individual tree detection. (**A**) UAV-derived point cloud; (**B**) Classified ground points; (**C**) Digital Terrain Model (DTM); (**D**) UAV-point cloud height normalization; (**E**) Canopy Height Model (CHM).

3. Results

The UAV-derived orthomosaic and CHM are presented in Figure 3. The UAV-derived CHM and LM algorithm proved to be very effective for detecting individual trees in the considered open canopy forest (Table 2; Figures 4 and 5). Importantly, FWS and SWS combinations were found to be a determining factor for the accuracy of ITD (Table 2). FWS of 3×3 and 5×5 are more accurate compared to 7×7 and 9×9. Lower SWS was found favorable towards ITD success rate as well. Larger SWS resulted in over estimation of tree tops. Finally, we did accuracy assessment on the best combination, which was 3×3 SWS and 3×3 FWS (Table 3).

Table 2. The highlighted grey color represents the best results, which were determined by comparing the number of trees detected (NTD) to the field-based tree inventory number (N). The closest values of NTD compared with N were selected as the best results.

Ref. (FID)	Ref. (N)	Fixed Tree Window Sizes (FWS)														
		3 × 3				5 × 5				7 × 7				9 × 9			
		Smoothing Window Sizes (SWS)														
		NF	3 × 3	5 × 5	7 × 7	NF	3 × 3	5 × 5	7 × 7	NF	3 × 3	5 × 5	7 × 7	NF	3 × 3	5 × 5	7 × 7
1	16	24	12	9	7	13	11	9	6	9	10	8	5	6	8	8	5
2	18	39	17	13	11	22	13	12	9	14	13	9	8	9	8	9	7
3	17	54	22	14	6	22	18	11	6	15	14	8	5	12	10	6	3
4	10	34	12	8	6	12	9	7	5	7	8	6	5	5	6	5	5
5	10	43	11	6	7	17	9	7	4	9	8	5	4	5	4	4	4
6	6	28	6	4	3	10	4	3	3	5	3	3	3	5	3	3	3
7	19	24	13	11	4	15	12	9	4	10	10	9	4	7	8	7	2
8	10	19	13	6	4	5	7	5	4	3	4	4	3	0	2	3	2
9	7	27	6	6	5	9	6	5	5	4	6	5	5	3	6	5	4
10	12	30	12	8	6	9	10	6	5	6	6	5	3	5	5	5	3
11	15	29	12	10	6	15	11	9	6	9	10	8	6	7	9	7	5
12	19	39	19	14	10	21	16	13	8	13	16	12	6	9	11	11	5
13	12	33	10	10	9	13	10	10	8	10	10	10	8	7	10	9	8
14	9	33	9	8	8	15	8	7	7	9	8	7	6	7	7	7	4
15	20	42	19	13	9	18	16	11	5	14	13	10	5	11	11	6	3
16	13	20	13	11	5	13	11	10	5	10	10	10	5	6	10	8	4
17	10	32	10	10	8	14	10	9	7	10	10	9	7	10	10	8	6
18	11	27	11	11	7	13	9	9	4	8	9	8	4	7	10	7	3
19	13	24	13	10	8	13	10	9	7	9	9	9	7	7	9	8	6
20	7	26	7	6	4	10	6	6	4	5	6	5	4	5	5	5	3
21	22	29	17	10	9	17	12	9	6	11	8	8	6	7	8	7	4
22	10	33	11	10	6	15	10	8	5	11	10	7	4	6	7	6	4
23	7	33	7	7	4	15	7	6	4	9	7	4	4	5	5	4	4
24	13	35	15	11	8	13	13	9	6	9	8	7	5	7	7	7	4
25	6	21	5	5	3	6	7	3	3	3	3	3	3	3	2	3	3
26	8	25	9	8	6	10	8	8	5	9	8	8	5	9	7	6	4
27	10	28	11	8	4	16	11	5	4	7	6	3	4	3	5	4	4
28	14	30	14	8	9	12	11	8	6	9	8	7	5	9	8	6	4
29	13	25	12	6	3	11	7	4	3	5	4	4	3	3	3	3	3
30	10	30	10	9	7	14	9	8	6	10	8	6	5	6	7	6	5
Total	367	916	358	270	192	408	301	235	160	262	253	207	147	191	211	183	124

FID: FeatureID; Ref.: reference number of tree per test plot (N); FWS: fixed treetop window size; SWS: fixed smoothing window size; NF: no filter applied.

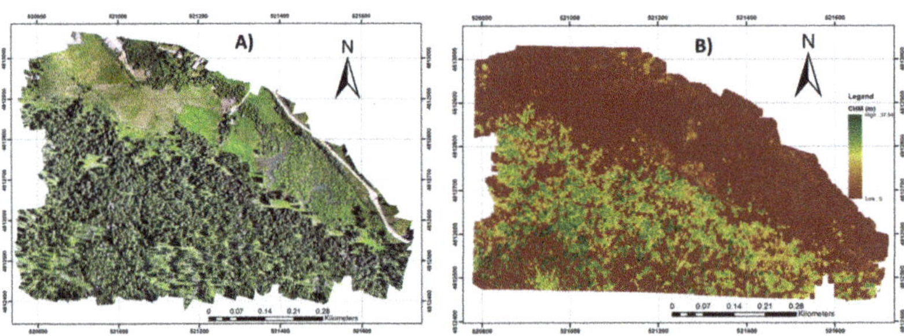

Figure 3. UAV-derived Orthomosaic (**A**) and Canopy Height Model (**B**).

For this study, the recall (r) had an overall value of 0.85 with a range of 0.68 to 1.00; overall value of precision (p) was 0.87 with a range of 0.69 to 1.00; and overall value of *F*-score was 0.86 and it varied from 0.73 to 0.95 (Table 3). Among the 367 reference trees used for this study, 358 (97.55%) trees were detected. In total, the algorithm missed 55 trees, and falsely detected 46 trees, indicating that under detection was higher than over detection.

Table 3. Accuracy assessment results of UAV-based individual tree detection based on False Positive (FP), False Negative (FN), True Positive (TP), recall (*r*), precision (*p*) and *F*-score (*F*) statistics parameters; where, Ref. (N) is the reference number of trees per test plot.

		Number of Trees						
Ref. (FID)	Ref. (N)	UAV	FP	FN	TP	*r*	*p*	*F*
1	16	12	0	4	12	0.75	1.00	0.86
2	18	17	1	2	16	0.89	0.94	0.91
3	17	22	6	1	16	0.94	0.73	0.82
4	10	12	2	0	10	1.00	0.83	0.91
5	10	11	2	1	9	0.90	0.82	0.86
6	6	6	1	1	5	0.83	0.83	0.83
7	19	13	0	6	13	0.68	1.00	0.81
8	10	13	4	1	9	0.90	0.69	0.78
9	7	6	1	2	5	0.71	0.83	0.77
10	12	12	3	3	9	0.75	0.75	0.75
11	15	12	1	4	11	0.73	0.92	0.81
12	19	19	2	2	17	0.89	0.89	0.89
13	12	10	0	2	10	0.83	1.00	0.91
14	9	9	1	1	8	0.89	0.89	0.89
15	20	19	2	3	17	0.85	0.89	0.87
16	13	13	1	1	12	0.92	0.92	0.92
17	10	10	2	2	8	0.80	0.80	0.80
18	11	11	3	3	8	0.73	0.73	0.73
19	13	13	1	1	12	0.92	0.92	0.92
20	7	7	1	1	6	0.86	0.86	0.86
21	22	17	0	5	17	0.77	1.00	0.87
22	10	11	1	0	10	1.00	0.91	0.95
23	7	7	1	1	6	0.86	0.86	0.86
24	13	15	2	0	13	1.00	0.87	0.93
25	6	5	0	1	5	0.83	1.00	0.91
26	8	9	1	0	8	1.00	0.89	0.94
27	10	11	3	2	8	0.80	0.73	0.76
28	14	14	2	2	12	0.86	0.86	0.86
29	13	12	1	2	11	0.85	0.92	0.88
30	10	10	2	2	8	0.80	0.80	0.80
Total	367	358	47	56	311	0.85	0.87	0.86

The observed and computed tree density in the study area from the UAV-derived CHM were 305 and 300 trees per hectare (TPH; trees·ha^{-1}), respectively. The most accurate results in the ITD were obtained primarily in test subplots with TPH ranging from 150 to 325 trees·ha^{-1} (Plot FID: 4, 13, 16, 19, 22, 24, 25, 26). On average, 93.2% of trees were detected correctly, with commission and omission errors limited to 8.7% and 6.8%, respectively, with a *F*-score of 0.92. In contrast, the algorithm detected only 81.4% of trees in subplots with TPH > 325 trees·ha^{-1} with a *F*-score of 0.83. The associated commission and omission errors were 14.4% and 18.6%, respectively.

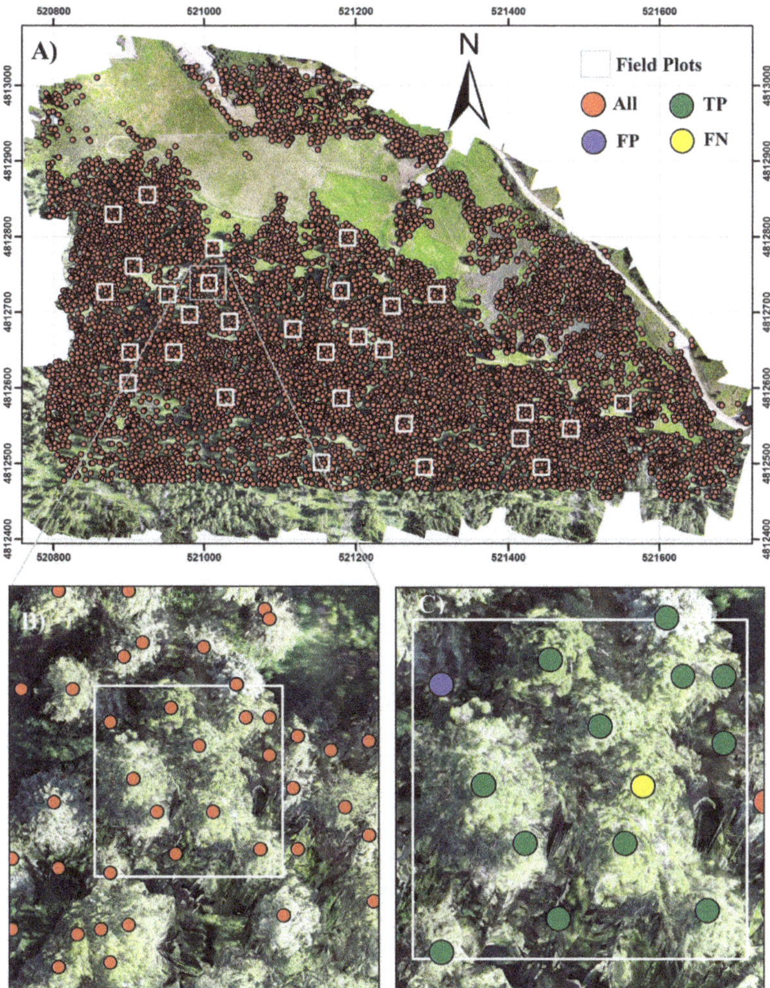

Figure 4. Individual tree detection from UAV-derived Canopy Height Model (CHM). Red dots represent the tree tops detected in the study area at stand level (**A**), and plot level (Plot 19) (**B**). Blue and yellow dots represent the commission (FP) and omission (FN) errors while the green dots represent the true positive trees (TP) at the plot 19 and (**C**).

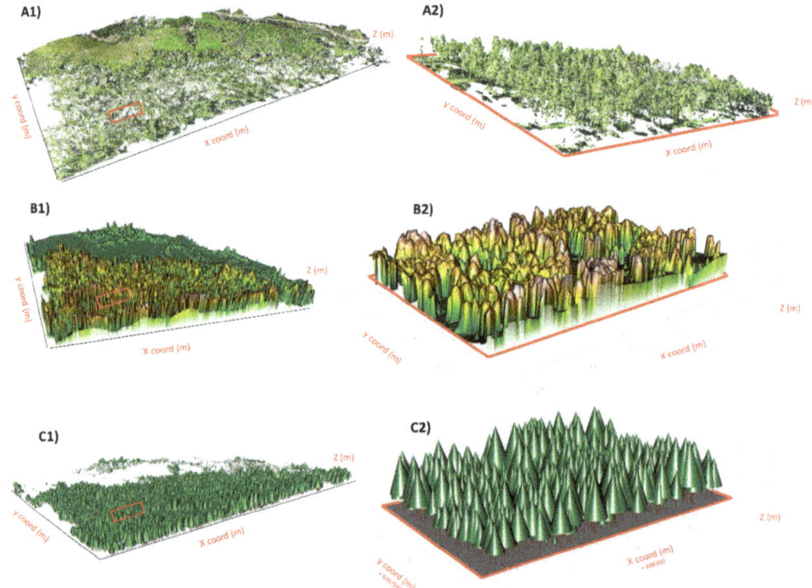

Figure 5. Illustration of individual tree detection at landscape (**A1–C1**) and plot (**A2–C2**) scales. (**A1,A2**) UAV-point cloud; (**B1,B2**) UAV-derived CHM; (**C1,C2**) Virtual forest in 3D.

4. Discussion

In the past decade, several studies have highlighted the potential for remote sensing in forestry. In particular, UAV equipped with consumer-grade onboard system camera and different sensors have been used to estimate tree counts, tree heights and crowns measurements due to its low cost and faster performance compared to traditional methods [22,23,69–74]. In this paper, we present a simplified framework for automated ITD from UAV-SfM derived CHM based on algorithms designed for LiDAR data processing. From the perspective of UAV-SfM remote sensing applications, this study can be recognized as a pioneering study for automatic ITD in open canopy mixed conifer forest, which have a comparative efficiency to LiDAR. The presented methodology has the potential to serve as a highly effective, affordable, and easy-to-use approach for ITD and therefore to support forest monitoring and inventory management. Because of the open canopy, we were able to collect sufficient ground points and perform operations similar to LiDAR; the derived data was found to be very detailed and of good quality. Herein, based on our results, it can be stated that UAV-SfM derived CHM coupled with LM algorithm is very capable of deriving tree counts with an acceptable accuracy ($F > 0.80$) in open canopy mixed conifer forests.

The accuracy of ITD in this study was found to be sensitive towards FWS and SWS combinations, and we noticed an inverse relationship between FWS and ITD, as reported by previous studies based on LiDAR data [7]. The most accurate results (i.e., results closest to the observed tree count) for SWS were found at 3×3 irrespective of FWS. Similar results were reported by Silva et al. (2016), where the 3×3 filter eliminated spurious local maxima caused due to irregularity of tree crowns and tree branches. In brief, a consistent FWS parameter can be seen as being more advantageous while performing ITD across a large area. Hence, we chose the best combination, which was 3×3 FWS and 3×3 SWS, for performing accuracy assessment. Even so, it should be borne in mind that the combinations of filter sizes and CHM filter conditions can be affected by forest types and tree size (which is correlated with Tree age) as well [75]. Depending on the intended usage, the users should always make a trade-off between commission and omission errors, and the overall efficiency of a particular window size can be viewed as a function of the object-resolution relationship which can serve as a first step towards the development of site specific tree identification models. [65]. However, more research pertaining to algorithmic development is needed to develop automated site specific adaptable window size estimations.

Even though we obtained the most accurate results for 3×3 FWS, we observed high commission error in all the cases and on average the algorithm missed around eight trees·ha^{-1}. As the tree detection is based on FWS, having a small FWS results in selection of non-existent trees or multiple trees for an individual tree crown [65]. This further implies that detection of larger trees would be easier as their crown consists of a higher number of pixels compared to the smaller trees. Hence, it is beneficial to have a rough estimate of the threshold tree size beforehand. This inverse relationship between canopy closure and suitable window sizes that we observed in this study was similar to previously reported studies [7,76]. Considering that, the higher success rate of 3×3 can be attributed to denser canopy over the study sites. This would also depend on the diameter and height of the tree species under consideration. In addition, proximity to neighboring trees, trees located under other trees, trees found in shadows or trees having low spectral contrast with respect to understory vegetation are also considered detrimental to tree detection [65]. The overall transferability of our framework can be expected to rely on multiple factors with openness of canopy as one of the most dominant. Hence, the algorithm needs to be continuously improved to adapt to different case scenarios.

In the past decade, several studies have highlighted the potential for using UAV for vegetation mapping on small and large scales [77–80]. However, research in this area is still in its infancy, especially for ITD [70]. Lim et al. [23] identified 11 of 13 trees, with one false positive, utilizing stacked RGB ortho-image and CHM segmentation. Trees not identified possessed small canopies which were not clearly defined from neighboring trees. Sperlich et al. [81] developed point clouds from UAV-based aerial photographs and achieved ITD accuracy of 87.68% within a search radius of 1 m using a pouring algorithm [25]. Kattenborn et al. [25] upgraded Sperlich's algorithm by geometrically classifying UAV-derived point clouds and detected individual palm trees with an overall mapping accuracy of 86.1% for a 9.4 ha study area and 98.2% for dense palm stands. Irrespective of tree detection techniques, heterogeneous stands with mixed and unmanaged trees offer more difficulty in selecting globally optimal segmentation parameters compared to homogenous stands [82]. Changing the crown overlapping threshold could be a viable strategy for determining emergent trees, as crown overlap is a major source for omission errors [83].

Similarly, airborne and terrestrial airborne LiDAR have also been explored for ITD, and have found to provide accurate results depending on forest conditions [84–89]. For instance, Silva et al. (2016), using airborne LiDAR data, found a similar accuracy (*F*-score > 80) for ITD in longleaf pine forest in Georgia, USA. Maas et al. [86] conducted numerous pilot studies using terrestrial LiDAR and reported that more than 97% of the trees could be detected correctly. Ritter et al. [87] proposed a two-stage density clustering approach for the automatic mapping of tree positions based on terrestrial lidar point cloud data sampled under limited sighting conditions. In their study, the authors detected tree positions in a mixed and vertically structured stand with an overall accuracy of 91.6%, and with omission- and commission error of only 5.7% and 2.7% respectively. Eysn et al. [89] used eight different ITD detection algorithms on airborne LiDAR data of different forest types of Alpine Space and automatically matched the ITD results to forest inventory data. While LiDAR is considered to be a well-suited technology for ITD, the acquisition cost of LiDAR data is still high, which makes the technology impractical for ITD in small areas and over time.

UAV imaging and terrestrial laser scanning systems are appropriate to survey small areas, while airborne laser scanning and hyperspectral systems are mostly used to survey large areas. The area sampled per time unit for these sensors depend of many factors, such as overlap percentage, flight speed, altitude, weather conditions and aim of the study [22,90,91]. UAV-imagery as well as photographs captured by manned aircrafts works differently from LiDAR technologies—whether it be the process of data extraction or be generation of metrics for assessing forest structure [92–96]. Even so, herein we are showing that the algorithm designed for LiDAR data processing can be also used for the purpose of processing UAV-SfM data ITD. Imaging technology has been found to be more reliable in capturing spectral information of the upper canopy while LiDAR—due to its ability to penetrate into the canopy—is more efficient for vertical stratification of vegetation layers and terrain associated with dense canopies [58,96]. Hence, in the case of close canopy forests, for accurate estimation of tree location and canopy attributes from UAV-SfM, external sources might need to be relied on for supplying DTMs [58,80,95].

While assessing the tree detection capability of our framework, tree location error was not estimated due to data limitations. Nevertheless, we anticipate a possible increase of errors with increasing density of stands, as reported in previous studies [31]. GPS errors associated with tree location estimation is another common source of uncertainty that needs further investigation [95]. Another weakness of this methodology can be attributed to the fact that irregular shapes of tree crowns can be a big problem in high density areas. As shadowing within and between crowns influences canopy reflectance, texture complexity, tree species variety, vertical stand structure and image quality can affect the delineation capacity of the algorithm [82,96–99]. The flight time of small UAVs is also shorter compared to LiDAR, which results in a relatively small sensed area. Even though flying at a higher altitude can be viewed as a remedy, it is detrimental towards the overall accuracy with respect to the resolution (ground sample distance) and it is restricted due to shorter signal range; it might create legal conflicts and privacy issues as well [40,100]. If high density point clouds are to be obtained, the process can become very time consuming. Also, environmental factors such as wind speed, lower visibility due to fog, shadows and temperature variations that can interfere in the efficacy of UAV operations should also be taken into consideration and addressed when working with forest remote sensing, especially for ITD [22,101,102]. Thus, the success rate of the method can be influenced by environmental conditions, existence of different forest types, particularly mixed species, and multilayered forests.

Modern forestry mostly requires forest information in digital format for maintaining a continuous workflow, and UAV based remote sensing offers a promising future in that regard [103,104]. In addition to that, ease of data collection, flexible control of spatial and temporal resolution, low operational costs, and safer work environment underpins the possibility of having a "UAV as a service" data collection market in the foreseeable future. The underlying technological advancement in multi-scale visual mapping, 3D digital modeling and time series analysis using SfM algorithms also empowers research in sectors outside forestry such as construction management [105], water contamination [106], archaeology [107], energy systems [108], computational biology [109] and habitat conservation [110]. Nevertheless, there are a lot of challenges associated with regulated, safe, and comprehensive applications of UAV remote sensing [9]. As multidisciplinary collaborations to promote the standardization of UAV remote sensing development are very limited, upcoming research expeditions should compare themselves with previous studies for systematically determining the optimal UAV remote sensing strategy for given forest lands [9,111]. Formulating methods to increase point density and developing strategies that optimize tree detection algorithms based on the characteristics of the point cloud can surely open new windows in UAV data analytics. Based on the findings presented in this study, future research directions should include species identification and evaluating the accuracy of estimating other tree-level characteristics such as DBH and crown area, which are important factors required for estimating biomass and stem volume.

5. Conclusions

In this study, we investigated the use of consumer grade cameras attached to UAVs and structure-from-motion algorithm for automatic ITD using a local-maxima based algorithm on UAV-derived Canopy Height Models (CHMs). Overall, this research study suggests that LM algorithm combined with proper fixed tree window sizes (FWS) and fixed smoothing window sizes (SWS) is capable of deriving tree counts with acceptable accuracy ($F > 0.80$) using UAV-derived CHM in open canopy forests. The proposed framework exposes the future potential of UAV-based photogrammetric point clouds for ITD and forest monitoring, and emphasizes that future research should focus on the estimation of individual tree attributes such as tree height, crown size and diameter, and thereby develop predictive models for estimating aboveground biomass and stem volume from UAV-imagery.

Acknowledgments: We would like to thank Joseph Roise, Travis Howell, Stacy Nelson, James McCarter, Siamak Khorram, Trevor Walker, Shalini Shankar, Timur Girgin, and the Department of Environmental Science and Forestry at North Carolina State University for providing the required training, support, facilities and resources to process, analyze and interpret the UAV data.

Author Contributions: All the authors have made substantial contribution towards the successful completion of this manuscript. They all have been involved in designing the study, drafting the manuscript and engaging in critical discussion. M.M., C.A.S, C.K. contributed with the methodological framework, data processing analysis and write up. P.J., G.C., A.C., A.H. and M.D. contributed to the interpretation, quality control and revisions of the manuscript. All authors read and approved the final manuscript.

Conflicts of Interest: The authors declare no conflict of interest.

References

1. Gatziolis, D.; Lienard, J.F.; Vogs, A.; Strigul, N.S. 3D tree dimensionality assessment using photogrammetry and small unmanned aerial vehicles. *PLoS ONE* **2015**, *10*, e0137765. [CrossRef] [PubMed]

2. Cubbage, F.; Roise, J.; Sutherland, R. The Proposed Sale of the Hofmann Forest: A Case Study in Natural Resource Policy. In *Forest Economics and Policy in a Changing Environment: How Market, Policy, and Climate Transformations Affect Forests—Proceedings of the 2016 Meeting of the International Society of Forest Resource Economics*; e-Gen. Tech. Rep. SRS-218; Department of Agriculture Forest Service: Asheville, NC, USA, 2016; p. 81.

3. Kwak, D.A.; Lee, W.K.; Lee, J.H.; Biging, G.S.; Gong, P. Detection of individual trees and estimation of tree height using LiDAR data. *J. For. Res.* **2007**, *12*, 425–434. [CrossRef]

4. Chen, Q.; Baldocchi, D.; Gong, P.; Kelly, M. Isolating individual trees in a savanna woodland using small footprint lidar data. *Photogramm. Eng. Remote Sens.* **2006**, *72*, 923–932. [CrossRef]

5. Strigul, N.; Pristinski, D.; Purves, D.; Dushoff, J.; Pacala, S. Scaling from trees to forests: Tractable macroscopic equations for forest dynamics. *Ecol. Monogr.* **2008**, *78*, 523–545. [CrossRef]

6. Strigul, N. *Individual-Based Models and Scaling Methods for Ecological Forestry: Implications of Tree Phenotypic Plasticity*; INTECH Open Access Publisher: Rijeka, Croatia, 2012.

7. Silva, C.A.; Hudak, A.T.; Vierling, L.A.; Loudermilk, E.L.; O'Brien, J.J.; Hiers, J.K.; Jack, S.B.; Gonzalez-Benecke, C.; Lee, H.; Falkowski, M.J.; et al. Imputation of Individual Longleaf Pine (*Pinus palustris* Mill.) Tree Attributes from Field and LiDAR Data. *Can. J. Remote Sens.* **2016**, *42*, 554–573. [CrossRef]

8. Gardner, T.A.; Barlow, J.; Araujo, I.S.; Ávila-Pires, T.C.; Bonaldo, A.B.; Costa, J.E.; Esposito, M.C.; Ferreira, L.V.; Hawes, J.; Hernandez, M.I.M.; et al. The cost-effectiveness of biodiversity surveys in tropical forests. *Ecol. Lett.* **2008**, *11*, 139–150. [CrossRef] [PubMed]

9. Tang, L.; Shao, G. Drone remote sensing for forestry research and practices. *J. For. Res.* **2015**, *26*, 791–797. [CrossRef]

10. Hansen, M.C.; Potapov, P.V.; Moore, R.; Hancher, M.; Turubanova, S.A.; Tyukavina, A.; Thau, D.; Stehman, S.V.; Goetz, S.J.; Loveland, T.R.; et al. High-resolution global maps of 21st-century forest cover change. *Science* **2013**, *342*, 850–853. [CrossRef] [PubMed]

11. Crowther, T.W.; Glick, H.B.; Covey, K.R.; Bettigole, C.; Maynard, D.S.; Thomas, S.M.; Smith, J.R.; Hintler, G.; Duguid, M.C.; Amatulli, G.; et al. Mapping tree density at a global scale. *Nature* **2015**, *525*, 201–205. [CrossRef] [PubMed]

12. Khorram, S.; van der Wiele, C.F.; Koch, F.H.; Nelson, S.A.; Potts, M.D. Remote Sensing: Past and Present. In *Principles of Applied Remote Sensing*; Springer International Publishing: Cham, Switzerland, 2016; pp. 1–20.

13. Roise, J.P.; Harnish, K.; Mohan, M.; Scolforo, H.; Chung, J.; Kanieski, B.; Catts, G.P.; McCarter, J.B.; Posse, J.; Shen, T. Valuation and production possibilities on a working forest using multi-objective programming, Woodstock, timber NPV, and carbon storage and sequestration. *Scand. J. For. Res.* **2016**, *31*, 674–680. [CrossRef]

14. Waser, L.T.; Baltsavias, E.; Ecker, K.; Eisenbeiss, H.; Ginzler, C.; Küchler, M.; Thee, P.; Zhang, L. High-resolution digital surface models (DSMs) for modelling fractional shrub/tree cover in a mire environment. *Int. J. Remote Sens.* **2008**, *29*, 1261–1276. [CrossRef]

15. Wallerman, J.; Bohlin, J.; Fransson, J.E. Forest height estimation using semi-individual tree detection in multi-spectral 3D aerial DMC data. In Proceedings of the 2012 IEEE International Geoscience and Remote Sensing Symposium, Munich, Germany, 22–27 July 2012.

16. Hudak, A.T.; Haren, A.T.; Crookston, N.L.; Liebermann, R.J.; Ohmann, J.L. Imputing forest structure attributes from stand inventory and remotely sensed data in western Oregon, USA. *For. Sci.* **2014**, *60*, 253–269. [CrossRef]

17. Hansen, E.H.; Gobakken, T.; Bollandsås, O.M.; Zahabu, E.; Næsset, E. Modeling aboveground biomass in dense tropical submontane rainforest using airborne laser scanner data. *Remote Sens.* **2015**, *7*, 788–807. [CrossRef]

18. Gholizadeh, A.; Mišurec, J.; Kopačková, V.; Mielke, C.; Rogass, C. Assessment of Red-Edge Position Extraction Techniques: A Case Study for Norway Spruce Forests Using HyMap and Simulated Sentinel-2 Data. *Forests* **2016**, *7*, 226. [CrossRef]

19. Zhang, Z.; Kazakova, A.; Moskal, L.M.; Styers, D.M. Object-based tree species classification in urban ecosystems using LiDAR and hyperspectral data. *Forests* **2016**, *7*, 122. [CrossRef]

20. Holmgren, J.; Persson, Å. Identifying species of individual trees using airborne laser scanner. *Remote Sens. Environ.* **2004**, *90*, 415–423. [CrossRef]

21. Silva, C.A.; Klauberg, C.; Hudak, A.T.; Vierling, L.A.; Jaafar, W.S.W.M.; Mohan, M.; Garcia, M.; Ferraz, A.; Cardil, A.; Saatchi, S. Predicting Stem Total and Assortment Volumes in an Industrial *Pinus taeda* L. Forest Plantation Using Airborne Laser Scanning Data and Random Forest. *Forests* **2017**, *8*, 254. [CrossRef]

22. Zarco-Tejada, P.J.; Diaz-Varela, R.; Angileri, V.; Loudjani, P. Tree height quantification using very high resolution imagery acquired from an unmanned aerial vehicle (UAV) and automatic 3D photo-reconstruction methods. *Eur. J. Agron.* **2014**, *55*, 89–99. [CrossRef]

23. Lim, Y.S.; La, P.H.; Park, J.S.; Lee, M.H.; Pyeon, M.W.; Kim, J.I. Calculation of Tree Height and Canopy Crown from Drone Images Using Segmentation. *J. Korean Soc. Surv. Geod. Photogramm. Cartogr.* **2015**, *33*, 605–614. [CrossRef]

24. Torres-Sánchez, J.; López-Granados, F.; Serrano, N.; Arquero, O.; Peña, J.M. High-throughput 3-D monitoring of agricultural-tree plantations with unmanned aerial vehicle (UAV) technology. *PLoS ONE* **2015**, *10*, e0130479. [CrossRef] [PubMed]

25. Kattenborn, T.; Sperlich, M.; Bataua, K.; Koch, B. Automatic Single Tree Detection in Plantations using UAV-based Photogrammetric Point clouds. *Int. Arch. Photogramm. Remote Sens. Spat. Inf. Sci.* **2014**, *40*, 139. [CrossRef]

26. Mlambo, R.; Woodhouse, I.H.; Gerard, F.; Anderson, K. Structure from Motion (SfM) photogrammetry with drone data: A low cost method for monitoring greenhouse gas emissions from forests in developing countries. *Forests* **2017**, *8*, 68. [CrossRef]

27. Miller, E.; Dandois, J.P.; Detto, M.; Hall, J.S. Drones as a Tool for Monoculture Plantation Assessment in the Steepland Tropics. *Forests* **2017**, *8*, 168. [CrossRef]

28. Zhang, J.; Hu, J.; Lian, J.; Fan, Z.; Ouyang, X.; Ye, W. Seeing the forest from drones: Testing the potential of lightweight drones as a tool for long-term forest monitoring. *Biol. Conserv.* **2016**, *198*, 60–69. [CrossRef]

29. Hung, C.; Bryson, M.; Sukkarieh, S. Multi-class predictive template for tree crown detection. *ISPRS J. Photogramm. Remote Sens.* **2012**, *68*, 170–183. [CrossRef]

30. Wallace, L.; Lucieer, A.; Watson, C.S. Evaluating tree detection and segmentation routines on very high resolution UAV LiDAR data. *IEEE Trans. Geosci. Remote Sens.* **2014**, *52*, 7619–7628. [CrossRef]

31. Gong, P.; Sheng, Y.; Biging, G.S. 3D model-based tree measurement from high-resolution aerial imagery. *Photogramm. Eng. Remote Sens.* **2002**, *68*, 1203–1212.

32. Song, C. Estimating tree crown size with spatial information of high resolution optical remotely sensed imagery. *Int. J. Remote Sens.* **2007**, *28*, 3305–3322. [CrossRef]

33. Sumnall, M.J.; Hill, R.A.; Hinsley, S.A. Comparison of small-footprint discrete return and full waveform airborne LiDAR data for estimating multiple forest variables. *Remote Sens. Environ.* **2016**, *173*, 214–223. [CrossRef]

34. Puliti, S.; Gobakken, T.; Ørka, H.O.; Næsset, E. Assessing 3D point clouds from aerial photographs for species-specific forest inventories. *Scand. J. For. Res.* **2017**, *32*, 68–79. [CrossRef]

35. Brandtberg, T.; Walter, F. Automated delineation of individual tree crowns in high spatial resolution aerial images by multiple-scale analysis. *Mach. Vis. Appl.* **1998**, *11*, 64–73. [CrossRef]

36. Wang, L.; Gong, P.; Biging, G.S. Individual tree-crown delineation and treetop detection in high-spatial-resolution aerial imagery. *Photogramm. Eng. Remote Sens.* **2004**, *70*, 351–357. [CrossRef]

37. Strecha, C.; Von Hansen, W.; Van Gool, L.; Fua, P.; Thoennessen, U. On benchmarking camera calibration and multi-view stereo for high resolution imagery. In Proceedings of the 2008 IEEE Conference on Computer Vision and Pattern Recognition, Anchorage, AK, USA, 23–28 June 2008; pp. 1–8.

38. Küng, O.; Strecha, C.; Beyeler, A.; Zufferey, J.C.; Floreano, D.; Fua, P.; Gervaix, F. The accuracy of automatic photogrammetric techniques on ultra-light UAV imagery. In Proceedings of the UAV-g 2011-Unmanned Aerial Vehicle in Geomatics, Zurich, Switzerland, 14–16 September 2011; No. EPFL-CONF-168806.

39. Remondino, F.; Barazzetti, L.; Nex, F.; Scaioni, M.; Sarazzi, D. UAV photogrammetry for mapping and 3d modeling—Current status and future perspectives. *Int. Arch. Photogramm. Remote Sens. Spat. Inf. Sci.* **2011**, *38*, C22. [CrossRef]

40. Tomaštík, J.; Mokroš, M.; Saloň, Š.; Chudý, F.; Tunák, D. Accuracy of Photogrammetric UAV-Based Point Clouds under Conditions of Partially-Open Forest Canopy. *Forests* **2017**, *8*, 151. [CrossRef]

41. James, M.R.; Robson, S. Straightforward reconstruction of 3D surfaces and topography with a camera: Accuracy and geoscience application. *J. Geophys. Res. Earth Surf.* **2012**, *117*. [CrossRef]

42. Fritz, A.; Kattenborn, T.; Koch, B. UAV-based photogrammetric point clouds—Tree stem mapping in open stands in comparison to terrestrial laser scanner point clouds. *Int. Arch. Photogramm. Remote Sens. Spat. Inf. Sci.* **2013**, *40*, 141–146. [CrossRef]

43. Haala, N.; Hastedt, H.; Wolf, K.; Ressl, C.; Baltrusch, S. Digital photogrammetric camera evaluation–generation of digital elevation models. *Photogramm.-Fernerkund.-Geoinf.* **2010**, *2010*, 99–115. [CrossRef] [PubMed]

44. Baltsavias, E.; Gruen, A.; Eisenbeiss, H.; Zhang, L.; Waser, L.T. High-quality image matching and automated generation of 3D tree models. *Int. J. Remote Sens.* **2008**, *29*, 1243–1259. [CrossRef]

45. Westoby, M.J.; Brasington, J.; Glasser, N.F.; Hambrey, M.J.; Reynolds, J.M. 'Structure-from-Motion' photogrammetry: A low-cost, effective tool for geoscience applications. *Geomorphology* **2012**, *179*, 300–314. [CrossRef]

46. James, M.R.; Robson, S. Mitigating systematic error in topographic models derived from UAV and ground-based image networks. *Earth Surface Proc. Landforms* **2014**, *39*, 1413–1420. [CrossRef]

47. Hernández-Clemente, R.; Navarro-Cerrillo, R.M.; Ramírez, F.J.R.; Hornero, A.; Zarco-Tejada, P.J. A novel methodology to estimate single-tree biophysical parameters from 3D digital imagery compared to aerial laser scanner data. *Remote Sens.* **2014**, *6*, 11627–11648. [CrossRef]

48. Dempewolf, J.; Nagol, J.; Hein, S.; Thiel, C.; Zimmermann, R. Measurement of Within-Season Tree Height Growth in a Mixed Forest Stand Using UAV Imagery. *Forests* **2017**, *8*, 231. [CrossRef]

49. Kwak, D.A.; Lee, W.K.; Lee, J.H. Predicting forest stand characteristics with detection of individual tree. In Proceedings of the MAPPS/ASPRS 2006 Fall Conference, San Antonio, TX, USA, 6–10 November 2006.

50. Larsen, M.; Eriksson, M.; Descombes, X.; Perrin, G.; Brandtberg, T.; Gougeon, F.A. Comparison of six individual tree crown detection algorithms evaluated under varying forest conditions. *Int. J. Remote Sens.* **2011**, *32*, 5827–5852. [CrossRef]

51. Lee, J.H.; Biging, G.S.; Fisher, J.B. An Individual Tree-Based Automated Registration of Aerial Images to Lidar Data in a Forested Area. *Photogramm. Eng. Remote Sens.* **2016**, *82*, 699–710. [CrossRef]

52. Descombes, X.; Pechersky, E. Tree Crown Extraction Using a Three State Markov Random Field. Ph.D. Thesis, INRIA, Rocquencourt, France, 2006.

53. Perrin, G.; Descombes, X.; Zerubia, J. *A Non-Bayesian Model for Tree Crown Extraction Using Marked Point Processes*; INRIA: Rocquencourt, France, 2006.

54. Gougeon, F.A. Automatic individual tree crown delineation using a valley-following algorithm and rule-based system. In Proceedings of the International Forum on Automated Interpretation of High Spatial Resolution Digital Imagery for Forestry, Victoria, BC, Canada, 10–12 February 1998; pp. 11–23.

55. Koch, B.; Heyder, U.; Weinacker, H. Detection of individual tree crowns in airborne lidar data. *Photogramm. Eng. Remote Sens.* **2006**, *72*, 357–363. [CrossRef]

56. R Core Team. *R: A Language and Environment for Statistical Computing*; R Foundation for Statistical Computing: Vienna, Austria; Available online: http://www.R-project.org (accessed on 15 October 2015).

57. AgiSoft, L.L.C. PhotoScan Professional Edition v.1.0.3. Available online: www.agisoft.ru (accessed on 3 October 2015).

58. Dandois, J.P.; Ellis, E.C. High spatial resolution three-dimensional mapping of vegetation spectral dynamics using computer vision. *Remote Sens. Environ.* **2013**, *136*, 259–276. [CrossRef]

59. Lucieer, A.; Turner, D.; King, D.H.; Robinson, S.A. Using an Unmanned Aerial Vehicle (UAV) to capture micro-topography of Antarctic moss beds. *Int. J. Appl. Earth Obs. Geoinf.* **2014**, *27*, 53–62. [CrossRef]

60. Turner, D.; Lucieer, A.; Wallace, L. Direct georeferencing of ultrahigh-resolution UAV imagery. *IEEE Trans. Geosci. Remote Sens.* **2014**, *52*, 2738–2745. [CrossRef]

61. Verhoeven, G. Taking computer vision aloft—Archaeological three-dimensional reconstructions from aerial photographs with photoscan. *Archaeol. Prospect.* **2011**, *18*, 67–73. [CrossRef]

62. Isenburg, M. LAStools—Efficient Tools for LiDAR Processing. Available online: lastools.org (accessed on 3 October 2015).

63. Kraus, K.; Pfeifer, N. Determination of terrain models in wooded areas with airborne laser scanner data. *ISPRS J. Photogramm. Remote Sens.* **1998**, *53*, 193–203. [CrossRef]

64. Silva, C.A.; Crookston, N.L.; Hudak, A.T.; Vierling, L.A. rLiDAR: An R Package for Reading, Processing and Visualizing LiDAR (Light Detection and Ranging) Data, Version 0.1. Available online: http://cran.rproject.org/web/packages/rLiDAR/index.html (accessed on 15 October 2015).

65. Wulder, M.; Niemann, K.O.; Goodenough, D.G. Local maximum filtering for the extraction of tree locations and basal area from high spatial resolution imagery. *Remote Sens. Environ.* **2000**, *73*, 103–114. [CrossRef]

66. Li, W.; Guo, Q.; Jakubowski, M.K.; Kelly, M. A new method for segmenting individual trees from the lidar point cloud. *Photogramm. Eng. Remote Sens.* **2012**, *78*, 75–84. [CrossRef]

67. Goutte, C.; Gaussier, E. A probabilistic interpretation of precision, recall and F-score, with implication for evaluation. In Proceedings of the European Conference on Information Retrieval, Compostela, Spain, 21–23 March 2005; Springer: Berlin/Heidelberg, Germany, 2005; pp. 345–359.

68. Sokolova, M.; Japkowicz, N.; Szpakowicz, S. Beyond accuracy, F-score and ROC: A family of discriminant measures for performance evaluation. In Proceedings of the Australasian Joint Conference on Artificial Intelligence, Auckland, New Zealand, 1–5 December 2008; Springer: Berlin/Heidelberg, Germany, 2008; pp. 1015–1021.

69. Puttock, A.K.; Cunliffe, A.M.; Anderson, K.; Brazier, R.E. Aerial photography collected with a multirotor drone reveals impact of Eurasian beaver reintroduction on ecosystem structure 1. *J. Unmanned Veh. Syst.* **2015**, *3*, 123–130. [CrossRef]

70. Koh, L.; Wich, S. Dawn of drone ecology: Low-cost autonomous aerial vehicles for conservation. *Trop. Conserv. Sci.* **2012**, *5*, 121–132. [CrossRef]

71. Paneque-Gálvez, J.; McCall, M.K.; Napoletano, B.M.; Wich, S.A.; Koh, L.P. Small drones for community-based forest monitoring: An assessment of their feasibility and potential in tropical areas. *Forests* **2014**, *5*, 1481–1507. [CrossRef]

72. Getzin, S.; Wiegand, K.; Schöning, I. Assessing biodiversity in forests using very high-resolution images and unmanned aerial vehicles. *Methods Ecol. Evol.* **2012**, *3*, 397–404. [CrossRef]

73. Felderhof, L.; Gillieson, D. Near-infrared imagery from unmanned aerial systems and satellites can be used to specify fertilizer application rates in tree crops. *Can. J. Remote Sens.* **2012**, *37*, 376–386. [CrossRef]

74. Wallace, L.; Watson, C.; Lucieer, A. Detecting pruning of individual stems using airborne laser scanning data captured from an unmanned aerial vehicle. *Int. J. Appl. Earth Obs. Geoinf.* **2014**, *30*, 76–85. [CrossRef]

75. Lindberg, E.; Hollaus, M. Comparison of methods for estimation of stem volume, stem number and basal area from airborne laser scanning data in a hemi-boreal forest. *Remote Sens.* **2012**, *4*, 1004–1023. [CrossRef]

76. Falkowski, M.J.; Smith, A.M.; Gessler, P.E.; Hudak, A.T.; Vierling, L.A.; Evans, J.S. The influence of conifer forest canopy cover on the accuracy of two individual tree measurement algorithms using lidar data. *Can. J. Remote Sens.* **2008**, *34* (Suppl. S2), S338–S350. [CrossRef]

77. Merino, L.; Caballero, F.; Martínez-de-Dios, J.R.; Maza, I.; Ollero, A. An unmanned aircraft system for automatic forest fire monitoring and measurement. *J. Intell. Robot. Syst.* **2012**, *65*, 533–548. [CrossRef]

78. Martínez-de Dios, J.R.; Merino, L.; Caballero, F.; Ollero, A. Automatic forest-fire measuring using ground stations and unmanned aerial systems. *Sensors* **2011**, *11*, 6328–6353. [CrossRef] [PubMed]

79. Ota, T.; Ogawa, M.; Shimizu, K.; Kajisa, T.; Mizoue, N.; Yoshida, S.; Takao, G.; Hirata, Y.; Furuya, N.; Sano, T.; et al. Aboveground biomass estimation using structure from motion approach with aerial photographs in a seasonal tropical forest. *Forests* **2015**, *6*, 3882–3898. [CrossRef]

80. Panagiotidis, D.; Abdollahnejad, A.; Surový, P.; Chiteculo, V. Determining tree height and crown diameter from high-resolution UAV imagery. *Int. J. Remote Sens.* **2017**, *38*, 2392–2410. [CrossRef]

81. Sperlich, M.; Kattenborn, T.; Koch, B.; Kattenborn, G. Potential of Unmanned Aerial Vehicle Based Photogrammetric Point Clouds for Automatic Single Tree Detection. Available online: http://www.dgpf.de/neu/Proc2014/proceedings/papers/Beitrag270.pdf (accessed on 15 January 2015).

82. La, H.P.; Eo, Y.D.; Chang, A.; Kim, C. Extraction of individual tree crown using hyperspectral image and LiDAR data. *KSCE J. Civ. Eng.* **2015**, *19*, 1078–1087. [CrossRef]

83. Zhou, J.; Proisy, C.; Descombes, X.; Le Maire, G.; Nouvellon, Y.; Stape, J.L.; Viennois, G.; Zerubia, J.; Couteron, P. Mapping local density of young Eucalyptus plantations by individual tree detection in high spatial resolution satellite images. *For. Ecol. Manag.* **2013**, *301*, 129–141. [CrossRef]

84. Eysn, L.; Hollaus, M.; Lindberg, E.; Berger, F.; Monnet, J.-M.; Dalponte, M.; Kobal, M.; Pellegrini, M.; Lingua, E.; Mongus, D.; et al. A benchmark of lidar-based single tree detection methods using heterogeneous forest data from the alpine space. *Forests* **2015**, *6*, 1721–1747. [CrossRef]

85. Astrup, R.; Ducey, M.J.; Granhus, A.; Ritter, T.; von Lüpke, N. Approaches for estimating stand-level volume using terrestrial laser scanning in a single-scan mode. *Can. J. For. Res.* **2014**, *44*, 666–676. [CrossRef]

86. Maas, H.G.; Bienert, A.; Scheller, S.; Keane, E. Automatic forest inventory parameter determination from terrestrial laser scanner data. *Int. J. Remote Sens.* **2008**, *29*, 1579–1593. [CrossRef]

87. Ritter, T.; Schwarz, M.; Tockner, A.; Leisch, F.; Nothdurft, A. Automatic Mapping of Forest Stands Based on Three-Dimensional Point Clouds Derived from Terrestrial Laser-Scanning. *Forests* **2017**, *8*, 265. [CrossRef]

88. Liang, X.; Hyyppä, J. Automatic stem mapping by merging several terrestrial laser scans at the feature and decision levels. *Sensors* **2013**, *13*, 1614–1634. [CrossRef] [PubMed]

89. Liang, X.; Litkey, P.; Hyyppa, J.; Kaartinen, H.; Vastaranta, M.; Holopainen, M. Automatic stem mapping using single-scan terrestrial laser scanning. *IEEE Trans. Geosci. Remote Sens.* **2012**, *50*, 661–670. [CrossRef]

90. White, J.C.; Wulder, M.A.; Vastaranta, M.; Coops, N.C.; Pitt, D.; Woods, M. The utility of image-based point clouds for forest inventory: A comparison with airborne laser scanning. *Forests* **2013**, *4*, 518–536. [CrossRef]

91. Whitehead, K.; Hugenholtz, C.H. Remote sensing of the environment with small unmanned aircraft systems (UASs), part 1: A review of progress and challenges. *J. Unmanned Veh. Syst.* **2014**, *2*, 69–85. [CrossRef]

92. Lin, Y.; Jiang, M.; Yao, Y.; Zhang, L.; Lin, J. Use of UAV oblique imaging for the detection of individual trees in residential environments. *Urban For. Urban Green.* **2015**, *14*, 404–412. [CrossRef]

93. White, J.C.; Stepper, C.; Tompalski, P.; Coops, N.C.; Wulder, M.A. Comparing ALS and image-based point cloud metrics and modelled forest inventory attributes in a complex coastal forest environment. *Forests* **2015**, *6*, 3704–3732. [CrossRef]

94. Penner, M.; Woods, M.; Pitt, D.G. A comparison of airborne laser scanning and image point cloud derived tree size class distribution models in boreal Ontario. *Forests* **2015**, *6*, 4034–4054. [CrossRef]

95. Lisein, J.; Pierrot-Deseilligny, M.; Bonnet, S.; Lejeune, P. A photogrammetric workflow for the creation of a forest canopy height model from small unmanned aerial system imagery. *Forests* **2013**, *4*, 922–944. [CrossRef]

96. Vauhkonen, J.; Korpela, I.; Maltamo, M.; Tokola, T. Imputation of single-tree attributes using airborne laser scanning-based height, intensity, and alpha shape metrics. *Remote Sens. Environ.* **2010**, *114*, 1263–1276. [CrossRef]

97. Wing, M.G.; Eklund, A.; John, S.; Richard, K. Horizontal measurement performance of five mapping-grade global positioning system receiver configurations in several forested settings. *West. J. Appl. For.* **2008**, *23*, 166–171.

98. Asner, G.P.; Heidebrecht, K.B. Spectral unmixing of vegetation, soil and dry carbon cover in arid regions: Comparing multispectral and hyperspectral observations. *Int. J. Remote Sens.* **2002**, *23*, 3939–3958. [CrossRef]

99. Wulder, M.A.; Dechka, J.A.; Gillis, M.A.; Luther, J.E.; Hall, R.J.; Beaudoin, A.; Franklin, S.E. Operational mapping of the land cover of the forested area of Canada with Landsat data: EOSD land cover program. *For. Chron.* **2003**, *79*, 1075–1083. [CrossRef]

100. Puliti, S.; Ørka, H.O.; Gobakken, T.; Næsset, E. Inventory of small forest areas using an unmanned aerial system. *Remote Sens.* **2015**, *7*, 9632–9654. [CrossRef]

101. Nevalainen, O.; Honkavaara, E.; Tuominen, S.; Viljanen, N.; Hakala, T.; Yu, X.; Hyyppä, J.; Saari, H.; Pölönen, I.; Imai, N.N.; et al. Individual tree detection and classification with UAV-based photogrammetric point clouds and hyperspectral imaging. *Remote Sens.* **2017**, *9*, 185. [CrossRef]

102. Dandois, J.P.; Olano, M.; Ellis, E.C. Optimal altitude, overlap, and weather conditions for computer vision UAV estimates of forest structure. *Remote Sens.* **2015**, *7*, 13895–13920. [CrossRef]

103. Birdal, A.C.; Avdan, U.; Türk, T. Estimating tree heights with images from an unmanned aerial vehicle. *Geomat. Nat. Hazards Risk* **2017**. [CrossRef]

104. Tang, L.; Shao, G.; Dai, L. Roles of digital technology in China's sustainable forestry development. *Int. J. Sustain. Dev. World Ecol.* **2009**, *16*, 94–101. [CrossRef]

105. Golparvar-Fard, M.; Peña-Mora, F.; Savarese, S. D4AR—A 4-dimensional augmented reality model for automating construction progress monitoring data collection, processing and communication. *J. Inf. Technol. Constr.* **2009**, *14*, 129–153.

106. Jat, P.; Serre, M.L. Bayesian Maximum Entropy space/time estimation of surface water chloride in Maryland using river distances. *Environ. Pollut.* **2016**, *219*, 1148–1155. [CrossRef] [PubMed]

107. Green, S.; Bevan, A.; Shapland, M. A comparative assessment of structure from motion methods for archaeological research. *J. Archaeol. Sci.* **2014**, *46*, 173–181. [CrossRef]

108. Azadeh, A.; Taghipour, M.; Asadzadeh, S.M.; Abdollahi, M. Artificial immune simulation for improved forecasting of electricity consumption with random variations. *Int. J. Electr. Power Energy Syst.* **2014**, *55*, 205–224. [CrossRef]

109. Murugesan, S.; Bouchard, K.; Chang, E.; Dougherty, M.; Hamann, B.; Weber, G.H. Multi-scale visual analysis of time-varying electrocorticography data via clustering of brain regions. *BMC Bioinform.* **2017**, *18*, 236. [CrossRef] [PubMed]

110. Johnston, A.N.; Moskal, L.M. High-resolution habitat modeling with airborne LiDAR for red tree voles. *J. Wildl. Manag.* **2017**, *81*, 58–72. [CrossRef]

111. Shahbazi, M.; Théau, J.; Ménard, P. Recent applications of unmanned aerial imagery in natural resource management. *GISci. Remote Sens.* **2014**, *51*, 339–365. [CrossRef]

 *forests*

Article

Estimation and Extrapolation of Tree Parameters Using Spectral Correlation between UAV and Pléiades Data

Azadeh Abdollahnejad *, Dimitrios Panagiotidis and Peter Surový

Department of Forest Management, Faculty of Forestry & Wood Sciences,
Czech University of Life Sciences (CULS), 165 00 Prague, Czech Republic;
panagiotidis@fld.czu.cz (D.P.); surovy@fld.czu.cz (P.S.)
* Correspondence: abdolahnejad@fld.czu.cz; Tel.: +420-774-844-679

Received: 20 December 2017; Accepted: 8 February 2018; Published: 11 February 2018

Abstract: The latest technological advances in space-borne imagery have significantly enhanced the acquisition of high-quality data. With the availability of very high-resolution satellites, such as Pléiades, it is now possible to estimate tree parameters at the individual level with high fidelity. Despite innovative advantages on high-precision satellites, data acquisition is not yet available to the public at a reasonable cost. Unmanned aerial vehicles (UAVs) have the practical advantage of data acquisition at a higher spatial resolution than that of satellites. This study is divided into two main parts: (1) we describe the estimation of basic tree attributes, such as tree height, crown diameter, diameter at breast height (DBH), and stem volume derived from UAV data based on structure from motion (SfM) algorithms; and (2) we consider the extrapolation of the UAV data to a larger area, using correlation between satellite and UAV observations as an economically viable approach. Results have shown that UAVs can be used to predict tree characteristics with high accuracy (i.e., crown projection, stem volume, cross-sectional area (CSA), and height). We observed a significant relation between extracted data from UAV and ground data with $R^2 = 0.71$ for stem volume, $R^2 = 0.87$ for height, and $R^2 = 0.60$ for CSA. In addition, our results showed a high linear relation between spectral data from the UAV and the satellite ($R^2 = 0.94$). Overall, the accuracy of the results between UAV and Pléiades was reasonable and showed that the used methods are feasible for extrapolation of extracted data from UAV to larger areas.

Keywords: downscaling; Pléiades imagery; unmanned aerial vehicle; stem volume estimation; remote sensing

---

1. Introduction

Technological advances in unmanned aerial vehicles (UAVs) have made it feasible to obtain high-resolution imagery and three-dimensional (3D) data for assessing tree attributes and forest monitoring. Methods of data acquisition with remotely-sensed aerial or satellite data at high spatial resolution have partially replaced conventional methods of field measurement for forest inventory purposes [1–4]. Repeated observation from modern satellites, as well as improvements in UAV technology, have contributed significantly to our understanding of the dynamics of complex ecosystems, particularly forests. Accurate quantification of tree basic parameters, such as height, crown diameter, and diameter at breast height (DBH), is essential for decision-making and planning. Modern techniques of remote sensing can provide accurate estimations of tree height and crown area characteristics at the individual level using a series of algorithms [5]. The individual tree identification (IDS) algorithm allows for the estimation of crown diameter [6], and the smoothing of the canopy height model (CHM) using local maxima techniques [7] can provide estimates of individual tree heights.

Volume and above ground biomass (AGB) of small spatial extent areas can be derived with the help of forest inventory data. However, field measurements are typically unbiased, time-consuming, and expensive. For spatially larger areas, modelling of volume and biomass requires the use of remote sensing information for practical purposes. Additionally, remote sensing techniques are able to improve the value of inventoried data with detailed coverage at affordable costs [8,9]. The strength of remote sensing approaches based on very high resolution (VHR) images using aerial imagery or downscaling methods (i.e., calibration of satellite images based on UAVs) is that they allow for the construction of high-quality 3D digital surface models (DSMs) that can be used to estimate several forest tree attributes, such as height, DBH, and crown diameter [3].

Landsat satellites have been used to predict volume and AGB, mainly because of their long-term data record and the favorable compromise between aerial coverage, spectral sensitivity, and spatial resolution [10]. Newer satellites (e.g., GeoEye-1, IKONOS, WorldView-2, and Pléiades) have significantly increased the potential for data acquisition at higher spatial resolutions (i.e., 2 m multispectral). However, satellites are still unable to provide the desired spatial resolution needed for forestry applications; UAVs, on the other hand, can provide higher resolution data and are frequently used to capture aerial images for planning purposes [11].

Only a limited number of studies have used optical VHR sensors for image matching and estimating forest parameters. In previous studies, the IKONOS, Cartosat-1, and Worldview-2 satellites were used for estimation of structural (e.g., height, DBH, stem volume) and textural (i.e., composition, AGB) metrics with acceptable accuracy [12–15].

However, a review of past studies found that few researchers have used the Pléiades satellite for forestry purposes [16] and no studies used it for AGB estimation. While previous studies using Pléiades emphasized combining spectral derivatives and textural metrics to image matched height metrics, these studies utilized combinations of either height and textural metrics or height metrics and spectral derivatives to improve the remote sensing estimation of forest parameters.

The main objectives of this study were to: (i) evaluate UAV performance to estimate key individual tree parameters, including height, crown diameter, cross-sectional area (CSA), and stem volume; and (ii) to extrapolate the extracted UAV data to larger areas, based on spectral correlation between UAV and satellite, as a practical method for detailed information collection across a large area that would not be economically feasible with ground-based assessments.

2. Materials and Methods

2.1. Study Area

The Doksy territory lies on the shores of Lake Mácha in Northern Bohemia in the Czech Republic (Figure 1). The lake is largely surrounded by dense forests covering an area of 300 km^2. Geologically, the area is characterized by sandstone pseudokarst in the late stages of development, and the soils are either sandy or peaty, with shallow, peaty basins over rocky sandstone hummocks and sporadic volcanic hills. We selected three evenly-aged (managed stands) 40 × 40 m experimental plots and sampled all trees in each plot to estimate the tree heights, DBH, crown diameters, and stem volume. All plots were located northeast of the city of Doksy (−718000, −991250 NW to −717000, −991950 SE in the local coordinate system S-JTSK/Krovak East North, Figure 1). The research was carried out in a 140-year-old *Pinus sylvestris* L. (Scots pine) monoculture natural stand established on sandy soils (68%). The vegetative period tends to be rather warm and dry. The mean annual air temperature is 7.3 °C and the average annual maximum temperature is 31.5 °C. The mean annual precipitation is 635 mm, with only 354 mm during the growing season.

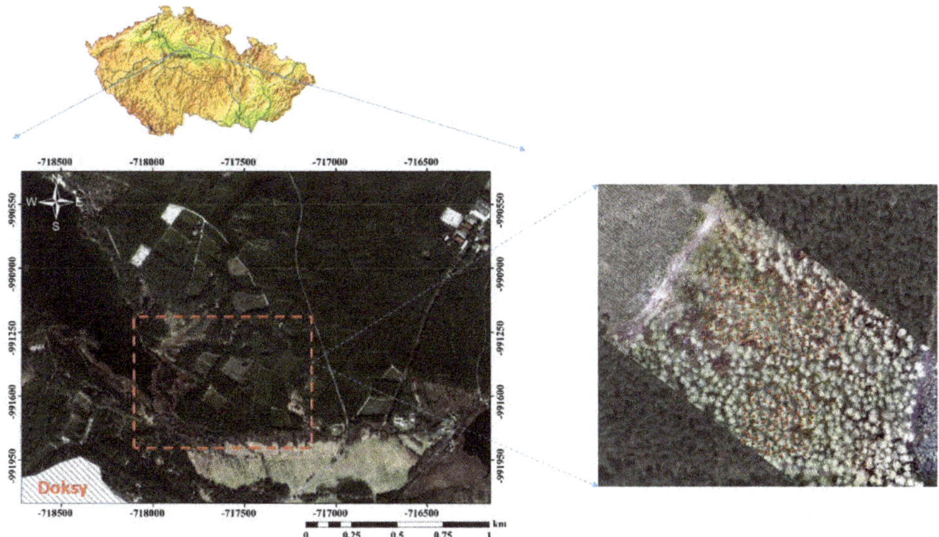

Figure 1. Location of the study area in the northern part of the Czech Republic in the local coordinate system S-JTSK/Krovak East North. Source: Google Earth (elaborated work in ArcGis for higher map detail).

2.2. Field Survey

For the acquisition of field measurements, we used field-map technology. Field-map is a software and hardware product which combines flexible real-time geographic information system (GIS) software field-map with electronic equipment of high accuracy for mapping and dendrometric measurement. For this study, the dendrometric device which we used was the TruPulse 360/B laser range finder by lnc. Centennial, Colorado, USA. For the purpose of this study, we used the TruPulse 360/B to measure (i) the positions of individual trees by distance based on the field triangulation approach, (ii) tree heights based on the functionality of the digital relascope, and (iii) the mean of the horizontal widths of the tree crown radius in four orientations (east, west, north, and south). The coordinate system was set to S-JTSK/Krovak East North (a local coordinate system that is mainly used in the vicinity of the Czech Republic). The whole data collection process is rather ergonomic since mapped data can be seen simultaneously in the monitor of the computer in the form of layers (either point, lines, or polygons). Each layer can have a number of attributes (i.e., tree positions, heights, etc.) which are stored in a fully relational database.

In addition, the DBH of trees (1.3 m above ground) was measured by a Häglof digital caliper (Häglof Sweden AB, Långsele, Sweden) in two perpendicular directions. Diameters were then transferred via bluetooth to the field-map computer and stored in the relative database. We considered the mean of measured diameters at breast height for computing the CSA. Finally, all the data was extracted by connecting the field-mapper with a universal serial bus (USB) to the personal computer (PC) for further processing. All field measurements were taken in November, 2015. In total, we sampled 223 live trees from all three plots.

2.3. Aerial and Satellite Imagery

The UAV data were acquired in March 2016. The platform model used was an octocopter SteadiDrone EI8HT, ready to fly (RTF) embedded with an RGB (red, green, blue) high-resolution camera. The camera was a Sony Alpha 6000 with an adjusted focal length of 25 mm. The octocopter needed approximately 7 min to complete a flight for each plot based on the predefined parameters (e.g., the number of waypoints) and the flight mode was set to semi-automatic. The flight path lines covered the entire study area and produced a set of images of the area. The octocopter was guided by a DJI (Dà-Jiāng Innovations Science and Technology Company, Shen Zhen, China) ground station, which is a global positioning system (GPS) flight planning and waypoint-based autopilot software. We performed three flights in total, one flight per plot, at a height of approximately 70 m above the ground with 80% frontal overlap and 70% side overlap.

In order to improve the accuracy of the 3D model, we also set up four ground control points (GCPs) randomly distributed within each plot; these points were measured using the Leica real-time kinematic (RTK) system, model RX1250XC with centimeter accuracy. Due to the low image quality, four of the 596 original images were excluded from the alignment process. During the alignment process, we set the accuracy to high for optimization of the final 3D model. We used Agisoft Photoscan© software (V 1.2.6, St. Petersburg, Russia) to construct the digital terrain model (DTM) and DSM from the 3D model with a cell size of 0.01 × 0.01 m. The reconstructed mesh of the 3D model was based on automatic classification on certain point classes through the triangulated irregular network (TIN) method. Due to the relatively open canopy in large parts of the study area, small bushes were often abundant in the understory, and, therefore, we classified them as ground points. For setting the parameters for the automatic classification, due to the presence of small bushes near the trees, we decreased the maximum angle from 15 (default value) to 11, the maximum distance from 4 to 1.5 m, while the cell size remained the same. All of the processing was conducted by one computer operator using an Intel® Core™ i7-6700K with a base clock of 4 GHz and 32 GB random access memory (RAM) running with the Windows 10 Professional Edition 64-bit operating system.

We used the Pléiades 1A satellite (launched 16 December 2012) to acquire the space-borne image. The image was taken 27 March 2016, and it had 20 bits/pixel dynamic range of acquisition. For this study, we used one frame with a total area of 25 km² (5 × 5 km). The image consisted of four multispectral bands: RGB, infrared (IR), and one panchromatic (PAN), as can be seen in Table 1. We used six GCPs for georeferencing the satellite image. For point acquisition, we used GPS RTK Leica model RX1250XC with a maximum error of two centimeters. The RTK correction was carried out by using a base/rover set, which sends and receives fast-rate over-the-air RTK data corrections, using the PDLGFU15 radio module.

Table 1. Pleiades-1A satellite sensor characteristics [17].

Imagery Products	Panchromatic: 50-cm resolution, black and white
	2-m multispectral (RGB—red, green, blue) Bundle: 50-cm black and white and 2-m multispectral
Spectral Bands	Panchromatic: 480–830 nm Blue: 430–550 nm Green: 490–610 nm Red: 600–720 nm Near Infrared: 750–950 nm
Image Location Accuracy	With ground control points: 1 m Without ground control points: 3 m (CE90)

Also, dark object correction was used to derive atmospheric optical information for radiometric normalization using the minimum digital number (DN) value of satellite images = water.

Normalized values of each band (RGB) from the UAV and satellite were then used to compare the spectral data (DN) between the UAV and Pléiades bands (separately) at the individual tree level using Equation (1):

$$Z_i = \frac{x_i - x_{min}}{x_{max} - x_{min}} \tag{1}$$

where Z_i describes the normalized data between 0 and 1, x_i describes the spectral data (for both the UAV and satellite), and $x_{\max}$ and $x_{\min}$ are the maximum and minimum value for each band, respectively. In addition, using the nearest neighbor method, we resampled all three multi-spectral bands from the satellite from 2 m to 1 cm to assign more weight to pixels that cover more crown area (Figure 2). For extracting the spectral data (UAV and satellite), we used zonal statistics in ArcGIS desktop V.10.4.1 (ESRI Inc., Redlands, CA, USA), with the crown area for each individual tree as the zonal layer. We considered the same weight for averaging the DN values of pixels within the zonal layer.

To evaluate the greenness of the detected trees from the UAV and eliminate the dry trees and gap areas, the normalized difference vegetation index (NDVI) was used as a detector index (Equation (2)). This index is usually used to determine the visible spectral response by defining the ratio of greenness per individual tree applied to satellite data [18]:

$$\text{NDVI} = \frac{NIR - R}{NIR + R} \tag{2}$$

where NIR stands for near-infrared and R refers to the red band.

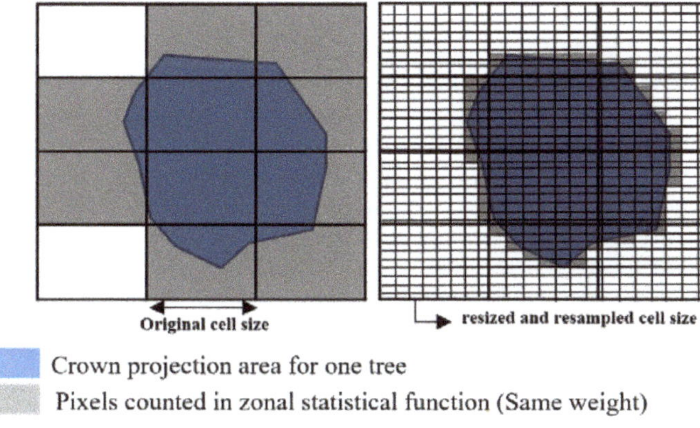

Figure 2. Example of resampled pixels, from 2 m to 1 cm.

2.4. Estimation of Height

To extract the height from the UAV, we computed the CHM, which was derived from the subtraction of the DSM from the DTM. The local maxima algorithm was then used to estimate the height; this algorithm enhances the maximum value within a specified kernel size. As a first step, we used the focal statistics tool in ArcGIS to identify the highest pixel value using the CHM as the input data layer. We performed a low-pass filter to reduce the noise effect and regulate the values of the smoothing window [19]. Among the several processing types we tested in different variances of radius using circular-shaped areas, the best results were at a kernel size with a radius of 1 m based on the average crown diameter derived from the ground measurements (Figure 3). For matching the pixel values, we used the conditional if/else statement on each of the input cells of CHM and focal statistics results, by entering the following command "Con ("CHM" = "focal statistics result", 1)" using the ArcGIS V. 10.4.1 (ESRI, Redlands, CA, United States) raster calculator.

This conditional tool performs an if/else statement on each input cell and it returns a binary layer with a value of zero assigned as no data and a value of one for data. Finally, the return value was the value when the CHM value equaled the focal statistics output.

Figure 3. The applied method of local maxima seeds for the derivation of tree heights (red crosses show the detected treetops by the local maxima approach).

2.5. Estimation of Crown Projection

For the extraction of the crown area, we used the method of inverse watershed segmentation (IWS), as proposed by Panagiotidis et al. [20]. To implement the IWS, we multiplied the CHM model by -1 so that treetops would appear as ponds and crowns as watersheds. Then, we created a flow direction layer for creating hydrological drainage basins [21]. We applied the ArcGIS reclassify tool on the CHM to create a Boolean layer comprised of two categories; we used a threshold value of 15 m as the classification delimiter. Essentially, that means that pixels with values above 15 m were assigned with a value of 1, and those that were lower than 15 m were assigned a value of 0. Afterwards, the CHM was converted to polygons; the center of each polygon was identified, the polygons were converted to lines, and the lines to points. This allowed us to assign points to the periphery of each polygon. Finally, we used the ArcGIS "zonal geometry as table" tool to define individual tree crown diameters.

2.6. Estimation of Cross-Sectional Area and Stem Volume

To estimate the CSA and volume of individual trees, the coefficient of determination (R^2) between the crown diameter and DBH was calculated. The formula of this coefficient was then used to measure the CSA, by replacing the measured crown size (ground) with the crown size extracted from the UAV. Once we had defined the height and CSA, we were able to estimate the stem volume from the UAV. Additionally, for modelling the stem volume, we considered the shape of tree stems as cylinders.

2.7. Extrapolation of Tree Attributes

We were able to calculate calibrated equations that can be used to determine tree attributes such as CSA and stem volume directly using the spectral information from Pléiades for each tree based on the following equations: linear relation between the UAV spectral data and tree parameters such as CSA and volume (Equation (3)) and linear relation between the UAV and Pléiades spectral data (Equation (4)). In details:

$$CSA = AX_1 + B \tag{3}$$

$$X_1 = CX_2 + D \tag{4}$$

where CSA is cross-section area, X_1 is UAV spectral data, and X_2 is satellite spectral data.

By replacing X_1 in Equation (3) with the Equation number (4), we will be able to measure the CSA directly by satellite spectral data using the below equation:

$$CSA = A(CX_2 + D) + B \tag{5}$$

2.8. Statistical Evaluation and Validation of Data

All statistical analyses were conducted in IBM SPSS V.24 (64-bit 2016) and Excel (Microsoft® Office). The linear regression was used to study correlation between the ground data and tree parameters predicted by the UAV.

Pearson correlation coefficient was computed to analyze the relationships between the spectral values of the UAV RGB bands and tree parameters derived from ground inventory and UAV in two different probability values ($p < 0.05$ and $p < 0.01$). In this study, due to the lack of an IR band in the UAV approach, we computed the vegetation index (VI) [22], green-red vegetation index (GRVI) [22], and visible atmospherically-resistant index, green (VARI g) [23] using the following equations:

$$VI = \frac{Green}{Red} \tag{6}$$

$$GRVI = \frac{(Green - Red)}{(Green + Red)} \tag{7}$$

$$VARI\ g = \frac{(Green - red)}{Green + Red - Blue} \tag{8}$$

3. Results

For the sake of simplicity, we divided the results into three parts. In the first part, we presented an overview of tree characteristics in our study area. In the second part, we mainly focused on the potential of the UAV platform deployed with an RGB camera to act as an accurate, alternative field measurement technique. In the third part, we tried to extrapolate the extracted data from the UAV to a larger forested area based on spectral correlation between Pléiades and the UAV.

3.1. General Evaluation of the Study Area

The evaluation of tree attributes showed similarities between the sample plots. However, there was a difference between the variability of sample plots, whereas plot 1 had a lower amount of variability compared with the other two plots (Table 2).

Table 2. Descriptive statistics of the three plots based on ground survey.

Sample	Index	Crown Projection (m²)	DBH (m)	CSA (m²)	Height (m)	Volume (m³)
Plot 1 N = 74	Mean	15.24	0.28	0.06	21.23	1.30
	Variability	30.61	0.15	0.06	7.00	1.30
	Std.	5.20	0.03	0.01	1.33	0.30
Plot 2 N = 72	Mean	14.71	0.28	0.06	23.03	1.43
	Variability	41.54	0.17	0.08	13.80	2.10
	Std.	7.75	0.04	0.02	2.24	0.46
Plot 3 N = 77	Mean	14.94	0.28	0.06	24.49	1.56
	Variability	28.67	0.17	0.08	8.20	2.13
	Std.	6.45	0.03	0.02	2.00	0.45
Total N = 223	Mean	14.97	0.28	0.06	22.96	1.43
	Variability	41.54	0.18	0.08	13.80	2.21
	Std.	6.51	0.03	0.01	2.32	0.42

Std. = standard deviation; DBH = diameter at breast height; CSA = cross-sectional area.

In addition, the NDVI showed that the mean greenness of trees ranged from 0.3 to 0.55 (Figure 4); this range is associated with shrubs-grasslands and temperate forest land-cover classes [24].

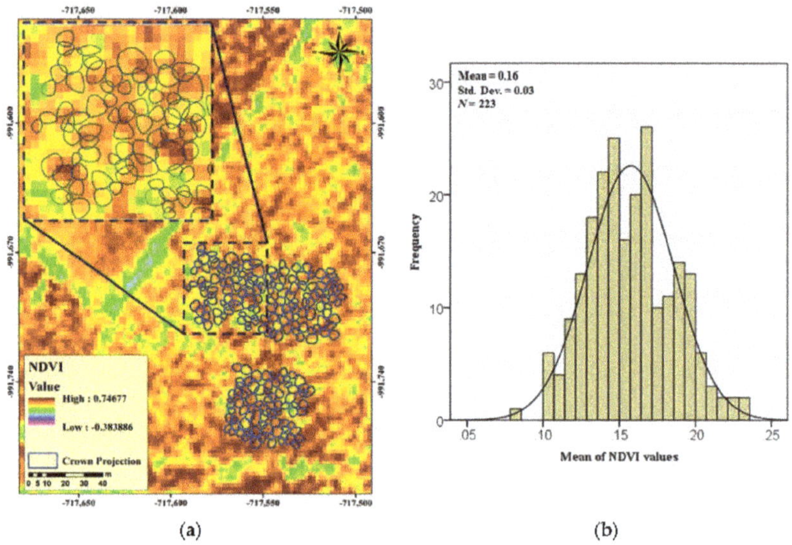

(a)   (b)

Figure 4. (**a**) The NDVI (normalized difference vegetation index) calculated for the study area and (**b**) the histogram of NDVI derived from Pléiades 1A satellite imagery.

3.2. UAV Performance

Linear regression (Figure 5) exhibited a strong relationship in the significant level of α = 0.05 (R^2 = 0.78) between the cross-section area and crown projection derived from ground data with a root mean square error percent (RMSE%) = 11.35 (Table 3; ID = 1). For the crown projections, a strong relationship (R^2 = 0.78; Figure 6a) between the ground and UAV data was observed with an RMSE% = 20.96 (Table 3; ID = 2). Based on these results, we were able to estimate cross-sectional areas from the UAV using adjusted R and RMSE% with an R^2 = 0.60 (Figure 6b) and RMSE% = 15.24 (Table 3; ID = 3). Additionally, the results of the height estimation showed strong correlation between the ground and extracted data from the UAV with an R^2 = 0.87 (Figure 6c) and RMSE% = 3.73 (Table 3; ID = 4). Finally, we calculated the stem volume based on data from the UAV and the ground. The comparison of the stem volume between the ground and UAV data showed significant correlation with R^2 = 0.71 (Figure 6d) and RMSE% = 15.88 (Table 3; ID = 5).

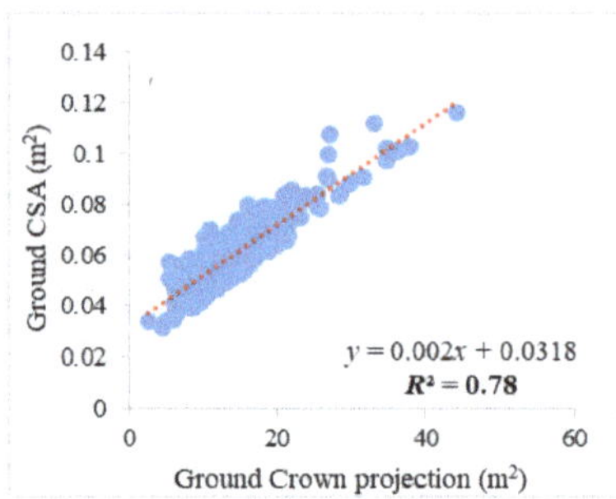

Figure 5. The correlation between crown projection and cross-sectional areas.

Table 3. Statistical summary table, where ID represents the simplicity of identification of each tested regression; N: number of observations; RMSE: root mean square error; and df: degrees of freedom. ID represents the simplicity of identification of each tested regression.

ID	N	RMSE	RMSE%	Bias	Bias %	df	*p*-Value
1	223	0.0069	11.35	0.0018	2.96	222	0.00
2	223	3.14	20.96	-	-	222	0.00
3	223	0.01	15.24	-	-	222	0.00
4	223	0.86	3.73	-	-	222	0.00
5	223	0.23	15.88	-	-	222	0.00

ID 1 indicates the relation between the ground CSA versus ground crown projection; ID 2 indicates the relation between the crown projection derived from UAV and ground data; ID 3 indicates the relation between the CSA derived from the UAV and ground data; ID 4 indicates the relation between the estimated height from the UAV and ground data; and ID 5 indicates the relation between the stem volume derived from the UAV and ground data.

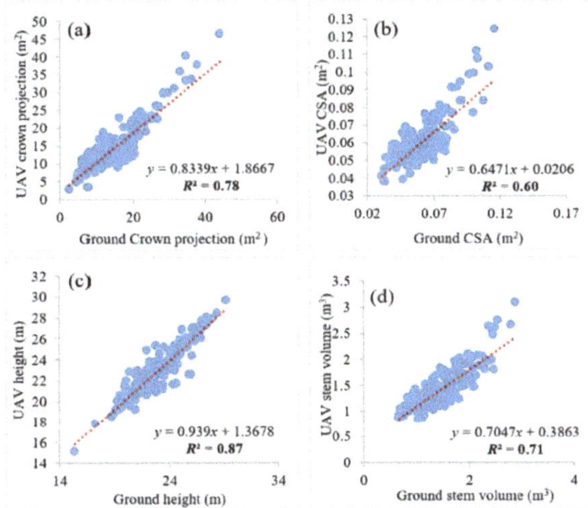

Figure 6. The correlation of (**a**) crown projection between the ground and UAV (unmanned aerial vehicle) data in m^2; (**b**) cross-sectional areas (CSA) between the ground and UAV data in m^2; (**c**) height between the ground and UAV data in m; and (**d**) stem volume between the ground and UAV data in m^3.

Moreover, the results of the correlation coefficient indicated that the most important descriptive statistic index for all models was the sum of the spectral data as can be seen in Table 4. Finally, our results showed that vegetation indicators that were computed using RGB bands had a significant correlation at the level of $\alpha = 0.01$ with the individual tree stem volume, as can be seen in Table 5.

Table 4. Pearson correlation between the UAV spectral descriptive data and tree attributes at the level of individual trees.

Tree Parameter	Min	Max	Mean	Std.	Sum	Median
CSA (Ground)	−0.208 **	0.133 *	0.224 **	0.160 *	0.864 **	0.229 **
Stem volume (Ground)	−0.200 **	0.131	0.239 **	0.180 **	0.795 **	0.255 **
CSA (UAV)	−0.140 *	0.0641	0.055	0.012	0.821 **	0.049
Stem volume (UAV)	−0.158 *	0.072	0.107	0.05	0.795 **	0.115

** Correlation is significant at the 0.01 level (two-tailed). * Correlation is significant at the 0.05 level (two-tailed).

Table 5. The Pearson correlation coefficient between the different vegetation indicators derived from UAV and tree attributes at the level of individual trees. VI: vegetation index; GRVI: green-red vegetation index; and VARI g: vegetation atmospherically resilient index, green.

Vegetation Index	CSA (Ground)	Stem Volume (Ground)	CSA (UAV)	Stem Volume (UAV)
VI	0.103	0.208 **	0.110	0.218 **
GRVI	−0.106	−0.210 **	−0.113	−0.221 **
VARI g	−0.106	−0.211 **	−0.112	−0.221 **

** Correlation is significant at the 0.01 level (two-tailed).

The selection process for the best independent variable for further analysis was based on the largest positive or negative correlations with the dependent variables. Based on our results, we chose the sum of pixel values as an independent variable to calculate the regression between the spectral data and tree attributes (Figure 7). Our results showed there was high correlation between the UAV main bands and tree attributes based on R^2 (Figure 7).

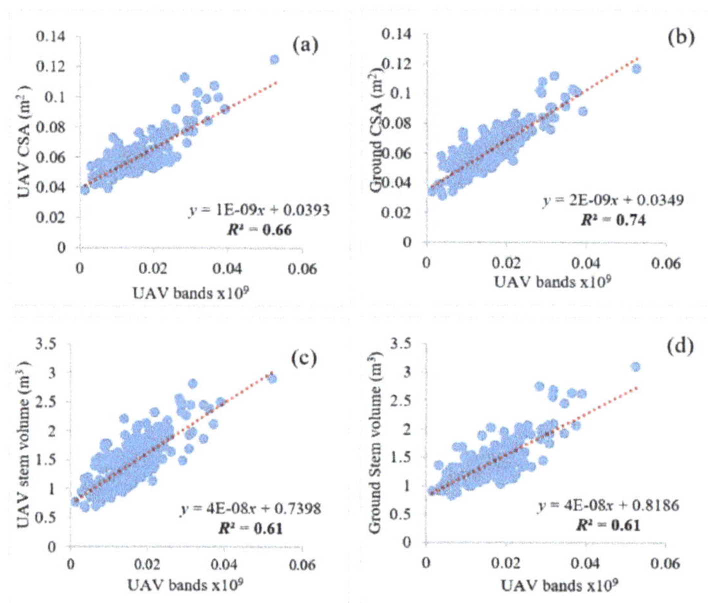

Figure 7. The relationship between spectral data of the UAV bands ($\times 10^9$) and tree characteristics: (**a**) UAV cross-section area in m²; (**b**) ground cross-section area in m²; (**c**) UAV stem volume in m³; and (**d**) ground stem volume in m³.

3.3. UAV-Pléiades Extrapolation

Overall, our findings indicated that there is a strong relationship between the UAV and Pléiades spectral data with an $R^2 = 0.94$ (Figure 8). Based on the regression model equations (Figures 7 and 8), we were able to calculate calibrated formulas that could determine tree attributes, such as CSA and stem volume (Table 6), directly by using the spectral information from Pléiades for each tree. Formulas presented in Table 6 were used to extrapolate the extracted data from the UAV to a larger area based on satellite spectral information.

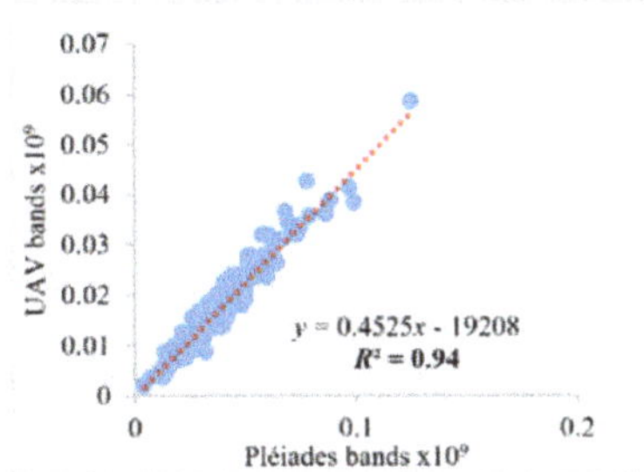

Figure 8. The relationship between spectral data between the UAV and Pléiades bands ($\times 10^9$).

Table 6. Basic errors of CSA and volume based on extrapolated data.

Index	CSA	Volume
RMSE	0.018	0.21
RMSE%	30.56	31.01
Bias	−0.015	−0.39

Our results showed that the computed formulas (9 and 10) could be used for the extrapolation of CSA and volume at the individual tree level with significant accuracy, as can be seen in Table 6. In addition, Figure 9 indicates that there were no significant differences between the mean of CSA and volume derived from UAV and ground data. Also, the same figure and Table 6 show that the extrapolation method has estimated the tree parameters with reasonable accuracy.

$$\text{CSA} = 1 \times 10^{-9} \, (0.4525x - 19208) + 0.0392 \tag{9}$$

$$\text{Volume} = 4 \times 10^{-8} \, (0.4525x - 19208) + 0.7398 \tag{10}$$

where the x is the spectral information derived from satellite imagery.

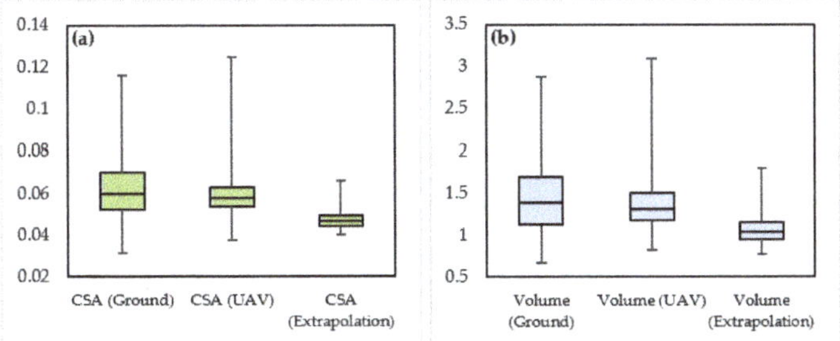

Figure 9. Box-and-whisker plots for comparison of the differences between the means of tree parameters: (**a**) CSA (m^2) (**b**) Volume (m^3) derived from different approaches. The medians of the measured values are marked by vertical lines inside the boxes. The ends of the boxes are the interquartile range (upper: Q3 and lower: Q1). The whiskers show the highest and lowest observations.

4. Discussion

Many studies have provided good examples of 3D model reconstructions of VHR DSMs from UAV-based imagery to systematically observe forest attributes, such as tree crown, tree height, and DBH [25,26]. Structural forest attributes are commonly extracted from the CHM using regression models to predict tree characteristics for forest inventory purposes [27]. Detailed CHMs from remote sensing data have recently gained more attention because they can be used to efficiently predict key forest parameters, such as tree height, crown projection, and stem volume. In this study, it was demonstrated that a consumer-grade camera deployed on a UAV platform generated a 3D scene of the entire study area, which was used to quantify single-tree parameters based on automatic and semi-automatic methods that can be used to support detailed construction or update existing forest inventories. Of course, this is just one of the basic reasons why UAVs are gradually replacing conventional field measurements. However, some applications require more accurate results in terms of absolute means, and this low-cost aerial approach would, therefore, require the collection of ground truth data to ensure this.

Based on the linear regression and RMSE results, we concluded that there is generally a high correlation between estimated (UAV) and observed data (ground) (Table 3; Figures 5 and 6). In this particular study, the methodological approach and the algorithms that were used to determine tree heights and crown diameters were influenced by the relative homogeneity of the study area; all trees were of similar age and had similar morphological characteristics (Table 2). Aslo, as it can be seen in Figure 9, the amount of CSA and volume for 50% of the trees distributed around the median, but the remaining 50% were distributed in a wider range.

In general, estimation of forest parameters in homogeneous forests is preferable for the application of algorithms, such as local maxima and IWS, because homogeneous forests allow for higher precision using a single kernel size method for smoothing the CHM [20,28,29]. It is evident that the UAV can be used to efficiently estimate heights and crown diameters (Figure 6a–c, and Table 3).

Since there was a high correlation between the crown and DBH from the ground measurements (Figure 5), and due to the fact that UAVs can be used to estimate both heights and crown diameters, there is a possibility for indirect measurement of stem diameters (Figure 6b and Table 3; ID 3) and volume (Figure 6d and Table 3; ID 5) [30].

Additionally, we evaluated the relationship between the tree parameters (CSA and stem volume of individual trees) and the spectral information that was extracted from the UAV imagery. Our results showed that there is a significant correlation between the CSA and stem volume and spectral/textural values derived from the UAV. These results are similar to the results of several studies that have emphasized the possibility of the estimation of tree/stand parameters, such as basal area (BA) and volume, using spectral (main bands) and textural (vegetation indicators) values of aerial/satellite images [31–33]. For the vegetation index (VI), a low value for this attribute implies lower density and stem volume, while higher values indicate dense canopy and higher stem volumes. This explains the positive correlation that we found between the VI, CSA, and stem volume in our study. This result is in accordance with the results of Wallner et al. [31].

The inverse relation between the stem volume and VARI g and GRVI indicators (Table 5) can be explained as follows: due to the fact that vegetation absorbs more red light and reflects more green light, a high value in the red band indicates less vegetation occurrence, which can be explained by the negative correlation between (i) VARI g and GRVI, and (ii) CSA and stem volume.

Finally, our results showed that spectral reflection of individual trees from both Pléiades and the UAV had a high correlation of more than 90% (Figure 8). Based on this finding, high-resolution satellite imagery can be used to extrapolate areas that cannot be reached with a UAV. However, the described methodology can be applied only in the case where the areas present similar structural characteristics. Although the extrapolation process in our study caused underestimation of tree parameters compared to that of ground data and UAV (Figure 9), the suggested methodology can be used for practical forestry and can open up new scientific areas of extrapolation methods to inspire other researchers in the forest community. In future work, we plan to use auxiliary data such as environmental data (edaphic, climatic, and topographic) in order to be able to enhance the methodology for homogeneous areas and to calibrate and assess the methodology in the case of diverse forested areas. For areas with higher variability, one suggestion is to divide the whole area into more homogenous areas, through the construction of homogenous groups or classes. We also assume that the use of hyperspectral sensors can lead to a better result with higher accuracy.

5. Conclusions

In this study, we proposed a method to test the performance of UAV image-based point clouds to accurately estimate tree attributes. For this purpose, detection algorithms based on high-quality CHM were used. To potentially improve the results of the 3D image reconstruction model and ensure the integrity of the results based on CHM, we used four GCPs, measured with RTK GPS. Many studies have previously treated estimates of tree parameters, such as tree crown delineation and treetop detection, as two separate procedures [28]. We extrapolated the estimated data from the UAV to a larger area based on a significant correlation (R-squared values and Pearson correlation percentages were greater than 0.90) between the spectral data from UAV and Pléiades. This study demonstrated that it is possible to use calibrated (linear regression) formulas (Table 6) to extrapolate data into larger forested areas (downscaling). Precision forestry is focusing on the use of high-resolution data to support site-specific tactical and operational decision-making (e.g., area productivity) over large forest areas. Therefore, the application of this study, as well as other similar studies, should be explored in the content of the European Common Policy to assess the full potential of these methods for covering larger forested areas.

Regarding the performance of remote sensing versus field measurements, based on our empirical data, we can conclude that the positive comparisons between reference ground measurements and remote sensing estimation of tree attributes confirmed the potential of the workflow process that can be applied as a quick and effective alternative technique to characterize forest tree parameters.

Acknowledgments: This research was supported by the project of the Internal Grant Agency (IGA) of the Faculty of Forestry and Wood Sciences, Czech University of Life Sciences (CULS) in Prague (No. A14/16), and by the Ministry of Agriculture of the Czech Republic Project (No. QJ1520037).

Author Contributions: Azadeh Abdollahnejad and Dimitrios Panagiotidis designed the experiment, wrote the manuscript, and led the image processing and evaluation. Azadeh Abdollahnejad performed the statistical analyses. Peter Surový and Dimitrios Panagiotidis acquired the UAV data. Peter Surový supervised the manuscript.

Conflicts of Interest: The authors declare no conflict of interest.

References

1. Järnstedt, J.; Pekkarinen, A.; Tuominen, S.; Ginzler, C.; Holopainen, M.; Viitala, R. Forest variable estimation using a high-resolution digital surface model. *J. Photogramm. Remote Sens.* **2012**, *74*, 78–84. [CrossRef]
2. Straub, C.; Stepper, C.; Seitz, R.; Waser, L.T. Potential of UltraCamX stereo images for estimating timber volume and basal area at the plot level in mixed European forests. *Can. J. For. Res.* **2013**, *43*, 731–741. [CrossRef]
3. White, J.C.; Wulder, M.A.; Vastaranta, M.; Coops, N.C.; Pitt, D.; Woods, M. The utility of image-based point clouds for forest inventory: A comparison with airborne laser scanning. *Forests* **2013**, *4*, 518–536. [CrossRef]
4. Stepper, C.; Straub, C.; Pretzsch, H. Using semi-global matching point clouds to estimate growing stock at the plot and stand levels: Application for a broadleaf-dominated forest in central Europe. *Can. J. For. Res.* **2014**, *45*, 111–123. [CrossRef]
5. Carleer, A.P.; Debeir, O.; Wolff, E. Assessment of Very High Spatial Resolution Satellite Image Segmentations. *Photogramm. Eng. Remote Sens.* **2005**, *71*, 1285–1294. [CrossRef]
6. Edson, C.; Wing, M.G. Airborne Light Detection and Ranging (LiDAR) for Individual Tree Stem Location, Height, and Biomass Measurements. *Remote Sens.* **2011**, *3*, 2494–2528. [CrossRef]
7. Popescu, S.C.; Wynne, R.H.; Nelson, R.F. Measuring Individual Tree Crown Diameter with Lidar and Assessing Its Influence on Estimating Forest Volume and Biomass. *Can. J. For. Res.* **2003**, *29*, 564–577. [CrossRef]

8.   Tomppo, E.; Olsson, H.; Ståhl, G.; Nilsson, M.; Hagner, O.; Katila, M. Combining national forest inventory field plots and remote sensing data for forest databases. *Remote Sens. Environ.* **2008**, *112*, 1982–1999. [CrossRef]

9.   McRoberts, R.E.; Cohen, W.B.; Næsset, E.; Stehman, S.V.; Tomppo, E.O. Using remotely sensed data to construct and assess forest attribute maps and related spatial products. *Scand. J. For. Res.* **2010**, *25*, 340–367. [CrossRef]

10.   Shao, Z.; Zhang, L. Estimating Forest Aboveground Biomass by Combining Optical and SAR Data: A Case Study in Genhe, Inner Mongolia, China. *Sensors* **2016**, *16*, 834. [CrossRef] [PubMed]

11.   Lehmann, J.R.K.; Nieberding, F.; Prinz, T.; Knoth, C. Analysis of Unmanned Aerial System-Based CIR Images in Forestry—A New Perspective to Monitor Pest Infestation Levels. *Forests* **2015**, *6*, 594–612. [CrossRef]

12.   Kayitakire, F.; Hamel, C.; Defourny, P. Retrieving forest structure variables based on image texture analysis and Ikonos-2 imagery. *Remote Sens. Environ.* **2006**, *102*, 390–401. [CrossRef]

13.   St-Onge, B.; Hu, Y.; Vega, C. Mapping the height and above-ground biomass of a mixed forest using LiDAR and stereo Ikonos images. *Int. J. Remote Sens.* **2008**, *29*, 1277–1294. [CrossRef]

14.   Ozdemir, I.; Karnieli, A. Predicting forest structural parameters using the image texture derived from worldview-2 multispectral imagery in a dryland forest, Israel. *Int. J. Appl. Earth Obs. Geoinf.* **2011**, *13*, 701–710. [CrossRef]

15.   Shamsoddini, A.; Trinder, J.C.; Turner, R. Pine plantation structure mapping usingWorldView-2 multispectral image. *Int. J. Remote Sens.* **2013**, *34*, 3986–4007. [CrossRef]

16.   Immitzer, M.; Stepper, C.; Böck, S.; Straub, C.; Atzberger, C. Forest ecology and management use of WorldView-2 stereo imagery and National Forest Inventory data for wall-to-wall mapping of growing stock. *For. Ecol. Manag.* **2016**, *359*, 232–246. [CrossRef]

17.   Astrium GEO-Information Services, Pléiades Imagery User Guide. Available online: http://www.cscrs.itu. edu.tr/assets/downloads/PleiadesUserGuide.pdf (accessed on 1 August 2012).

18.   Surový, P.; Ribeiro, N.A.; Pereira, J.S.; Yoshimoto, A. Estimation of Cork Production Using Aerial Imagery. *Rev. Árvore* **2015**, *39*, 853–861. [CrossRef]

19.   Pitkänen, J.; Maltamo, M.; Hyyppä, J.; Yu, X. Adaptive Methods for Individual Tree Detection on Airborne Laser Based Canopy Height Model. In *Proceedings of ISPRS Working Group VIII/2: "Laser-Scanners for Forest and Landscape Assessment"*; Theis, M., Koch, B., Spiecker, H., Weinacker, H., Eds.; University of Freiburg: Freiburg, Germany, 2004; pp. 187–191.

20.   Panagiotidis, D.; Abdollahnejad, A.; Surový, P.; Chiteculo, V. Determining tree height and crown diameter from high-resolution UAV imagery. *Int. J. Remote Sens.* **2016**, *38*, 1–19. [CrossRef]

21.   Wannasiri, W.; Nagai, M.; Honda, K.; Santitamont, P.; Miphokasap, P. Extraction of Mangrove Biophysical Parameters Using Airborne LiDAR. *Remote Sens.* **2013**, *5*, 1787–1808. [CrossRef]

22.   Tucker, C.J. Red and Photographic Infrared Linear Combinations for Monitoring Vegetation. *Remote Sens. Environ.* **1979**, *8*, 127–150. [CrossRef]

23.   Gitelson, A.; Stark, R.; Grits, U.; Rundquist, D.; Kaufman, Y.; Derry, D. Vegetation and Soil Lines in Visible Spectral Space: A Concept and Technique for Remote Estimation of Vegetation Fraction. *Int. J. Remote Sens.* **2002**, *23*, 2537–2562. [CrossRef]

24.   Arulbalaji, P.; Gurugnanam, B. Evaluating the Normalized Difference Vegetation Index Using Landsat Data by Envi in Salem District, Tamilnadu, India. *Int. J. Dev. Res.* **2014**, *4*, 1844–1846.

25.   Baltsavias, E.; Gruen, A.; Eisenbeiss, H.; Zhang, L.; Waser, L.T. High-Quality Image Matching and Automated Generation of 3D Tree Models. *Int. J. Remote Sens.* **2008**, *29*, 1243–1259. [CrossRef]

26.   Dandois, J.P.; Ellis, E.C. Remote Sensing of Vegetation Structure Using Computer Vision. *Remote Sens.* **2010**, *2*, 1157–1176. [CrossRef]

27.   Næsset, E. Predicting Forest Stand Characteristics with Airborne Scanning Laser Using a Practical Two-Stage Procedure and Field Data. *Remote Sens. Environ.* **2002**, *80*, 88–99. [CrossRef]

28.   Wang, L.; Gong, P.; Biging, G.S. Individual Tree-Crown Delineation and Treetop Detection in High-Spatial-Resolution Aerial Imagery. *Photogramm. Eng. Remote Sens.* **2004**, *70*, 351–357. [CrossRef]

29.   Jakubowski, M.K.; Li, W.; Guo, Q.; Kelly, M. Delineating Individual Trees from Lidar Data: A Comparison of Vector- and Raster-based Segmentation Approaches. *Remote Sens.* **2013**, *5*, 4163–4186. [CrossRef]

30. Tuominen, S.; Balazs, A.; Saari, H.; Pölönen, I.; Sarkeala, J.; Viitala, R. Unmanned aerial system imagery and photogrammetric canopy height data in area-based estimation of forest variables. *Silva Fenn.* **2015**, *49*, 1348. [CrossRef]
31. Wallner, A.; Elatawneh, A.; Schneider, T.; Knoke, T. Estimation of forest structural information using rapideye satellite data. *Forestry* **2015**, *88*, 96–107. [CrossRef]
32. Straub, C.; Weinacker, H.; Koch, B. A comparison of different methods for forest resource estimation using information from airborne laser scanning and CIR orthophotos. *Eur. J. For. Res.* **2010**, *129*, 1069–1080. [CrossRef]
33. Heiskanen, J. Estimating aboveground tree biomass and leaf area index in a mountain birch forest using ASTER satellite data. *Int. J. Remote Sens.* **2006**, *27*, 1135–1158. [CrossRef]

 *forests*

Article

Effects of Tree Trunks on Estimation of Clumping Index and LAI from HemiView and Terrestrial LiDAR

Yunfei Bao [1,*], Wenjian Ni [2], Dianzhong Wang [1], Chunyu Yue [1], Hongyan He [1] and Hans Verbeeck [3]

[1]   Beijing Institute of Space Mechanics and Electricity, No. 104, Road Youyi, Beijing 100094, China; drgnw@163.com (D.W.); ycy1893@163.com (C.Y.); yesterday75@163.com (H.H.)
[2]   State Key Laboratory of Remote Sensing Science, Institute of Remote Sensing and Digital Earth, Chinese Academy of Sciences, Beijing 100101, China; niwj@radi.ac.cn
[3]   CAVELab Computational and Applied Vegetation Ecology, Faculty of Bioscience Engineering, Ghent University, 9000 Ghent, Belgium; Hans.Verbeeck@ugent.be
*   Correspondence: byf_rs@163.com; Tel.: +86-10-6811-4759

Received: 25 January 2018; Accepted: 13 March 2018; Published: 15 March 2018

Abstract: Estimating clumping indices is important for determining the leaf area index (LAI) of forest canopies. The spatial distribution of the clumping index is vital for LAI estimation. However, the neglect of woody tissue can result in biased clumping index estimates when indirectly deriving them from the gap probability and LAI observations. It is difficult to effectively and automatically extract woody tissue from digital hemispherical photos. In this study, a method for the automatic detection of trunks from Terrestrial Laser Scanning (TLS) data was used. Between-crown and within-crown gaps from TLS data were separated to calculate the clumping index. Subsequently, we analyzed the gap probability, clumping index, and LAI estimates based on TLS and HemiView data in consideration of woody tissue (trunks). Although the clumping index estimated from TLS had better agreement ($R^2 = 0.761$) than that from HemiView, the change of angular distribution of the clumping index affected by the trunks from TLS data was more obvious than with the HemiView data. Finally, the exclusion of the trunks led to a reduction in the average LAI by ~19.6% and 8.9%, respectively, for the two methods. These results also showed that the detection of woody tissue was more helpful for the estimation of clumping index distribution. Moreover, the angular distribution of the clumping index is more important for the LAI estimate than the average clumping index value. We concluded that woody tissue should be detected for the clumping index estimate from TLS data, and 3D information could be used for estimating the angular distribution of the clumping index, which is essential for highly accurate LAI field measurements.

Keywords: clumping index; leaf area index; trunk; terrestrial LiDAR; HemiView

1. Introduction

Leaf area index (LAI) is an important factor in describing ecosystem structure and function. It not only relates to photosynthetic and respiration activities, but also plays a dominant role in the reflectance characteristics of vegetated land surfaces [1]. In many studies, passive optical remote sensing images from satellites and airborne platforms have been used to retrieve LAI [2–4]. Light Detection and Ranging (LiDAR) is an active optical sensing technique and offers an alternative to estimate forest structure in 3D. It has therefore been widely used to estimate forest structure parameters [5–9], especially the LAI parameters [10–15].

However, the calibration of satellite-based and airborne-based LAI maps requires adequate ground truth points. The fast and accurate extraction of canopy structure parameters in the field is very important for ground-based LAI estimates, which are important to evaluate the accuracy of airborne LAI estimates. However, in some studies, the LAI has often been replaced with the retrieved plant area index (PAI), which includes the leaf area index, wood area index, and other component area indices. This does not benefit the validation of retrieval results from satellite images. Currently, ground-based LAI measurements can be performed through two major techniques: direct and indirect measurements [16,17]. Direct (Destructive) LAI determination is not compatible with long-term monitoring of the spatial and temporal dynamics of leaf area development [18]. Optical indirect techniques are the methods most commonly used in validation studies due to the speed and ease of LAI measurements over large areas. Optical techniques are based on the measurements of light transmittance through canopies [17,19] and have been implemented using multiple commercial optical instruments including the LAI-2000 plant canopy analyzer (LI-COR, Lincoln, NE, USA), AccuPAR (Decagon Devices, Inc., Pullman, WA, USA), Tracing radiation and Architecture of Canopies system (TRAC, 3rd Wave, Nepean, Ontario, Canada), and digital hemispherical photographs [17]. The accuracy of these indirect LAI measurements is often limited by leaf clumping, woody-to-total area ratio, and illumination conditions.

The clumping effect is an important factor complicating indirect LAI measurements. This index at different scales quantifies the spatial pattern of leaf distribution and is a transforming factor for calculating the true LAI from the effective LAI. However, the estimation of the clumping index can be affected by woody material. There are typically two types of clumping index retrieval methods: the finite length averaging method (hereafter LX) [20], and the gap size distribution method (hereafter CC) [21]. Due to the limitations of the two methods [22,23], a new method was developed by combining the gap size distribution and logarithmic methods for LAI estimation (hereafter CLX) [23]. However, the clumping index results from the three methods notably differ. The CC method does not show radical changes of the clumping index value outside zenith angles ranging from 30° to 60°, which is in contrast to the other two methods. Based on hemispherical photography, the clumping index increases with the view zenith angle in both boreal and temperate forests [23–25]. Although the angular dependence of clumping is an important characteristic in determining the clumping index, a limited number of clumping indices have been estimated within a narrow and moderate range of zenith angles including 30–80°, 54–63° and 57.5° [24,26,27]. Given the different ranges of the view zenith angle, different clumping index values were retrieved from three instruments in an open savanna ecosystem using the same methods as CLX or CC [28]. This indicates that estimating the angular distribution of the clumping index from hemispherical photography within a stand level remains challenging [23,28]. Meanwhile, several clumping index values covering a wide range of view zenith angles are required to calculate hemispherical average clumping index values, where the clumping index changes depending on the view zenith angle [28]. To quantify and interpret the clumping index of the forest canopy, several questions need to be addressed including the range of the view zenith angle and how the clumping index changes with the view zenith angle. Additionally, during LAI retrieval, gap probability and clumping index estimates are affected by the quantity of woody tissue. For deciduous forests, the wood area index is generally estimated during leafless periods and subtracted from the total PAI [29]. However, this method is not suitable for coniferous forests. Although the contribution of woody tissue to gap probability and LAI estimation is obvious, it remains undetermined in the case of indirect measurements in coniferous forests [30]. This problem is even bigger in Mediterranean forests or other semi-arid forests where trees and shrubs usually have small leaves and the importance of trunks should be proportionally more important. A few studies have aimed to quantify the effect of woody tissue on the gap probability and clumping index estimation by developing an image analysis approach [31,32], which can still be affected by imaging conditions and image processing techniques.

Terrestrial Laser Scanning (TLS) has been used as an active optical sensing tool to estimate forest structure parameters from field measurements [9,33]. 3D space information from TLS is useful for the estimation of gap distribution and tree structure. Recent studies have proven that TLS offers high performance estimates of gap probability and LAI [22,34,35]; however, very few studies have addressed the impact of the clumping index in these LAI estimates. Moorthy et al. (2011) calculated the clumping index using TLS data based on the CC method and an obvious improvement in LAI estimates was achieved in comparison to field estimates from LAI-2000 [36]. The clumping index was estimated by using the gap size information derived from images of gap probabilities yielded by the Echidna Validation Instrument (EVI), which is a TLS method with dual wavelengths [24]. The TLS scanner can produce a 'pseudo' Normalized Difference Vegetation Index (NDVI) image, which is beneficial for the detection of woody material. In their study [24], woody structures were considered in clumping index estimates and identified using a threshold of the ratio of total power to reflected pulse width. García et al. evaluated the potential of TLS to estimate the clumping index in different vegetation types based on the spatial distribution of the returns, the gap distribution and the gap size distribution, and proved that TLS data could be used to estimate the clumping index [37]. Given the importance of the spatial distribution of the clumping index, the angular distribution of woody tissue and its effect on the spatial distribution of the clumping index needs to be estimated. Meanwhile, most of the woody tissue is trunk, which has a greater effect on the clumping index than other components. To solve this problem, we introduced a method that could be used to exclude the trunk from terrestrial LiDAR data and we analyzed the influence of the trunk on the gap probability, clumping index, and LAI at all view zenith angles.

In this study, we first extracted the tree trunk point clouds from terrestrial LiDAR data. Within-crown and between-crown gap probabilities were calculated to estimate the clumping index and LAI. The angular dependence of the gap probability, clumping index, and LAI were analyzed before and after the exclusion of tree trunks. At the same time, we compared the performance of TLS with that of digital hemispherical photos with respect to the angular distribution of the clumping index and LAI and the analysis of trunk effects. We addressed the following research questions: (1) How does the clumping index change within the whole range of view zenith angles? (2) How strong is the influence of tree trunks on the canopy gap probability, clumping index, and LAI at different view zenith angles? (3) How different are the canopy gap probability, clumping index, and LAI estimates derived from TLS data and HemiView photos?

2. Materials

2.1. Study Area Description

The study area is located in the Qilian Mountains within the Gansu Province in Western China ($38°19'$–$38°29'$ N, $100°12'$–$100°20'$ E), as shown in Figure 1. The elevation varies from 2500 to 3800 m above sea level. This area is cold and dry with low precipitation and presents the typical features of a temperate continental mountainous climate. The atmospheric circulation in this area is controlled by the Mongolia anticyclone in winter. Influenced by climate and terrain, the prevalent vegetation types in the study area are mountainous pastures and forests. The dominant vegetation types include two tree species, the Qinghai Spruce (*Picea crassifolia* Kom.) and Qilian Juniper (*Sabina przewalskii* Kom.), as well as grassland. The vegetation density varies depending on the terrain, soil, water, and climate factors. In this study, the coniferous tree species Qinghai spruce was selected as the study object.

2.2. Sampling Design and Field Measurements

A 100×100 m sample plot was selected in the study area. The plot was selected as the major area for field measurements (Figure 1). The ground in the plot was relatively flat with a slope of about five degrees. The plot was dominated by the same tree species, the Qinghai spruce. This plot was divided into 16 subplots, with each subplot having a size of 25×25 m. Measurement points were located in the center of each subplot (Figure 2) and TLS and HemiView (Delta-T devices Ltd., Cambridge, UK) were used to acquire data at the 13 subplots (red solid circles in Figure 1).

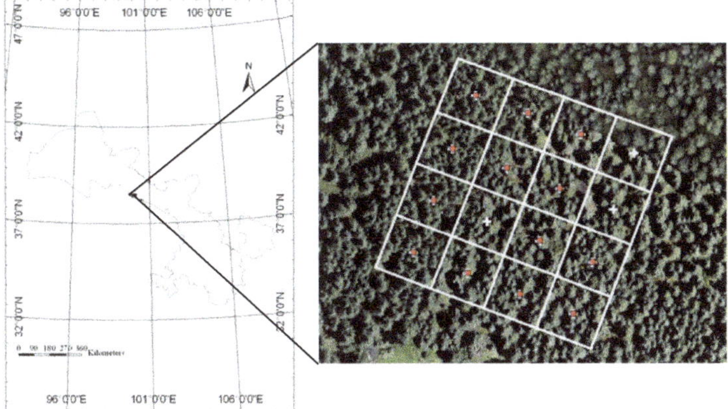

Figure 1. Study area and distribution of measurement points (base map is an aerial photo), located within the Qilian Mountains in Gansu Province, Western China. The sixteen white solid crosses represent the locations of the centers of the subplots and the thirteen red solid circles represent the locations at which data were measured.

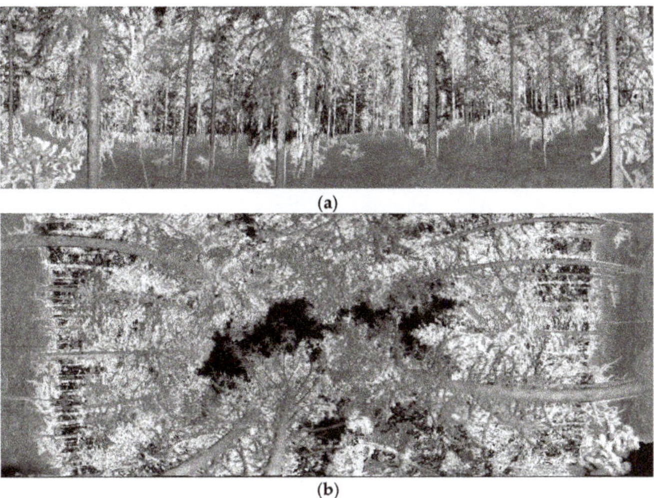

Figure 2. Intensity of vertical and tilt scanning of the terrestrial laser scanner. (**a**) Vertical mount; and (**b**) Tilt mount. The black areas indicate where no points were recorded.

The terrestrial laser scanner RIEGL LMS-Z360i (RIEGL, Horn, Austria) was used in this study. The configuration of RIEGL LMS-Z360i used in the field is shown in Table 1. The scanner is based on the principle of time-of-flight measurement with short infrared laser pulses (900 nm). The first and last returns could be recorded, and the first returns of points were used in our study. When the laser scanner was vertically mounted on a tripod located at the center of the subplot, the vertical range of scanning (line scanning angle) was $-50°\sim40°$ and the horizontal range of scanning (frame scanning angle) was $360°$ (Table 1) The vertical angle of $40°\sim90°$ needed to be obtained by tilt scanning of TLS at the same position to cover the upper hemisphere view. The TLS data from the vertical and tilt scan at the same position were used for the detection of trunks and the estimation of gap probability in the study. Figure 2 shows the intensities obtained using the vertical and tilt mounts of the terrestrial laser scanner.

Table 1. Configuration of the LMS-360i laser scanner.

Parameters	Characteristics/Value
Scanning mechanism	Rotating/oscillating
Measurement principle	Single-shot time-of-flight measurement
Target detection modes	First, last, or alternating target
Laser wavelength	900 nm
Measurement range	1~200 m
Laser pulse repetition rate	24,000 Hz
Laser beam divergence	$\leq$2 mrad (focused to infinity)
Measurement resolution	5 mm
Line scan angle range	$-50°{\sim}+40°$
Frame scan angle range	$0°{\sim}360°$
Line angle step width	0.008°
Frame angle step width	0.01°

In addition, orientation information can be acquired using a compass and spirit level. Three angles of the vertical and tilt scans were obtained using this method including the roll, pitch, and yaw angle. A compass was employed to orient the vertical and tilt scans of the laser scanner and to measure the yaw angle around the scanner's Z-axis. The roll angle of the X-axis of the scanner and the pitch angle of the Y-axis of the scanner were determined with an electronic spirit level. As the intersection angle between the vertical and tilt mount was 90°, we obtained a tilt mount calibration matrix from the three angles based on the method in [38]. The data acquired at the tilt mount can be transformed into its corresponding vertical mount coordinates using the tilt mount calibration matrix.

We took hemispherical photographs as field samples using HemiView (Delta-T Device, Cambridge, UK) including a self-levelling mount and a Nikon Coolpix 8400 with FC-E9 adapter lens (Nikon, Tokyo, Japan). The camera and lens were mounted and levelled with a bubble level before use. The measurements were performed under overcast conditions or conditions of diffuse skylight. A hemispherical photograph provides a permanent record and is a valuable source of information for canopy gaps. The photographs were processed with digital hemispherical photography (DHP) software to derive the effective LAI. Hemispherical photographs were acquired at a height of 0.8 m for each station in this plot. An example is shown in Figure 3. Similar to the case of the terrestrial LiDAR, the measurement locations in the plot are shown in Figure 1.

Figure 3. Example of a hemispherical photograph.

3. Methods

3.1. Gap Probability Theory

Beer's Law, which relates the absorption of light to the properties of particles [39], was used to relate the LAI to the gap probability of the canopy. The original formulation of Beer's Law assumes the random distribution of light-intercepting elements in the pathway of penetrating beams. However, leaves in natural forest canopies are clumped within crowns instead of being randomly distributed. Therefore, Nilson introduced the clumping index $\Omega(\theta)$ to the relationship between the LAI and gap probability [40]:

$$P(\theta) = \exp\left(\frac{-L_e G(\theta)}{\cos(\theta)}\right) = \exp\left(\frac{-L\Omega(\theta)G(\theta)}{\cos(\theta)}\right),\tag{1}$$

where θ is the view zenith angle; $G(\theta)$ is the leaf projection function, namely the G-function, which quantifies the projection coefficient of unit foliage area on a plane perpendicular to the view direction [41]; and $\Omega(\theta)$ is the clumping index, which quantifies the degree of the deviation of foliage spatial distribution from the random case, and it is a value between 0 and 1 (the higher value denotes less clumping degree). L_e and L are the effective LAI and true LAI, respectively. Random distribution of the elements (no clumping) was assumed for the retrieval of L_e while L could be calculated after the clumping index was taken into account. In fact, L refers to the plant area index, which includes the LAI and wood area index if it has not been corrected for the influence of woody tissue on $P(\theta)$. Although several instruments have been used to perform clumping index retrievals such as the Tracing Radiation and Architecture of Canopies (TRAC) instrument, hemispherical photography, and Echidna Validation Instrument [24], the effect of woody tissue on the clumping index and LAI needs to be accounted for. In this study, we compared the results from the HemiView data with those from the TLS data with and without woody tissue.

3.2. HemiView

An important step in measuring the clumping index of forests is the separation of between-crown gaps from within-crown gaps. To derive the gap fraction from hemispherical images, each image must be separated into areas of gaps and plant tissue using an image classification algorithm. The mean gap probability can be obtained by dividing the average number of pixels of all gaps ($\overline{gt}$) by the total number of pixels A of each image at each view zenith angle. Thus, the effective LAI is calculated by:

$$L_e = -\frac{\ln\left(\frac{\overline{gt}}{A}\right)\cos(\theta)}{G(\theta)},\tag{2}$$

A method was proposed for calculating the clumping index from digital cover photography (DCP), which offers the advantage of separately determining different gaps in the canopy including between-crown and within-crown gaps [32]. The between-crown gap probability can be derived from the average number of pixels of the between-crown gap $\overline{gl}$ and the total number of pixels A. at each view zenith angle. Therefore, the LAI can be calculated when the clumping effect can be considered explicitly [32,42]:

$$L = -\left(1 - \frac{\overline{gl}}{A}\right)\frac{\ln\left(\frac{\overline{gt}-\overline{gl}}{A-\overline{gl}}\right)\cos(\theta)}{G(\theta)},\tag{3}$$

where $\left(1 - \frac{\overline{gl}}{A}\right)$ is the crown cover; and $\overline{gl}$ is the average number of pixels of the larger gap of the image, which is generally the gap between adjacent crowns. The calculation of the clumping index is as follows:

$$\Omega(\theta) = \frac{L_e}{L} = \frac{\ln\left(\frac{\overline{gt}}{A}\right)}{\ln\left(\frac{\overline{gt}-\overline{gl}}{A-\overline{gl}}\right)}\left(\frac{1}{1-\frac{\overline{gl}}{A}}\right),\tag{4}$$

Based on the above-mentioned methods, we should successively retrieve the gap probability, clumping index, and LAI. However, before retrieving the gap probability, it is essential to extract the tree trunks and between-crown gaps from the images. The separation was performed using Adobe Photoshop and digital hemispherical photography (DHP) software. The shape features of objects can be used to identify the trunk tissue in an image. The separation of the trunk area from other elements of trees in this study was conducted with object-based image analysis using the eCognition software (Trimble, Munich, Germany) [32]. The advantage of this method is that it allows for the use of object mean values for the image instead of separately applying a threshold to each pixel. The threshold values for trunk tissue detection depend on the object size classes. Objects were only classified as trunk tissue when they were not obscured by leaves.

The gaps were classified using thresholds of the object's average brightness and blue difference [32]. Hemispherical photos were analyzed using Adobe Photoshop 10 (Adobe, San Jose, CA, USA) to separate the between-crown and within-crown gaps as follows: large gaps between the tree crowns in each photo were selected using the 'Magic wand' tool with the SHIFT key held down and the total number of pixels of large gaps was recorded using the histogram. Then, the hemispherical sphere was divided into segments along the zenithal and horizontal directions separately using DHP software. The mean gap probability $\overline{P(\theta)}$ and between-crown gap probability $\overline{P_b(\theta)}$ of ten view zenith angles ($\theta = 4.5°, 13.5°, 22.5°, \ldots, 85.5°$) were then calculated based on the DHP software. Subsequently, the clumping index and LAI were derived based on Equations (3) and (4).

In addition to the analysis of the influence of the trunks on the gap fraction, clumping index, and LAI, the relative bias was used as the evaluation index. The relative biases of the effect on the gap fraction, clumping index, and LAI were defined as:

$$\varepsilon P(\theta) = \frac{\overline{P_{in}(\theta)} - \overline{P_{ex}(\theta)}}{\overline{P_{in}(\theta)}},$$

$$\varepsilon CI(\theta) = \frac{\overline{CI_{in}(\theta)} - \overline{CI_{ex}(\theta)}}{\overline{CI_{in}(\theta)}}, \tag{5}$$

$$\varepsilon L(\theta) = \frac{\overline{L_{in}(\theta)} - \overline{L_{ex}(\theta)}}{\overline{L_{in}(\theta)}}.$$

where $P_{in}(\theta)$, $CI_{in}(\theta)$ and $L_{in}(\theta)$ are the gap probability, clumping index, and LAI including trunks, respectively. Here, $P_{ex}(\theta)$, $CI_{ex}(\theta)$ and $L_{ex}(\theta)$ are the gap probability, clumping index, and LAI excluding trunks, respectively.

3.3. Terrestrial Laser Scanner

3.3.1. Gap Probability Calculation

The directional canopy gap fraction was obtained from the data of a terrestrial laser scanner [33]. The gap fraction was derived from the terrestrial laser scanner data based on the ratio of the number of laser beams passing through the canopy to the total number of beams emitted into the canopy. A laser scanner model was developed to determine the number and direction of all shots in a scan [43]. The interception of laser beams and objects can be expressed as polar coordinates (r, θ, ϕ), which can be determined by the range, zenith angle θ, and azimuth angle ϕ. We subdivided the view zenith and azimuth angles based on the minimum resolution angle of the laser scanner. The view zenith angle θ (from 0° to 90°) was divided into m. subangles and the azimuth angle ϕ (from 0° to 360°) was divided into n subangles. If a laser beam did not intercept an object, the gap fraction of the angle voxel was '1'; if the laser beam intercepted an object, the gap fraction of the angle voxel was '0'.

In addition, the forest canopy structure was reconstructed using a 3D arrangement of voxels created from all LiDAR shots and the gap fraction was calculated from the voxels [44]. The gap fraction of each 5° zenith angle was calculated as:

$$P(\theta_k) = \frac{18 \times \sum_{i=5k-5}^{5k} \sum_{j=1}^{360} P(\theta_i, \varphi_j)}{90 \times 360} = \frac{1}{5} \sum_{i=5k-5}^{5k} \left(\frac{\sum_{j=1}^{360} P(\theta_i, \varphi_j)}{360} \right), \tag{6}$$

where k is the number of divisions from 1 to 18; and $P(\theta_k)$ is the mean gap fraction of each 5° zenith angle. The azimuth angle was divided into 360 1° steps and the view zenith angle from 0° to 90° was divided into 18 5° steps.

3.3.2. Between-Crown Gap Separation

To estimate the clumping index and LAI, between-crown gaps were separated from within-crown gaps. As the TLS records the gap fraction and distance of objects in each direction, LiDAR data at different distances were processed for calculating the gap fraction, and it was possible to distinguish within-crown and between-crown gaps based on the change in size and distance of connected gap segments. Between-crown gaps consist of many large connected gaps, which form relatively large and open spaces. In contrast, within-crown gaps are broken gaps that are distributed among canopy leaves and disconnected. The change of large connected gaps was analyzed from the farthest to the nearest distance. If the size of the connected gap remained unchanged at all distances, the path lengths of the laser beams did not intercept any leaves. With reference to Zhao et al.'s method [24], a threshold value was chosen to identify between-crown gaps and how they changed with distance:

$$S_{th} = \pi \left(l \times \frac{n \times d\theta}{2} \right)^2, \tag{7}$$

where l is the distance at which a single laser beam arrives; $d\theta$ is the beam divergence for a single laser beam; and n is an integer number.

If the area of the gap fraction was larger than the threshold value, the gap was classified as a between-crown gap. If it was smaller than the threshold value, the gap was considered to be a within-crown gap. Once the between-crown gaps were distinguished from the within-crown gaps, the corresponding total gap fractions and between-crown gap fractions for each zenith angle could be calculated.

3.3.3. Clumping Index Calculation

The effective LAI can be described using Equation (1). If only the total gap probability within gaps is known, the assumption must be tailored to randomly distributed leaf elements. The effective LAI can be derived from:

$$L_e = -\frac{\ln \left(\overline{P(\theta)} \right) \cos(\theta)}{G(\theta)}, \tag{8}$$

where $\overline{P(\theta)}$ is the canopy mean gap probability, which can be estimated from terrestrial LiDAR point cloud data [33].

After separating the between-crown gap probability using the method described in Section 3.3.2, the LAI can be calculated using a method similar to that used in [32,39]:

$$L = -(1 - \overline{P_b(\theta)}) \frac{\ln \left(\frac{\overline{P(\theta)} - \overline{P_b(\theta)}}{1 - \overline{P_b(\theta)}} \right) \cos(\theta)}{G(\theta)}, \tag{9}$$

where $\overline{P_b(\theta)}$ is the between-crown gap probability.

Hence, the clumping index can be directly calculated by dividing the effective LAI by the true LAI:

$$\Omega(\theta) = -\left(\frac{1}{1 - \overline{P_b(\theta)}}\right) \frac{\ln\left(\overline{P_b(\theta)}\right)}{\ln\left(\frac{\overline{P(\theta)} - \overline{P_b(\theta)}}{1 - \overline{P_b(\theta)}}\right)}, \tag{10}$$

3.3.4. Trunk Detection

To estimate the impact of woody tissue (trunk) on the LAI, the point cloud returned from tree trunks should first be identified. The central idea is that there are many points on the trunk surface towards the TLS and that no point is on the opposite side due to the shading effect of trunks [45]. When the laser scans the trunk of a tree, the laser shots are occluded by the trunk with no points in the longer range. The detailed algorithm for the identification of tree trunks can be described as follows. First, we identified the 'trunk angle' along the azimuth direction based on these characteristics (as shown in Figure 4). We defined the trunk angle as a range of angles within which the trunk point cloud between ground and a height of 1.3 m occurs when they are projected in the x-y plane. In a plot with 25 m × 25 m size, we searched for points from far away to near the azimuth angle and saved this angle if there were no points beyond a certain distance. Second, after identifying the 'trunk angle', we extracted the points in this range of the 'trunk angle' from all points based on the azimuth angles. Finally, we confirmed the positions of the trunks based on the point density in vertical direction by setting a threshold to remove other non-trunk points. The threshold value could be established based on the points' distance and configuration of the laser scanner. This way, we could extract the point cloud of the tree trunks based on the 'trunk angle' characteristics. However, leaves obscured by the trunk should be considered in the calculation of LAI. We assumed that the distribution of the obscured leaves was the same as the distribution of the other leaves. We then calculated the mean gap fraction for each view zenith angle after removing the point cloud of the trunks. The clumping index and LAI were also retrieved from the TLS data.

To analyze the effect of trunks on the gap fraction, clumping index, and LAI based on TLS data and compare them with results from the HemiView data, the relative biases were calculated using Equation (5) and interpolated with several angles from HemiView.

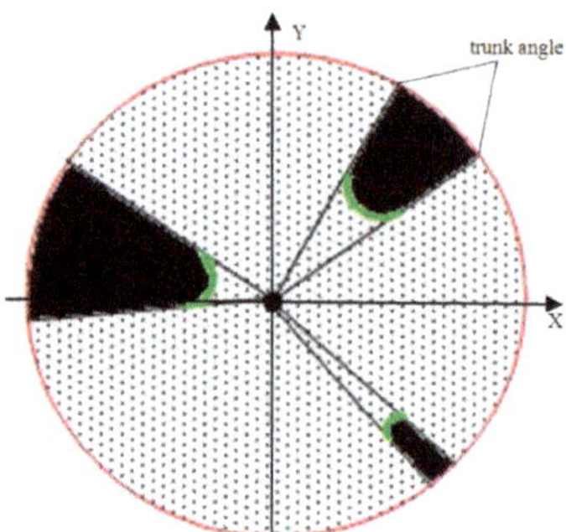

Figure 4. Sketch of Terrestrial Laser Scanning (TLS) scanning in forests. The center of the circle denotes the position of the laser scan; the green curve denotes the cross section of the trunk surface; and the black zone denotes no point cloud; top view.

4. Results and Discussion

4.1. Canopy Gap Fraction Distribution

The terrestrial laser scanner data of our sample plots were selected to evaluate their potential for the estimation of gap fractions. For a comparison with the results from the HemiView, twenty of the directional gap fractions within each 5° from the TLS were averaged as the single gap fraction value for the corresponding angle from 0° to 90°. As shown in Figure 5, the gap fractions derived from the hemispherical photography and terrestrial laser scanner showed good consistency, but the latter was lower than the former in the zenith angle range from 10° to 70°. We found that the gap probabilities from TLS and HemiView were affected by the trunks and increased after excluding them. The detailed results of the effect will be analyzed in Section 4.4. Based on Figure 5b, the trunks had a greater effect at some view zenith angles (from 15° to 55°) than at other zenith angles for HemiView, while the trunks at most view zenith angles had a similar effect on the gap probabilities for TLS. Due to the active TLS detection technique, we could automatically identify trunks from point cloud data and remove them. However, we could not determine if other trees were located behind the identified trunks from the LiDAR data. Therefore, the gap probabilities at most view zenith angles changed after removing the trunks from the point cloud. For HemiView, Photoshop and DHP software were used to process photos, then the trunk pixels could be manually selected based on the background of sky light (Figure 3). At small view zenith angles, the trunks are too thin to be identified from bright background. Furthermore, it is also difficult to determine if there is another tree located behind the identified trunks at larger view zenith angles due to the dense forest background. Therefore, the trunk surface area at these angles, to a large part, could not be removed.

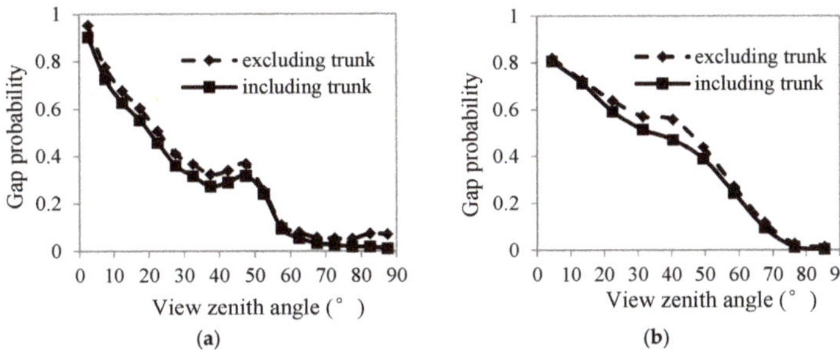

Figure 5. Angular distribution of the gap fraction distribution. (**a**) TLS; and (**b**) HemiView.

4.2. Clumping Index

In general, the estimated clumping indices increased with increasing view zenith angle (Figure 6). The clumping index significantly changed at larger view zenith angles after removing the trunk points from the point cloud. This indicated that the influence of the trunks on the clumping index was smaller at smaller view zenith angles than at larger angles. For HemiView, the effect of trunks on the clumping index between 30° and 70° was about 10–30% more than at other angles (Figure 6b), which was similar to the influence on the gap probability. However, the clumping index decreased with the removal of trunks, which differed from the TLS results. In addition, the clumping index of both TLS and HemiView showed a similar behavior, which was observed in several other studies [23,30]. Rye et al. asserted that heterogeneous ecosystem-scale tree distribution patterns caused different tree clustering at different zenith angles [28]. From the viewpoint of 3D space, the trunk distribution was clumped, not uniform. While projected to 2D images, trunks at larger view zenith angles showed uniformity. Thus, the clumping index from HemiView at larger view zenith angles was higher than at lower view zenith angles and monotonically increased with increasing view zenith angle.

In addition, we also estimated the hemispherical average clumping index from two types of data after removing the trunks (Figure 7). The results obtained using TLS and HemiView were highly correlated ($R^2 = 0.761$). The results also showed that the clumping index results from TLS were higher than those from HemiView in the case of more clumping, while the opposite result was obtained in the case of less clumping. Although good agreement has been achieved between EVI images and hemispherical photos using the 2D image processing method [24], we could be confident that the dimension of data from HemiView and TLS and their processing method also affects the estimation of the clumping index, especially the effect of removing trunks.

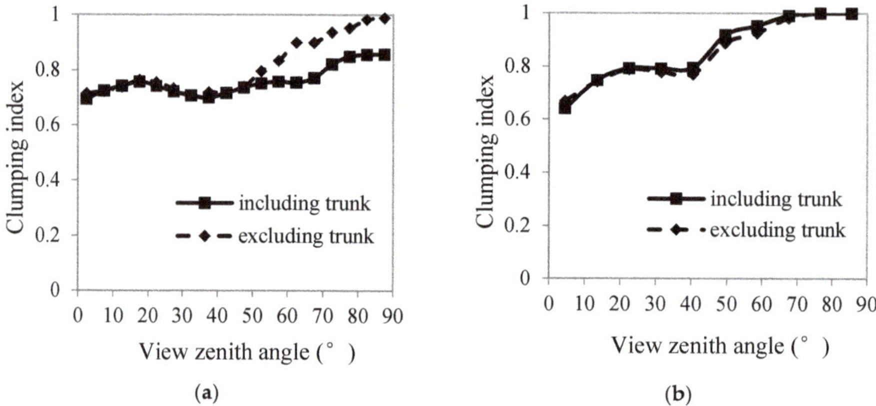

Figure 6. Angular distribution of the clumping index: (**a**) TLS; and (**b**) HemiView.

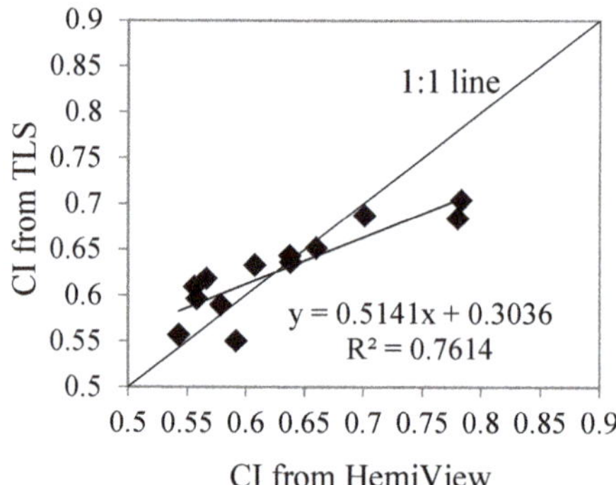

Figure 7. Comparison of the hemispherical clumping index between TLS and HemiView.

4.3. Leaf Area Index

Based on the methods proposed in Section 2, the effective LAI and LAI in the study area were estimated from the TLS data and HemiView images. The angular distributions of LAI from TLS and HemiView have a similar shape (Figure 8); however, the trunks have a 3–25% effect on the LAI, and the effect varies with the view zenith angle. The exclusion of trunks yielded an ~19.6% lower LAI based on the TLS data, while the exclusion of trunks from the HemiView data yielded an ~8.9% lower LAI on average. Due to the exclusion of trunks from the TLS data, both the decrease of the gap probability and increase of the clumping index reduced the estimated LAI value. In contrast, the exclusion of trunks from the HemiView data and the associated decrease of the gap probability reduced the retrieved

LAI value and the decrease of the clumping index resulted in a LAI increase. Therefore, the exclusion of trunks had less effect on the LAI from HemiView than on that from TLS. The magnitude of the influence of trunks on the LAI is therefore not only affected by the proportion of tree distribution, but by its patterns and measuring methods. In addition, we compared the effective LAI and LAI of 13 measurement sites derived from TLS with that obtained from HemiView (Figure 9a,b) and found that they had good correlation, though the correlation coefficient of HemiView and TLS decreased from 0.724 to 0.613 after introducing the clumping index. Meanwhile, we found that the results from the TLS data were generally higher than those derived from the HemiView images and we assumed it was due to the higher gap fractions over a 30°–70° view zenith angle determined based on the HemiView data (Figure 5), which was stated in [29].

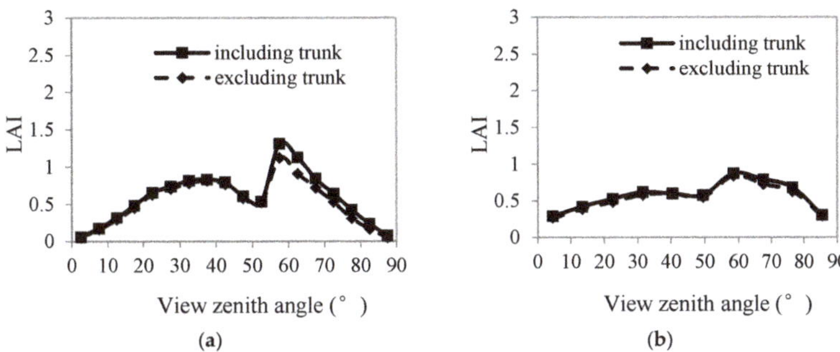

Figure 8. Angular distribution of Leaf Area Index (LAI). (**a**) TLS; and (**b**) HemiView.

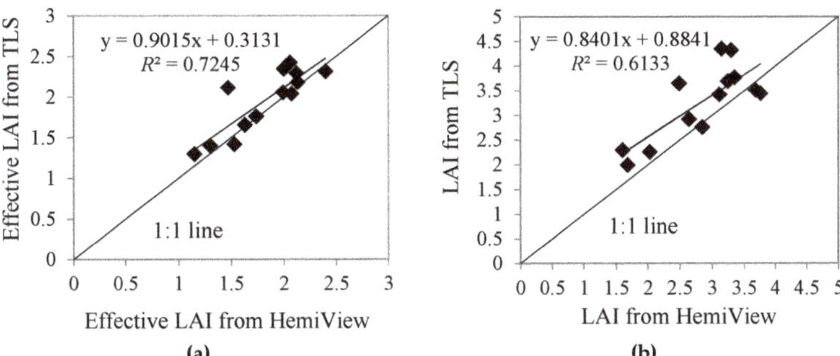

Figure 9. Comparison of the effective LAI (**a**), and LAI (**b**) between TLS and HemiView.

4.4. Tree Trunk Effects

The influence of tree trunks on the gap probability, clumping index, and LAI was analyzed. The effect of the trunks on the gap probability increased with the increasing view zenith angle (Figure 10a). The trunk surface area in the sensor view field increased with the increasing view zenith angle. In this case, εP was <20% when the view zenith angle was <60°. When the angle was >60°, the relative bias of the gap probability increased rapidly. The results indicated that trunks had less effect on larger gap probabilities but affected smaller gap probabilities more. The trunks had a stronger effect on εCI based on TLS when compared with HemiView, especially between 50° and 90° (Figure 10b,c). We analyzed the effect of the trunks on the clumping index by comparing Figure 6a,b and Figure 10b, and found that the clumping index was more affected by trunks when the proportion of the trunk surface area at a certain view zenith angle was close to the gap fraction segment at the corresponding angle such as for the view zenith angles of 60–90° in Figure 6a and 30–60° in Figure 6b.

Thus, εCI was relatively larger at the corresponding angle (Figure 10b). However, relative to εP, εL was small in that it was not only proportional to the logarithm of $(1 - \varepsilon P)$ but was inversely proportional to the logarithm of the gap probability including trunks. These results showed that the gap probability, clumping index, and LAI were indeed affected by trunks, especially the zenith angular distribution of the trunk surface area.

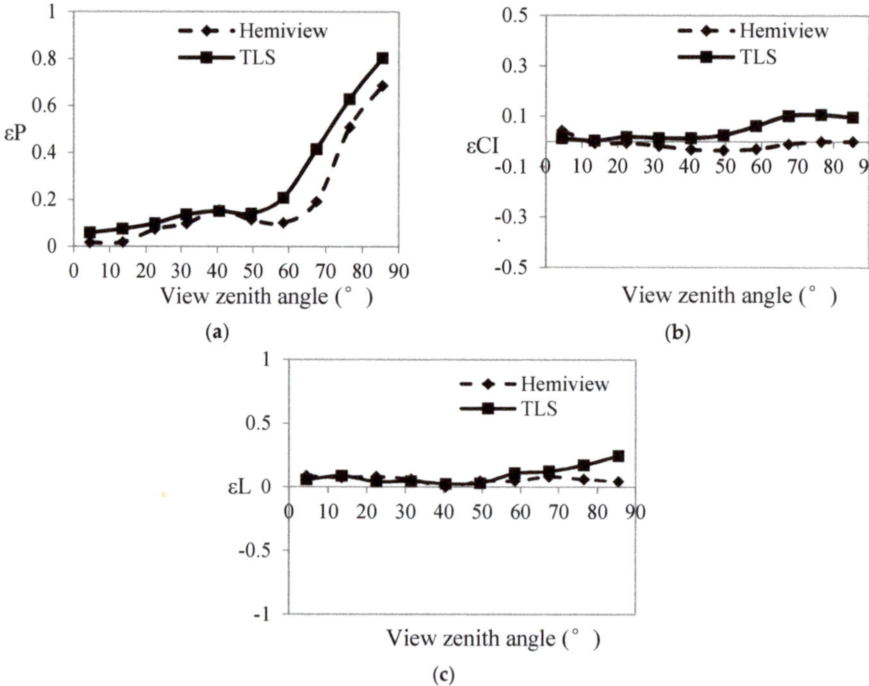

Figure 10. Comparison of the relative bias (εP, εCI, εL) between TLS and HemiView. (**a**) Gap probability (P); (**b**) Clumping index (CI); and (**c**) LAI (*L*).

5. Conclusions

In this study, we employed angularly distributed gap probability observations to derive the angularly distributed clumping index and angular distribution of LAI from TLS and HemiView data. In addition, we analyzed the effect of trunks on the gap probability, clumping index, and LAI estimates. We concluded the following. (1) Compared with the results of the clumping index and LAI from HemiView, the results from TLS proved that it had the potential to estimate the clumping index, either the angular distribution or hemispherical average value; (2) Based on the comparison of the results from TLS with those from HemiView, trunks had different effects on the gap probability, clumping index, and LAI. Although the LAI from HemiView was lower than the TLS derived LAI, the effect of trunks on LAI from HemiView was also less than the corresponding effect for TLS due to the different effects of trunks on the gap probability and clumping index. Additionally, the bias of clumping index had more effect on LAI estimates than the bias of gap probability; (3) The spatial distribution (angular distribution) of the clumping index was more important for accurate LAI estimates than the average clumping index value, especially for the clumping index at a larger view zenith angle [46]. Although the average clumping index values from HemiView and TLS were close, different LAI estimates originated from the angular distribution of the clumping index, which is seriously affected by the 3D distribution of trunks; (4) Trunk detection is important to estimate other forest structure parameters.

From a comparison to the TLS data process, we could easily detect the trunks from the 2D hemi image and calculate their area, but the position distribution of trunks could not be detected from 2D images. The clumping index is a structure parameter that is relative to depth information, so it could be affected by the 3D distribution of trunks. Furthermore, 3D information from TLS could be helpful for detecting the spatial distribution of trunks and the clumping index, which may improve the retrieved LAI accuracy. Therefore, different tree densities and distributions detected from TLS data and their effects on the clumping index estimate should be the focus of further studies.

We considered the trunk's effect on the gap probability, clumping index, and LAI but did not remove the trunk's effect on the G-function in the gap probability model. We hypothesized that the separate treatment of leaf and wood projection functions could be a key point for the improvement of the accuracy of indirect LAI retrievals via the application of the gap probability model [42]. At the same time, further validation with respect to different types and densities of forests is required.

Acknowledgments: This work was funded by the National Science Foundation of China (Grant Nos. 41401410, 41611530544 and 41401411). The authors thank Feilong Lin and Zhuo Fu for providing HemiView data. We are also grateful for Kim Calder and Sruthi Moorthy who gave useful suggestions.

Author Contributions: Yunfei Bao and Wenjian Ni conceived, designed, and performed the experiments; Yunfei Bao analyzed the TLS data, and Dianzhong Wang analyzed the HemiView data; Chunyu Yue and Hongyan He contributed analysis tools; Yunfei Bao wrote the paper; and Hans Verbeeck reviewed and revised the paper.

Conflicts of Interest: The authors declare no conflict of interest. The funding sponsors had no role in the design of the study; in the collection, analyses, or interpretation of data; in the writing of the manuscript, and in the decision to publish the results.

References

1. Chen, J.M.; Leblanc, S.G. A four-scale bidirectional reflectance model based on canopy architecture. *IEEE Trans. Geosci. Remote Sens.* **1997**, *35*, 1316–1337. [CrossRef]
2. Weiss, M.; Baret, F. Evaluation of canopy biophysical variable retrieval performances from the accumulation of large swath satellite data. *Remote Sens. Environ.* **1999**, *70*, 293–306. [CrossRef]
3. Cohen, W.B.; Maiersperger, T.K.; Gower, S.T.; Turner, D.P. An improved strategy for regression of biophysical variables and Landsat ETM+ data. *Remote Sens. Environ.* **2003**, *84*, 561–571. [CrossRef]
4. Kötz, B.; Schaepman, M.; Morsdorf, F.; Bowyer, P.; Itten, K.; Allgöwer, B. Radiative transfer modeling within a heterogeneous canopy for estimation of forest fire fuel properties. *Remote Sens. Environ.* **2004**, *92*, 332–344. [CrossRef]
5. Bao, Y.F.; Cao, C.X.; Zhang, H.; Chen, E.X.; He, Q.S.; Huang, H.B.; Li, Z.Y.; Li, X.W.; Gong, P. Synchronous estimation of DTM and fractional vegetation cover in forested area from airborne LIDAR height and intensity data. *Sci. China Ser.* **2008**, *E51*, 176–187. [CrossRef]
6. Oukoulas, S.; Blackburn, G.A. Mapping individual tree location, height and species in broadleaved deciduous forest using airborne LIDAR and multi-spectral remotely sensed data. *Int. J. Remote Sens.* **2005**, *26*, 431–455. [CrossRef]
7. Solberg, S.; Næsset, E.; Hanssen, K.H.; Christiansen, E. Mapping defoliation during a severe insect attack on Scots pine using airborne laser scanning. *Remote Sens. Environ.* **2006**, *102*, 364–376. [CrossRef]
8. Wagner, W.; Hollaus, M.; Briese, C.; Ducic, V. 3D vegetation mapping using small-footprint full-waveform airborne laser scanners. *Int. J. Remote Sens.* **2008**, *29*, 1433–1452. [CrossRef]
9. Watt, P.; Donoghue, D. Measuring forest structure with terrestrial laser scanning. *Int. J. Remote Sens.* **2005**, *26*, 1437–1446. [CrossRef]
10. Cao, C.X.; Bao, Y.F.; Chen, W.; Tian, R.; Dang, Y.F.; Li, L.; Li, G.H. Extraction of forest structural parameters based on the intensity information of high-density airborne light detection and ranging. *J. Appl. Rem. Sens.* **2012**, *6*, 063533.
11. Fu, Z.; Wang, J.D.; Song, J.L.; Zhou, H.M.; Pang, Y.; Chen, B.S. Estimation of forest canopy leaf area index using MODIS, MISR, and LiDAR observations. *J. Appl. Rem. Sens.* **2011**, *5*, 053530. [CrossRef]
12. Jensen, J.L.R.; Humes, K.S.; Vierling, L.A.; Hudak, A.T. Discrete return Lidar-based prediction of leaf area index in two conifer forests. *Remote Sens. Environ.* **2008**, *112*, 3947–3957. [CrossRef]

13. Morsdorf, F.; Kötz, B.; Meier, E.; Itten, K.I.; Allgöwer, B. Estimation of LAI and fractional cover from small footprint airborne laser scanning data based on gap fraction. *Remote Sen. Environ.* **2006**, *104*, 50–61. [CrossRef]

14. Riaňo, D.; Vallddares, F.; Condés, S.; Chuvieco, E. Estimation of leaf area index and covered ground from airborne laser scanner (Lidar) in two contrasting forests. *Agric. For. Meteorol.* **2004**, *124*, 269–275. [CrossRef]

15. Richardson, J.J.; Moskal, L.M.; Kim, S.H. Modeling approaches to estimate effective leaf area index from aerial discrete-return LiDAR. *Agric. For. Meteorol.* **2009**, *149*, 1152–1160. [CrossRef]

16. Gower, S.T.; Kucharik, C.J.; Norman, J.M. Direct and indirect estimation of leaf area index, fAPAR, and net primary production of terrestrial ecosystems. *Remote Sens. Environ.* **1999**, *70*, 29–51. [CrossRef]

17. Jonckheere, I.; Fleck, S.; Nackaerts, K.; Muys, B.; Coppin, P.; Weiss, M.; Baret, F. Review of methods for in situ leaf area index determination. Part I. Theories, sensors and hemispherical photography. *Agric. For. Meteorol.* **2004**, *121*, 19–35. [CrossRef]

18. Chason, J.W.; Baldocchi, D.D.; Hustona, M.A. Comparison of direct and indirect methods for estimating forest canopy leaf-area. *Agric. For. Meteorol.* **1991**, *57*, 107–128. [CrossRef]

19. Weiss, M.; Baret, F.; Smith, G.J.; Jonckheere, I.; Coppin, P. Review of methods for in situ leaf area index (LAI) determination—Part II: Estimation of LAI, errors and sampling. *Agric. For. Meteorol.* **2004**, *121*, 37–53. [CrossRef]

20. Lang, A.R.G.; Xiang, Y. Estimation of leaf area index from transmission of direct sunlight in discontinuous canopies. *Agric. For. Meteorol.* **1986**, *41*, 179–186. [CrossRef]

21. Chen, J.M.; Cilhar, J. Quantifying the effect of canopy architecture on optical measurements of leaf area index using two gap size analysis methods. *IEEE Trans. Geosci. Remote Sens.* **1995**, *33*, 777–787. [CrossRef]

22. Leblanc, S.G. Correction to the plant canopy gap-size analysis theory used by the Tracing Radiation and Architecture of Canopies instrument. *Appl. Opt.* **2002**, *41*, 7667–7670. [CrossRef] [PubMed]

23. Leblanc, S.G.; Chen, J.M.; Fernandes, R.; Deering, D.W.; Conley, A. Methodology comparison for canopy structure parameters extraction from digital hemispherical photography in boreal forests. *Agric. For. Meteorol.* **2005**, *129*, 187–207. [CrossRef]

24. Zhao, F.; Strahler, A.H.; Schaaf, C.L.; Yao, T.; Yang, X.Y.; Wang, Z.S.; Schull, M.A.; Román, M.O.; Woodcock, C.E.; Olofsson, P.; et al. Measuring gap fraction, element clumping index and LAI in Sierra Forest Stands using a full-waveform ground-based Lidar. *Remote Sens. Environ.* **2012**, *125*, 73–79. [CrossRef]

25. Chen, J.M. Optically-based methods for measuring seasonal variation of leaf area index in boreal conifer stands. *Agric. For. Meteorol.* **1996**, *80*, 173–176. [CrossRef]

26. Kucharik, C.J.; Norman, J.M.; Gower, S.T. Characterization of radiation regimes in nonrandom forest canopies: Theory, measurements, and simplified modeling approach. *Tree Physiol.* **1999**, *19*, 695–706. [CrossRef] [PubMed]

27. Jonckheere, I.; Muys, B.; Coppin, P. Allometry and evaluation of in situ optical LAI determination in Scots pine: a case study in Belgium. *Tree Physiol.* **2005**, *25*, 723–732. [CrossRef] [PubMed]

28. Ryu, Y.; Sonnentag, O.; Nilson, T.; Vargas, R.; Kobayashi, H.; Wenk, R.; Baldocchi, D.D. How to quantify tree leaf area index in a heterogeneous savanna ecosystem: A multi-instrument and multi-model approach. *Agric. For. Meteorol.* **2010**, *150*, 63–76. [CrossRef]

29. Ryu, Y.; Verfaillie, J.; Macfarlane, C.; Kobayashi, H.; Sonnentag, O.; Vargas, R.; Ma, S.; Baldocchi, D.D. Continuous observation of tree area index at ecosystem scale using upward-pointing digital cameras. *Remote Sens. Environ.* **2012**, *126*, 116–125. [CrossRef]

30. Woodgate, W.; Disney, M.; Armston, J.D.; Jones, S.D.; Suarez, L.; Hill, M.J.; Wilkes, P.; Soto-Berelov, M.; Haywood, A.; Mellor, A. An improved theoretical model of canopy gap probability for leaf area index estimation in woody ecosystems. *For. Ecol. Manag.* **2015**, *358*, 303–320. [CrossRef]

31. Woodgate, W.; Armston, J.D.; Disney, M.; Jones, S.D.; Suarez, L.; Hill, M.J.; Wilkes, P.; Soto-Berelov, M. Quantifying the impact of woody material on leaf area index estimation from hemispherical photography using 3D canopy simulations. *Agric. For. Meteorol.* **2016**, *226–227*, 1–12. [CrossRef]

32. Piayda, A.; Dubbert, M.; Werner, C.; Correia, A.V.; Pereira, J.S.; Cuntz, M. Influence of woody tissue and leaf clumping on vertically resolved leaf area index and angular gap probability estimates. *For. Ecol. Manag.* **2015**, *340*, 103–113. [CrossRef]

33. Bao, Y.F. Estimation of canopy gap fraction based on multi scanning data from terrestrial laser scanner. In Proceedings of the International Geoscience and Remote Sensing Symposium, Munich, Germany, 22–27 July 2012; pp. 6479–6482.

34. Danson, F.M.; Gaulton, R.; Armitage, R.P.; Disney, M.; Gunawan, O.; Lewis, P.; Pearson, G.; Ramirez, A.F. Developing a dual-wavelength full-waveform terrestrial laser scanner to characterize forest canopy structure. *Agric. For. Meteorol.* **2014**, *198–199*, 7–14. [CrossRef]

35. Jupp, D.L.B.; Culvenor, D.S.; Lovell, J.L.; Newnham, G.J.; Strahler, A.H.; Woodcock, C.E. Estimating forest LAI profiles and structural parameters using a ground-based laser called Echidna (R). *Tree Physiol.* **2009**, *29*, 171–181. [CrossRef] [PubMed]

36. Moorthy, I.; Miller, J.R.; Berni, J.A.J.; Zarco-Tejada, P.; Hu, B.; Chen, J. Field characterization of olive (*Olea europaea* L.) tree crown architecture using terrestrial laser scanning data. *Agric. For. Meteorol.* **2011**, *151*, 204–214. [CrossRef]

37. García, M.; Gajardo, J.; Riaño, D.; Zhao, K.; Martín, P.; Ustin, S. Canopy clumping appraisal using terrestrial and airborne laser scanning. *Remote Sens. Environ.* **2015**, *161*, 78–88. [CrossRef]

38. Ni, W.J.; Sun, G.Q.; Guo, Z.F.; Huang, H.B. A method for the registration of multiview range images acquired in forest areas using a terrestrial laser scanner. *Int. J. Remote Sens.* **2011**, *32*, 9769–9787. [CrossRef]

39. Beer, A. Bestímmung der absorption des rothenlichts in farbigen Flüssigkeiten. *Ann. Phys. Chem.* **1852**, *86*, 78–88. [CrossRef]

40. Nilson, T. A theoretical analysis of the frequency of gaps in plant stands. *Agric. For. Meteorol.* **1971**, *8*, 25–38. [CrossRef]

41. Ross, J. *The Radiation Regime and Architecture of Plant Stands*; Junk Publishers: Hague, The Netherlands, 1981.

42. Macfarlance, C.; Arndt, S.K.; Livesley, S.J.; Edgar, A.C.; White, D.A.; Adams, M.A.; Eamus, D. Estimation of leaf area index in eucalypt forest with vertical foliage, using cover and fullframe fisheye photography. *Forest Ecol. Manag.* **2007**, *242*, 756–763. [CrossRef]

43. Danson, F.M.; Hetherington, D.; Morsdorf, F.; Koetz, B.; Allgower, B. Forest canopy gap fraction from terrestrial laser scanning. *IEEE Trans. Geosci. Remote Sens. Lett.* **2007**, *4*, 157–160. [CrossRef]

44. Takeda, T.; Oguma, H.; Sano, T.; Yone, Y.; Fujinuma, Y. Estimating the plant area density of a Japanese larch (*Larix kaempferi* Sarg.) plantation using a ground-based laser scanner. *Agric. For. Meteorol.* **2008**, *148*, 428–438. [CrossRef]

45. Bao, Y.F.; Cao, C.X.; Ni, W.J.; Li, Z.Y.; Li, X.W. Study on the method for detection of a single tree based on ground-based LiDAR. In Proceedings of the Conference Remote Sensing of China, Hangzhou, China, 27–31 August 2010; pp. 07201–07207. (In Chinese)

46. Woodgate, W.; Armston, J.D.; Disney, M.; Suarez, L.; Jones, S.D.; Hill, M.J.; Wilkes, P.; Soto-Berelov, M. Validating canopy clumping retrieval methods using hemispherical photography in a simulated Eucalypt forest. *Agric. For. Meteorol.* **2017**, *247*, 181–193. [CrossRef]

 *forests*

Article

Automatic Mapping of Forest Stands Based on Three-Dimensional Point Clouds Derived from Terrestrial Laser-Scanning

Tim Ritter [1,*], Marcel Schwarz [1], Andreas Tockner [1], Friedrich Leisch [2] and Arne Nothdurft [1]

[1] Department of Forest- and Soil Science, Institute of Forest Growth, University of Natural Resources and Life Sciences (BOKU), Vienna 1180, Austria; marcel.schwarz@boku.ac.at (M.S.); andreas.tockner@students.boku.ac.at (A.T.); arne.nothdurft@boku.ac.at (A.N.)

[2] Department of Landscape, Spatial and Infrastructure Sciences, Institute of Applied Statistics and Computing, University of Natural Resources and Life Sciences (BOKU), Vienna 1180, Austria; friedrich.leisch@boku.ac.at

* Correspondence: tim.ritter@boku.ac.at; Tel.: +43-1-47654-91414

Received: 6 July 2017; Accepted: 21 July 2017; Published: 25 July 2017

Abstract: Mapping of exact tree positions can be regarded as a crucial task of field work associated with forest monitoring, especially on intensive research plots. We propose a two-stage density clustering approach for the automatic mapping of tree positions, and an algorithm for automatic tree diameter estimates based on terrestrial laser-scanning (TLS) point cloud data sampled under limited sighting conditions. We show that our novel approach is able to detect tree positions in a mixed and vertically structured stand with an overall accuracy of 91.6%, and with omission- and commission error of only 5.7% and 2.7% respectively. Moreover, we were able to reproduce the stand's diameter in breast height (DBH) distribution, and to estimate single trees DBH with a mean average deviation of ± 2.90 cm compared with tape measurements as reference.

Keywords: terrestrial laser scanning; forest inventory; density-based clustering

---

1. Introduction

Catalogues of key attributes surveyed in forest inventories have broadened [1] in the context of redefined goals of sustainable forest management [2,3]. Whereas traditional sampling protocols were mainly designed to provide precise information on the timber growing stock, modern survey networks have been developed towards multi-purpose forest inventories which also consider aspects of biodiversity and carbon sequestration [4,5]. In order to fulfil increased information needs and to compensate for extra expenses associated with additional field work, remote sensing techniques have been successfully adopted in forest inventory practice [6,7].

Compared to large scale forest inventories, information needs are much higher in intensive long-term forest monitoring plots, originally established to study the effects of atmospheric pollution [8] and now providing valuable data for climate change impact analyses [9]. However, evaluation of long-term forest growth trends related to changing climate requires in-depth understanding of the mechanisms behind inter-tree competition and species mingling. Competition among neighboring trees mainly depends on the inter-tree distances formed by the overall spatial tree pattern in a forest stand. Thus, mapping of exact tree positions can be regarded as crucial task of the field work associated with forest monitoring.

Traditionally, a full census of long-term observation plots is performed using easy-to-handle devices that combine a compass and a laser range finder, like the computer-aided field data collection system FieldMap (IFER - Monitoring and Mapping Solutions Ltd., Jilove u Prahy, Czech Republic) [10]. Nevertheless, the mapping of all tree positions in a complete forest stand is time consuming and cost-intensive. In addition, recorded tree locations, especially in rough terrain, often have high measurement errors because position- distance- and angular-errors propagate in consequence of multiple traverses subsequently aligned over longer distances [11].

Alternative methods for the collection of precise position data exist using new sensor techniques, such as airborne digital imagery [12–14], terrestrial laser scanning (TLS) [15,16] and airborne laser scanning (ALS) [15,17]. Whereas airborne digital imagery and ALS have been successfully used for data collection on large areas and single layer forest stands, they are limited in providing data for single trees, like diameter and height estimates, especially in diverse, multi-layer stands and in stands with dense understory and tree regeneration [18–21], unless they are combined with terrestrial measurements [7,22–24]. However, a rapid and automated measurement of objects in the three-dimensional space and the creation of 3D-point clouds with millions to billions of points in millimeter-resolution, is possible using TLS [16]. Thus, tree parameters such as tree positions, stem numbers, diameters and heights can principally be derived from TLS data [25,26].

However, obtaining tree parameters from 3D-point clouds is challenging, and requires development of new methods that can automatically process point cloud data, and aggregate them to comprehensible statistics. Another important task is to assess the accuracy of these methods. Especially the accuracy of tree positions extracted from 3D-point cloud data is a crucial point in this task, as it provides the foundation for further calculations of tree parameters.

In this paper, we propose (i) an approach for the detection of tree positions based on two-stage density-based clustering applied to TLS point clouds, and (ii) an algorithm for tree diameter measurement under limited sighting conditions of sample trees. Data for this study were collected in a 4.08 ha forest stand located in the Austrian pre-Alps. A full census measured with the FieldMap system serves as reference. Dense 3D-point cloud data were collected via multi-scan TLS. Our method for detecting tree positions is evaluated against the full census reference data and by means of manifold statistics. Our approach for tree diameter measurement is evaluated against tape measurements. We hypothesize that our methods outperform existing approaches and can be easily adopted in forest monitoring practice in conjunction with TLS.

2. Data and Methods

2.1. Experimental Stand

The survey was conducted in a 4.08 ha stand located in the training forest of the University of Natural Resources and Life Sciences, Vienna (BOKU) near Forchtenstein in the Lower-Austrian pre-Alps. The test site has sub montane altitude ranging from 608 m to 632 m a.s.l. (above sea level), average slope of 15% and a northern exposition. Climate is continental with average annual temperature of 6.5 °C and mean annual precipitation of 796 mm (unpublished data from the Heuberg weather station [27]). Geological bedrock is mainly silicate dominated by gneiss, mica schist and quartzite and soil type is a brown earth of varying depth [28]. The test stand is characterized by a high vertical structural diversity with respect to tree heights, including spatially irregularly distributed understory and regeneration. It also has a high species diversity comprising different conifers and deciduous tree species. Overstory trees are approx. 110 year old and are composed of 38% spruce (*Picea abies*), 38% beech (*Fagus sylvatica*), 15% fir (*Abies alba*), 5% pine (*Pinus sylvestris*), 4% larch (*Larix decidua*) and less than 1% of other broadleaf species. Natural regeneration occurs in irregularly arranged clusters consisting of approx. 90% beech and 10% fir.

2.2. Data Collection

In summer 2015, a full census of live trees was performed using the FieldMap system. All live trees with a diameter at breast height (DBH) greater than or equal to 10 cm (measured with diameter tape) were recorded. In total, 1789 trees (684 spruce, 259 fir, 65 larch, 91 pine, 678 beech and 12 other broadleaf species) were surveyed, this represents a stem density of 438 trees ha^{-1}. DBH ranged from

10.0 cm to 67.3 cm, with mean 35.0 cm and median 36.3 cm. Tree heights were measured using a Vertex IV ultrasonic hypsometer [29] and had a range from 5.4 m to 40.9 m, with mean of 27.8 m and median of 29.7 m.

Positions for all 1789 trees were recorded as x-coordinate and y-coordinate in a local FieldMap coordinate reference system. For a transformation to a global coordinate reference system, 48 permanently marked reference points were recorded in the local FieldMap coordinate reference system and simultaneously georeferenced via differential GPS in WGS84. FieldMap coordinates were used as reference for the evaluation of our new tree position detection algorithm and diameter measured with girth tapes served as reference for our automatic tree diameter measurement routine.

In winter 2015/16, a FARO (FARO Technologies Inc., Lake Mary, Florida, USA) Focus^{3D} X330 device [30] was used for a full terrestrial laser scan of the test stand. Scanning was performed in a multi-scan mode with in total 117 scans obtained from different positions. Scanning positions were approximately regularly spaced with a mean distance of approx. 20 m between two neighboring positions. Exact scanning positions were defined in the field, guided by a visual assessment of the local sighting conditions. The device's scan quality parameter was set to 4x, producing a moderate noise reduction, and the resolution was set to $r = 6.136$ mm/10 m, i.e., $\frac{1}{4}$ of the maximum possible resolution ($r_{max} = 1.534$ mm/10 m). These settings resulted in a scanning time of approx. 11 min for every single scan. Nearby each scanning position, six Styrofoam balls with a diameter of 15 cm were placed on top of monopods in approx. 1 m height, so that three Styrofoam balls were visible from every two neighboring scan positions. Single scan raw data were aligned by means of the Styrofoam balls, using the software FARO Scene 6.0 [31]. The aligned scans were then further processed to obtain a 3D-point cloud of the whole forest stand in a local x-, y- and z-coordinate reference system. The cut-off distance for each scan in FARO Scene was set to 45 m. The resulting point cloud had a size of 2.84×10^9 points. The local x-, y- and z-coordinates of 20 out of the total of 48 artificial reference points were manually extracted from the TLS point cloud.

2.3. Coordinate Transformation

Based on the reference points, both, the TLS point cloud coordinates as well as the tree coordinates recorded with the FieldMap system were transformed into WGS84 using a linear model with intercept β_0, regression coefficient β_1 and rotation angle φ (Equations (1) and (2)), the corresponding estimates of the model parameters are given in Table 1.

$$x_{global} - x_{local} = \beta_0 + \beta_1 \times \left(x \times \cos\left(\frac{\varphi}{360}2\pi\right) + y \times \sin\left(\frac{\varphi}{360}2\pi\right) \right), \tag{1}$$

$$y_{global} - y_{local} = \beta_0 + \beta_1 \times \left(-x \times \sin\left(\frac{\varphi}{360}2\pi\right) + y \times \cos\left(\frac{\varphi}{360}2\pi\right) \right), \tag{2}$$

Table 1. Parameter estimates for the coordinate transformation of the local coordinates from the FieldMap survey and from the terrestrial laser scanning (TLS) to WGS84.

	β_0 (m)	β_1	φ (°)
FieldMap x-coordinate	−3612.23	1.003	3.24
TLS x-coordinate	−3329.37	1.000	8.25
FieldMap y-coordinate	282,336.70	1.004	3.24
TLS y-coordinate	282,345.70	1.000	8.25

FieldMap coordinates were rotated by 3.24°, due to magnetic declination. TLS coordinates were rotated by 8.25°, as the orientation of the first scanner position is assumed to represent an Azimuth of 0° in FARO Scene. The stretching of the FieldMap coordinates ($\beta_1 > 1$) indicates the measuring inaccuracy of the device.

The resulting root mean squared deviation (RMSD) between the differential GPS data and the transformed coordinates was 1.02 m for the FieldMap data (based on 48 reference points) and 1.84 m for the TLS data (based on 20 reference points) respectively. The resulting RMSD between the TLS and the FieldMap data was 2.18 m (based on 20 reference points).

2.4. Point Cloud Processing

Please note that a step-by-step overview of our workflow and the applied software functions and parameters can be found in the appendix (Table A1).

For processing of the point cloud we used the LAStools (rapidlasso GmbH, Gilching, Germany) software [32]. To achieve a higher performance with LAStools during data processing, every second point was removed from the point cloud, finally resulting in a thinned cloud of 1.42×10^9 points. The thinned point cloud was rotated and shifted according to procedure explained above. The complete transformed point cloud was then split into 1087 quadratic tiles of 10 m edge length and with an additional buffer of 2 m width on each side. Each tile was then processed with the LAStools noise filter. Subsequently, all points were classified into ground and non-ground points using the "lasground" function. The non-ground points were normalized relative to the DEM derived from ground points; the ground points were discarded from the data. After the above described filtering and classification, the tiles were re-merged to form a complete point cloud for the entire test stand. Before the application of our new algorithms for stem position detection and diameter measurement, the complete point cloud was split into 155 tiles of 25 m × 25 m edge length and with an additional 5 m buffer on each side. These tiles were exported in .xyz format, so that further data analysis could be easily processed by the statistical software R (R Foundation for Statistical Computing, Vienna, Austria) [33].

2.5. Two-Step Clustering for the Detection of Tree Positions

A novel two-step clustering approach was developed for the detection of tree positions. In the first step the 3D-point cloud of each tile was stratified into different vertical strata by means of the normalized z-coordinates, and cluster centroids were detected within each layer. In step two, cluster centroids obtained from step one were projected onto a plane by simply discarding their z-coordinates, and a further clustering was applied to the projected xy-coordinates of cluster centroids from step one. The rationale behind our approach is that stage-one clustering does not only detect the centers of tree stems but also finds local point density maxima at locations where thicker branches and understory vegetation exist. As a tree trunk has generally a vertical orientation, stage-one clustering therefore results in multiple cluster centroids in different horizontal layers nearby the position of the stem center. Second-stage clustering of first-stage centroids works then as noise filter and discards first-stage centroids with lower local point density at positions of branches and small understory trees.

In both steps, we used a density-based clustering algorithm presented by Rodriguez and Laio [34] and implemented in the R-package densityClust [35]. The clustering algorithm is solely based on the distance between data points. Cluster centers are defined as local maxima in the density of data points. The algorithm is based on the assumptions that (i) cluster centers are surrounded by areas of lower local density, and that (ii) cluster centers are located at relatively large distance from points with larger local density [34]. For each data point i the local density ρ_i and the distance δ_i from points of higher density is calculated depending on the distances d_{ij} between data points. The local density ρ_i equals the number of data points that are closer to point i than the cut off distance d_c. The distance δ_i is calculated as the minimum distance between point i and any other point with higher density, $\delta_i = \min_{j:\rho_j > \rho_i} (d_{ij})$. δ_i exceeds the typical nearest neighbor distance for points that are local or global maxima in the density, so cluster centers are recognized as points, having anomalously large values of δ_i [34].

An appropriate cut off distance value d_c for a given distance matrix can be estimated using the R package densityClust [35] and requires values for the parameters neighbourRateLow and neighbourRateHigh. Thereby, the average number of points having a maximum distance d_c to a cluster centroid lies between neighbourRateLow and neighbourRateHigh.

The local density ρ_i and the distance values δ_i, are calculated, from a distance matrix using a Gaussian kernel for smoothed density estimation. For estimating the local values of ρ_i and δ_i, either the cut off distance d_c estimated from the distance matrix, or a pre-defined value is used.

Finally, local estimates for ρ_i and δ_i are used to detect the local density maxima in the point data and assigns the remaining points to one of these maxima, upper and lower thresholds for ρ and δ have to be provided [35].

2.5.1. Stage-One Clustering

The point cloud of each tile (25 m × 25 m with an additional 5 m buffer) was stratified into horizontal layers, each of which having a vertical extent of 21 cm, and with an overlap of 7 cm between two neighboring layers. The bottom of the lowest layer was placed at 0.04 meters above normalized zero height and the top of the highest layer at 16 meter height. This procedure resulted in 228 horizontal layers. In each of these layers 5000 points were randomly sampled, and if a stratum contained less than 5000 points the complete set of points was selected. Layers with less than 100 points were excluded from the cluster analysis.

In each layer the sample points were clustered around centroids, which represented local point density maxima d_c was estimated for each layer, depending on the corresponding distance matrix. Cut-off distances obtained from the previous sub-step were then used to calculate the local density and distance values ρ_i and δ_i. Finally, cluster centroids in each subsample from each layer were searched.

2.5.2. Stage-Two Clustering

In the second step of our two-step clustering approach cluster centroids from step one were clustered once again. For each tile, the cluster centroids from all layers were merged and projected onto one single horizontal plane. In the second-stage clustering, the cut off distance was manually set to $d_c = 0.04$ m and a prior estimation to derive d_c was therefore not necessary. Parameter estimates for the local density and distance parameters ρ_i and δ_i were then used in the final cluster search. The xy-coordinates of the centroids from stage-two clustering were finally used as estimates for tree positions. Neighboring position estimates having a distance of less than or equal to 50 cm were treated as a single tree position and merged with the R-package spatstat [36].

Trials have shown that further subsampling from stage-one clusters prior to stage-two clustering achieved enhanced results. We thus tested manifold subsamples from stage-one clusters defined by a restricted range of z-coordinates. Hereby, a lower threshold of z-coordinates from the sequence with minimum 0 m, maximum 10 m and step width 0.5 m was tested, and the upper threshold was chosen from the sequence with minimum 4 m, maximum 16 m and step width 0.5 m. Subsamples were only considered which had minimum vertical distance of 3.5 m between the lower and the upper threshold. Thus, in total 125 variants of vertical subsampling from stage-one clusters were tested, each of which having a different lower and upper limit of stage-one cluster z-coordinates.

2.6. Assignment of Tree Locations

As explained in section coordinate transformation, measurement error propagated along traverses when FieldMap system was used for the manual mapping of tree positions. Thus, in order to assess performance of our clustering approach, coordinates from the FieldMap system were matched with TLS data. The matching of tree coordinates from the two different measurement systems (FieldMap vs. TLS) was carried out using a sub-pattern assignment algorithm implemented in the R-package spatstat [36]. Automatically detected tree positions, which could not be assigned to an existing tree measured in the FieldMap campaign, were manually double checked in the TLS point cloud. In some cases extra IDs were given to positions where the cluster algorithm proposed a tree location, which was not recorded with FieldMap but visible in the point cloud. This exclusively happened for standing deadwood that was not measured with FieldMap.

As the point cloud was not exactly truncated to the stand borders, trees were also detected in neighboring stands. To exclude those trees outside the area of interest from further analysis, a boundary around the border trees of the FieldMap survey was constructed, using an α-convex hull [37], i.e., a generalization of the convex hull, that was created around the tree coordinates with an α-value of 30, using the alphahull package [38] implemented in R. An edge of the hull is drawn between two tree positions if a closed disk of radius $1/\alpha$ exists that has the property that the tree positions lie on its boundary. Additionally, a small buffer was added to the α-convex hull with a dilation radius of 1 m to consider the tree radii. All trees detected outside the resulting hull were neglected. The final assignment of predicted tree positions to corresponding approved coordinates of existing trees was done using the function pppdist() from R-package spatstat. The function allows to assign two point matrices with x- and y-coordinates by minimizing the average Euclidean distance between the matched points by a sub-pattern assignment [36].

The quality of the tree detection was determined by the overall accuracy

$$acc_{\text{total}}[\%] = 100\% - (o[\%] + c[\%]), \tag{3}$$

Thereby, o is the omission error, i.e., the percentage of trees which were not found by the two-step clustering algorithm (false negative). c is the commission error, i.e., the percentage of predicted trees, to which an existing tree could not be assigned (false positive). The detection rate d_r is given by $d_r[\%] = 100\% - o[\%]$ and the assignment rate a_r is $a_r[\%] = 100\% - c[\%]$.

To investigate whether and to which extend edge effects occurred at the border of the stand, the above described performance measures were calculated for inner sub-windows of the entire test stand, from which three buffer-zones of width 1, 5 and 10 m have been subtracted (erosion 1 m, 5 m and 10 m respectively).

2.7. Measurement of DBH

For the measurement of DBH, a circular window of 0.8 m radius was constructed around every identified tree position, and a layer of 30 cm vertical extent (from 1.15 m to 1.45 m above the ground-level) was subset from that window. 3D-points were then projected onto the horizontal plane, resulting in a two-dimensional point cloud with xy-coordinates. In a first step, a circle was fitted to the 2D-point cloud, using the circular cluster method of Müller and Garlipp [39], which is based on edge identification by redescending M-estimators (these are non-decreasing near the origin, but decreasing toward 0 far from the origin). The method is implemented in the R-package edci [40]. We used a dense grid of 25×25 starting points (i.e., possible circle centers) with a distance of 3.2 cm between two neighboring starting points, and 5 starting radii per starting point providing in total 3125 starting values (i.e., combinations of radius and centerpoint) for the optimization algorithm. The initial diameter estimate (obtained from the Müller and Garlipp algorithm [39]) is henceforward referred to as DBH_1. In a second step, all data points with distance of greater than or equal to 5 cm to the circular arc associated with DBH_1 were removed. Another circle was then fitted to the remaining point cloud using the least-squares-based algorithm of Chernov [41], implemented in the R-package "conicfit" [42]. The diameter measurement resulting from step two is referred to as DBH_2. In a third step, all points having a distance greater than or equal to 2.5 cm to the circular arc associated with DBH_2 were removed, and another circle with diameter DBH_3 was fitted to the remaining point cloud, using the same method as for DBH_2.

In a fourth step, an ellipse was fitted to the same points as DBH_3, using the method of Fitzgibbon et al. [43], which is extremely robust and efficient, as it incorporates the ellipticity constraint into the normalization factor, and as it can be solved naturally by a generalized eigensystem. The method is implemented in the R-package conicfit [42]. We computed the sum of the ellipse's two radii and refer to it as DBH_4; as well as twice the quadratic mean of these radii, which we refer to as DBH_5.

Due to dense understory vegetation, the visibility of the stem in 1.3 m height was sometimes limited and the DBH-measurement failed. Therefore, a second layer of 30 cm vertical extent was selected from 2 m above the DBH (i.e., from 3.15 m to 3.45 m above the ground-level), and five diameters ($UD_1, \ldots, UD_5$) were measured analogously to the DBH-measurement . We then estimated a taper form correction constant of tc = 2.62 cm ($\hat{=}$ 1.31 cm m^{-1}) from the median of the differences between the DBH-fits and the corresponding UD-fits, which can be applied to predict DBH from the upper diameter.

Our procedure generally achieved up to ten diameter measurements per tree ($DBH_1, \ldots, DBH_5$ and $UD_1, \ldots, UD_5$), from which one measure was selected as final DBH estimate using the following decision rules:

1. Remove all DBH estimates larger than 80 cm ($DBH_x > 80$ cm $\lor$ UD_x + tc > 80 cm). Rationale: according to a priori knowledge, trees with larger DBH do not exist in the test stand.
2. Remove UD estimates larger than the corresponding DBH-estimate ($UD_x > DBH_x$). Rationale: Diameter increasing with increasing height is illogical.
3. Remove all DBH_x, having values larger than the corresponding UD_x value plus twice the taper form correction constant ($DBH_x > UD_x + 2$ tc). Rationale: A taper of more than 2.62 cm m^{-1} is highly unlikely.
4. Given that three or more DBH-estimates, and three or more UD-estimates remain for a specific tree after steps 1–3: Remove all DBH-estimates for that tree if their standard deviation is larger than the standard deviation of the corresponding UD-estimates. Rationale: According to a visual inspection of the fitted circles and ellipses, a high variation between the fits indicates much noise near the stem, resulting in poor fits.
5. Neglect DBH_4 and DBH_5 as well as UD_4 and UD_5 (diameters measured by ellipse fitting) if the semi-minor axis is smaller than 80% of the semi-major axis. Rationale: According to visual inspection of the fitted ellipses, strong flattening indicates much noise near the stem, resulting in poor fits.
6. Select the first available measure from the sequence DBH_4, DBH_5, DBH_1, DBH_2, DBH_3, UD_4, UD_5, UD_1, UD_2, UD_3 after application of steps 1 until 5. Rationale: Diameter measurement in 1.3 m height is favored over measurement in 3.3 m height, as no assumptions regarding tc must be made, and diameter estimation via ellipse-fits is favored over estimates from circle-fits, as ellipses may fit circular and elliptical stem cross-sections as well.
7. Apply the taper form correction constant of 2.62 cm m^{-1} if an UD estimate is finally selected.

If the procedure outlined above does not provide a DBH estimate (either due to convergence failure of the optimization algorithms, or because all measures were removed as consequence of the seven decision rules), the tree location proposed by the two-stage clustering algorithm is treated as false detection, and the erroneous tree location is therefore removed from further analysis.

3. Results

3.1. Two-Step Clustering for the Detection of Tree Positions

Analyses showed that omission error of the two-step clustering approach strongly depends on the position of the lower limit of the subset layer, which is used for subsampling from stage-one clusters prior to stage-two clustering. In contrast, position of the upper limit had no relevance (Figure 1A). If a lower limit of less than or equal to 1.5 m was used, omission error was higher than 50%; but if a lower limit of 2 m height was applied, omission error decreased to less than 5%. With a lower limit increasing from 2 m to 10 m, omission error slightly increased up to approx. 10%.

In contrast, the commission error of clustering-based tree detection strongly depends on both the upper and lower limit of the layer (Figure 1B). Commission error increased continuously with increasing upper limit. Commission error decreased by approx. 5 percentage points with lower limit increasing from 0 m to 1.5 m. With lower limit increasing from 1.5 m to 2 m, commission error increased by approx. 2.5 percentage points, and with any further increase of the lower limit beyond 2 m, commission error decreased.

The overall accuracy is less than 40% for a lower limit of less than 2 m and became greater than 70% for higher positions of the upper limit (Figure 1C). A further increase of the layer's lower limit of more than 4 m has no significant effect anymore. Overall accuracy is highest for an upper border of 7.5 m and decreases by approx. one percentage point per each 0.5 m increase of the upper limit's position.

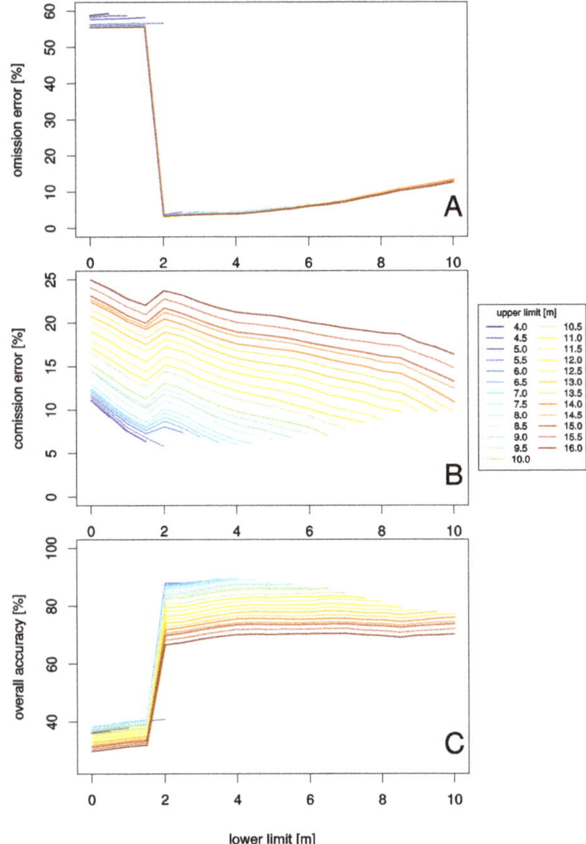

Figure 1. Accuracy of two-step clustering. (**A**): Omission error, (**B**): Commission error, (**C**): Overall accuracy.

In summary, subsampling from stage-one cluster centroids with a horizontal layer having lower limit at 4 m height and upper limit at 7.5 m height showed best performance and was therefore used for our further analysis.

3.2. Assignment of Tree Locations and DBH Estimation

DBH estimation was applied to the subset point clouds around 1778 locations proposed as tree positions from two-stage clustering. Estimation performed successfully in 1762 cases, which were only regarded as proposed tree locations; the other 16 positions were removed from the stem list. In total 1686 positions were assigned to FieldMap tree positions meaning that 76 out of 1762 proposed trees had no correspondence in the FieldMap data achieved by the traditional survey. 27 out of the 76 unassigned tree positions were identified as standing deadwood through the visual inspection in the point cloud, and 49 unassigned locations were hence regarded as false positive detections. In summary, 1713 out of 1816 trees (1789 "FieldMap trees" plus 27 standing dead trees) were correctly detected, 103 existing trees were missed, and 49 locations were falsely classified as tree position (Table 2). This represents an overall accuracy of 91.6%, an omission error of 5.7% and a commission error of 2.7%.

Table 2. Number of trees sampled through the FieldMap survey and number of trees detected and missed by means of terrestrial laser scanning (TLS).

	Spruce	Fir	Larch	Pine	Beech	Other Broad Leaf	NA	Total
FieldMap sampled	684	259	65	91	678	12		1789
FieldMap corrected	684	259	65	91	678	12	27	1816
correct detections	664	239	62	87	624	10	27	1713
false positive detections							49	49
missed tree positions	20	20	3	4	54	2		103

The mean absolute deviation between the TLS-based automatic DBH estimates and the girth tape measurements is 2.90 cm, the corresponding RMSD is 4.95 cm, and the corresponding bias is 0.37 cm. While the correlation coefficient between the measurements and estimates is high ($r = 0.90$), some extreme outliers exist, especially for trees having a small manually measured DBH (Figure 2).

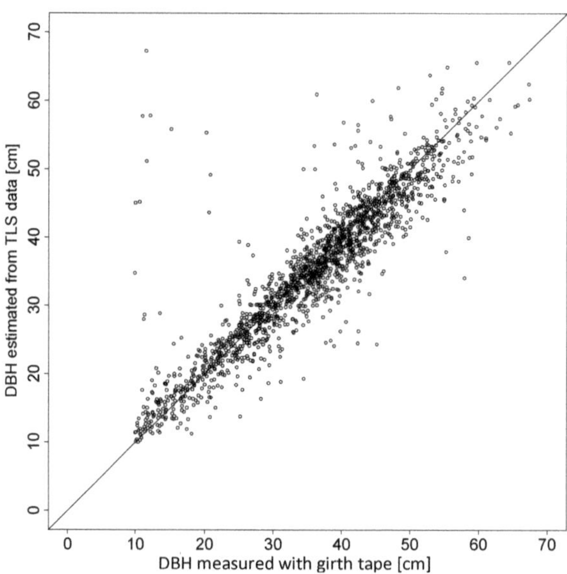

Figure 2. Comparison of DBH values measured manually with girth tape, and those estimated automatically from the TLS data.

The empirical cumulative distribution functions (ECDF) and density functions of DBH values obtained from the TLS measurements, and those measured with girth tape in the field, are presented in Figure 3, corresponding descriptive statistics are presented in Table 3. According to a Kolmogorov-Smirnov-test, the two distributions do not differ significantly ($p \approx 0.13$), i.e., despite the relatively high RMSD, the new algorithms yielded a stand diameter distribution that is statistically equal to the one derived from the field map survey.

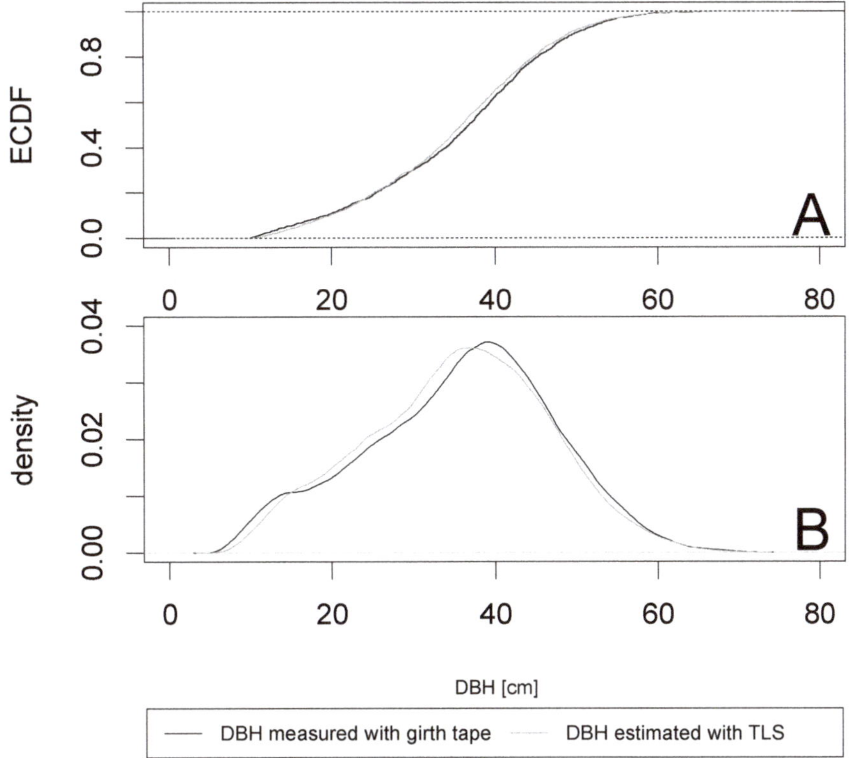

Figure 3. Distribution of measured and fitted DBH-values. (**A**): ECDF = empirical cumulative distribution function (**B**): density = Epanechnikov kernel density estimation.

Table 3. Descriptive statistics of TLS based DBH estimates and girth tape measurements.

	Min. (cm)	1st Quartile (cm)	Median (cm)	3rd Quartile (cm)	Max. (cm)	Mean (cm)	SD (cm)
Girth tape measurements	10.0	28.0	36.3	43.5	67.3	35.0	11.4
TLS based estimates	10.0	27.2	35.6	43.0	67.2	34.9	11.3

The basal area of the experimental stand was calculated as 45.7 m^2 ha^{-1} and 46.9 m^2 ha^{-1} based on TLS and FieldMap survey data respectively, this corresponds to a deviation of 1.2 m^2 ha^{-1} or 2.6%.

3.3. Stand Mapping and Erosion

The map of tree positions (Figure 4) suggests that an edge effect exists and trees were more likely missed nearby the stand border. Moreover, the assumption that the detection rate might be higher for bigger trees seems intuitively reasonable.

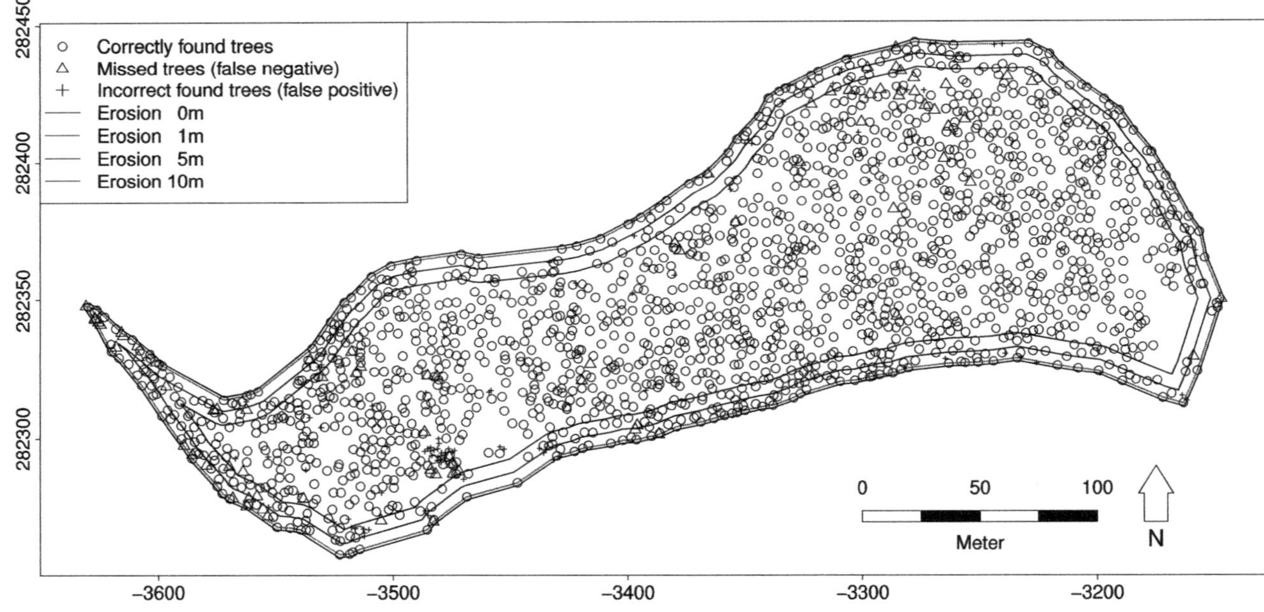

Figure 4. Result of final tree mapping, WGS84 coordinates are in (m). Erosion indicates negative buffer-zones of 0 m, 1 m, 5 m, and 10 m width respectively that were used to account for possible edge effects.

Figure 5 presents overall accuracy, commission error and omission error for an inner sub-window after removal of an outer buffer of varying size (different line types in Figure 5) and if only trees were considered having a DBH larger than a specific threshold (along the *x*-axis in Figure 5). Omission error and commission error decrease with increasing buffer size, and hence, overall accuracy increases with increasing buffer size. Omission error and commission error almost continuously decrease, and overall accuracy increases, with a DBH threshold increasing from 10 cm to 20 cm. Performance results do not change for any thresholds larger than 22 cm. When compared to the entire stand, application of a buffer of 10 m width increases overall accuracy by approx. 2 percentage points. Overall accuracy had its maximum of 97.2%, if a 10 m buffer was applied and if only trees were considered having a DBH larger than or equal to 21 cm.

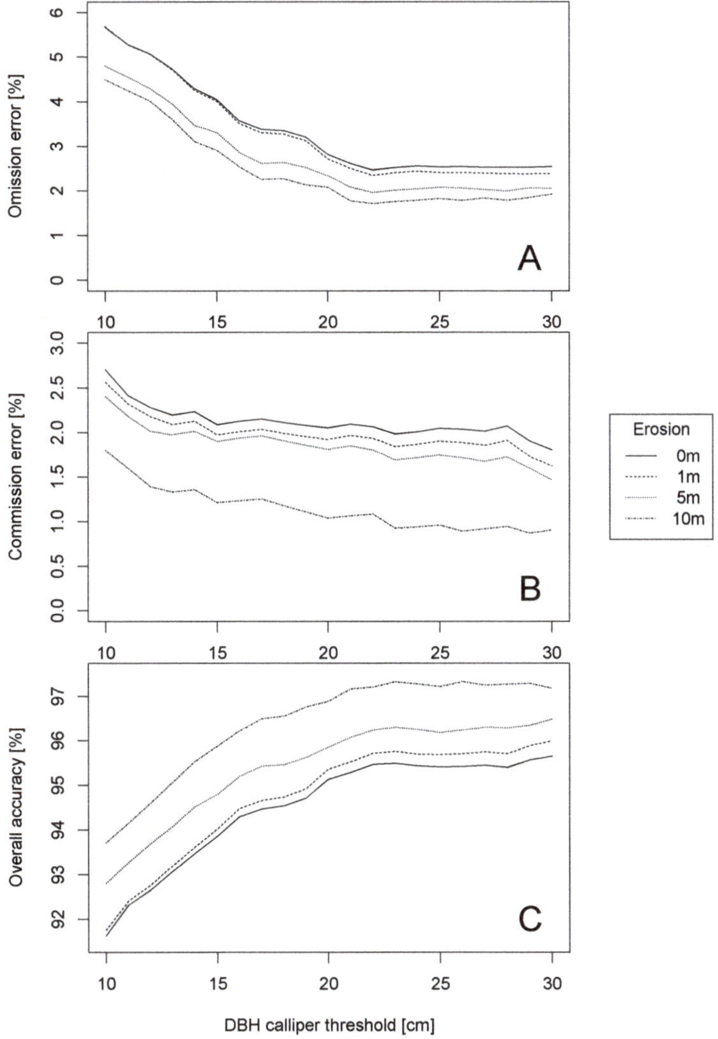

Figure 5. Effect of erosion and calliper threshold on the accuracy (lower limit = 4 m, upper limit = 7.5 m) (**A**): Omission error, (**B**): Commission error, (**C**): Overall accuracy.

4. Discussion

4.1. Subset from Point Clouds with Lower and Upper Limit

We made trials to find possible optima for a lower as well as an upper limit of a horizontal layer defining a subset from the entire point cloud on which the two-stage clustering algorithm for tree detection is applied. Because prior knowledge of existing tree locations was used in the optimization, general application of this procedure is not feasible when tree positions are predicted only from TLS data. However, results from our trial showed that the optimal layer for finding trees in point clouds is the space above the natural regeneration or ground vegetation and below the crowns. Thus, the lower limit should be set to a value high enough, so that most of the noise introduced by ground vegetation and natural regeneration is omitted. That is the case for a lower border of approx. 3 m in our study; further upwards shifts of the lower border had only small effects on the overall accuracy (Figure 1C). The upper border should be set to a value high enough, so that enough sublayers exist, and low enough, so that noise from tree crowns is minimized. The upper limit of the subset layer was set to 7.5 m, which corresponds to the 12.3%-quantile of the overstory crown base heights.

We would thus recommend that the field crews should sample some heights from understory vegetation and natural regeneration as well as from crown base heights of the overstory. As field crews have a break of approx. 11 min during every single scan (i.e., the time the devices needs to complete the scan), these additional measurements would consume no extra working time.

Trees with forked trunks might result in double detections if the bifurcation is below the upper limit of the horizontal layer. We therefore recommend that a bifurcation is treated as the crown base height during the sampling.

4.2. Transferability to Other Stands

Our novel methodology was so far only tested by means of a single but large and complex experimental stand. Transferability of the methods and results to other situations will be therefore examined in our future work. Our method (and TLS in general) is restricted in so far as stems as well as the reference markers must be visible from the scanner position. This might become problematic, e.g., in very dense pole stands or in tropical forests. However, feasibility of our approach was demonstrated in a complex scenario, with different tree species in a highly vertically structured stand and with presence of dense natural regeneration. It can be thus expected that the presented methodology should successfully work also in most forest landscapes in the temperate and boreal zones.

4.3. Estimation of DBH and Basal Area

The algorithm for automatic DBH estimation relies on prior knowledge of the maximum DBH in the stand, thus it is not feasible if only TLS data is available. However, field crews can use the 11 min scanning time to measure the DBH of the largest tree in sight. Based on these measurements and an additional safety margin of, e.g., 20%, the necessary information of the maximum DBH in the stand can be obtained without extra working time.

Despite the relatively high RMSD between the TLS-based automatic DBH estimates and the girth tape measurements, stand diameter distributions derived from the two approaches are statistically equal (Figure 3). The deviation between the basal area calculated from TLS and FieldMap survey data respectively, was only as 1.2 m^2 ha^{-1} (i.e., 2.5%).

4.4. Verification of Automatically Detected Tree Positions

Due to the relatively long traverses required by the field work, tree coordinates from both the manual registration with FieldMap and the automatic TLS-based approach were affected by measurement error. Moreover, the TLS based approach derived the positions in a height of more than 4 m, so that for slanting trees there is a deviation to the position of the trunk base. Thus, a software algorithm had to be applied to map the point pattern derived from one measurement system to the pattern from the other system. As we had no prior experience with the applied automatic tree pattern algorithm, we have visually checked the mapping results in the 3D point cloud for all existing as well as proposed tree locations.

4.5. Edge Effects

As overall accuracy increased when results were evaluated on an inner sub-window, edge effects must have existed. One explanation is that the scanner could not be placed optimally nearby the stand borders, because restrictions were given by the steep terrain on embankments alongside the forest roads, which surround the experimental stand. Another possible explanation is the higher density of understory vegetation at the edges of the stand, due to the higher availability of light near the ground in this area. Moreover, TLS data collection had been aborted due to sudden rain falls. Therefore, the two final scans could not be completed. Consequently, there is a region with low TLS point density in the very northern part of the experimental stand, where the detection rate was very low.

4.6. Comparison with Other Studies

4.6.1. Tree Detection

In existing literature on automatic tree detection and mapping the overall accuracy and the detection rate are commonly used as performance measure. Our method achieved an overall accuracy between 91.6% for all trees having a DBH greater than or equal to 10 cm and 97.2% for trees with DBH greater than or equal to 21 cm standing on an inner sub-window after removal of an 10 m wide buffer zone; these coincide with detection rates of 94.3% and 98.2%, respectively.

The highest detection rate so far provided in existing literature is 100% reported by Maas et al. [44]; the corresponding overall accuracy is 92.8%. However, those results are based on one single 0.07 ha sample plot, containing only 14 trees (all of them 140 year old beeches), representing a very low tree density of only 200 trees ha^{-1}. Other studies reported overall accuracies (the metric is called "completeness" in that reference) between 86.9% and 89.0% for dense but homogeneous bamboo stands [45] and detection rates of 90.4% in boreal coniferous forests [26] (the metric is called "recall" in that reference). Comparison of our multi-scan TLS based overall accuracies with measures obtained by single-scan TLS or airborne laser scanning (ALS) may be criticized as unfair, because the latter data are associated with lower measurement intensity. Nevertheless, the comparison is useful, as it demonstrates future capability of high resolution Lidar technology applied in forest monitoring practice in general.

Single-scan TLS is a time and cost efficient approach, however, it is limited to small sample plots, as the detection accuracy drastically decreases with increasing scanner to tree distance [46–48]. While distance sampling based approaches are able to correct for the non-detection, regarding estimates of the number of trees and the corresponding volume per ha [46,47], the observed tree pattern is thinned and remains incomplete. Thus, single-scan TLS cannot be applied for the mapping of entire forest stands.

Liang et al. [16] provide a comprehensive overview of detection rates reported in former studies. Hereby, single scan TLS generally yielded low detection rates between 22.0% and 72.3% [48–55]. However, one reference [44] reported extraordinary high detection rates between 86.7% and 100% (with corresponding overall accuracy of 73.3–96.5%) on three circular 0.07 ha plots in low-density forests.

ALS is even more time efficient than single scan TLS, as very large areas can be sampled in a very short time. However, the possible accuracy of ALS based tree detection is still insufficient for mapping experimental stands. As an example, Eysn et al. [56] reported omission errors between 42% and 65% and commission errors between 22% and 104%, while Dalponte et al. [57] reported overall accuracies (the metric is called accuracy index in that reference) between 30% and 44%.

4.6.2. DBH Measurement

Existing studies on DBH measurement from TLS have reported RMSD between measured and fitted DBH values ranging from 0.7 cm to 7.0 cm, and a corresponding bias between −1.6 cm and 1.6 cm [44,48,51,55]. This is on par with the results we could achieve with our new approach (RMSD = 4.95 cm, and bias = 0.38 cm). Our results did not outperform those from other existing applications, but our proposed approach has a relevant extra feature, as DBH can be estimated even under limited sighting conditions, when it cannot be measured directly.

5. Conclusions

We showed that TLS in multi-scan mode can be successfully applied for the automatic mapping of forest stands. The overall accuracy of up to 97.2% justifies operational use of TLS in forest monitoring.

Due to the fact that the highest overall accuracy was achieved for trees with DBH greater than or equal to 21 cm, we are confident that TLS can be applied in forest mensuration practice in the near future, because these larger trees represent the majority of growing stock volume.

Stem maps derived from TLS-based tree detection can be further used to collect additional information (e.g., tree species, health status, etc.) or to correct erroneous tree detections and DBH fits during traditional field surveys. Thus, TLS may not completely substitute traditional fieldwork, but will contribute to drastically enhance labor efficiency: TLS related field work took approx. one person-week, whereas the FieldMap survey took 24 person-weeks. Despite the need for additional working time for co-registration of the scans, TLS offers a great cost saving potential.

The complete workflow is based on established software products, there was no need for novel developments of programs or interfaces. Raw scanner data are manually co-registered in FARO Scene and the resulting point cloud is exported in standard *xyz*-format. Thus, data obtained with other instruments or co-registered within other software products, can also be used for further processing. All scripts for further processing are written in R, and the LASTools are directly invoked by system commands from within these R-scripts. We successfully tested our workflow under Windows (Microsoft Corporation, Redmond, Washington, USA) and GNU/Linux (Free Software Foundation, Boston, Massachusetts, USA) operating systems.

Long-term storage of the TLS point-cloud datasets should carefully be planned, as the required amount of storage capacity increases by six orders of magnitude compared to traditional measurements (the point-cloud of our experimental stand needs approx. 114 GB of storage capacity, whereas the FieldMap survey data just need 163 KB). To ensure long term data availability, we strongly recommend the use of standard and open data formats (like *xyz*-format), even though the necessary storage capacity is higher than for proprietary formats.

Another possible application of the new methods is change detection on already mapped stands. Using the automated tree detection and point assignment algorithms, it should be possible to detect the drop out of trees caused by harvesting or natural hazards, this would be a great tool for permanent monitoring of experimental stands.

Acknowledgments: The authors are grateful to Josef Gasch, head of the BOKU Forest Demonstration Centre, for his kind support and cooperativeness during the installation of the test site. The authors appreciate much the careful field work of Josef Paulic, Rudolph Feichter and Magdalena Höhne. The authors acknowledge the comments and suggestions provided by two anonymous reviewers.

Author Contributions: A.N. and T.R. designed the experiment; T.R. and M.S. performed the T.L.S. related field work and analyzed the data; A.N. and F.L. supported data analysis; all authors developed the tree detection algorithm; A.T., A.N., M.S. and T.R. programmed the tree detection algorithm; T.R. and A.N. developed and programmed the D.B.H. measuring algorithm; T.R. and M.S. drafted the manuscript.

Conflicts of Interest: The authors declare no conflict of interest.

Table A1. Step-by-step workflow and applied software functions and parameters.

Step No.	Step/Sub-Step		Software	Package/Function		Parameters
1	Co-registration of scans		FARO Scene			
2	Export in .xyz format					
3	Import data			txt2las		
4	Coordinate transformation			las2las		- rotate_xy −8.25 0 0 - reoffset −3329.372 282345.7 0
5	Thinning the point cloud			lasthin		- keep_every_nth 2
6	Splitting point cloud into 1087 tiles			lastile		- tile_size 10 - buffer 2
7	Filtering		LAStools	lasnoise		- step 0.35 - isolation 850
8	Differentiate between ground points and non-ground points. Normalize relative to the DEM			lasground		- step 1 - spike 0.4 - bulge 0.5 - offset 0.1 - replace z
9	Re-merge the tiles and remove the ground points			lastile		- drop_classification 2
10	Splitting point cloud into 155 tiles			lastile		- tile_size 25 - buffer 5
11	Export in txt format			las2txt		
12	Import data			data.table	fread()	
13a	Stage one clustering	Estimate cut off distance			estimateDc()	NeighbourhoodRateLow = 0.005 NeighbourhoodRateHigh = 0.01
13b		Calculate local density		densityClust	densityClust()	Dc = [estimated dc from estimateDc()]
13c		Find cluster centroids			findClusters()	rho = 2 delta = 0.5
14a	Stage two clustering	Calculate local density		densityClust	densityClust()	Dc = 0.04
14b		Find cluster centroids			findClusters()	rho = 2 delta = 0.5
14c		Join clusters with a distance of less than 50 cm	R	spatstat	connected.ppp()	R = 0.5
15a	Assignment of tree locations	Construct α-convex hull		alphahull	ahull()	Alpha = 30
15b		Point assignment		spatstat	pppdist()	Cutoff = 1.5
16a	Diameter estimation	DBH		edci	circMclust()	Nx = 25 ny = 25 nr = 5
				conicfit	LMcircleFit	
					EllipseDirectFit	
16b		Upper Diameter		edci	circMclust()	Nx = 25 ny = 25 nr = 5
				conicfit	LMcircleFit	
					EllipseDirectFit	

References

1. Köhl, M.; Magnussen, S.S.; Marchetti, M. *Sampling Methods, Remote Sensing and GIS Multiresource Forest Inventory*; Springer-Verlag: Berlin/Heidelberg, Germany, 2006; ISBN 978-3-540-32572-7.
2. MCPFE. Background information for improved Pan-European indicators for sustainable forest management. In Proceedings of the Ministerial Conference on the Protection of Forests in Europe (MCPFE) Liaison Unit, Vienna, Austria, 7–8 October 2002.

3. MCPFE. State of Europe's Forests 2003: The MCPFE report on sustainable forest management in Europe. In Proceedings of the Ministerial Conference on the Protection of Forests in Europe (MCPFE) Liaison Unit, Vienna, Austria, 28–30 April 2003.

4. IUFRO. *Guidelines for Designing Multipurpose Resource Inventories: A Project of IUFRO Research Group 4.02.02*; Lund, H.G., Ed.; IUFRO World Series: Vienna, Austria, 1998; Volume 8.

5. Kenning, R.S.; Ducey, M.J.; Brissette, J.C.; Gove, J.H. Field efficiency and bias of snag inventory methods. *Can. J. For. Res.* **2005**, *35*, 2900–2910. [CrossRef]

6. Næsset, E. Predicting forest stand characteristics with airborne scanning laser using a practical two-stage procedure and field data. *Remote Sens. Environ.* **2002**, *80*, 88–99. [CrossRef]

7. Maltamo, M.; Bollandsås, O.M.; Gobakken, T.; Næsset, E. Large-scale prediction of aboveground biomass in heterogeneous mountain forests by means of airborne laser scanning. *Can. J. For. Res.* **2016**, *1144*, 1138–1144. [CrossRef]

8. De Vries, W.; Vel, E.; Reinds, G.J.; Deelstra, H.; Klap, J.M.; Leeters, E.E.J.M.; Hendriks, C.M.A.; Kerkvoorden, M.; Landmann, G.; Herkendell, J.; et al. Intensive monitoring of forest ecosystems in Europe 1. Objectives, set-up and evaluation strategy. *For. Ecol. Manag.* **2003**, *174*, 77–95. [CrossRef]

9. Mellert, K.H.; Deffner, V.; Küchenhoff, H.; Kölling, C. Modeling sensitivity to climate change and estimating the uncertainty of its impact: A probabilistic concept for risk assessment in forestry. *Ecol. Modell.* **2015**, *316*, 211–216. [CrossRef]

10. IFER. Field-Map—Tool Designed for Computer Aided Field Data Collection. Available online: http://www.fieldmap.cz/ (accessed on 28 February 2017).

11. Field, H.L. *Landscape Surveying*; Cengage Learning: Delmar, CA, USA, 2012; ISBN 1111310602.

12. Culvenor, D.S. TIDA: An algorithm for the delineation of tree crowns in high spatial resolution remotely sensed imagery. *Comput. Geosci.* **2002**, *28*, 33–44. [CrossRef]

13. Pouliot, D.; King, D.; Bell, F.; Pitt, D. Automated tree crown detection and delineation in high-resolution digital camera imagery of coniferous forest regeneration. *Remote Sens. Environ.* **2002**, *82*, 322–334. [CrossRef]

14. Ke, Y.; Quackenbush, L.J. A review of methods for automatic individual tree-crown detection and delineation from passive remote sensing. *Int. J. Remote Sens.* **2011**, *32*, 4725–4747. [CrossRef]

15. Van Leeuwen, M.; Nieuwenhuis, M. Retrieval of forest structural parameters using LiDAR remote sensing. *Eur. J. For. Res.* **2010**, *129*, 749–770. [CrossRef]

16. Liang, X.; Kankare, V.; Hyyppä, J.; Wang, Y.; Kukko, A.; Haggrén, H.; Yu, X.; Kaartinen, H.; Jaakkola, A.; Guan, F.; et al. Terrestrial laser scanning in forest inventories. *ISPRS J. Photogramm. Remote Sens.* **2016**, *115*, 63–77. [CrossRef]

17. Næsset, E.; Gobakken, T.; Holmgren, J.; Hyyppä, H.; Hyyppä, J.; Maltamo, M.; Nilsson, M.; Olsson, H.; Persson, Å.; Söderman, U. Laser scanning of forest resources: the nordic experience. *Scand. J. For. Res.* **2004**, *19*, 482–499. [CrossRef]

18. Kwak, D.-A.; Lee, W.-K.; Lee, J.-H.; Biging, G.S.; Gong, P. Detection of individual trees and estimation of tree height using LiDAR data. *J. For. Res.* **2007**, *12*, 425–434. [CrossRef]

19. Ferraz, A.; Bretar, F.; Jacquemoud, S.; Gonçalves, G.; Pereira, L.; Tomé, M.; Soares, P. 3-D mapping of a multi-layered Mediterranean forest using ALS data. *Remote Sens. Environ.* **2012**, *121*, 210–223. [CrossRef]

20. Kaartinen, H.; Hyyppä, J.; Yu, X.; Vastaranta, M.; Hyyppä, H.; Kukko, A.; Holopainen, M.; Heipke, C.; Hirschmugl, M.; Morsdorf, F.; et al. An international comparison of individual tree detection and extraction using airborne laser scanning. *Remote Sens.* **2012**, *4*, 950–974. [CrossRef]

21. Vauhkonen, J.; Ene, L.; Gupta, S.; Heinzel, J.; Holmgren, J.; Pitkanen, J.; Solberg, S.; Wang, Y.; Weinacker, H.; Hauglin, K.M.; et al. Comparative testing of single-tree detection algorithms under different types of forest. *Forestry* **2012**, *85*, 27–40. [CrossRef]

22. Maack, J.; Lingenfelder, M.; Weinacker, H.; Koch, B. Modelling the standing timber volume of Baden-Württemberg—A large-scale approach using a fusion of Landsat, airborne LiDAR and National Forest Inventory data. *Int. J. Appl. Earth Obs. Geoinf.* **2016**, *49*, 107–116. [CrossRef]

23. Deo, R.K.; Froese, R.E.; Falkowski, M.J.; Hudak, A.T. Optimizing variable radius plot size and LiDAR resolution to model standing volume in conifer forests. *Can. J. Remote Sens.* **2016**, *42*, 428–442. [CrossRef]

24. Scrinzi, G.; Clementel, F.; Floris, A. Angle count sampling reliability as ground truth for area-based LiDAR applications in forest inventories. *Can. J. For. Res.* **2015**, *45*, 506–514. [CrossRef]

25. Srinivasan, S.; Popescu, S.C.; Eriksson, M.; Sheridan, R.D.; Ku, N.-W.; Waser, L.T.; Wynne, R.H.; Thenkabail, P.S. Terrestrial laser scanning as an effective tool to retrieve tree level height, crown width, and stem diameter. *Remote Sens.* **2015**, *7*, 1877–1896. [CrossRef]

26. Yang, B.; Dai, W.; Dong, Z.; Liu, Y. Automatic forest mapping at individual tree levels from terrestrial laser scanning point clouds with a hierarchical minimum cut method. *Remote Sens.* **2016**, *8*, 372. [CrossRef]

27. Hager, H. Die Klimastationen im lehrforst der universität für bodenkultur. *Allg. Forstzeitung* **1980**, *91*, 54–57.

28. Glatzel, G.; Sieghardt, M. Die Böden des Lehrforstes. *Allg. Forstztg.* **1980**, *91*, 53–54.

29. Haglöf Measurement Solutions in Forest and Field. Available online: http://www.haglofcg.com/index.php/en/products/instruments/height/341-vertex-iv (accessed on 17 March 2017).

30. FARO Laser Scanner FARO Focus3D—Overview—3D Surveying. Available online: http://www.faro.com/en-us/products/3d-surveying/faro-focus3d/overview (accessed on 28 February 2017).

31. FARO FARO Laser Scanner Software—SCENE—Overview. Available online: http://www.faro.com/en-us/products/faro-software/scene/overview (accessed on 28 February 2017).

32. Isenburg, M. LAStools—Efficient LiDAR Processing Software (version 160429, Academic). Available online: https://rapidlasso.com/lastools/ (accessed on 28 February 2017).

33. R Development Core Team. *R: A Language and Environment for Statistical Computing*, R Version 3.3.2; R Foundation for Statistical Computing: Vienna, Austria, 2016.

34. Rodriguez, A.; Laio, A. Clustering by fast search and find of density peaks. *Science* **2014**, *344*, 1492–1496. [CrossRef] [PubMed]

35. Pedersen, T.L.; Hughes, S. *Densityclust: Clustering by Fast Search and Find of Density Peaks*, R package version 0.2.1; 2016. Available online: https://cran.r-project.org/package=densityClust (accessed on 15 June 2017).

36. Baddeley, A.; Rubak, E.; Turner, R. *Spatial Point Patterns: Methodology and Applications with R*; Chapman and Hall/CRC Press: London, UK, 2015.

37. Edelsbrunner, H.; Kirkpatrick, D.; Seidel, R. On the shape of a set of points in the plane. *IEEE Trans. Inf. Theory* **1983**, *29*, 551–559. [CrossRef]

38. Pateiro-Lopez, B.; Rodriguez-Casal, A. *Alphahull: Generalization of the Convex Hull of a Sample of Points in the Plane*, R package version 2.1; 2016. Available online: https://cran.r-project.org/package=alphahull (accessed on 15 June 2017).

39. Müller, C.H.; Garlipp, T. Simple consistent cluster methods based on redescending M-estimators with an application to edge identification in images. *J. Multivar. Anal.* **2005**, *92*, 359–385. [CrossRef]

40. Garlipp, T. Edci: Edge Detection and Clustering in Images. 2016. Available online: https://CRAN.R-project.org/package=edci (accessed on 17 June 2017).

41. Chernov, N. *Circular and Linear Regression: Fitting Circles and Lines by Least Squares*; CRC Press: Taylor & Francis Group, Boca Raton, Florida, USA, 2011; ISBN 9781439835906.

42. Gama, J.; Chernov, N. *Conicfit: Algorithms for Fitting Circles, Ellipses and Conics Based on the Work by Prof. Nikolai Chernov*, R package version 1.0.4; 2015. Available online: https://CRAN.R-project.org/package=conicfit (accessed on 17 June 2017).

43. Fitzgibbon, A.W.; Pilu, M.; Fisher, R.B. Direct least squares fitting of ellipses. *IEEE Trans. Pattern Anal. Mach. Intell.* **1996**, *21*, 476–480. [CrossRef]

44. Maas, H.-G.; Bienert, A.; Scheller, S.; Keane, E. Automatic forest inventory parameter determination from terrestrial laser scanner data. *Int. J. Remote Sens.* **2008**, *29*, 1593–1879. [CrossRef]

45. Xia, S.; Wang, C.; Pan, F.; Xi, X.; Zeng, H.; Liu, H. Detecting stems in dense and homogeneous forest using single-scan TLS. *Forests* **2015**, *6*, 3923–3945. [CrossRef]

46. Astrup, R.; Ducey, M.J.; Granhus, A.; Ritter, T.; von Lüpke, N. Approaches for estimating stand-level volume using terrestrial laser scanning in a single-scan mode. *Can. J. For. Res.* **2014**, *44*, 666–676. [CrossRef]

47. Ducey, M.J.; Astrup, R. Adjusting for nondetection in forest inventories derived from terrestrial laser scanning. *Can. J. Remote Sens.* **2013**, *39*, 410–425.

48. Olofsson, K.; Holmgren, J.; Olsson, H. Tree stem and height measurements using terrestrial laser scanning and the RANSAC algorithm. *Remote Sens.* **2014**, *6*, 4323–4344. [CrossRef]

49. Thies, M.; Spiecker, H. Evaluation and future prospects of terrestrial laser scanning for standardized forest inventories. *Int. Arch. Photogramm. Remote Sens. Spat. Inf. Sci.* **2004**, *36*, 192–197.

50. Strahler, A.H.; Jupp, D.L.; Woodcock, C.E.; Schaaf, C.B.; Yao, T.; Zhao, F.; Yang, X.; Lovell, J.; Culvenor, D.; Newnham, G.; et al. Retrieval of forest structural parameters using a ground-based lidar instrument (Echidna®). *Can. J. Remote Sens.* **2008**, *34*, 426–440. [CrossRef]
51. Brolly, G.; Kiraly, G. Algorithms for stem mapping by means of terrestrial laser scanning. *Acta Silv. Lign. Hung* **2009**, *5*, 119–130.
52. Murphy, G.E.; Acuna, M.A.; Dumbrell, I. Tree value and log product yield determination in radiata pine (Pinus radiata) plantations in Australia: comparisons of terrestrial laser scanning with a forest inventory system and manual measurements. *Can. J. For. Res.* **2010**, *40*, 2223–2233. [CrossRef]
53. Lovell, J.L.; Jupp, D.L.B.; Newnham, G.J.; Culvenor, D.S. Measuring tree stem diameters using intensity profiles from ground-based scanning lidar from a fixed viewpoint. *ISPRS J. Photogramm. Remote Sens.* **2011**, *66*, 46–55. [CrossRef]
54. Liang, X.; Litkey, P.; Hyyppa, J.; Kaartinen, H.; Vastaranta, M.; Holopainen, M. Automatic stem mapping using single-scan terrestrial laser scanning. *IEEE Trans. Geosci. Remote Sens.* **2012**, *50*, 661–670. [CrossRef]
55. Liang, X.; Hyyppä, J. Automatic stem mapping by merging several terrestrial laser scans at the feature and decision levels. *Sensors* **2013**, *13*, 1614–1634. [CrossRef] [PubMed]
56. Eysn, L.; Hollaus, M.; Lindberg, E.; Berger, F.; Monnet, J.-M.; Dalponte, M.; Kobal, M.; Pellegrini, M.; Lingua, E.; Mongus, D.; et al. A benchmark of lidar-based single tree detection methods using heterogeneous forest data from the alpine space. *Forests* **2015**, *6*, 1721–1747. [CrossRef]
57. Dalponte, M.; Reyes, F.; Kandarre, K.; Gianelle, D. Delineation of individual tree crowns from ALS and hyperspectral data: A comparison among four methods. *Eur. J. Remote Sens.* **2015**, *48*, 365–382. [CrossRef]

 *forests*

Article

Estimation of Forest Biomass Patterns across Northeast China Based on Allometric Scale Relationship

Xiliang Ni [1,†], **Chunxiang Cao** [1,*], **Yuke Zhou** [2,†], **Lin Ding** [1], **Sungho Choi** [3], **Yuli Shi** [4], **Taejin Park** [3], **Xiao Fu** [5], **Hong Hu** [6] and **Xuejun Wang** [7]

[1] State Key Laboratory of Remote Sensing Science, Institute of Remote Sensing and Digital Earth, Chinese Academy of Sciences, Beijing 100101, China; nixl@radi.ac.cn (X.N.); dinglin@radi.ac.cn (L.D.)
[2] State Key Laboratory of Resources and Environmental Information System, Institute of Geographical Sciences and Natural Resources Research, Chinese Academy of Sciences, Beijing 100101, China; zyk@lreis.ac.cn
[3] Department of Earth and Environment, Boston University, 675 Commonwealth Avenue, Boston, MA 02215, USA; schoi@bu.edu (S.C.); parktj@bu.edu (T.P.)
[4] School of Remote Sensing, Nanjing University of Information Science and Technology, Nanjing 210044, China; ylshi.nuist@gmail.com
[5] College of Applied Sciences and Humanities of Beijing Union University, Beijing 100083, China; fuxiao@buu.edu.cn
[6] Haihe basin Soil and Water Conservation Monitor Centre, Tianjing 300171, China; huhong-2004@163.com
[7] Survey Planning and Design Institute, State Forest Administration of China, Beijing 100714, China; wangxuejun320@126.com
* Correspondence: caocx@radi.ac.cn; Tel.: +86-010-6483-6205
† These authors contributed equally to this work.

Received: 30 June 2017; Accepted: 1 August 2017; Published: 8 August 2017

Abstract: This study develops a modeling framework for utilizing the large footprint LiDAR waveform data from the Geoscience Laser Altimeter System (GLAS) onboard NASA's Ice, Cloud, and Land Elevation Satellite (ICESat), Moderate Resolution Imaging Spectro-Radiometer (MODIS) imagery, meteorological data, and forest measurements for monitoring stocks of total biomass (including aboveground biomass and root biomass). The forest tree height models were separately used according to the artificial neural network (ANN) and the allometric scaling and resource limitation (ASRL) tree height models which can both combine the climate data and satellite data to predict forest tree heights. Based on the allometric approach, the forest aboveground biomass model was developed from the field measured aboveground biomass data and the tree heights derived from two tree height models. Then, the root biomass should scale with the aboveground biomass. To investigate whether this approach is efficient for estimating forest total biomass, we used Northeast China as the object of study. Our results generally proved that the method proposed in this study could be meaningful for forest total biomass estimation (R^2 = 0.699, RMSE = 55.86).

Keywords: forest aboveground biomass; root biomass; tree heights; GLAS; artificial neural network; allometric scaling and resource limitation

---

1. Introduction

As the principal part of terrestrial ecosystems, forest ecosystems hold approximately 80% of the terrestrial aboveground and below-ground biomass, and play very important roles in the global carbon cycle and climate change [1–3]. Since the temperate and boreal forests play a crucial role as atmospheric CO2 sinks, more attention has been paid to climatic warming in mid-and high-latitudes than in low-latitudes [4–6]. Many studies have shown that the lack of accurate forest biomass maps has generated a large uncertainty in forest carbon stocks of temperate and boreal forest regions [7–9]. The accurate estimation of forest biomass is not only necessary for improving the estimation of carbon pools, but also very important for forest management and understanding the response to climate change [10–13]. At the same time, the estimation of root biomass is suggested to be helpful for improving the estimation of terrestrial carbon stocks [14–16]. A large number of biomass estimation methods that involve extrapolation of biomass measurements in sufficient number of field sample plots can achieve very good estimation results for small forest areas [17,18]. However, there are some problems related to obtaining reliable forest biomass estimation results in large scale studies, because of the lack of field data, inconsistency of data collection methods, and less consideration of root biomass [19]. The forest biomass estimation based on satellite remote sensing is considered as a fairly advantageous approach in large scale studies because of wide and synoptic data [4,20,21]. As there is no remote sensing instrument developed to have the capacity of measuring biomass directly, the ground inventory data is required for building relationships between remote sensing parameters and biomass. Although passive optical satellite data can be used for biomass estimation across a variety of spatial and temporal scales due to its inexpensive technology, the low saturation level of vegetation spectral bands has had a serious effect on biomass estimation [22–24]. LiDAR is often more accurate in predicting forest structural parameters because its signal does not saturate in high-biomass forests [25,26]. Especially, airborne LiDAR can obtain the forest biomass estimation results with high precision. However, the high acquisition costs of large observation data bring limitations of its application at large scale region [27,28]. The Geoscience Laser Altimeter System (GLAS), onboard the Ice, Cloud, and land Elevation Satellite (ICESat), is a spaceborne LiDAR system which can provide the global recording of full waveforms over large footprints [29,30]. The GLAS data can be used for estimating forest tree heights and biomass by associating field plot data, because it has the capability of obtaining the vertical structural information. However, relatively sparse GLAS footprints could not provide the continuous forest biomass distribution image without the help of other remote sensing datasets [30–32]. Therefore, it is necessary that multi-sensor datasets synergy is considered to estimate forest biomass at the regional scale.

The northeast part of China (NE China) is the most important forest region in China. NE China possesses the largest contiguous forest land area in China, accounting for 40% of total country forest biomass [33,34]. The forest of NE China was exploited much later than eastern and southern parts of the country. It has experienced drastic climatic warming since the 1980s, which brought a significant growth rate in forest biomass and productivity [35,36]. These changes also suggested that it is necessary to obtain the accurate forest biomass estimation.

In this study, we explored the capabilities of satellite remote sensing for mapping forest biomass in NE China region. The forest tree heights were estimated using artificial neural network (ANN) method and the allometric scaling and resource limitation (ASRL) approach by combining the climate data and satellite data, including ICESat/GLAS data, MODIS land surface products, and some ancillary datasets [37,38]. The potential information on root-shoot biomass allocation was studied using the field biomass measurements, and the best forest biomass estimation model was established.

2. Materials and Methods

2.1. Study Area

The research area of this study is NE China, which is defined here to extend longitudinally from 115°37′ E to 135°05′ E and latitudinally from 38°43′ N to 53°34′ N, with a total area of 1.66 × 10^6 km^2 (Figure 1). NE China in this study mainly includes Liaoning (LN), Jilin (JL), and Heilongjiang (HLJ) provinces and four leagues (an administrative division) in Eastern Inner Mongolia Autonomous region (IMA) [16].

The study area encompasses all the major forest types in Northeast Asia. The major forest types mainly include cold temperate coniferous forest, temperate coniferous and broad-leaved forests. The forested land area in NE China is about of 50.5 million ha, accounting for the largest area of natural forests in the country. The elevation in most part of NE China is below 400 m (Figure 1). The DaXing'anling Mountain, Xing'anling Mountain and Changbai Mountain have relatively high elevation. The precipitation in NE China generally ranges from 300 mm to1000 mm because of the monsoon climate of medium latitudes. The annual mean temperature ranged from −4 to 11.5 °C.

Three forest zones are covered from south to north in NE China, including warm temperate deciduous broadleaf forest zone, temperate coniferous and broadleaf mixed forest zone, and boreal forest zone [16,39].

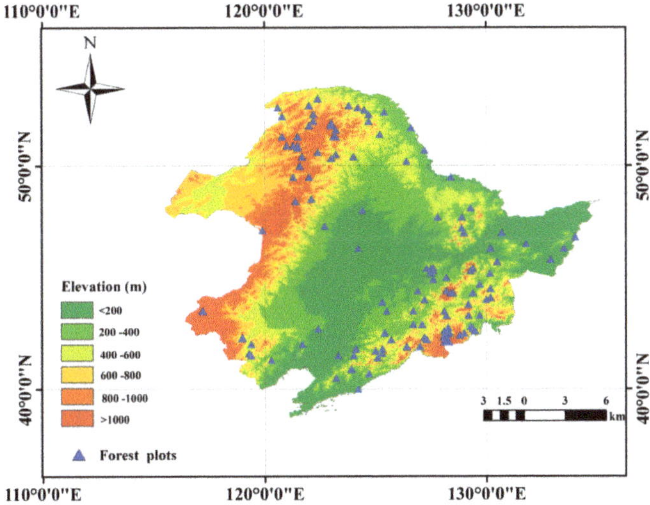

Figure 1. The study area and the biomass data collected in this study. These data are distributed well across the forest regions of the study area.

2.2. Datasets

2.2.1. Climate Data

The climatic variables used as the inputs of the two tree heights models in this study include precipitation, temperature, relative humidity, wind speed, and incoming solar radiation. The precipitation and temperature were the inputs of ANN tree heights model. These five climatic variables which determine the available and evaporative flow rate of trees were inputs of ASRL model. All these climate data were obtained from the China Meteorological Data Sharing Service System (CMDSSS) which provides the annual meteorological records [40,41].

In this study area, the observations spanning over recent 30 years (1978–2007) from total 118 meteorological stations were collected, comprised of annual total precipitation, average temperature, relative humidity, wind speed, and solar radiation time. We further interpolated these climatic variables to generate the continuous gridded maps of annual average climatic variables at 1km spatial resolution by using ordinary kriging [42].Then, we averaged the continuous gridded maps of annual average precipitation and temperature from 2004 to 2006 in order to obtain mean precipitation and temperature maps during 2004–2006 as inputs of ANN tree heights model, respectively. Similarly, the mean values of five annual average climatic variables during 1978–2007 were obtained to be the inputs of ASRL model.

2.2.2. Land Surface Reflectance

The MODIS nadir bidirectional reflectance distribution function adjusted reflectance (NBAR) product which is named as MCD43B4 provides the land surface reflectance data at a spatial resolution of 1 km [43]. The NBAR product represents the best characterization of surface reflectance over a 16-day period. In this study, we selected the seven spectral bands of MCD43B4 data in growing season from 2004 to 2006 as the inputs of ANN tree heights model. The seven bands of MCD43B4 data used in this study include three visible bands (460, 555, and 659 nm) and four near-infrared bands (865, 1240, 1640, and 2130 nm). The MCD43B4 product can cover the study region by selecting the images with the scene orbiter h25v03, h25v04, h26v03, h26v04, h27v04, and h27v05.

2.2.3. Ancillary Data

In this study, there are two types of ancillary data. The first set of ancillary data was used as the inputs of ASRL-based tree height model, and consisted of the Advanced Space borne Thermal Emission and Reflection Radiometer (ASTER) Global DEM (GDEM V2, spatial resolution of 30 m), and the eight-day composites of the post-processed Moderate Resolution Imaging Spectro radiometer (MODIS) leaf area index (LAI) products (1 km spatial resolution) [43]. We further calculated the mean value of LAI in summer (June to August, the approximate growing season) for the time period from 2003 to 2006.The mean value of LAI was calculated by using the Equation (1):

$$LAI_{avg} = \frac{\sum_{i=1}^{n} \sum_{j=1}^{m} LAI_{8-day,i,j}}{m \times n} \tag{1}$$

Here, LAI_{avg} is the average LAI (June to August months from 2003 to 2006) and $LAI_{8-day,i,j}$ is the eight-day LAI product for j^{th} week ($m = 12$) of ith year ($n = 4$).

In addition, the second set of ancillary data was used to identify forested lands, and consisted of land cover (LC) and vegetation continuous field (VCF) [44]. We selected the International Geosphere-Biosphere Programme (IGBP) of LC product (MCD12Q1, 500 m grid) and tree cover percent of VCF product (MOD44B, 250 m grid) of NE China for the year 2005 [45]. The land cover belonging to one of the five forest classes per the IGBP, and consisting of more than 50% tree cover per the VCF product were defined as forest land.

2.2.4. Tree Height and Biomass Measurements

GLAS-Derived Tree Heights

GLAS-Derived Tree Heights were used as the standard forest tree heights for training the ANN model and optimizing the parameters of ASRL based tree height model in this study.

The ICESat/GLAS is designed to obtain characteristics of the earth surface structures in three dimensions with high accuracy [26,38].

The Release-33 of GLAS laser altimetry data available from the National Snow and Ice Data Center [46] was used in this study. GLAS waveform data provide information on land elevation and vegetation cover within its ellipsoidal footprints about 72 diameters at about 170 m interval along the sub-satellite track [47,48]. National Snow and Ice Data Center released 15 GLAS products.

Amongst various altimetry products of the Release-33, we selected GLA14 product (GLAS Level-2 Land Surface Altimetry product, National Snow and Ice Data Center (NSIDC), Boulder, CO, USA) for the maximum tree height retrieval from May to October (2003–2006).

The GLAS waveform data are generally affected by the following factors: atmospheric forward scattering, signal saturation, background noise (low cloud), and the topographic slope gradient effects. In order to obtain the best GLAS data waveform for deriving accurate tree heights, the GLAS footprints screening is needed. At the same time, GLAS footprints over non-forest should be filtered. The GLAS footprints processing steps for selecting valid GLAS waveforms can be depicted as follows [37,38,47,48].

Firstly, the GLAS data of the approximate growing season from 2003 to 2006 were considered. Then, the GLAS data were further screened by applying the atmospheric forward scattering filters, signal saturation filters, background noise level correction filters, land cover mask conditions filters, and topographic slope gradient filter. Based on the procedure of GLAS footprints filtering, the best GLAS data waveform was selected.

In order to obtain the most accurate GLAS tree heights, the topographic correction approach of GLAS-derived tree heights was used according to the Equation (2) [37,38,41,47–50]:

$$H_{GLAS} = \left(D_{SigBegOff} - D_{gpCntRngOff} \right) - \frac{d_{GLAS} \times tan\theta}{2} \tag{2}$$

where H_{GLAS} is the GLAS-derived tree heights, $D_{SigBegOff}$ represents the location the GLAS full-waveformbeginning signal, and $D_{gpCntRngOff}$ represents the location of the ground peak, d_{GLAS} is the GLAS footprint size, and θ is the topographic slope [40,48–52].

Field-Measured Tree Biomass

In this study, 515 field measured tree biomass measurement plots were compiled from NE China region. All these plots were compiled from literature by Wang [16]. For each plot, the total tree biomass, shoot (stem, branch and leaf) biomass and root biomass were calculated.

In these plots, 85 plots were sampled by Wang [16]. The shoot and root biomass for the plots were estimated with DBH (and tree height) using allometric relation-ships developed by Zhu [53]. In addition, there were 161 plots obtained from the data base of Luo [54], and the rest of other plots were collected from 59 sources [16]. Since there are different numbers of effective records on the total biomass, shoot biomass and root biomass in Wang's database, the plots which include all the useful records on the total biomass, shoot biomass, and root biomass were considered. Finally, 432 effective plots with valuable information of total biomass, shoot biomass and root biomass, were selected from the 515 plots to build and validate the biomass model. The detailed statistics information of these records is shown in Table 1.

Table 1. The statistical information of the field-measured biomass.

Statistics	Shoot Biomass(t/ha)	Root Biomass(t/ha)	Total Biomass (t/ha)
Maximum Value	369.1	106	432.4
Minimum Value	8.8	1.9	10.7
Mean Value	113.1044715	26.62682927	139.7052846
Variance	6088.720266	389.2960119	9226.870054

2.3. Methodology

We developed the approach of estimating forest biomass by combining multisource satellite data, meteorological data and field measured forest plot data.

Our analysis consisted of two main parts, (a) predicting the tree heights by using two tree heights models; (b) developing the biomass model for estimating forest biomass in study region (Figure 2).

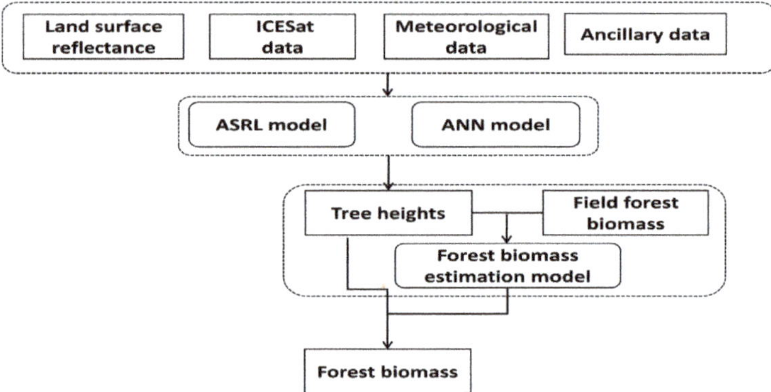

Figure 2. The approach total flowchart of this study.

This section describes tree height estimation methods used in this study and forest biomass modeling.

2.3.1. Tree Height Estimation Methods

Artificial Neural Network (ANN) Tree Heights Model Approach

The artificial neural network (ANN) tree heights model was proposed by Ni [38] for mapping the forest tree heights over China continent. We employed the ANN tree height model to obtain the tree heights in NE China. The inputs of ANN tree heights model consisted of 11 parameters, including climate variables (temperature, precipitation), ancillary data (land cover class, vegetation cover fraction), and seven multispectral bands of MODIS NBAR data. With the help of back-propagation (BP) process algorithm, the ANN tree height models was trained by using the GLAS-derived tree heights. The training and validation data pairs for each pixel of study region were selected according to the most similar climatic condition factor discipline according to described by Ni [38]. For each target pixel, one ANN-based tree heights model was trained by using 15 pairs of training data, while five pairs of validation data were used to prevent the over-fitting of the model. When the ANN-based tree heights model was trained well, it was used for estimating the forest canopy height over the target pixel.

Allometric Scaling and Resource Limitations (ASRL) Model Approach

We employed the ASRL model approach developed by Sungho et al. [41] to obtain the tree heights in NE China. The ASRL is a mechanism model based on a combination of the allometric scaling laws and local resource availability [41]. It can predict the potential maximum tree heights. A key hypothesis of the ASRL model is that a tree maintains its evaporative flow by collecting sufficient resources (water and light) while satisfying its basal metabolic needs, in turn, limiting maximum growth. The mechanism has a quantitative expression by the simple inequality ($Q\_p \geq Q\_e \geq Q\_0$). Here, the Q_p means the available flow rate, Q_e refers to the actual flow rate of a tree, and Q_0 represents the basic or required metabolic flow rate of a tree. Shi et al. [55] and Sungho et al. [41] proposed the parametric optimization methods of the ASRL model in order to obtain the actual tree heights. Shi et al. [55] recommend a parametric optimization, which can iteratively adjust three ASRL parameters to minimize disparities between the reference and modeled heights, and help improving the overall prediction accuracy. In this study, we applied the same processes to optimize ASRL model and obtain the actual tree heights in NE China.

2.3.2. Forest Biomass Modeling

Zianis and Mencuccini [56] developed an allometric approach to estimate aboveground forest biomass (M) by regarding M and corresponding tree diameter at breast height (DBH). The method can be described from the following allometric equation [34] (Equation (3)):

$$M = a \times DBH^b \tag{3}$$

where *a* is the allometric intercept and *b* is the allometric exponent. Based on the scale relationship between tree height (*H*) and DBH, the aboveground treebiomass estimation equation can be rewritten as Equation (4):

$$M = a \times H^b + c \tag{4}$$

where *c* is a constant, *a* is the allometric intercept and *b* is the allometric exponent.

In this study, we have used the field measured biomass data and forest tree heights which were calculated by averaging the ANN-derived tree heights and ASRL tree heights to build the least-square regression model to obtain the coefficients of Equation (4).

In order to obtain the forest root biomass and the total biomass, we developed the allometric relationship equation based on the field-measured tree aboveground biomass and root biomass. Figure 3 shows the fitted allometric relationship equations described in Equations (3) and (4).

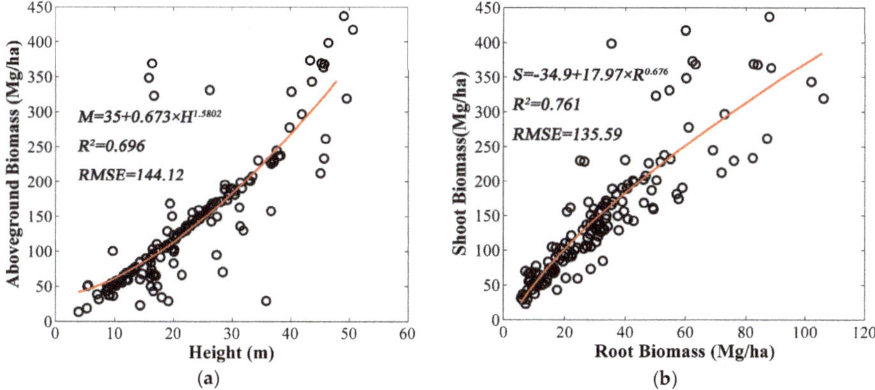

Figure 3. The allometric relationships of forest tree heights, aboveground biomass and root biomass. (**a**) The allometric relationship between field aboveground biomass and height; and (**b**) the allometric relationship between the field aboveground biomass and root biomass.

3. Results

3.1. Canopy Heights Mapping in NE China by Two Tree Height Methods

Following the artificial neural network (ANN) tree height model, a contiguous map of canopy heights with a spatial resolution of 1 km in NE China region was generated. Figure 4a shows the tree height map derived from the ANN model in NE China.

Similarly, a contiguous map of the tree heights over NE China was generated from the optimized ASRL mode by using the gridded climate and ancillary data mentioned in Section 2.2. This map (Figure 4b) closely represents the spatial distribution of tree heights(mean = 24.37, SD = 9.35) in NE China regionwith the spatial resolution of 1 km, showing a high correspondence to the spatial pattern of ANN tree heights (mean = 26.66 m, SD = 10.13).In order to investigate the accuracy of the two tree heights maps, we tested the tree heights by comparing with the GLAS-derived tree heights according to the validation methods from Ni [37] and Ni [38], respectively. The validation results ($R^2 = 0.811$, RMSE = 4.79; $R^2 = 0.664$, RMSE = 5.24) demonstrate the effectiveness of the ANN- and ASRL-based tree heights models.

From the estimation result of the ANN and ASRL tree heights in NE China, we can see that relatively tall trees were growing in the eastern research areas. The trees distributed in Northwestern China show obviously lower heights than those in eastern regions.

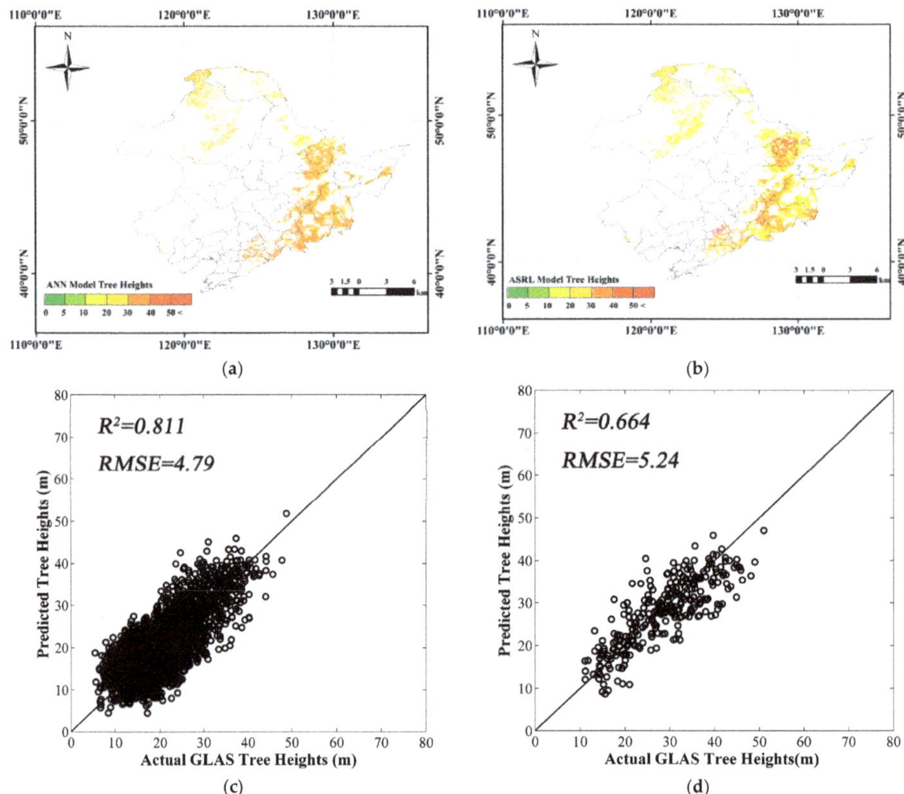

Figure 4. The forest tree heights based on ANN and ASRL methods respectively. (**a**) ANN tree heights; (**b**) ASRL tree heights; (**c**) ANN based tree heights validation; (**d**) ASRL based tree heights validation.

3.2. Forest Biomass Estimation

We firstly estimated the forest aboveground biomass by using the allometric approach described in Section 2.3.2. Based on the forest aboveground biomass model described in Figure 3a, a contiguous map of forest aboveground biomass at a spatial resolution of 1 km was generated (Figure 5a).

In addition, the forest root biomass was estimated according to the relationship equation described as Figure 3b. The total biomass was calculated by summing the forest aboveground biomass and root biomass. The root biomass and total biomass map can be seen from the Figure 5c,e.

The estimation result shows that the forest aboveground biomass, root biomass, and total biomass have the same density distribution trend in the research region. The forests in the eastern research region have relatively large forest biomass density. The low biomass density was distributed in northwestern research region. The parts of the southeastern NE China (Eastern Liaoning Province) show the largest biomass density in the research area.

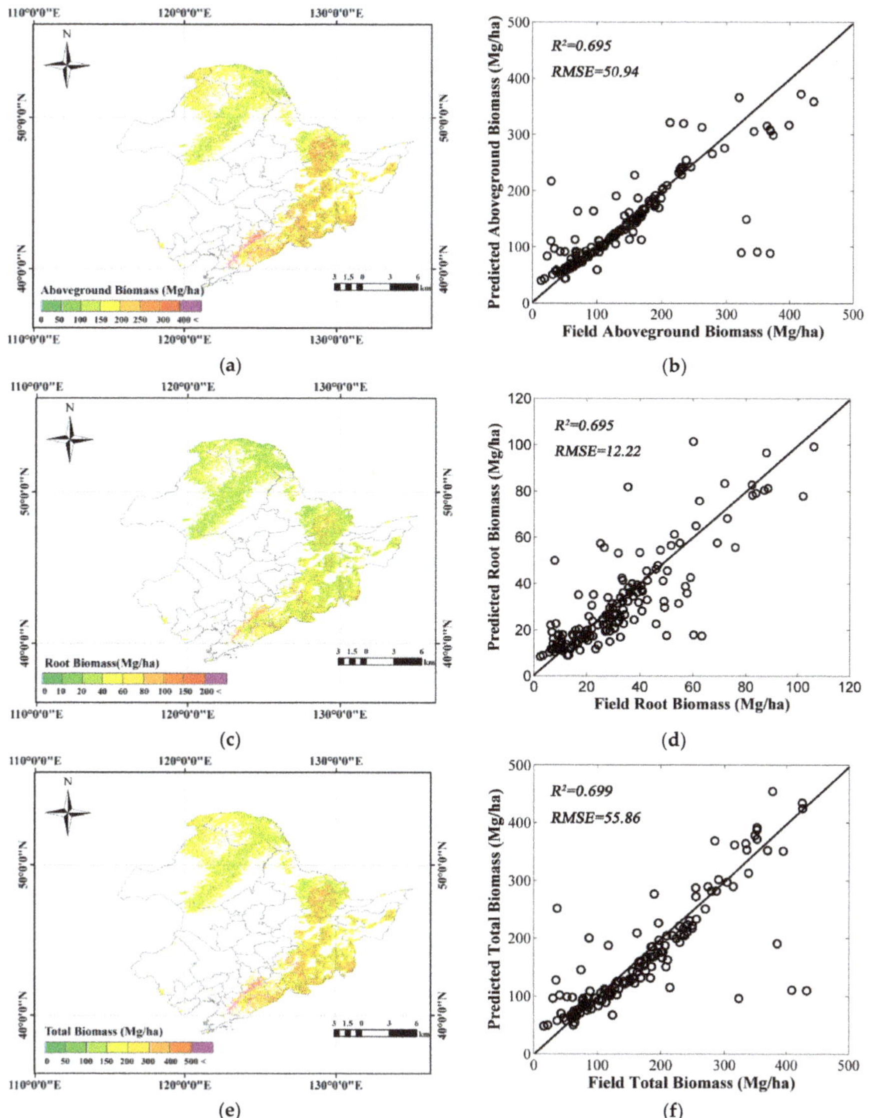

Figure 5. The estimation and validation results of forest aboveground biomass, root biomass, and total biomass in NE China region.(**a**) Forest aboveground biomass; (**b**) estimation validation of aboveground biomass; (**c**)forest root biomass; (**d**) estimation validation of root biomass; (**e**) forest total biomass; and (**f**) estimation validation of total biomass.

4. Discussion

A method to assess the forest biomass in NE China forests region was developed by combining satellite LiDAR data, optical remote sensing data, and field measurements. Due to the great difficulties in measuring root biomass for forest total biomass estimation, it is the primary method for estimating root biomass by building the allomeric relationship with aboveground biomass [54].

Based on the field measured forest aboveground biomass and root biomass, we obtained a relatively good allometric model for predicting root biomass ($R^2 = 0.761$) in the study area. However, there are some factors that influence the precision of root biomass prediction from forest aboveground biomass, including climate factors, and forest types. Especially at large scale, varying climate factors and forest types might cause the different allometric relationships between aboveground biomass and root biomass. Similarly, allometric models of aboveground biomass estimation from tree heights are also influenced by the climate factors and forest types.

In this case, different allometric models are necessary in the ecological zones which differ according to the climate factors and forest types. Generally, these allometric models would be optimized in these ecological zones. Due to the lack enough fields measured forest biomass and low resolution forest types data, in this study, we just used the united model which ignored the climate factors and forest types.

As shown in Figure 5, most of predicted and field biomass has good linear relationship, while there is a small quantity of points scattered out across the isoline. Although the validation result of forest biomass shows that our allometric metrics are efficient, some uncertainties occurred in the validation (Figure 5). The scattered points might be caused from the combination of several influenced factors, such as forest biomes, different measurement rules of field biomass data from various studies, and the propagation of errors from tree heights and biomass estimation models. The following points could be improved for higher precision of the forest biomass estimation in the future. Firstly, more effective geostatistics interpolation methods should be used for obtaining finer climate factor raster data, especially in the mountain areas which lack enough meteorological stations. Secondly, it is necessary to collect more field measured forest plots data, including the tree heights, aboveground biomass, root biomass, and total biomass. Finally, the allometric models for estimating aboveground biomass from tree heights and predicting root biomass from aboveground biomass should be respectively developed in the ecological zones which are divided based on the climate data and the forest type data.

5. Conclusions

We used two tree height models to obtain the forest tree heights map of NE China by combining GLAS data, meteorological data and optical remote sensing data. Based on the collected field measured forest plots data, we developed the allometric models for predicting the aboveground biomass and root biomass. Then, the forest total biomass map was generated by summing the aboveground biomass and root biomass.

The assessment of the three forest biomass map using field measured forest plots data was performed separately. At the forest sites level, high correlation appeared in the estimation results of aboveground biomass and total biomass ($R^2 = 0.695$, RMSE = 50.94; $R^2 = 0.699$, RMSE = 55.86). The estimation results of root biomass showed, relatively, a slightly lower correlation with the field measured data ($R^2 = 0.695$, RMSE = 12.22). In summary, this study demonstrated that forest aboveground biomass and root biomass can be estimated with relatively high precision by combining satellite LiDar data, MODIS data, and climate data in regional scale.

This study highlights the potential to estimate forest total biomass in conjunction with novel and efficient methods, such as ANN tree heights metrics, ASRL-based tree height models and allometric scaling relationships among tree heights, forest aboveground biomass, and root biomass. Combinations of these techniques were able to quantify the forest structure parameters and biomass. Application of the approach proposed in this study at the national scale would provide an opportunity to understand carbon sinks of all forest land in China. These relatively fine-scaled, spatially-explicit forest biomass maps can provide critical information for forest carbon cycle studies and forest resource management.

Acknowledgments: The authors would like to thank the three anonymous reviewers whose comments significantly improved this manuscript. This study was partially funded by the Special Fund for Forest Scientific Research in the Public Welfare (grant no. 201504323), the National Key Research and Development Program of China (grant no. 2016YFB0501505), the Special Fund for the Ecological Assessment of Three Gorges Project

Forests **2017**, *8*, 288

(grant no. 0001792015CB5005), the National NaturalScience Foundation (2016, grant no.41601478), the National Key R and D Program (2016YFC0500103) of China, and the Key Programs of the Chinese Academy of Sciences (Grand No.KZZD-EW-TZ-17).

Author Contributions: The analysis was performed by Xiliang Ni. All authors contributed with ideas, writing, and discussions.

Conflicts of Interest: The authors declare no conflict of interest.

References

1. Houghton, R.A.; Hall, F.; Goetz, S.J. Importance of biomass in the global carbon cycle. *J. Geophys. Res.* **2009**, *114*, G00E03. [CrossRef]

2. Chi, H.; Sun, G.; Huang, J.; Guo, Z.; Ni, W.; Fu, A. National Forest Aboveground Biomass Mapping from ICESat/GLAS Data and MODIS Imagery in China. *Remote Sens.* **2015**, *7*, 5534–5564. [CrossRef]

3. Deo, R.K.; Russell, M.B.; Domke, G.M.; Andersen, H.-E.; Cohen, W.B.; Woodall, C.W. Evaluating Site-Specific and Generic Spatial Models of Aboveground Forest Biomass Based on Landsat Time-Series and LiDAR Strip Samples in the Eastern USA. *Remote Sens.* **2017**, *9*, 598. [CrossRef]

4. Myneni, R.B.; Dong, J.; Tucker, C.J.; Kaufmann, R.K.; Kauppi, P.E.; Liski, J.; Zhou, L.; Alexeyev, V.; Hughes, M.K. A large carbon sink in the woody biomass of northern forests. *Proc. Natl. Acad. Sci. USA* **2001**, *98*, 14784–14789. [CrossRef] [PubMed]

5. Schimel, D.S.; House, J.I.; Hibbard, K.A.; Bousquet, P.; Ciais, P.; Peylin, P.; Braswell, B.H.; Apps, M.J.; Baker, D.; Bondeau, A.; et al. Recent patterns and mechanisms of carbon exchange by terrestrial ecosystems. *Nature* **2001**, *414*, 169–172. [CrossRef] [PubMed]

6. Serreze, M.C.; Walsh, J.E.; Iii, F.S.C.; Osterkamp, T.; Dyurgerov, M.; Romanovsky, V.; Oechel, W.C.; Morison, J.; Zhang, T.; Barry, R.G. Observational evidence of Recent change in the northern high-latitude environment. *Clim. Chang.* **2000**, *46*, 159–207. [CrossRef]

7. Neigh, S.R.; Nelson, R.F.; Ranson, K.J.; Margolis, H.A.; Montesano, P.M.; Sun, G.; Kharuk, V.; Næsset, E.; Wulder, M.A.; Andersen, H. Taking stock of circum boreal forest carbon with ground measurements, airborne and space borne LiDAR. *Remote Sens. Environ.* **2013**, *137*, 274–287. [CrossRef]

8. Goodale, C.L.; Apps, M.J.; Birdsey, R.A.; Field, C.B.; Heath, L.S.; Houghton, R.A.; Jenkins, J.C.; Kohlmaier, G.H.; Kurz, W.; Liu, S.R.; et al. Forest carbon sinks in the Northern Hemisphere. *Ecol. Appl.* **2002**, *12*, 891–899. [CrossRef]

9. Fang, J.Y.; Brown, S.; Tang, Y.H.; Naruurs, G.-J.; Wang, X.P.; Shen, H.H. Overestimated biomass carbon pools of the northern mid-and high latitude forests. *Clim. Chang.* **2006**, *74*, 355–368. [CrossRef]

10. LeToan, T.; Quegan, S.; Davidson, M.J.; Balzter, H.; Paillou, P.; Papathanassiou, K.; Plummer, S.; Rocca, F.; Saatchi, S.; Shugart, H.; et al. The BIOMASS mission: Mapping global forest biomass to better understand the terrestrial carbon cycle. *Remote Sens. Environ.* **2011**, *115*, 2850–2860. [CrossRef]

11. Hudak, A.T.; Strand, E.K.; Vierling, L.A.; Byrne, J.C.; Eitel, J.U.H.; Martinuzzi, S.; Falkowski, M.J. Quantifying aboveground forest carbon pools and fluxes from repeat LiDAR surveys. *Remote Sens. Environ.* **2012**, *123*, 25–40. [CrossRef]

12. Barbosa, J.M.; Melendez-Pastor, I.; Navarro-Pedreno, J.; Bitencourt, M.D. Remotely sensed biomass over steep slopes: An evaluation among successional stands of the Atlantic Forest, Brazil. *ISPRS J. Photogramm.* **2014**, *88*, 91–100. [CrossRef]

13. Houghton, R.A. Above ground forest biomass and the global carbon balance. *Glob. Chang. Biol.* **2005**, *11*, 945–958. [CrossRef]

14. Cairns, M.A.; Brown, S.; Helmer, E.H.; Baumgardner, G.A. Root biomass Allocation in the world's upland forests. *Oecologia* **1997**, *111*, 1–11. [CrossRef] [PubMed]

15. Mokany, K.; Raison, R.J.; Prokushkin, A.S. Critical analysis of root: Shoot ratios in terrestrial biomes. *Glob. Chang. Biol.* **2005**, *11*, 1–13. [CrossRef]

16. Wang, X.; Fang, J.; Zhu, B. Forest biomass and root-shoot allocation in northeast China. *For. Ecol. Manag.* **2008**, *255*, 4007–4020. [CrossRef]

17. Chave, J.; Andalo, C.; Brown, S.; Cairns, M.A.; Chambers, J.Q.; Eamus, D.; Fölster, H.; Fromard, F.; Higuchi, N.; Kira, T.; et al. Tree allometry and improved estimation of carbon stocks and balance in tropical forests. *Oecologia* **2005**, *145*, 87–99. [CrossRef] [PubMed]

18. Boudreau, J.; Nelson, R.F.; Margolis, H.A.; Beaudoin, A.; Guindon, L.; Kimes, D.S. Regional aboveground forest biomass using airborne and space borne LiDAR in Quebec. *Remote Sens. Environ.* **2008**, *112*, 3876–3890. [CrossRef]

19. Guo, Z.F.; Chi, H.; Sun, G. Estimating forest aboveground biomass using HJ-1 Satellite CCD and ICES at GLAS waveform data. *Sci. China Earth Sci.* **2010**, *53*, 16–25. [CrossRef]

20. Hayashi, M.; Saigusa, N.; Oguma, H.; Yamagata, Y. Forest canopy height estimation using ICESat/GLAS data and error factor analysis in Hokkaido, Japan. *ISPRS J. Photogramm.* **2013**, *81*, 12–18. [CrossRef]

21. Ram, D.; Matthew, R.; Grant, D.; Hans-Erik, A.; Warren, C.; Christopher, W.; Michael, J.F.; Warren, B.C. Using Landsat Time-Series and LiDAR to Inform Aboveground Forest Biomass Baselines in Northern Minnesota, USA. *Can. J. Remote Sens.* **2017**, *43*, 28–47.

22. Gibbs, H.K.; Brown, S.; Niles, J.O.; Foley, J.A. Monitoring and estimating tropical forest carbon stocks: Making REDD a reality. *Environ. Res. Lett.* **2007**, *2*, 1–13. [CrossRef]

23. Englhart, S.; Keuck, V.; Siegert, F. Aboveground biomass retrieval in tropical forests—The potential of combined X- and L-band SAR data use. *Remote Sens. Environ.* **2011**, *115*, 1260–1271. [CrossRef]

24. Pflugmacher, D.; Cohen, W.B.; Kennedy, R.E.; Yang, Z.Q. Using Landsat-derived disturbance and recovery history and lidar to map forest biomass dynamics. *Remote Sens. Environ.* **2013**, *151*, 124–137. [CrossRef]

25. Blair, J.B.; Rabine, D.L.; Hofton, M.A. The Laser Vegetation Imaging Sensor: A medium-altitude, digitisation-only, airborne laser altimeter for mapping vegetation and topography. *ISPRS J. Photogramm.* **1999**, *54*, 115–122. [CrossRef]

26. Abshire, J.B.; Sun, X.L.; Riris, H.; Sirota, J.M.; McGarry, J.F.; Palm, S.; Yi, D.H.; Liiva, P. Geoscience Laser Altimeter System (GLAS) on the ICESat mission: On-orbit measurement performance. *Geophys. Res. Lett.* **2005**, *32*. [CrossRef]

27. Zolkos, S.G.; Goetz, S.J.; Dubayah, R.A. Meta-analysis of terrestrial aboveground biomass estimation using lidar remote sensing. *Remote Sens. Environ.* **2013**, *128*, 289–298. [CrossRef]

28. Avitabile, V.; Baccini, A.; Friedl, M.A.; Schmullius, C. Capabilities and limitations of Landsat and land cover data for aboveground woody biomass estimation of Uganda. *Remote Sens. Environ.* **2012**, *117*, 366–380. [CrossRef]

29. Schutz, B.E.; Zwally, H.J.; Shuman, C.A.; Hancock, D.; DiMarzio, J.P. Overview of the ICESat Mission. *Geophys. Res. Lett.* **2005**, *32*. [CrossRef]

30. Sun, G.; Ranson, K.J.; Guo, Z.; Zhang, Z.; Montesano, P.; Kimes, D. Forest biomass mapping from lidar and radar synergies. *Remote Sens. Environ.* **2011**, *115*, 2906–2916. [CrossRef]

31. Koch, B. Status and future of laser scanning, synthetic aperture radar and hyperspectral remote sensing data for biomass assessment. *ISPRS J. Photogramm.* **2010**, *65*, 581–590. [CrossRef]

32. Baccini, A.; Laporte, N.; Goetz, S.J.; Sun, M.; Dong, H. A first map of tropical Africa's aboveground biomass derived from satellite imagery. *Environ. Res. Lett.* **2008**, *3*, 1–9. [CrossRef]

33. Fang, J.Y.; Chen, A.P.; Peng, C.H.; Zhao, S.Q.; Ci, L.J. Changes in forest biomass carbon storage in China between 1949 and 1998. *Science* **2001**, *292*, 2320–2322. [CrossRef] [PubMed]

34. Deo, R.K. Modeling and Mapping of aboveground Biomass and Carbon Sequestration in the Cool Temperature Forest of North-East China. Master's Thesis, International Institution for Geo-Information Science and Earth Observation Enschede, Enschede, The Netherlands, 2008.

35. Corfee-Morlot, J.; Maslin, M.; Burgess, J. Global warming in the public sphere. *Philos. Trans. R. Soc.A—Math. Phys. Eng. Sci.* **2007**, *365*, 2741–2776. [CrossRef] [PubMed]

36. Piao, S.; Fang, J.; Zhou, L.; Ciais, P.; Zhu, B. Variations in satellite-derived phenology in China's temperate vegetation. *Glob. Chang. Biol.* **2006**, *12*, 672–685. [CrossRef]

37. Ni, X.; Park, T.; Choi, S.; Shi, Y.; Cao, C.; Wang, X.; Lefsky, M.A.; Simard, M.; Myneni, R.B. Allometric scaling and resource limitations model of tree heights: Part 3. Model optimization and testing over continental China. *Remote Sens.* **2014**, *6*, 3533–3553. [CrossRef]

38. Ni, X.; Zhou, Y.; Cao, C.; Wang, X.; Shi, Y.; Park, T.; Choi, S.; Myneni, R.B. Mapping Forest Canopy Height over Continental China Using Multi-Source Remote Sensing Data. *Remote Sens.* **2015**, *7*, 8436–8452. [CrossRef]

39. Zhou, Y.L. *Geography of the Vegetation in Northeast China*; Science Press: Beijing, China, 1997.

40. China Meteorological Data Sharing Service System. Available online: http://cdc.cma.gov.cn/ (accessed on 15 March 2013).

41. Choi, S.; Ni, X.; Shi, Y.; Ganguly, S.; Zhang, G.; Duong, H.V.; Lefsky, M.A.; Simard, M.; Saatchi, S.S.; Lee, S. Allometric scaling and resource limitations model of tree heights: Part 2. Site based testing of the model. *Remote Sens.* **2013**, *5*, 202–223. [CrossRef]
42. Olea, R.A. *Geostatistics for Engineers and Earth Scientists*; Springer: New York, NY, USA, 1999.
43. Yuan, H.; Dai, Y.J.; Xiao, Z.Q.; Ji, D.Y.; Wei, S.G. Reprocessing the MODIS Leaf Area Index products for land surface and climate modelling. *Remote Sens. Environ.* **2011**, *115*, 1171–1187. [CrossRef]
44. Available online: https://lpdaac.usgs.gov/dataset_discovery/modis/modis_products_table/mod44b_v006 (accessed on 6 August 2017).
45. Román, M.O.; Schaaf, C.B.; Woodcock, C.E.; Strahler, A.H.; Yang, X.; Braswell, R.H.; Curtis, P.; Davis, K.J.; Dragoni, D.; Goulden, M.L.; et al. The MODIS (Collection V005) BRDF/albedo product: Assessment of spatial representativeness over forested landscapes. *Remote Sens. Environ.* **2009**, *113*, 2476–2498. [CrossRef]
46. Available online: https://icesat.gsfc.nasa.gov/icesat/ (accessed on 6 August 2017).
47. Simard, M.; Pinto, N.; Fisher, J.B.; Baccini, A. Mapping forest canopy height globally with spacebornelidar. *J. Geophys. Res.-Biogeosci.* **2011**, *116*. [CrossRef]
48. Lefsky, M.A. A global forest canopy height map from the moderate resolution imaging spectroradiometer and the geoscience laser altimeter system. *Geophys. Res. Lett.* **2010**, *37*. [CrossRef]
49. Lee, S.; Ni-Meister, W.; Yang, W.; Chen, Q. Physically based vertical vegetation structure retrieval from ICESat data: Validation using LVIS in White Mountain National Forest, New Hampshire, USA. *Remote Sens. Environ.* **2011**, *115*, 2776–2785. [CrossRef]
50. Neuenschwander, A.L.; Urban, T.J.; Gutierrez, R.; Schutz, B.E. Characterization of ICESat/GLAS waveforms over terrestrial ecosystems: Implications for vegetation mapping. *J. Geophys. Res. Biogeosci.* **2008**, *113*. [CrossRef]
51. Zhang, G.; Ganguly, S.; Nemani, R.R.; White, M.A.; Milesi, C.; Hashimoto, H.; Wang, W.; Saatchi, S.; Yu, Y.; Myneni, R.B. Estimation of forest aboveground biomass in California using canopy height and leaf area index estimated from satellite data. *Remote Sens. Environ.* **2014**, *151*, 44–56. [CrossRef]
52. Ni, X.L.; Shi, Y.L.; Choi, S.H.; Cao, C.X.; Myneni, R.B. Estimation of tree heights using remote sensing data and an allometric scaling and resource limitations (ASRL) model. In Proceedings of the 2012 IEEE International Geoscience and Remote Sensing Symposium (IGARSS), Munich, Germany, 22–27 July 2012; pp. 7248–7251.
53. Zhu, B. Carbon Stocks of Main Forest Ecosystems in Northeast China. Master's Thesis, Peking University, Beijing, China, 2005.
54. Luo, T.X. Patterns of net primary productivity for Chinese major forest types and their mathematical models. Ph.D. Thesis, Chinese Academy of Sciences, Beijing, China, 1996.
55. Shi, Y.; Choi, S.; Ni, X.; Ganguly, S.; Zhang, G.; Duong, H.V.; Lefsky, M.A.; Simard, M.; Saatchi, S.S.; Lee, S. Allometric scaling and resource limitations model of tree heights: Part 1. Model optimization and testing over continental USA. *Remote Sens.* **2013**, *5*, 284–306. [CrossRef]
56. Zianis, D.; Mencuccini, M. On simplifying allometric analyses of forest biomass. *For. Ecol. Manag.* **2004**, *187*, 311–332. [CrossRef]

 *forests*

Article

Analysis of Global LAI/FPAR Products from VIIRS and MODIS Sensors for Spatio-Temporal Consistency and Uncertainty from 2012–2016

Baodong Xu [1,2,3], Taejin Park [2,*], Kai Yan [2], Chi Chen [2], Yelu Zeng [4], Wanjuan Song [2], Gaofei Yin [5], Jing Li [1,3], Qinhuo Liu [1,3], Yuri Knyazikhin [2] and Ranga B. Myneni [2]

[1]   State Key Laboratory of Remote Sensing Science, Institute of Remote Sensing and Digital Earth, Chinese Academy of Sciences, Beijing 100101, China; xubd@radi.ac.cn (B.X.); lijing01@radi.ac.cn (J.L.); liuqh@radi.ac.cn (Q.L.)
[2]   Department of Earth and Environment, Boston University, Boston, MA 02215, USA; kaiyan.earthscience@gmail.com (K.Y.); chenchi@bu.edu (C.C.); wandasong9703@gmail.com (W.S.); jknjazi@bu.edu (Y.K.); ranga.myneni@gmail.com (R.B.M.)
[3]   University of Chinese Academy of Sciences, Beijing 100049, China
[4]   Department of Global Ecology, Carnegie Institution for Science, Stanford, CA 94305, USA; zengyelu@163.com
[5]   Institute of Mountain Hazards and Environment, Chinese Academy of Sciences, Chengdu 610041, China; gaofeiyin@imde.ac.cn
*   Correspondence: parktj@bu.edu; Tel.: +1-617-893-1988; Fax: +1-617-353-8399

Received: 8 December 2017; Accepted: 29 January 2018; Published: 1 February 2018

Abstract: The operational Moderate Resolution Imaging Spectroradiometer (MODIS) Leaf Area Index (LAI) and Fraction of Photosynthetically Active Radiation absorbed by vegetation (FPAR) algorithm has been successfully implemented for Visible Infrared Imager Radiometer Suite (VIIRS) observations by optimizing a small set of configurable parameters in Look-Up-Tables (LUTs). Our preliminary evaluation showed reasonable agreement between VIIRS and MODIS LAI/FPAR retrievals. However, there is a need for a more comprehensive investigation to assure continuity of multi-sensor global LAI/FPAR time series, as the preliminary evaluation was spatiotemporally limited. In this study, we use a multi-year (2012–2016) global LAI/FPAR product generated from VIIRS and MODIS to evaluate for spatiotemporal consistency. We also quantify uncertainty of the product by utilizing available ground measurements. For both consistency and uncertainty evaluation, we account for variations in biome type and temporal resolution. Our results indicate that the LAI/FPAR retrievals from VIIRS and MODIS are consistent at different spatial (i.e., global and site) and temporal (i.e., 8-day, seasonal and annual) scales. The estimate of mean discrepancy (-0.006 ± 0.013 for LAI and -0.002 ± 0.002 for FPAR) meets the stability requirement for long-term LAI/FPAR Earth System Data Records (ESDRs) from multi-sensors as suggested by the Global Climate Observing System (GCOS). It is noteworthy that the rate of retrievals from the radiative transfer-based main algorithm is also comparable between two sensors. However, a relatively larger discrepancy over tropical forests was observed due to reflectance saturation and an unexpected interannual variation of main algorithm success was noticed due to instability in input surface reflectances. The uncertainties/relative uncertainties of VIIRS and MODIS LAI (FPAR) products assessed through comparisons to ground measurements are estimated to be 0.60/42.2% (0.10/24.4%) and 0.55/39.3% (0.11/26%), respectively. Note that the validated LAI were only distributed in low domains (~2.5), resulting in large relative uncertainty. Therefore, more ground measurements are needed to achieve a more comprehensive evaluation result of product uncertainty. The results presented here generally imbue confidence in the consistency between VIIRS and MODIS LAI/FPAR products and the feasibility of generating long-term multi-sensor LAI/FPAR ESDRs time series.

Keywords: VIIRS; leaf area index (LAI); Fraction of Photosynthetically Active Radiation absorbed by vegetation (FPAR); MODIS; consistency; uncertainty; evaluation

1. Introduction

Leaf Area Index (LAI) and Fraction of Photosynthetically Active Radiation absorbed by vegetation (FPAR) are well-known as two essential variables to describe the exchange of fluxes of energy, mass (e.g., water, nutrients, and carbon dioxide) and momentum between the surface and atmosphere [1]. LAI, which is generally defined as one-sided green leaf area per unit ground area in broadleaf canopies and as the projected needle leaf area in coniferous canopies [2], is extensively used to characterize the structure and function of vegetation [3]. FPAR measures the fraction of radiation that leaves absorb in the 0.4–0.7 µm spectrum and it thus evaluates the energy absorption capacity of a canopy [4]. For effective use in most global models of climate, biogeochemistry and ecology, the long-term LAI/FPAR products have been generated from satellite remote sensing [2,5].

In particular, the ground-breaking Earth Observing System (EOS) Moderate Resolution Imaging Spectroradiometer (MODIS) sensors onboard Terra and Aqua satellites have provided an opportunity for opening a new horizon of global LAI/FPAR products [2]. The latest version (Collection 6, C6) of global LAI/FPAR products from MODIS (since February 2000) is freely available through the Land Processes Distributed Active Archive Center (LP DAAC), and is widely used by the scientific, public, and private user communities [5–7]. However, both Terra and Aqua MODIS have far exceeded their design life, and they will be likely terminated in the early 2020s [8]. In the context of MODIS termination, the Visible/Infrared Imager Radiometer Suite (VIIRS) instrument onboard the Suomi National Polar-orbiting Partnership (S-NPP) and Joint Polar Satellite System (JPSS) has inherited the scientific roles of MODIS to provide the long-term Earth System Data Records (ESDRs) [9]. Thus, developing consistent LAI/FPAR dataset from these new sensors to continue the MODIS ESDR is a high priority.

Due to sensor-specific spatial resolution and spectral band composition, it is unreliable to directly apply the MODIS algorithm to VIIRS observations [10,11]. With theoretical basis of "canopy spectral invariants" that permits an accurate decoupling of the structural and radiometric components of modeled and measured spectral Bi-directional Reflectance Factors (BRFs) [12–14], the MODIS algorithm has been successfully implemented with VIIRS by optimizing a small set of configurable parameters in Look-Up-Tables (LUTs) [11,15]. The preliminary evaluation shows reasonable agreement (global mean difference is 0.024 and 0.029 in January and July of 2015, respectively) between LAI/FPAR retrievals from VIIRS and MODIS [11]. Nevertheless, a comprehensive assessment of the VIIRS product using multi-year and global retrievals is still needed to assure the continuity with the MODIS time series.

Here, we have generated VIIRS LAI/FPAR products that cover the first five-years (2012–2016) of the S-NPP mission era, and we used them to achieve the following objectives: (1) to evaluate the spatiotemporal consistency between the latest version of VIIRS (Version 1, V1) and MODIS (C6) products; (2) to quantify the uncertainty of the VIIRS and MODIS products using available ground measurements; and (3) to understand the observed inconsistency between LAI/FPAR retrievals from two sensors. This paper is organized as follows. Section 2 describes the LAI/FPAR retrieval algorithm and its optimization for VIIRS sensor. Section 3 introduces the data and methods used in this study. Section 4 presents the evaluation results of VIIRS products. Section 5 discusses the reasons for the inconsistency between MODIS and VIIRS products and future developments of VIIRS product evaluation. Finally, Section 6 provides the concluding remarks on this study.

2. LAI/FPAR Retrieval Algorithm

2.1. Algorithm Description

The heritage algorithm, developed for MODIS, includes BRFs in the red and Near Infra-Red (NIR) bands, their uncertainties, sun-sensor geometry and a biome classification map. The biome map is required here as a priori-knowledge to reduce the number of unknown parameters of the "ill-posed" inversion problem [2]. The operational LAI/FPAR algorithm includes a main algorithm that is based on a three-dimensional (3D) Radiative Transfer (RT) equation. By describing the photon transfer process, this algorithm links surface spectral BRFs to both structural and spectral parameters of the vegetation canopy and soil [16,17]. Given atmosphere-corrected surface spectral BRFs and their uncertainties, the algorithm finds all candidates of LAI and FPAR by comparing observed and modeled BRFs within biome-specified thresholds of uncertainties that are stored in LUTs. Then, the mean and standard deviation of all LAI/FPAR candidates are reported as the retrieval and its dispersion, respectively. The main algorithm may fail to localize a solution if uncertainties of input BRFs are larger than specified threshold values or due to cloud effects or too low sun/view zenith angles. In this case of a failed solution, the back-up algorithm based on the LAI/FPAR-NDVI relationships for each biome is used to retrieve LAI/FPAR with relatively poor quality [2,13]. Finally, the LAI/FPAR product is generated over the composition period (8-day/4-day) using the daily products. In detail, the compositing algorithm includes two steps: (1) the main algorithm retrievals are first selected (if none available, back-up retrievals are selected); (2) the maximum FPAR and the corresponding LAI are selected as the final product value.

2.2. Generation of VIIRS-Specific LUTs

Although VIIRS was designed to be quite similar to MODIS, the differences between these two sensors cannot be ignored [18,19]. For example, the relative spectral responses (RSRs) of MODIS and VIIRS sensors are similar at NIR bands but quite different at red bands (VIIRS has a broader bandwidth than MODIS) [11]. Moreover, VIIRS has a wider swath of 3000 km with a view angle between ±56.28° than MODIS's swath of 2330 km with a view angle between ±55° [20], resulting in near-constant resolution of VIIRS from the nadir to the edge of the scan [21]. Additionally, the native instantaneous field of view (IFOV) of VIIRS and MODIS red/NIR bands are 375 m and 250 m, respectively. All discrepancies between VIIRS and MODIS have potential errors for their measured BRFs. Therefore, the generation of consistent LAI/FPAR products from MODIS and VIIRS requires parameterizations that account for sensor-specific features.

The theory of "canopy spectral invariants" provides the necessary parameterizations using a small set of measurable variables that describe the spectral response of vegetation to incident radiation at a range of spatial scales, including: (1) spectrally invariant canopy interceptances, recollision probability, directional escape probability, and their respective hemispherically averaged values; (2) spectrally varying ground reflectance and (3) the single-scattering albedo [12–14]. This theory is applied to adapt the LAI/FPAR algorithm to the spatial scale of VIIRS surface reflectance measurements. The spectrally invariant parameters for the different vegetation types in the VIIRS-LUTs were derived using the procedure described in [22]. The spectrally varying ground reflectance was taken from the soil reflectance model developed by [23,24], including 29 typical patterns of effective soil reflectance ranging from bright to dark [15]. Note that the uncertainty of soil reflectance was also considered in generation of VIIRS LUTs to well represent the global conditions [13]. The scale and spectral dependence of the single scattering albedo were exploited to account for VIIRS spatial resolution, spectral bandwidths and radiometric response [25–27]. This theory is the fundamental principle of the LAI/FPAR retrieval algorithm because knowing the invariants of the canopy and the single scattering albedo (SSA) of an average phytoelement (e.g., leaf, steam) at any wavelength makes it possible to reconstruct the radiation field of the canopy at any wavelength. The spectral invariant parameters permit decoupling of the structural and radiometric components of any optical sensor signal, which is the theoretical foundation of optimizing configurable parameters to achieve inter-sensor consistency in multi-sensor LAI/FPAR retrievals. Therefore, based on the theory of "canopy spectral invariants", VIIRS-specific LUTs have been developed and incorporated. The theoretical and technical details are described in [11,15].

3. Data and Method

3.1. VIIRS and MODIS LAI/FPAR

Testing and integrating the operational VIIRS LAI/FPAR product are currently under way and its completion is expected in the spring of 2018. The VIIRS LAI/FPAR product will be generated at 500-m spatial resolution and an 8-day time step over a sinusoidal grid. The access to the VIIRS LAI/FPAR product can be obtained from the product's user guide (https://viirsland.gsfc.nasa.gov/PDF/VIIRS_LAI_FPAR_UserGuide_V1.1.pdf). Here, we used the individual Science Computing Facility (SCF) (Boston University, Boston, MA, USA) to generate a multi-year VIIRS LAI/FPAR product spanning from 2012 to 2016. The operational production system, which is identical to S-NPP Land Science Investigator-led Processing Systems (LandSIPS), was embedded in the SCF, and the optimized LUTs for VIIRS sensor described in Section 2.2 were read for production. The VIIRS Level 2G (L2G) surface reflectance product, called VNP09GA, is composed of all available surface reflectance observations for a given day over a set of tiles with global coverage [28]. By calculating observation scores based on quality assurance (QA) and geometry information (i.e., the basis of high observation coverage, low sensor angle, the absence of cloud, cloud shadow, and aerosol loading), the algorithm produces the intermediate surface reflectance product that contains the best quality observation (i.e., VNP15IP). Then, it accounts for both the best quality observations and the biome map (MCDLCHKM, Figure S1) to produce daily LAI/FPAR retrievals (VNP15A1). With the compositing strategy described in Section 2.1, LAI/FPAR products in the 8-day interval (VNP15A2H) can be obtained. Note that all processes for producing 5-year global VIIRS LAI/FPAR from our SCF are completely identical to the operational production system in S-NPP LandSIPS.

The latest version (C6) of the MODIS LAI/FPAR product was generated and released to the public in 2015. The standard MODIS C6 LAI/FPAR product suite is at 500-m spatial resolution and includes LAI/FPAR retrievals from Terra MODIS, Aqua MODIS, and Terra MODIS + Aqua MODIS Combined [7]. The product suite has two types of temporal compositing periods (i.e., 8 and 4 days) and provides approximately 18- and 16-year records of global land surface from Terra (2000–present) and Aqua (2002–present) MODIS, respectively. The operational production workflow of MODIS is similar to that of the VIIRS product. All the MODIS product suites (MOD15A2H, MYD15A2H, MCD15A2H and MCD15A3H) adopt the same LAI/FPAR retrieval algorithm but different sensors (MOD for Terra and MYD for Aqua) and compositing methods (8-day for A2H and 4-day for A3H). Therefore, LAI/FPAR retrievals for all the MODIS products should be very similar as long as they have comparable observing conditions [29]. To verify this assumption, we compared VIIRS product with different MODIS products and found the results were very similar, that is, different MODIS product suites show similar LAI/FPAR retrievals. As Suomi-NPP is an afternoon-overpassing satellite (13:30 local time), which is identical to MODIS onboard Aqua, we used Aqua MODIS retrieval (MYD15A2H) as a fair counterpart of VIIRS product. Also, the same compositing strategy based on 8-day interval is another reason to use MYD15A2H to compare with the VIIRS LAI/FPAR product. All MODIS LAI/FPAR product suites are available via LP DAAC (https://lpdaac.usgs.gov/dataset_discovery/modis/modis_products_table) and the details of MYD15A2H can be found in [30].

Both VIIRS and MODIS LAI/FPAR products provide 6 scientific data sets (SDS): (1) Fpar; (2) Lai; (3) FparLai_QC; (4) FparExtra_QC; (5) FparStdDev; and (6) LaiStdDev. The first two SDSs are retrievals of respective FPAR and LAI, and the last two layers are the standard deviation of all candidates from solving the "ill-posed" inversion problem. In addition to LAI/FPAR values, the products contain the corresponding Quality Control (QC) data in the third and fourth layers, and the users are advised to consult the quality flags when using these products [31]. The key indicator of the quality of retrievals is the algorithm path, which distinguishes the following five categories: (1) main algorithm without saturation; (2) main algorithm with saturation; (3) back-up algorithm due to bad geometry (Backup-G); (4) back-up algorithm due to other problems (Backup-O), such as the cloud contamination, snow coverage, etc.; and (5) not produced due to invalid BRFs. In addition to the algorithm path, the QC layers, including the "FparLai_QC" and "FparExtra_QC" data sets, provide information about the presence of clouds, aerosols, and snow, inherited from input reflectance products. Overall, VIIRS and MODIS LAI/FPAR products have very similar characteristics, including the similar input BRFs at red and NIR bands, the similar process chains which consider the good quality of observations by calculating the observation score based on the quality flag (clouds, snow, cloud shadow, etc.) from reflectance products, the same input land cover maps (MCDLCHKM) and the same spatiotemporal resolution (500-m/8-day).

3.2. Evaluation of Continuity between VIIRS and MODIS

The evaluation of continuity between VIIRS and MODIS retrievals is prioritized in this study to build a long time series of LAI/FPAR ESDR that is independent of sensor observation. We evaluated the spatiotemporal continuity between VIIRS and MODIS products at two different scales: global and site scale. Continuity is defined as satisfying the following condition, $|V-M| < E$, where V is the VIIRS estimate of LAI/FPAR, M is the MODIS estimate of LAI/FPAR, and E is an accuracy specification in absolute LAI/FPAR units. In this study, the accuracy specifications for LAI and FPAR are less than 0.25 LAI and 0.02 FPAR, respectively. This specification also meets the long-term ESDR stability requirement suggested by the Global Climate Observing System (GCOS) [32]. However, several issues have to be addressed to achieve a reasonable comparison between VIIRS and MODIS products. First, product geometrical characteristics, including geolocation uncertainties, point spread function (PSF), and projection system, should be considered because the comparison needs to be performed over the same support area [33]. Many studies proposed that the comparison of products over a much larger area than their native spatial resolution is an effective way to reduce potential geolocation errors and PSF impacts [3,34,35]. A good practice suggested by the Committee on Earth Observation Satellites (CEOS) Land Product Validation (LPV) subgroup shows that the product can be evaluated over the area ranged from 3 km × 3 km to 10 km × 10 km [3,36–38]. To understand the product performance over different spatial scales, the VIIRS LAI/FPAR product was evaluated over 10 km × 10 km at global scale and 3 km × 3 km at site scale, respectively. Second, the quality control was performed for VIIRS and MODIS products to exclude pixels contaminated by clouds, shadow, cirrus, and snow [31]. The detailed selection of valid pixels from both VIIRS and MODIS products based on "FparLai_QC" and "FparExtra_QC" layers is shown in Figure S2. Note that only retrievals from the main algorithm were used in our study due to the low accuracy of back-up algorithm retrievals (Section 2.1). For both 10 km × 10 km and 3 km × 3 km areas, the mean LAI/FPAR was calculated only if more than 60% of pixels fell within this area. Additionally, the global land cover map (see Figure S1) was used to identify the vegetated pixel and the specific vegetation type in the analysis of the products.

At the global scale, the aggregated LAI/FPAR retrievals were compared at different temporal scales (8-day, seasonal and annual). We introduced the Benchmark Land Multisite Analysis and Intercomparison of Products (BELMANIP) network, designed to represent the global surface types and conditions, to reduce the geolocation uncertainty and sampling bias [39] at the site scale. The latest version (2.1) of the network contains 445 sites that are located on relatively flat and homogeneous areas within a 10 km × 10 km domain (see Figure S3). Thus, this version is widely used in the product intercomparison practice [31,36]. Similarly, we selected the valid LAI/FPAR retrievals for both VIIRS and MODIS in a 3 km × 3 km area to compare their spatiotemporal consistency using a series of statistical metrics (Bias, Root Mean Square Error (RMSE), relative RMSE (rRMSE) and coefficient of determination (R^2)). Additionally, the temporal continuity and consistency of the products were also evaluated by biome type.

As a key indicator of the quality of retrievals, the algorithm path of each pixel is stored in the LAI/FPAR product. By comparing the retrieval rate of different algorithm paths at the global scale, we can evaluate the overall quality of the product from VIIRS and MODIS. The main algorithm outputs retrievals at high precision in the case of low LAI and at moderate precision when LAI is high and surface reflectance has low sensitivity to LAI. Low-precision retrievals are obtained from the empirical back-up algorithm when the main algorithm fails. An indicator called "Retrieval Index (RI)" that characterizes the proportion of the good quality and high precision retrievals was adopted in our study. The rate of main algorithm retrievals (including retrievals of the main algorithm without saturation, and the main algorithm with saturation) in the 10 km × 10 km area was calculated at the global scale as Equation (1). We investigated RI to show global spatial coverage of the main algorithm retrievals changing with observation dates for the VIIRS product, and also to evaluate whether the proportions of good quality retrievals between VIIRS and MODIS are comparable.

$$RI = \frac{number\ of\ retrieved\ pixels\ via\ main\ algorithm}{number\ of\ total\ vegetated\ pixels} \times 100\% \tag{1}$$

3.3. Uncertainty Quantification

The uncertainty is defined as the RMSE between products and ground measurements or ground-measurements-derived reference maps. For this purpose, as most ground measurements were collected during the early EOS era (mostly from 2000 to 2008), here we first synthesized all available in-situ LAI/FPAR measurements over the globe during the VIIRS era (2012–present). Three different field campaign projects were active across 28 sites during the VIIRS era, and those include ImagineS [40], Heihe Watershed Allied Telemetry Experimental (HiWater) [41,42], and Huailai [43]. In particular, the ImagineS project, which is designed to support European Copernicus Global Land Service, has conducted a series of single date or multi-temporal field campaigns over 23 sites around the world during 2013–2016 (http://fp7-imagines.eu). Because of the spatial scale mismatch between ground measurements (tens of meters) and products (500-m), the "bottom-up" approach proposed by the Committee on Earth Observation Satellites (CEOS) Land Product Validation (LPV) subgroup is adopted in this practice [37,44]. The strategy is designed to correlate the scale of in situ biophysical measurements (i.e., LAI and FPAR) to that of the remote sensing product using finer-resolution images to bridge their scale gaps [45]. It is based on a two-stage sampling strategy that (1) uses multiple elementary sampling units (ESUs) to capture the variability across the extent (~3 km × 3 km) of a site and (2) repeats measurements within each ESU to capture the variability within the pixel (~30 m) of high-resolution imagery. Then, an empirical transfer function was established between the ground measurements and the spectral values from the high-resolution imagery to produce LAI/FPAR reference maps. The generated LAI/FPAR maps can be aggregated to match the product pixel grid and are used as the benchmark to validate products [46]. The standard deviation of LAI/FPAR from reference map was also calculated over the 3 km × 3 km region. We also matched the observation date between ground measurements and product at native temporal resolution (8-day). To reduce the geolocation uncertainties caused by different projection systems and point spread functions (PSF), the LAI/FPAR reference maps were averaged over the 3 km ×3 km area for product assessment [33,36]. We collected 68 LAI reference maps and 57 FPAR reference maps from all available ground sites (see Figure S3). However, to reduce unexpected error sources from land surface heterogeneity, we adopted the criterion named information entropy proposed by [31]. The information entropy (E) defined as Equation (2) is to keep only sites with a homogeneous land surface condition.

$$E = -\sum_{i=1}^{11} P_i \times log_2 P_i \tag{2}$$

where ith denotes the specific land cover type from MODIS land cover product, P_i indicates the proportion of the area covered by the ith land cover type. The reference map was selected if the E of this site is greater than 1 [31]. Note that the percentage of valid pixels in 3 km × 3 km area should also be greater than 60% as defined in Section 3.2. The detailed information of each reference map is shown in Table S1. Finally, a total of 26 LAI and 24 FPAR reference maps were used in this study. Additionally, the GCOS identified target requirement uncertainties/relative uncertainties of max (0.5, 20%) for LAI and max (0.05, 10%) for FPAR [32], which will also be utilized as the criterion to evaluate the uncertainty of VIIRS and MODIS products.

4. Results

4.1. Spatiotemporal Consistency between VIIRS and MODIS

4.1.1. Global Scale

Figure 1 depicts the 8-day global LAI/FPAR time series of VIIRS and MODIS from 2012 to 2016. Both VIIRS and MODIS tightly well capture the LAI and FPAR seasonality, with two distinct peaks every year caused by the hemispherically different seasons. The LAI and FPAR in boreal summer (i.e., July) are much greater (about 2.0 LAI and 0.51 FPAR) than those (about 1.5 LAI and 0.42 FPAR) in boreal winter (i.e., January) because of larger proportion of global vegetation in northern hemisphere. In particular, both retrievals from VIIRS and MODIS agree very well with long-term time series, with a mean difference of -0.006 ± 0.013 for LAI and -0.002 ± 0.002 for FPAR. The LAI/FPAR difference between VIIRS and MODIS meets the stability requirement of multi-sensor product time-series suggested by GCOS [32]. The standard deviation (std.) of the difference illustrates a slight stochastic difference between VIIRS and MODIS products. The seasonal change of std. of difference is also observed with higher values (0.46 for LAI and 0.060 for FPAR) in January and lower values (0.36 for LAI and 0.045 for FPAR) in July. This is because in January, the LAI/FPAR differences are lower in most regions but larger in tropical areas, resulting in larger std. of difference, as shown in Figure 2.

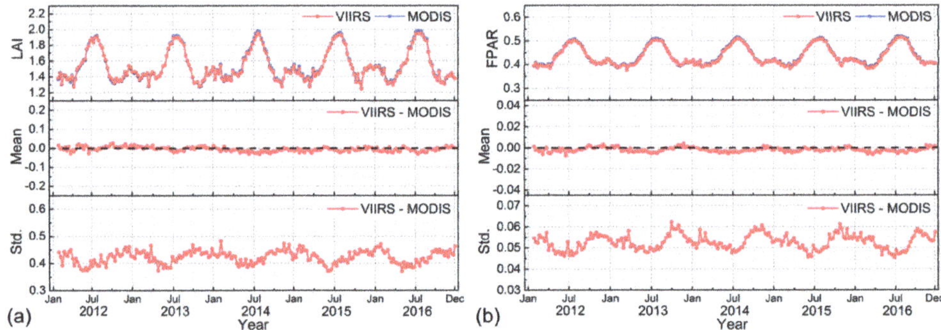

Figure 1. The global mean (**a**) LAI and (**b**) FPAR of VIIRS and MODIS at 8-day interval during 2012–2016. The top panels of (**a**,**b**) show the seasonal variation of VIIRS and MODIS LAI/FPAR retrievals, while the bottom two panels show the mean and standard deviation (std.) of difference (the difference was first calculated in each 10 km × 10 km grid, and then the mean and std. were calculated for all grids) between the two products (VIIRS-MODIS). The mean difference indicates whether VIIRS and MODIS products have systematic errors, while the std. of difference shows the dispersion of errors between VIIRS and MODIS products. Also see Figure S4 for the evaluation between VIIRS and MODIS LAI/FPAR products at the native spatial resolution (500-m) for each biome.

Details of the comparison at the seasonal scale can be found in Table 1. The mean and standard deviation (std.) of product difference were calculated at a 3-month interval (March–May (MAM), June–August (JJA), September–November (SON) and December–February (DJF)). Both LAI and FPAR show the largest differences in JJA, which are primarily due to the largest LAI and FPAR values in this period (Figure 1). Nevertheless, the std. presents the lowest magnitude in JJA, showing the similar differences at the global scale. Although the differences between VIIRS and MODIS change in different seasons, the differences present low mean and std. values. Results show that the LAI/FPAR difference between products can also satisfy the stability requirement of a long-term dataset for different seasons. Additionally, the mean and std. of LAI/FPAR difference also depend on the biome type (see Table S2). From biome-specific investigations, we do not find obvious systematic over- or under-estimation across biomes. Forest biomes (B5–B8) present relatively larger LAI/FPAR differences than non-forest biomes (B1–B4) in each season. Among all biomes, the needleleaf forest (B7–B8) shows the largest absolute mean difference between products, while the evergreen broadleaf forest (B5) presents the largest std. of LAI/FPAR difference. As shown in Table 1, most biomes show the largest mean difference in JJA among four seasons except for the needleleaf forest (B7–B8), which shows the largest differences in DJF. This is likely due to the snow coverage and the large solar zenith angle in the high latitude areas in DJF, introducing larger uncertainty in surface reflectances. The possible reasons for LAI/FPAR inconsistencies between VIIRS and MODIS products will be further discussed in Section 5.1.

Table 1. The mean and standard deviation (std.) of LAI/FPAR difference between VIIRS and MODIS products (VIIRS-MODIS) at the seasonal scale during 2012–2016.

Product	Statistical Indicator	Overall	MAM	JJA	SON	DJF
LAI	Mean	−0.008	−0.007	−0.014	−0.003	−0.006
	Std.	0.313	0.300	0.284	0.327	0.348
FPAR	Mean	−0.003	−0.004	−0.004	−0.001	−0.002
	Std.	0.036	0.035	0.033	0.038	0.039

The "MAM", "JJA", "SON" and "DJF" stand for "March–May", "June–August", "September–November" and "December–February", respectively.

In detail, the global spatial patterns of mean LAI and FPAR difference maps between VIIRS and MODIS products at two typical observation dates (17–24 January and 12–19 July) for the five-year period (2012–2016) are presented in Figure 2. The detailed spatial distribution of LAI/FPAR differences shows no observed systematic bias between VIIRS and MODIS products. The overall differences of LAI and FPAR are within ±0.25 (percentage: 83%) and ±0.02 (percentage: 70%), respectively, with larger differences over the dense forests. The invalid LAI/FPAR values at high latitudes in the northern hemisphere in January are mainly due to the snow coverage, polar night, and large solar zenith angle [11]. For LAI retrievals (Figure 2a,b), most regions in the northern hemisphere have a slightly larger difference in July than in January. However, the LAI difference in the equatorial region, e.g., Amazonia and central Africa, shows an opposite pattern due to the growing season of vegetation in January. From the spatial pattern comparison of LAI differences in January and July at the global scale, LAI differences depend on the LAI magnitudes, with larger LAI differences in high LAI domains. The possible reasons will be discussed in Section 5.1. We also observed similar spatial distribution of FPAR differences from Figure 2c,d.

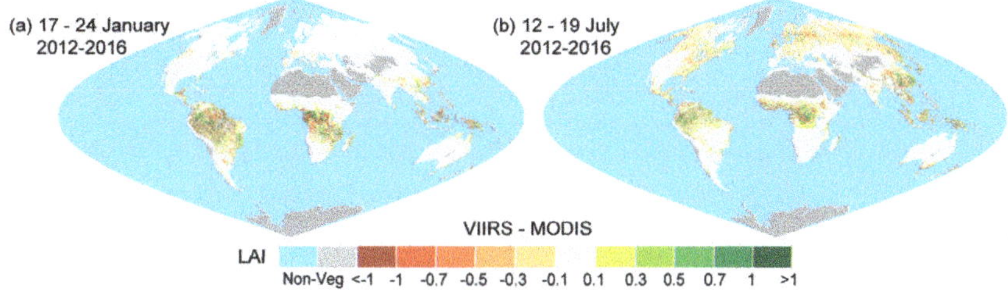

Figure 2. *Cont.*

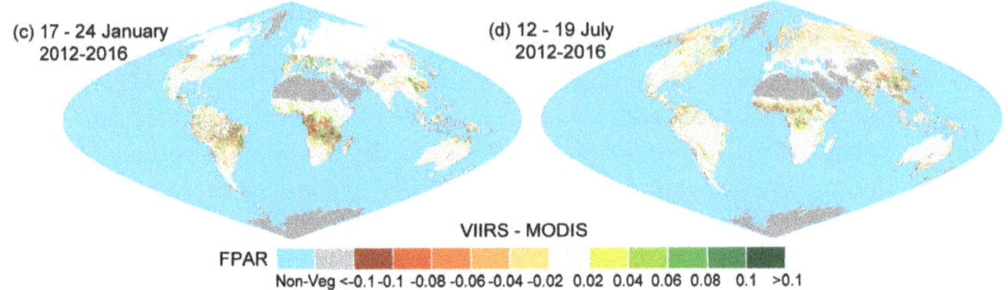

Figure 2. The global spatial pattern of mean (**a**,**b**) LAI difference and (**c**,**d**) FPAR difference (VIIRS-MODIS) in each 10 km × 10 km grid from 2012 to 2016 at two observation dates (17–24 January and 12–19 July). The white color indicates invalid LAI/FPAR values calculated from QC layers as mentioned in Section 3.2. An equal-area sinusoidal projection is used here.

As VIIRS will continue the role of MODIS to provide LAI/FPAR records, it is necessary to evaluate whether VIIRS can inherit MODIS for the long-term analysis of vegetation dynamics (i.e., LAI trend over several years [47,48]) at global scale. Here, we additionally investigate how well VIIRS captures the interannual variability of MODIS retrievals by splitting three large regions: the Northern Hemisphere (25° N–90° N), the equatorial region (25° N–25° S) and the Southern Hemisphere (25° S–90° S). The LAI anomaly, defined as a certain year of mean LAI value minus mean LAI value over 2012–2016, is used to quantify the interannual variability of products. A large LAI anomaly indicates that LAI changes considerably compared with the mean LAI over different years. As shown in Figure 3, VIIRS shows quite similar interannual variation for each region to those of MODIS. However, the magnitude of LAI anomalies is slightly different, especially in the equatorial region (Figure 3b). The variation (~0.03) of MODIS LAI anomaly in the equatorial region is much larger than that for VIIRS (~0.01) from 2013 to 2014, resulting in the opposite variation of 2014 at global scale (Figure 3d). This is likely because VIIRS reflectance is easier to be saturated due to the sensor's intrinsic limitation (VIIRS reflectance is generally larger in the NIR band and lower in the red band [11]). The saturated reflectance over dense canopy in tropical forests can only provide limited information for LAI retrievals [49], that is, the real vegetation change cannot be shown if the reflectance is saturated, resulting in the small variation of VIIRS LAI retrievals over different years.

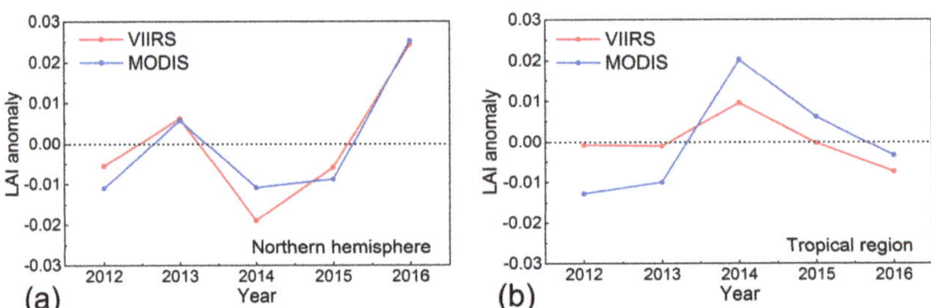

Figure 3. *Cont.*

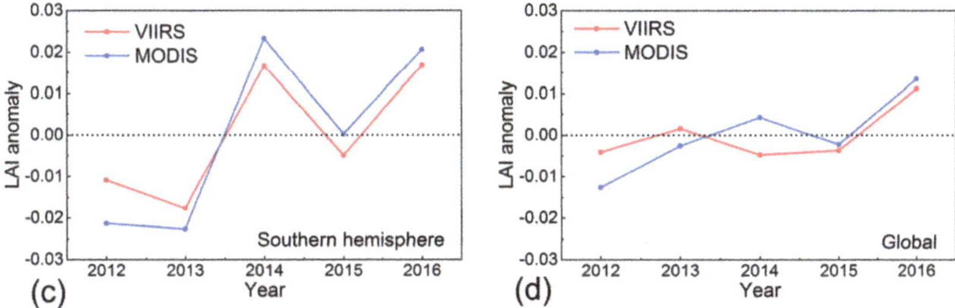

Figure 3. Time-series of annual LAI anomalies over 2012–2016 for VIIRS (red line) and MODIS (blue line). The values in figure (**a**–**d**) stand for the LAI anomaly in the Northern Hemisphere (25° N–90° N), equatorial region (25° N–25° S), Southern Hemisphere (25° S–90° S) and global vegetated area, respectively.

4.1.2. Site Scale

The VIIRS and MODIS products are further compared at the site scale for each biome over all BELMANIP-2.1 sampling sites. Figures 4 and 5 show the density scatter plots between VIIRS and MODIS for all 8-day LAI and FPAR retrievals from 2012 to 2016, respectively. The statistical results (R^2, Bias and RMSE) of all pixels indicate good consistency between VIIRS and MODIS in both LAI/FPAR retrievals. For LAI retrievals (Figure 4), non-forest biomes (i.e., B1–B4) have better agreement between products than forest biomes, which can also be found for FPAR retrievals (Figure 5). The best agreement is achieved over shrubs (Bias = 0.009, RMSE = 0.060) and broadleaf crops (Bias = −0.001, RMSE = 0.029) for LAI and FPAR retrievals, respectively. VIIRS retrievals from most biomes, except evergreen broadleaf forest (B5), capture more than 85% of the variations ($R^2 > 0.85$) observed in MODIS retrievals. The case of dense evergreen broadleaf forest shows less agreement in both LAI ($R^2 = 0.62$, RMSE = 0.69) and FPAR ($R^2 = 0.47$, RMSE = 0.048), which is not surprising because of sub-optimal quality of retrievals due to reflectance saturation for dense forest. In detail, small variations in reflectances can result in large variation in LAI under the condition of saturation [49]. Additionally, another possible reason of the inconsistency between VIIRS and MODIS retrievals for evergreen broadleaf forest will be discussed in Section 5.1.

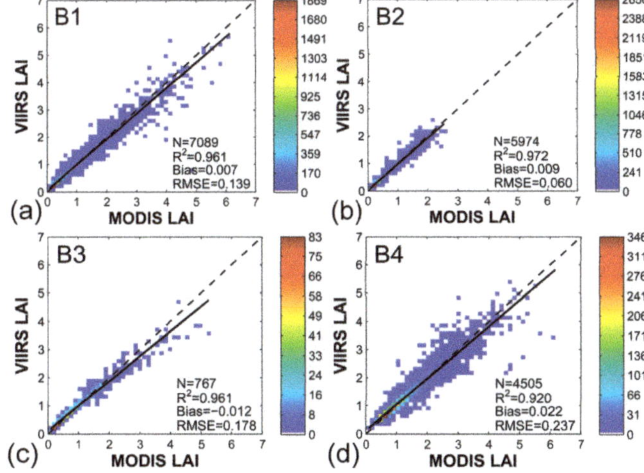

Figure 4. *Cont.*

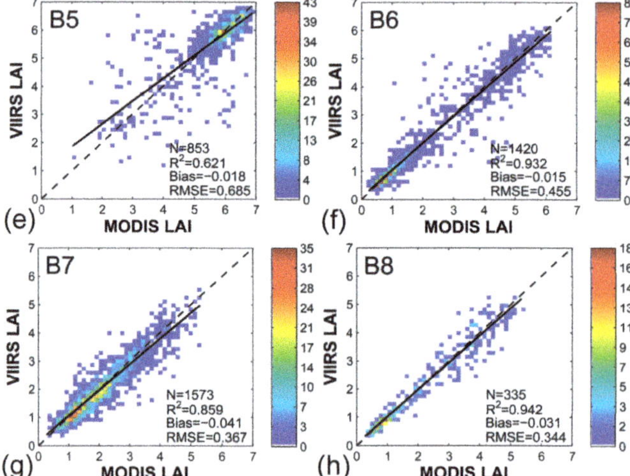

Figure 4. The biome-specific comparison between VIIRS and MODIS LAI retrievals at the site scale from 2012 to 2016. Multiple validation sites adopted from Benchmark Land Multi-site Analysis and Intercomparison of Products 2.1 (BELMANIP-2.1) are used for this comparison. The solid black line is the linear fit for all pixels and the colorbar shows the density of pixels falling at each grid. The (**a–h**) stands for biome 1–8, respectively. The eight biomes are (B1) grasses and cereal crops, (B2) shrubs, (B3) broadleaf crops, (B4) savannas, (B5) evergreen broadleaf forest, (B6) deciduous broadleaf forest, (B7) evergreen needleleaf forest, (B8) deciduous needleleaf forest.

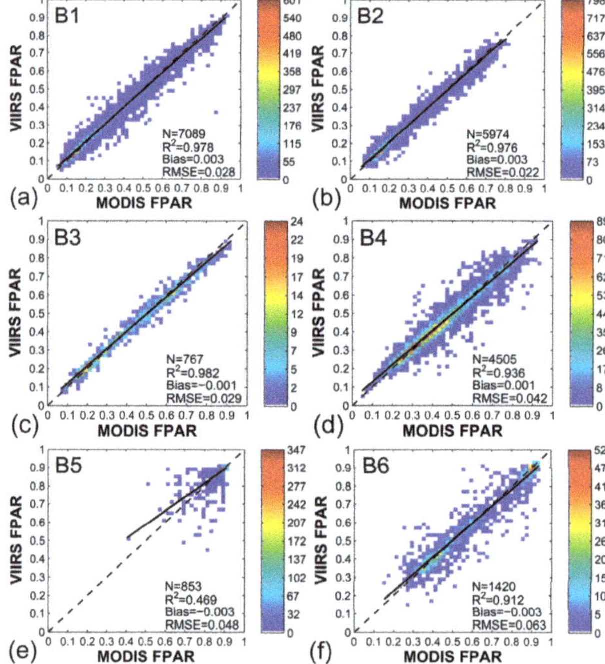

Figure 5. *Cont.*

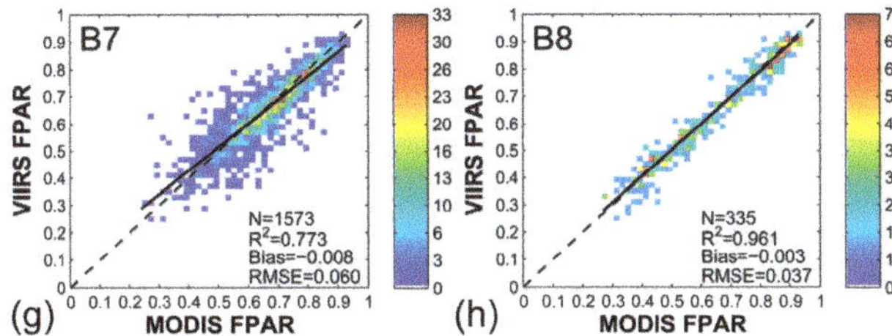

Figure 5. The (**a–h**) stands for biome 1–8, respectively. The eight biomes are (B1) grasses and cereal crops, (B2) shrubs, (B3) broadleaf crops, (B4) savannas, (B5) evergreen broadleaf forest, (B6) deciduous broadleaf forest, (B7) evergreen needleleaf forest, (B8) deciduous needleleaf forest.

Figure 6 shows the seasonal trajectory of VIIRS and MODIS LAI/FPAR over four example sites (grasses/cereal crops (B1), shrubs (B2), deciduous broadleaf forest (B6) and deciduous needleleaf forest (B8)) for the period from 2012 to 2016. All available ground measurements are additionally plotted as a reference. Both products shown in Figure 6 achieve a good temporal consistency across all biomes and comparable spatial coverage (i.e., missing data rate). Each site shows distinct characteristics of seasonal variations. For example, non-forest biomes (Figure 6a,b) exhibit smaller seasonal amplitude, less than 2 for LAI and 0.4 for FPAR, while forest biomes (Figure 6c,d) reveal a higher LAI (about 4 LAI unit) and FPAR (about 0.5 FPAR unit) seasonality. In addition, the algorithm implemented for both sensors agrees quite well with the ground measurements, indicating VIIRS and MODIS can capture the temporal trajectory of true LAI/FPAR values.

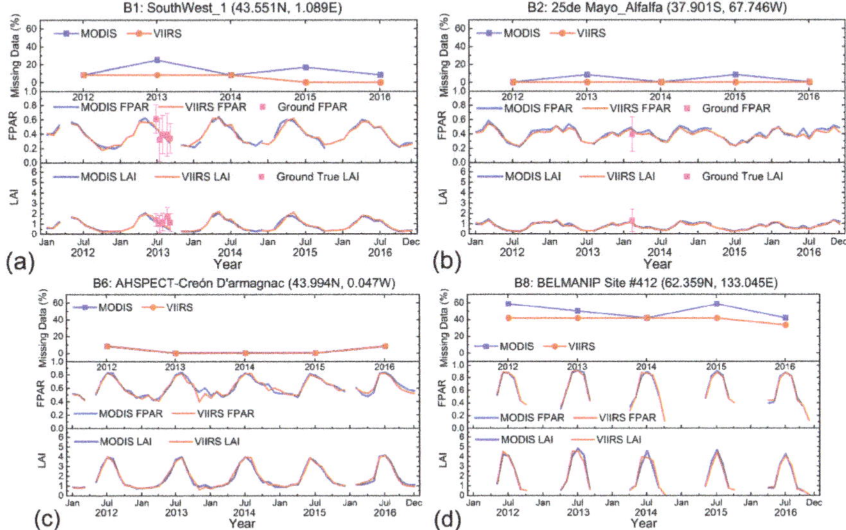

Figure 6. Temporal comparison of VIIRS and MODIS LAI/FPAR over four selected example sites. Monthly LAI and FPAR for the time period 2012–2016 are shown here. Red and blue lines indicate VIIRS and MODIS, respectively. The upper panel shows the missing data rate in each year and the following two panels show the seasonal variation of LAI/FPAR products with collected ground measurements. The error bar of the ground measurement was derived from the standard deviation of LAI/FPAR over the specified 3 km × 3 km region. Panel (**a–d**) are for grasses/cereal crops (B1), shrubs (B2), deciduous broadleaf forest (B6) and deciduous needleleaf forest (B8) cases, respectively.

4.2. Spatial Coverage

The VIIRS RI and the Difference of RI (DRI) between VIIRS and MODIS were calculated at the global scale from 2012 to 2016 to ensure their consistency of spatial coverage. For the sake of brevity, here we only present the results from two 8-day compositional dates (i.e., 17–24 January and 12–19 July) representing the respective boreal winter and summer season. Figure 7 shows seasonally different biome-specific retrieval success (i.e., RI) and RI difference (DRI) between VIIRS and MODIS. The global RIs of VIIRS in January and July are 58.8% and 85.1%, respectively. In general, the RI of forest biomes (B5–B8) is lower than that of non-forest biomes (B1–B4) for both two seasons. Both sensors give a lower RI in January for most biomes (see Figure S5), especially for needleleaf forest (B7–B8, <20%). The lower RI is partly because most needleleaf forests are situated in the high latitude or altitude regions (see Figure S1) where they are often covered by snow in January. In addition, a larger solar zenith angle at high latitudes in January is the most important reason for low needleleaf forest RI (See Figure S5c,d). The evergreen broadleaf forests (B5) in the equatorial region have comparable RIs in January and July, and relatively lower RIs than other biomes due to the impact of cloudy condition throughout the year.

From the comparison with calculated MODIS RIs, we find that VIIRS completely resembles their seasonal variations of each algorithm pathway including main and backup (Backup-G and Backup-O) retrievals (see details in Figure S5). As shown in Figure 7, the proportion of DRI ranging from −5% to 5% at global scale is 84.7% in January and 77.2% in July. For individual biome types, both VIIRS and MODIS RIs agree well in January for each biome type showing a low mean DRI (~1%). However, we also observed a relatively large mean DRI (~5%) from the comparison in July, especially for forest biomes. The reason for this inconsistency of RIs will be discussed in Section 5.1. The standard deviation (std.) of DRI shows that VIIRS has a stable DRI (~4%) with MODIS for shrubs (B2), while also presents a relatively unstable DRI (~9%) for evergreen broadleaf forest (B5) over five years.

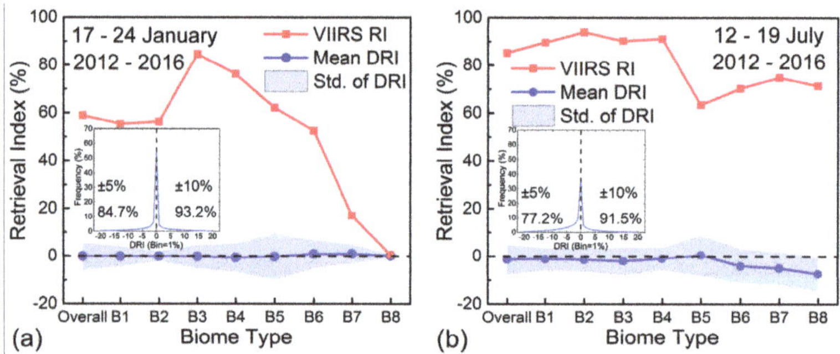

Figure 7. The retrieval index (RI) of VIIRS main algorithm (red line), the mean Difference of RI (DRI) (blue line) and the standard deviation (blue shadow) of DRI (the difference of RI was first calculated in each 10 km × 10 km grid, and then the mean and std. were calculated for all grids) between VIIRS and MODIS products (VIIRS-MODIS) for different biome types on (**a**) 17–24 January and (**b**) 12–19 July from 2012 to 2016. The inset figure in each panel shows the histogram of DRI between VIIRS and MODIS products.

4.3. Uncertainty Assessment

The uncertainty of both VIIRS and MODIS LAI/FPAR products was evaluated using the synthesized ground measurements, as shown in Figure 8. In general, our comparison reveals that VIIRS and MODIS show comparable uncertainties of LAI (VIIRS = 0.60, MODIS = 0.55) and FPAR (VIIRS = 0.10, MODIS = 0.11) retrievals. As the result of MODIS resembles that of VIIRS, here, we use VIIRS as a case to explain the general pattern of the results. For LAI, both ground measurements corrected (hereafter, true LAI) or uncorrected (hereafter, effective LAI) for clumping effect are introduced in this study but separately analyzed. Since the clumping effect was taken into account in the LAI retrieval algorithm, VIIRS displays better agreement with true LAI (Bias = 0.17, RMSE = 0.60, rRMSE = 42.2%) than with effective LAI (bias = 0.47, RMSE = 0.79, rRMSE = 66.0%) as expected. Specifically, VIIRS shows an important overestimation tendency for the effective LAI case (Bias = 0.47) because the ground-measured effective LAI is generally lower than true LAI [50]. As LAI was only distributed on relatively lower domains (<2.5), the RMSE is better than the rRMSE to show LAI product accuracy based on the accuracy requirement (max(0.5, 20%)) from GCOS. For FPAR, most scatters are distributed within 0–0.2 difference, indicating an overestimation tendency of VIIRS FPAR (Bias = 0.08). Nevertheless, VIIRS FPAR retrieval displays good agreement with ground measured FPAR in terms of R^2 (=0.83), RMSE (=0.10) and rRMSE (=24.4%), indicating that VIIRS can well capture the FPAR variation (the dispersion of VIIRS FPAR is low). In addition, we can see that both VIIRS and MODIS products cannot totally meet the target accuracy requirements suggested by GCOS, with 66.7% LAI and 37.5% FPAR pixels less than the LAI (max (0.5, 20%)) and FPAR (max (0.05, 10%)) specific criteria, respectively. However, it should be noted that the uncertainty of LAI/FPAR retrievals comes from both VIIRS products and other sources including uncertainties of reference maps and a mismatch in the spatial and temporal domains, despite that efforts were made to reduce such sources of uncertainty. Additionally, we should also note that the distribution of measurements is not sufficient enough because only small number of measurements and no forest biomes are included. Adding more ground measurements, especially from forest biomes, may achieve a more robust evaluation result of VIIRS LAI/FPAR product. Further discussions will be followed in Section 5.2.

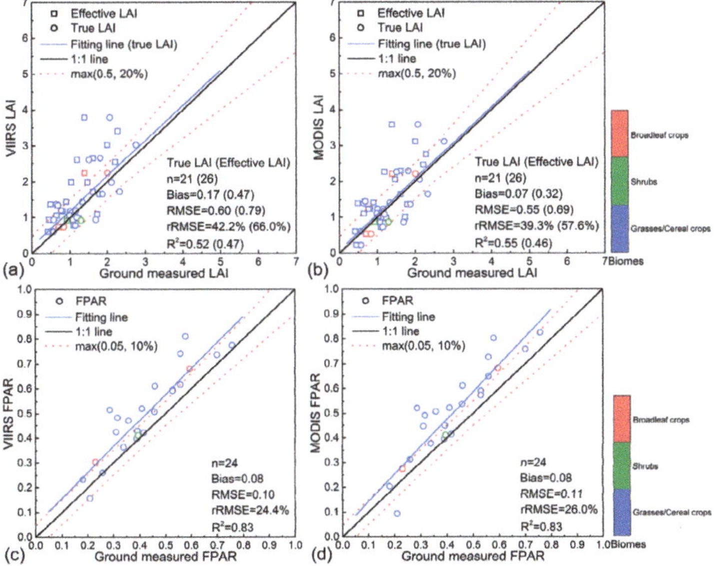

Figure 8. Comparison between ground measurements and products. (**a**,**b**) for VIIRS and MODIS LAI and (**c**,**d**) for FPAR. The 3 km × 3 km sites dominated by different biome types are depicted by different colors. Circles (squares) in panel (**a**,**b**) represent ground LAI measurements corrected (not corrected) for the clumping effect. The blue and black solid lines represent the overall fit and 1:1 line, respectively. The red dashed lines show the GCOS specifications boundaries (max (0.5, 20%) for LAI and max (0.05, 10%) for FPAR). Note that LAI was only distributed in relatively lower domains (~2.5), the RMSE is better than the rRMSE to show LAI product accuracy based on the accuracy requirement (max (0.5, 20%)) from GCOS.

5. Discussion

5.1. Understanding Inconsistency between VIIRS and MODIS

From our comprehensive evaluation presented in Section 4, we can conclude that the implemented LAI/FPAR algorithm on the VIIRS can generate highly consistent LAI/FPAR retrievals with those of MODIS at multi-spatiotemporal scales. However, as we observed some minor discrepancies between VIIRS and MODIS products; here, we investigate the possible reasons to better understand observed inconsistencies. The saturation frequency and dispersion of the retrieved LAI distribution (DLAI) are two elements by which the quality of the retrieval can be assessed. The accuracy of the retrieval decreases under conditions of saturation, that is, the reflectance data do not contain accurate information about the surface [13,49]. DLAI is defined as the standard deviation of all LAI candidates in LUTs enabling to reproduce input surface reflectances under various canopy and soil combinations (recall that the algorithm is solving the ill-posed inverse problem) [13,15]. Thus, in theory, the lower the DLAI value, the higher the accuracy of the algorithm as DLAI represents the degree of difficulty to localize a single solution.

In this manner, Figure 9 shows important relationships between observed difference and reliability of retrievals (i.e., saturation rate and DLAI). Note that all quantities in Figure 9 are based on 8-day global products from 2012 to 2016. First, Figure 9a shows how the saturation rate behaves as a function of LAI. Unsurprisingly, the saturation rate increases with increasing LAI and thus we can infer the accuracy of the retrievals is decreasing. In the same vein, DLAI increases with increasing LAI but DLAI tend to decrease at certain LAI (about 5 for non-forest and 4.5 for forest). This is because DLAI is not only related to LAI (higher LAI yields higher DLAI), but also depends on the number of available solutions in LUTs (Section 2.1), i.e., the nature of the standard deviation. Under the relationship of saturation rate and DLAI to LAI retrieval, we found that observed discrepancy between VIIRS and MODIS tends to increase with an increasing saturation rate and DLAI. We highlight that the discrepancy is abruptly increased when the saturation rate shows a sharp rise in higher LAI, namely, the reliability of retrieval determines the degree of consistency between two sensors. Additionally, we should also note the LAI difference (Bias = 0.02, RMSE = 0.69) for evergreen broadleaf forest (EBF) is not negligible (see Figure 4e); thus, a special caution is needed for this biome. Observing land surface over tropical regions where EBF mostly situated is a classical challenge in optical remote sensing field due to cloudy or high aerosol atmospheric conditions [51,52]. Therefore, this difference we observed here can be likely explained by different observing conditions in two sensors. For instance, although both Suomi-NPP and Aqua are afternoon-overpassing satellites, different orbital path and selection of daily best input observation may result in a dissimilar input surface reflectance. The impact of such discrepancy is expected to increase in highly varying atmospheric conditions over densely vegetated tropical forests. Another possible reason is QA which is slightly different between VIIRS and MODIS. For instance, Platnick et al. (2016) reported that cloud detection and flagging its mask in VIIRS is limited by absence of CO_2 and H_2O absorption channels [53], in turn such discrepancy results in unlike LAI/FPAR retrievals.

In addition, we also observed the relatively large RI difference in July (see Figure 7), which is likely explained by the reported instability of input surface reflectance for VIIRS [8]. The results from [8] show larger deviations of spectral adjustment coefficients to make MODIS-like VIIRS surface reflectances are observed for the initial years of VIIRS operation, while relatively better stability is observed for later years. As VIIRS LUTs have been developed using BRFs in 2015 based on the time-invariant basis, unstable input surface reflectances over different years may induce unexpected interannual variability of RIs. Therefore, the DRI between VIIRS and MODIS is less than 1% in 2015, whereas it shows relatively larger difference during 2012–2014 (see Figure S6). Here, we should note that RI differences of January and non-forest are smaller than July and forest biomes (Figure S6). The lower RI values are caused by the algorithm path, which is also sensitive to the high LAI values. More specifically, the domain of the main algorithm retrievals is more limited in dense vegetation with higher NIR reflectance and lower red reflectance (see Figure 10a in [7]), which introduces lower RI due to the interannual variability of surface reflectance.

There are additional minor factors that influence the inter-consistency of the LAI/FPAR product: compositing and off-nadir view zenith angle (VZA), which are from the intrinsic limitation of VIIRS and MODIS sensors. For both VIIRS and MODIS, the algorithm first accounts for 8 daily surface reflectances and produces daily LAI/FPAR. Based on the maximum FPAR compositing strategy, a daily retrieval is selected as 8-day product value. When selected VIIRS daily retrieval is separate from MODIS daily retrieval, the degree of inconsistency between VIIRS and MODIS may increase as [36] shows similar pattern in surface reflectance. For the off-nadir VZA case, MODIS pixels increase by a factor of six from nadir to edge of scan, whereas VIIRS restricts the pixel to an approximate twofold increase using an onboard aggregation scheme [11]. Therefore, a reduced effective spatial resolution of MODIS with the increase of VZA may introduce additional source of inconsistency between VIIRS and MODIS products [54].

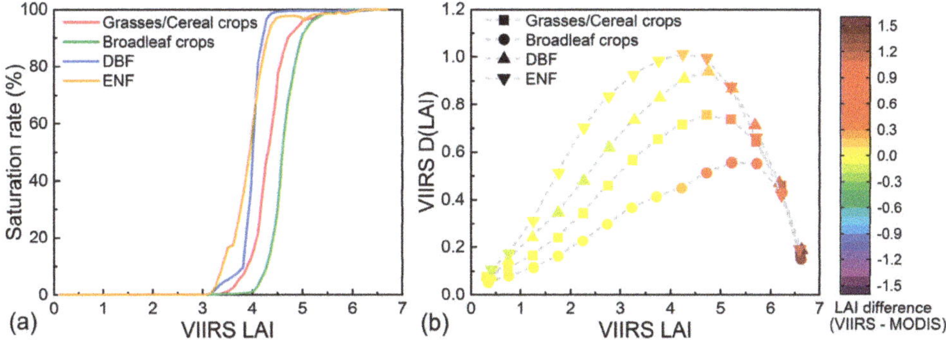

Figure 9. (**a**) The variation of saturation rate with VIIRS LAI retrievals; (**b**) The relationship among VIIRS LAI (x-axis), DLAI (y-axis) and the difference (colorbar) between VIIRS and MODIS LAI retrievals using all composition periods from 2012 to 2016. "DBF" and "ENF" represent "deciduous broadleaf forest" and "evergreen needleleaf forest", respectively.

5.2. Limitation and Future Direction

This study investigates the spatiotemporal consistency between S-NPP VIIRS and EOS MODIS LAI/FPAR products. Additionally, the uncertainty of VIIRS products was evaluated using available ground measurements. However, it should be noted that the quantified uncertainty is limited by an availability of the ground measurements (i.e., spatiotemporal and biome-specific coverage). As shown in Figure 8, we can know that the evaluation of VIIRS product uncertainty only achieved stage 1 ("Product accuracy is assessed from a small set of locations and time periods by comparison with in-situ or other suitable reference data." https://viirsland.gsfc.nasa.gov/Val_overview.html) based on CEOS validation stage hierarchy. Thus, adding more ground measurements from spatiotemporally and biome-specific well-distributed field campaigns are required. To achieve a robust evaluation result, we plan to synthesize in-situ reference data from globally distributed research networks (i.e., FLUXNET, Chinese Ecosystem Research Network (CERN), Terrestrial Ecosystem Research Network (TERN), et al.) by deploying a robust upscaling strategy developed by [55] to improve the spatiotemporal coverage of validation. These networks are not originally intended for validation purposes, but they have been continuously collecting in-situ LAI or FPAR data. These ground data have not been frequently utilized for assessing moderate resolution remote sensing products due to lack of spatial representativeness arising from the scale-mismatch problem. However, a new approach called Grading and Upscaling of Ground Measurements (GUGM) is able to resolve this scale mismatch issue and maximize the use of time-series of ground measurements from these networks. As the approach enables to provide frequent time-series of ground measurements, it would also provide a unique opportunity to evaluate the temporal performance of the LAI/FPAR products [56], allowing better understanding of the structure of the uncertainties and their evolution along different seasonal or annual contexts [33,57]. Therefore, more ground measurements covering the VIIRS product time-period (2012–present), will be collected from the global network of sites and used to better quantify the uncertainties, particularly the performance of time-series attached to the VIIRS product in a future study. This direction will help us to achieve a higher validation stage for VIIRS LAI/FPAR product.

In this study, we focused on evaluating continuity and consistency between retrievals from VIIRS and MODIS, and on quantifying uncertainty via direct validation approach. In general, "indirect validation" comparing VIIRS LAI/FPAR products with products from other moderate resolution sensors can afford one to identify the respective strengths and weaknesses of the products, thus leading to improvements in the next version of products, and to improve the confidence in research-quality of VIIRS products. Moreover, the intercomparison with other similar products is also necessary to achieve stage 2 and stage 3 defined from CEOS validation stage hierarchy ("Spatial and temporal consistency of the product and consistency with similar products has been evaluated over globally representative locations and time periods") [37]. Therefore, we will further analyze VIIRS products using the intercomparison with other global moderate resolution LAI/FPAR products such as GEOV1 [5], GLASS [6] and so on. A series of metrics and qualitative checks proposed by [36] will be adopted to assess the continuity, consistency and precision between VIIRS and other products.

In addition to the provisional results reported in [11,15], our multi-year evaluation clearly corroborates that the implemented heritage MODIS LAI/FPAR algorithm and a physically proven algorithm for achieving inter-sensor consistency allow the generation of multi-sensor LAI/FPAR ESDRs. Our results also satisfy the requirement of stability implying the potential of generating MODIS-like VIIRS LAI/FPAR ESDRs. In light of our S-NPP VIIRS LAI/FPAR experience and knowledge on developing a multi-sensor product, we are confident that we will be able to generate seamless and continuous VIIRS products from JPSS programs with minimal effort. This future plan will build a foundation of continuing sequential JPSS satellite programs, which include already budgeted JPSS-1 to JPSS-4 extending a more than 30+ years of long-term ESDRs.

6. Conclusions

It is critical that LAI/FPAR products are generated with high accuracy and precision, but more importantly, they must be produced with consistent algorithms across different sensor platforms in order to maintain a continuous and well-characterized data record. This is especially important for the VIIRS LAI/FPAR product on SNPP, which bridges the gap between NASA's EOS satellites and the next generation JPSS platforms. Thus, to evaluate the spatiotemporal consistency between VIIRS and MODIS LAI/FPAR products, multi-year time series of VIIRS and MODIS LAI/FPAR retrievals have been retrieved and their consistency is evaluated in this study. For the spatiotemporal consistency, the global mean differences between VIIRS and MODIS products at a native temporal composite (i.e., 8-day) or seasonally integrated are less than -0.006 for LAI and -0.002 for FPAR, respectively, indicating a reasonably high consistency between two sensors. These estimates meet the stability requirement for long-term LAI/FPAR ESDRs from multi-sensors suggested by GCOS. Generally the performance of the LAI/FPAR consistency for non-forest biomes is better than forest biomes. Additionally, the evaluation also showed a similar interannual variation and a comparable main algorithm retrieval rate from VIIRS with respect to those from MODIS during the study period (2012–2016), although several sources for inconsistency (e.g., stability of input surface reflectance) were discussed. Twenty-one true LAI, 26 effective LAI and 24 FPAR ground measurements with high reliability were used to validate both VIIRS and MODIS LAI/FPAR products. The results indicated that LAI retrievals from both sensors are closer to true LAI than effective LAI because of the clumping correction in the algorithm. We found a comparable LAI uncertainty/relative uncertainty of VIIRS (RMSE = 0.60/42.2%) with respect to that of MODIS (RMSE = 0.55/39.3%) and both VIIRS (RMSE = 0.10/24.4%) and MODIS

(RMSE = 0.11/26.0%) showed an overestimation of FPAR. This validation practice also supports the feasibility of producing consistent retrievals from VIIRS and MODIS. However, it also should be noted that more accurate ground measurements and better representation of different vegetation types in different LAI/FPAR ranges are required to refine the uncertainty evaluation of VIIRS LAI/FPAR product. Overall, this study imbues the confidence that S-NPP VIIRS and EOS MODIS can generate a continuous and well-characterized LAI/FPAR ESDRs. This important achievement plays an important role to bridge the gap between NASA's EOS satellites and the next generation of JPSS platforms.

Supplementary Materials: The following are available online at www.mdpi.com/1999-4907/9/2/73/s1, Figure S1: Global land cover map used in this study, Figure S2: The quality control based on "FparLai_QC" and "FparExtra_QC" layers for VIIRS and MODIS products under study, Table S1: Characteristics of the available ground sites and aggregated LAI_effective/LAI_true/FPAR values over 3 km × 3 km area, Figure S3: Spatial distribution of the BELMANIP-2.1 network and ground measurements for VIIRS LAI/FPAR product assessment, Table S2: The mean and standard deviation of LAI/FPAR difference between VIIRS and MODIS products (VIIRS-MODIS) at seasonal scale for each biome type from 2012 to 2016, Figure S4: The examples for the comparison between VIIRS and MODIS products at the native spatiotemporal (500-m/8-day) resolution, Figure S5: The annual variation of global retrieval rate of different algorithm paths for eight biome types (B1–B8 are same as in Figure S1), Figure S6: The annual variation of mean difference of retrieval index (DRI, calculated from VIIRS-MODIS) for non-forest and forest biomes in January and July from 2012 to 2016.

Acknowledgments: We sincerely appreciate all help from the MODIS and VIIRS science team members and validation collaborators. This study is supported by "Global LAI-FPAR Earth System Data Records from Suomi VIIRS to Extend the EOS MODIS Time Series" (NNX14AP80A) and "Maintenance and Evaluation of Collection 6 Terra and Aqua MODIS LAI/FPAR Products" (NNX14AI71G). This work is also funded by National Key R&D Program of China (2017YFA0603001), National Natural Science Foundation of China (41601403, 41671374, 41701401).

Author Contributions: Baodong Xu, Taejin Park, Yuri Knyazikhin, Ranga B. Myneni conceived and designed the experiments; Baodong Xu, Taejin Park, Kai Yan, Chi Chen and Wanjuan Song performed the experiments; Baodong Xu and Taejin Park discussed the results. All authors contributed to the writing of the paper.

Conflicts of Interest: The authors declare no conflict of interest.

References

1. Sellers, P.J.; Dickinson, R.E.; Randall, D.A.; Betts, A.K.; Hall, F.G.; Berry, J.A.; Collatz, G.J.; Denning, A.S.; Mooney, H.A.; Nobre, C.A.; et al. Modeling the exchanges of energy, water, and carbon between continents and the atmosphere. *Science* **1997**, *275*, 502–509. [CrossRef] [PubMed]

2. Myneni, R.B.; Hoffman, S.; Knyazikhin, Y.; Privette, J.L.; Glassy, J.; Tian, Y.; Wang, Y.; Song, X.; Zhang, Y.; Smith, G.R.; et al. Global products of vegetation leaf area and fraction absorbed par from year one of MODIS data. *Remote Sens. Environ.* **2002**, *83*, 214–231. [CrossRef]

3. Garrigues, S.; Lacaze, R.; Baret, F.; Morisette, J.T.; Weiss, M.; Nickeson, J.E.; Fernandes, R.; Plummer, S.; Shabanov, N.V.; Myneni, R.B.; et al. Validation and intercomparison of global leaf area index products derived from remote sensing data. *J. Geophys. Res.* **2008**, *113*, G02028. [CrossRef]

4. Fensholt, R.; Sandholt, I.; Rasmussen, M.S. Evaluation of MODIS LAI, FAPAR and the relation between FAPAR and NDVI in a semi-arid environment using in situ measurements. *Remote Sens. Environ.* **2004**, *91*, 490–507. [CrossRef]

5. Baret, F.; Weiss, M.; Lacaze, R.; Camacho, F.; Makhmara, H.; Pacholcyzk, P.; Smets, B. GEOV1: LAI and FAPAR essential climate variables and FCOVER global time series capitalizing over existing products. Part 1: Principles of development and production. *Remote Sens. Environ.* **2013**, *137*, 299–309. [CrossRef]

6. Xiao, Z.; Liang, S.; Wang, J.; Chen, P.; Yin, X.; Zhang, L.; Song, J. Use of general regression neural networks for generating the glass leaf area index product from time-series MODIS surface reflectance. *IEEE Trans. Geosci. Remote Sens.* **2014**, *52*, 209–223. [CrossRef]

7. Yan, K.; Park, T.; Yan, G.; Chen, C.; Yang, B.; Liu, Z.; Nemani, R.; Knyazikhin, Y.; Myneni, R. Evaluation of MODIS LAI/FPAR product collection 6. Part 1: Consistency and improvements. *Remote Sens.* **2016**, *8*, 359. [CrossRef]

8. Skakun, S.; Justice, C.O.; Vermote, E.; Roger, J.-C. Transitioning from MODIS to VIIRS: An analysis of inter-consistency of NDVI data sets for agricultural monitoring. *Int. J. Remote Sens.* **2018**, *39*, 971–992. [CrossRef]

9. Justice, C.O.; Román, M.O.; Csiszar, I.; Vermote, E.F.; Wolfe, R.E.; Hook, S.J.; Friedl, M.; Wang, Z.; Schaaf, C.B.; Miura, T.; et al. Land and cryosphere products from Suomi NPP VIIRS: Overview and status. *J. Geophys. Res. Atmos.* **2013**, *118*, 9753–9765. [CrossRef] [PubMed]

10. Ganguly, S.; Samanta, A.; Schull, M.A.; Shabanov, N.V.; Milesi, C.; Nemani, R.R.; Knyazikhin, Y.; Myneni, R.B. Generating vegetation leaf area index earth system data record from multiple sensors. Part 2: Implementation, analysis and validation. *Remote Sens. Environ.* **2008**, *112*, 4318–4332. [CrossRef]

11. Yan, K.; Park, T.; Chen, C.; Xu, B.; Song, W.; Yang, B.; Zeng, Y.; Liu, Z.; Yan, G.; Knyazikhin, Y.; et al. Generating global products of LAI and FPAR from SNPP-VIIRS data: Theoretical background and implementation. *IEEE Trans. Geosci. Remote Sens.* **2018**, *56*, 1–19. [CrossRef]

12. Huang, D.; Knyazikhin, Y.; Dickinson, R.E.; Rautiainen, M.; Stenberg, P.; Disney, M.; Lewis, P.; Cescatti, A.; Tian, Y.; Verhoef, W.; et al. Canopy spectral invariants for remote sensing and model applications. *Remote Sens. Environ.* **2007**, *106*, 106–122. [CrossRef]

13. Knyazikhin, Y.; Martonchik, J.V.; Myneni, R.B.; Diner, D.J.; Running, S.W. Synergistic algorithm for estimating vegetation canopy leaf area index and fraction of absorbed photosynthetically active radiation from MODIS and MISR data. *J. Geophys. Res. Atmos.* **1998**, *103*, 32257–32275. [CrossRef]

14. Knyazikhin, Y.; Schull, M.A.; Stenberg, P.; Mõttus, M.; Rautiainen, M.; Yang, Y.; Marshak, A.; Latorre Carmona, P.; Kaufmann, R.K.; Lewis, P.; et al. Hyperspectral remote sensing of foliar nitrogen content. *Proc. Natl. Acad. Sci. USA* **2013**, *110*, E185–E192. [CrossRef] [PubMed]

15. Park, T.; Yan, K.; Chen, C.; Xu, B.; Knyazikhin, Y.; Myneni, R.B. VIIRS Leaf Area Index (LAI) and Fraction of Photosynthetically Active Radiation Absorbed by Vegetation (FPAR) Product Algorithm Theoretical Basis Document (ATBD). Available online: https://viirsland.gsfc.nasa.gov/Products/NASA/LAI_FparESDR.html (accessed on 10 November 2017).

16. Myneni, R.B.; Ross, J. *Photon-Vegetation Interactions: Applications in Optical Remote Sensing and Plant Ecology*; Springer-Verlag: New York, NY, USA, 1991.

17. Ross, J. *The Radiation Regime and Architecture of Plant Stands*; Springer Science & Business Media: Medford, MA, USA, 1981.

18. Hillger, D.; Kopp, T.; Lee, T.; Lindsey, D.; Seaman, C.; Miller, S.; Solbrig, J.; Kidder, S.; Bachmeier, S.; Jasmin, T.; et al. First-light imagery from Suomi NPP VIIRS. *Bull. Am. Meteorol. Soc.* **2013**, *94*, 1019–1029. [CrossRef]

19. Liu, H.; Remer, L.A.; Huang, J.; Huang, H.-C.; Kondragunta, S.; Laszlo, I.; Oo, M.; Jackson, J.M. Preliminary evaluation of S-NPP VIIRS aerosol optical thickness. *J. Geophys. Res. Atmos.* **2014**, *119*, 3942–3962. [CrossRef]

20. Xiong, X.; Butler, J.; Chiang, K.; Efremova, B.; Fulbright, J.; Lei, N.; McIntire, J.; Oudrari, H.; Sun, J.; Wang, Z.; et al. VIIRS on-orbit calibration methodology and performance. *J. Geophys. Res. Atmos.* **2014**, *119*, 5065–5078. [CrossRef]

21. Wolfe, R.E.; Lin, G.; Nishihama, M.; Tewari, K.P.; Tilton, J.C.; Isaacman, A.R. Suomi NPP VIIRS prelaunch and on-orbit geometric calibration and characterization. *J. Geophys. Res. Atmos.* **2013**, *118*, 508–511, 521. [CrossRef]

22. Ganguly, S.; Schull, M.A.; Samanta, A.; Shabanov, N.V.; Milesi, C.; Nemani, R.R.; Knyazikhin, Y.; Myneni, R.B. Generating vegetation leaf area index earth system data record from multiple sensors. Part 1: Theory. *Remote Sens. Environ.* **2008**, *112*, 4333–4343. [CrossRef]

23. Baret, F.; Jacquemoud, S.; Hanocq, J.F. The soil line concept in remote sensing. *Remote Sens. Rev.* **1993**, *7*, 65–82. [CrossRef]

24. Jacquemoud, S.; Baret, F.; Hanocq, J.F. Modeling spectral and bidirectional soil reflectance. *Remote Sens. Environ.* **1992**, *41*, 123–132. [CrossRef]

25. Lewis, P.; Disney, M. Spectral invariants and scattering across multiple scales from within-leaf to canopy. *Remote Sens. Environ.* **2007**, *109*, 196–206. [CrossRef]

26. Schull, M.A.; Knyazikhin, Y.; Xu, L.; Samanta, A.; Carmona, P.L.; Lepine, L.; Jenkins, J.P.; Ganguly, S.; Myneni, R.B. Canopy spectral invariants, Part 2: Application to classification of forest types from hyperspectral data. *J. Quant. Spectrosc. Radiat. Transf.* **2011**, *112*, 736–750. [CrossRef]

27. Smolander, S.; Stenberg, P. Simple parameterizations of the radiation budget of uniform broadleaved and coniferous canopies. *Remote Sens. Environ.* **2005**, *94*, 355–363. [CrossRef]

28. Vermote, E.F. VIIRS/NPP Surface Reflectance Daily L2G Global 1 km and 500 m Sin Grid v001 [Data Set]. NASA EOS MODIS Land Processes DAAC. Available online: https://doi.org/10.5067/viirs/vnp09ga.001 (accessed on 15 November 2017).

29. Yang, W.; Shabanov, N.V.; Huang, D.; Wang, W.; Dickinson, R.E.; Nemani, R.R.; Knyazikhin, Y.; Myneni, R.B. Analysis of leaf area index products from combination of MODIS terra and aqua data. *Remote Sens. Environ.* **2006**, *104*, 297–312. [CrossRef]

30. Myneni, R.B.; Knyazikhin, Y.; Park, T. MYD15A2H MODIS/Aqua Leaf Area Index/FPAR 8-Day L4 Global 500m Sin Grid v006 [Data Set]. NASA EOS MODIS Land Processes DAAC. Available online: https://doi.org/10.5067/modis/myd15a2h.006 (accessed on 10 September 2017).

31. Yan, K.; Park, T.; Yan, G.; Liu, Z.; Yang, B.; Chen, C.; Nemani, R.; Knyazikhin, Y.; Myneni, R. Evaluation of MODIS LAI/FPAR product collection 6. Part 2: Validation and intercomparison. *Remote Sens.* **2016**, *8*, 460. [CrossRef]

32. Global Climate Observing System (GCOS). Systematic Observation Requirements for Satellite-Based Products for Climate, 2011 Update, Supplemental Details to the Satellite-Based Component of the Implementation Plan for the Global Observing System for Climate in Support of the UNFCCC (2010 Update). Available online: https://library.wmo.int/opac/doc_num.php?explnum_id=3710 (accessed on 9 September 2017).

33. Weiss, M.; Baret, F.; Garrigues, S.; Lacaze, R. LAI and FAPAR CYCLOPES global products derived from VEGETATION. Part 2: Validation and comparison with MODIS collection 4 products. *Remote Sens. Environ.* **2007**, *110*, 317–331. [CrossRef]

34. Fernandes, R.; Butson, C.; Leblanc, S.; Latifovic, R. Landsat-5 TM and landsat-7 ETM+ based accuracy assessment of leaf area index products for Canada derived from SPOT-4 VEGETATION data. *Can. J. Remote Sens.* **2003**, *29*, 241–258. [CrossRef]

35. Tan, B.; Woodcock, C.E.; Hu, J.; Zhang, P.; Ozdogan, M.; Huang, D.; Yang, W.; Knyazikhin, Y.; Myneni, R.B. The impact of gridding artifacts on the local spatial properties of MODIS data: Implications for validation, compositing, and band-to-band registration across resolutions. *Remote Sens. Environ.* **2006**, *105*, 98–114. [CrossRef]

36. Camacho, F.; Cernicharo, J.; Lacaze, R.; Baret, F.; Weiss, M. GEOV1: LAI, FAPAR essential climate variables and fcover global time series capitalizing over existing products. Part 2: Validation and intercomparison with reference products. *Remote Sens. Environ.* **2013**, *137*, 310–329. [CrossRef]

37. Fernandes, R.; Plummer, S.; Nightingale, J.; Baret, F.; Camacho, F.; Fang, H.; Garrigues, S.; Gobron, N.; Lang, M.; Lacaze, R.; et al. Global Leaf Area Index Product Validation Good Practices. Version 2.0. In *Best Practice for Satellite-Derived Land Product Validation (p. 76): Land Product Validation Subgroup (WGCV/CEOS)*; Schaepman-Strub, G., Román, M., Nickeson, J., Eds.; Committee on Earth Observation Satellites: Maryland, MD, USA, 2014; pp. 1–78. [CrossRef]

38. Verger, A.; Baret, F.; Weiss, M. Performances of neural networks for deriving LAI estimates from existing CYCLOPES and MODIS products. *Remote Sens. Environ.* **2008**, *112*, 2789–2803. [CrossRef]

39. Baret, F.; Morissette, J.T.; Fernandes, R.A.; Champeaux, J.L.; Myneni, R.B.; Chen, J.; Plummer, S.; Weiss, M.; Bacour, C.; Garrigues, S.; et al. Evaluation of the representativeness of networks of sites for the global validation and intercomparison of land biophysical products: Proposition of the ceos-belmanip. *IEEE Trans. Geosci. Remote Sens.* **2006**, *44*, 1794–1803. [CrossRef]

40. Camacho, F.; Lacaze, R.; Latorre, C.; Baret, F.; De la Cruz, F.; Demarez, V.; Di Bella, C.; García-Haro, J.; González-Dugo, M.P.; Kussul, N. Collection of ground biophysical measurements in support of copernicus global land product validation: The imagines database. In Proceedings of the EGU General Assembly, Vienna, Austria, 17–22 April 2015; Geophysical Research Abstracts, 17 EGU2015-2209-1.

41. Li, X.; Cheng, G.; Liu, S.; Xiao, Q.; Ma, M.; Jin, R.; Che, T.; Liu, Q.; Wang, W.; Qi, Y.; et al. Heihe watershed allied telemetry experimental research (HiWATER): Scientific objectives and experimental design. *Bull. Am. Meteorol. Soc.* **2013**, *94*, 1145–1160. [CrossRef]

42. Zhao, J.; Li, J.; Liu, Q.; Fan, W.; Zhong, B.; Wu, S.; Yang, L.; Zeng, Y.; Xu, B.; Yin, G. Leaf area index retrieval combining HJ1/CCD and Landsat8/OLI data in the Heihe River Basin, China. *Remote Sens.* **2015**, *7*, 6862. [CrossRef]

43. Zeng, Y.; Li, J.; Liu, Q.; Qu, Y.; Huete, A.; Xu, B.; Yin, G.; Zhao, J. An optimal sampling design for observing and validating long-term leaf area index with temporal variations in spatial heterogeneities. *Remote Sens.* **2015**, *7*, 1300. [CrossRef]

44. Yang, W.; Tan, B.; Huang, D.; Rautiainen, M.; Shabanov, N.V.; Wang, Y.J.; Privette, J.L.; Huemmrich, K.F.; Fensholt, R.; Sandholt, I.; et al. Modis leaf area index products: From validation to algorithm improvement. *IEEE Trans. Geosci. Remote Sens.* **2006**, *44*, 1885–1898. [CrossRef]

45. Tan, B.; Hu, J.N.; Zhang, P.; Huang, D.; Shabanov, N.; Weiss, M.; Knyazikhin, Y.; Myneni, R.B. Validation of moderate resolution imaging spectroradiometer leaf area index product in croplands of Alpilles, France. *J. Geophys. Res. Atmos.* **2005**, *110*, D01107. [CrossRef]

46. Morisette, J.T.; Baret, F.; Privette, J.L.; Myneni, R.B.; Nickeson, J.E.; Garrigues, S.; Shabanov, N.V.; Weiss, M.; Fernandes, R.A.; Leblanc, S.G.; et al. Validation of global moderate-resolution LAI products: A framework proposed within the CEOS land product validation subgroup. *IEEE Trans. Geosci. Remote Sens.* **2006**, *44*, 1804–1817. [CrossRef]

47. Jiang, C.; Ryu, Y.; Fang, H.; Myneni, R.; Claverie, M.; Zhu, Z. Inconsistencies of interannual variability and trends in long-term satellite leaf area index products. *Glob. Chang. Biol.* **2017**, *23*, 4133–4146. [CrossRef] [PubMed]

48. Zhu, Z.; Piao, S.; Myneni, R.B.; Huang, M.; Zeng, Z.; Canadell, J.G.; Ciais, P.; Sitch, S.; Friedlingstein, P.; Arneth, A.; et al. Greening of the earth and its drivers. *Nat. Clim. Chang.* **2016**, *6*, 791–795. [CrossRef]

49. Shabanov, N.V.; Huang, D.; Yang, W.Z.; Tan, B.; Knyazikhin, Y.; Myneni, R.B.; Ahl, D.E.; Gower, S.T.; Huete, A.R.; Aragao, L.E.O.C.; et al. Analysis and optimization of the MODIS leaf area index algorithm retrievals over broadleaf forests. *IEEE Trans. Geosci. Remote Sens.* **2005**, *43*, 1855–1865. [CrossRef]

50. Fang, H.; Wei, S.; Liang, S. Validation of MODIS and Cyclopes LAI products using global field measurement data. *Remote Sens. Environ.* **2012**, *119*, 43–54. [CrossRef]

51. Huete, A.; Didan, K.; Miura, T.; Rodriguez, E.P.; Gao, X.; Ferreira, L.G. Overview of the radiometric and biophysical performance of the MODIS vegetation indices. *Remote Sens. Environ.* **2002**, *83*, 195–213. [CrossRef]

52. Kobayashi, H.; Dye, D.G. Atmospheric conditions for monitoring the long-term vegetation dynamics in the Amazon using normalized difference vegetation index. *Remote Sens. Environ.* **2005**, *97*, 519–525. [CrossRef]

53. Platnick, S.; Cechini, M.; Boller, R.; Schmaltz, J.; Manoharan, S.; Amarasinghe, N.; Levy, R.I. MODIS atmosphere discipline team: C6 status II. MODAWG: MODIS-VIIRS product continuity for cloud mask, cloud-top and optical properties status. In Proceedings of the MODIS/VIIRS 2016 Science Team Meeting, Silver Spring, MD, USA, 6–10 June 2016. Available online: https://modis.gsfc.nasa.gov/sci_team/meetings/201606/presentations/plenary/platnick.pdf (accessed on 9 January 2018).

54. Pahlevan, N.; Sarkar, S.; Devadiga, S.; Wolfe, R.E.; Román, M.; Vermote, E.; Lin, G.; Xiong, X. Impact of spatial sampling on continuity of MODIS-VIIRS land surface reflectance products: A simulation approach. *IEEE Trans. Geosci. Remote Sens.* **2017**, *55*, 183–196. [CrossRef]

55. Xu, B.; Li, J.; Park, T.; Liu, Q.; Zeng, Y.; Yin, G.; Zhao, J.; Fan, W.; Yang, L.; Knjazikhin, Y.; et al. An integrated method for validating long-term leaf area index products using global networks of site-based measurements. *Remote Sens. Environ.* **2018**, accepted.

56. Claverie, M.; Vermote, E.F.; Weiss, M.; Baret, F.; Hagolle, O.; Demarez, V. Validation of coarse spatial resolution LAI and FAPAR time series over cropland in southwest France. *Remote Sens. Environ.* **2013**, *139*, 216–230. [CrossRef]

57. Barr, A.G.; Black, T.A.; Hogg, E.H.; Kljun, N.; Morgenstern, K.; Nesic, Z. Inter-annual variability in the leaf area index of a boreal aspen-hazelnut forest in relation to net ecosystem production. *Agric. For. Meteorol.* **2004**, *126*, 237–255. [CrossRef]

Article

A Comparison of Simulated and Field-Derived Leaf Area Index (LAI) and Canopy Height Values from Four Forest Complexes in the Southeastern USA

John S. Iiames [1,*], Ellen Cooter [2], Donna Schwede [2] and Jimmy Williams [3]

[1] U.S. Environmental Protection Agency, National Exposure Research Laboratory, Exposure Methods and Measurements Division, 109 T.W. Alexander Drive, Research Triangle Park, NC 27711, USA

[2] U.S. Environmental Protection Agency, National Exposure Research Laboratory, Atmospheric Modeling and Analysis Division, 109 T.W. Alexander Drive, Research Triangle Park, NC 27711, USA; Cooter.ellen@epa.gov (E.C.); Schwede.donna@epa.gov (D.S.)

[3] Texas A & M University, Agri-Life Research, Temple, TX 76502, USA; jwilliams@brc.tamus.edu

* Correspondence: Iiames.john@epa.gov; Tel.: +1-919-541-3039

Received: 3 November 2017; Accepted: 25 December 2017; Published: 12 January 2018

Abstract: Vegetative leaf area is a critical input to models that simulate human and ecosystem exposure to atmospheric pollutants. Leaf area index (LAI) can be measured in the field or numerically simulated, but all contain some inherent uncertainty that is passed to the exposure assessments that use them. LAI estimates for minimally managed or natural forest stands can be particularly difficult to develop as a result of interspecies competition, age and spatial distribution. Satellite-based LAI estimates hold promise for retrospective analyses, but we must continue to rely on numerical models for alternative management analysis. Our objective for this study is to calculate and validate LAI estimates generated from the USDA Environmental Policy Impact Climate (EPIC) model (a widely used, field-scale, biogeochemical model) on four forest complexes spanning three physiographic provinces in Virginia and North Carolina. Measurements of forest composition (species and number), LAI, tree diameter, basal area, and canopy height were recorded at each site during the 2002 field season. Calibrated EPIC results show stand-level temporally resolved LAI estimates with R^2 values ranging from 0.69 to 0.96, and stand maximum height estimates within 20% of observation. This relatively high level of performance is attributable to EPIC's approach to the characterization of forest stand biogeochemical budgets, stand history, interspecies competition and species-specific response to local weather conditions. We close by illustrating the extension of this site-level approach to scales that could support regional air quality model simulations.

Keywords: LAI; leaf area index; EPIC; simulation; satellite; MODIS; biomass; evaluation; southern U.S. forests

1. Introduction

The status and dynamics of vegetation leaf area, often reported in terms of leaf area index (LAI), can be a critical determinant of regional air and water quality [1]. LAI is commonly used as a surrogate of photosynthetically active area when photosynthesis is the principle process controlling chemical exchange between the atmosphere and underlying land surfaces. Leaf area influences the sequestration of carbon from carbon emissions [2], the removal of pollutant species through deposition [3], the biogenic emission of volatile organic compounds (BVOC) that contribute to tropospheric ozone formation [4], and the emission of greenhouse gases [5]. Temporally resolved leaf area and canopy heights for natural forest stands are critical inputs for process-based meteorological models. Leaf area estimates contribute to the calculation of surface evapotranspiration and albedo, and canopy height largely determines surface roughness, which contributes to mechanical mixing of the atmosphere [6]. Calculation of gaseous air pollutant deposition velocity (V_d) frequently requires values for leaf area and surface roughness [7–9]. The Clean Air Status and Trends Network (CASTNET), for instance, measures atmospheric concentrations and then estimates water vapor, ozone (O_3), sulfur dioxide (SO_2), and nitric acid (HNO_3) fluxes using the Multilayer Model (MLM) [8,10,11]. The MLM inputs a generalized annual LAI time-series developed from measured values of maximum LAI for each plant species and typical phenology. Estimates of deposition velocity calculated by MLM were seen to be highly sensitive to LAI time-series parameters [12] with differences in V_d of about 25% for sulfur dioxide and nitric acid and greater than 60% for ozone. On a regional scale, the United States Environmental Protection Agency's (US EPA) Community Multiscale Air Quality Model (CMAQ) [8] relies on output from the Weather Research Model (WRF) [13] and the Pleim-Xiu (PX) land surface scheme [14]. PX LAI estimates are based on deep soil temperature, an LAI response function based on soil temperature and a specified minimum and maximum LAI for each land use classification which results in known biases [15].

LAI inputs into air quality applications that rely on model estimates of chemical flux include periodic in situ point sampled and Light Detection and Ranging (LIDAR) LAI measurements, static look-up values, and satellite-derived LAI values. A number of studies have reported that simulating both accurate leaf- and canopy-scale fluxes is not possible when leaf-scale fluxes are scaled using an inaccurate LAI [16–19]. While more temporally and spatially detailed LAI inputs are likely to improve model estimates of meteorological conditions, biogenic emissions, and O_3 precursor estimates, interspecies competition, age and spatial distribution make temporally resolved LAI estimates for minimally managed or natural forest stands particularly difficult to develop. Satellite-based LAI estimates hold promise for retrospective analyses, e.g., [20,21], but we are still learning how best to make use of these data, and they will never be available for future meteorological alternative management applications. Therefore, we must continue to supplement temporally resolved, remotely sensed vegetative leaf area estimates with numerical model estimates.

The objective of this study is to evaluate the capacity of a biogeochemical agroecosystems model to generate more biologically (process) representative, accurate and temporally resolved forest stand-level LAI estimates than existing numerical methods. We calibrate the USDA Environmental Policy Integrate Climate (EPIC) model from LAI estimates evaluated at four mixed forest stands in the southeastern U.S. We then demonstrate the practical areal application of this approach to represent mixed-age, mixed forest stands that could support nitrogen flux estimates for air quality assessment and supplement remotely sensed observations.

2. Materials and Methods

2.1. Site Descriptions

The near-lab study site for the US EPA Office of Research and Development is the Albemarle-Pamlico Basin (APB), located in central-to-northern North Carolina and southern Virginia. The APB has a drainage area of 738,735 km^2 and includes three physiographic provinces: mountain, piedmont and coastal plain, ranging in elevation from 1280 m to sea level. The APB sub-basins include the Albemarle-Chowan, Roanoke, Pamlico, and Neuse River basins; all draining into the second largest estuarine system within the continental United States. Four 1 km^2 LAI calibration/validation sites were chosen within the APB representing all three physiographic provinces: 1. Appomattox (Upper Piedmont), 2. Hertford (Coastal Plain), 3. Fairystone (Upper Piedmont-Mountain), 4. Umstead (Lower Piedmont) (Figure 1). General descriptions of these sites are provided below.

2.1.1. Appomattox

The Appomattox field site is located in Campbell County, Virginia (37.219° N, −78.879° W) approximately 15.5 km south-southwest of Appomattox, Virginia (Figure 1 and Table 1). This Upper Piedmont region has a range in elevation from 165 m to 215 m above mean sea level. The area is a mixture of rural agricultural fields and managed (Loblolly pine—*Pinus taeda*) and unmanaged (oak-hickory) forest stands. The MeadWestvaco Corporation, a supporter of the Sustainable Forestry Initiative, permitted sampling access to the US EPA for LAI research on 505 hectares in 2002. Land cover (LC) percentages within the 1 km^2 area include: (1) coniferous (Total—67.6% (unthinned—30.6%, thinned —37.0%)); (2) deciduous (21.7%); and (3) other vegetation (harvested—10.6%). The modeled and in situ LAI comparison uses the 30.6 hectare unthinned pine stand; a 100% planted loblolly pine dominant-codominant crown class (15.9 m canopy height) underlain with a variety of understory deciduous species (Supplementary Materials Figures S1 and S2 and Table S1). The dominant soil types consist of very deep, well-drained, moderately permeable soils (Georgville and Tatum loams) on 2% to 15% slopes.

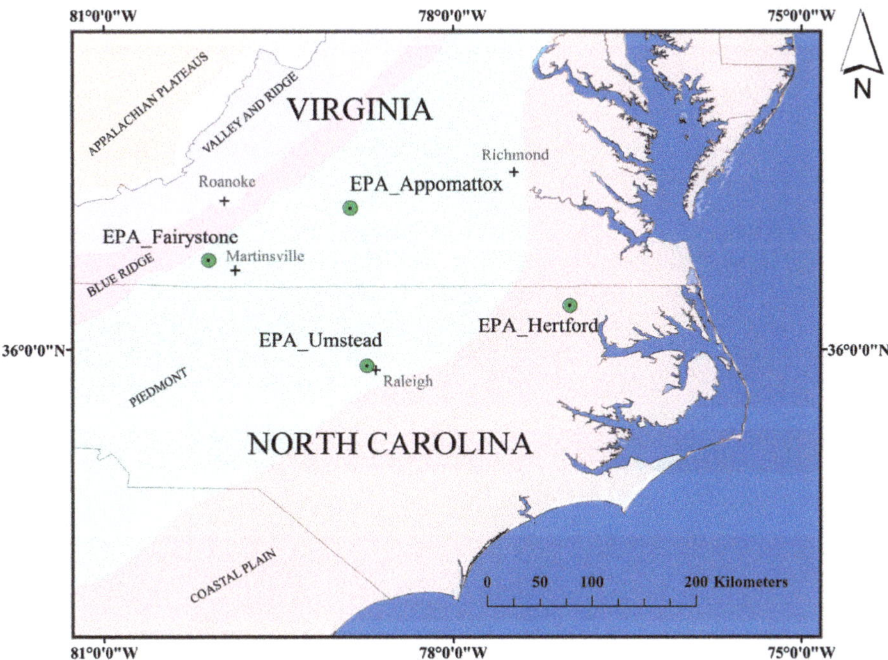

Figure 1. Four LAI field site locations in Virginia and North Carolina, USA.

2.1.2. Hertford

The Hertford site is located in Hertford County, North Carolina (36.383° N, −77.001° W), approximately 5.8 km west-southwest of Winton, North Carolina (Figure 1). This coastal plain site is 8–10 m above mean sea level (0% slope) with a moderately well drained thermic Aquic Hapludult soil type (Craven fine sandy loam). This site is dominated by 19 year-old planted loblolly pine, and was managed by International Paper, Inc. LC percentages within the 1 km^2 area include: (1) coniferous (72.3%); (2) deciduous (17.8%); and (3) other vegetation (agriculture—9.9%). The modeled and in situ LAI comparison was done only within the 19 year-old managed pine stand (Supplementary Materials Figures S3 and S4 and Table S1).

2.1.3. Fairystone

The Fairystone site is located in both Henry and Patrick Counties in Virginia (36.771° N, −80.092° W), approximately 20.7 km west-northwest of Martinsville, Virginia (Figure 1). This Blue Ridge eastern foothills site is 409 m above mean sea level and is defined by well drained mesic Typic Hapludult soil types (Fairystone and Littlejoe Series). The Virginia Department of Inland Game and Fisheries manage this upland hardwood site (100.0% deciduous) maintaining habitat suitable for turkeys, deer and a variety of small game and nongame wildlife (Supplementary Materials Figures S5 and S6 and Table S1).

2.1.4. Umstead

The North Carolina Parks and Recreation Department manages the Umstead site, which is located 12.5 km northwest of Raleigh, NC (William B. Umstead State Park) (Figure 1). This 2258-hectare park is a mixture of mature deciduous and coniferous tree species, with an average age approximately 75–85 years old (Supplementary Materials Figures S7 and S8 and Table S1). This Piedmont site ranges in elevation between 75.3 m and 120.5 m above mean sea level ($\bar{x}$ = 94.0 m), with slopes ranging between 0.1% and 25.0% ($\bar{x}$ = 6.2%) and is defined by 13 variant sandy loam soil types. Land-cover percentages within the 1 km^2 area include: (1) deciduous (77.8%) and (2) coniferous (22.2%) forest types.

2.2. In Situ Measurements

Forest stand characteristics were collected for all four validation sites over the 2002 growing season including (1) species type, and (2) structural (crown closure, height, diameter, LAI) and (3) stocking attributes (basal area and trees per hectare). On each site, the sampling design includes the establishment of the primary sampling unit, the quadrat (Figure 2B). Each quadrat is a 100 m × 100 m grid delineated by five parallel 100 m east-to-west (E–W) oriented transects labeled L1–L5—each transect line is separated 20 m apart in a north south (N–S) direction, with L1 representing the northern-most transect. Interspersed between these transects are 25 'point sampling' locations in a N–S and E–W grid network, separated 20 m apart, with the northwestern most point established between L1 and L2 transects at a 10-m inset from the origin of L1 and L2. The secondary sampling unit was three intersecting 50 m or 100 m transects at 90°, 135°, and 225° (Figure 2B). These sub-plots were located to augment data collected on the primary sampling unit(s), the quadrat (Figure 2C). Only one quadrat (Q1) was laid-out on three of the four research sites (Appomattox, Hertford, and Umstead), with four quadrats (Q1–Q4) established on the Fairystone site. Quadrat replicates were applied only at Fairystone to account for terrain and aspect differences prevalent in this upper piedmont region. Sub-plots were applied to account for multiple forest types within 1-km area (i.e., Appomattox).

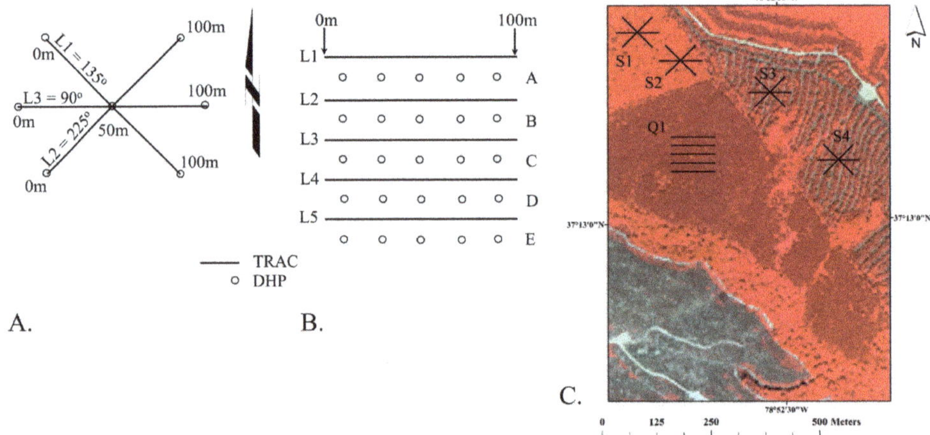

Figure 2. Plot design for (**A**) 50-m or 100-m sub-plot and (**B**) 100 m × 100 m quadrat. Distribution of quadrat and sub-plots on Appomattox research site (**C**) is illustrated by Q1 and S1–4. Note: DHP—Digital Hemispherical Photography; TRAC—Tracing Radiation and Architecture of Canopies analyzer.

Measurements of the forest structural attributes height (m) and diameter (cm) were made at all sites within the 100 m $\times$ 100 m quadrat areas using a point sampling method using a basal area 10-factor prism for trees larger than 5 cm in diameter at breast height (DBH), with breast height 1.37 m above the base of the tree. Stocking values, expressed as trees per hectare (TPH) and basal area per hectare (BA/H) were calculated from the point sampling data. BA is the cross-sectional area of a tree at 1.37 m above the tree base per unit area [22]. Three plots within each quadrat were sampled for understory components (stems less than 5 cm DBH) using a 4.57 m radius fixed area sampling method. Canopy closure, defined as the percent obstruction of the sky by canopy elements, was estimated using a Geographic Resources Solutions (GRS) Densitometer (http://www.BenMeadows.com/) along transects 1, 3, and 5 in each quadrat, taking obstruction/no obstruction readings by canopy height class (i.e., understory, intermediate, dominate) every two meters.

We define LAI here as one-half the total green leaf area per unit ground surface [23]. LAI estimates are based on the indirect in situ optical LAI estimation technique combining measurements from the Tracing Radiation and Architecture of Canopies analyzer (TRAC) optical sensor and digital hemispherical photography (DHP) in deciduous and coniferous forest stands. Parameters assessed within the TRAC-DHP method included the element clumping index (Ω_E) derived from the TRAC instrument, the effective leaf area index (L_e) measured with DHP, and the needle-to-shoot area index (λ_E) and the woody-to-total index (α) measured in the field and in the laboratory. LAI calculations use these characteristics as inputs into the modified Beer-Lambert light extinction function [24] (1):

$$LAI = (1 - \alpha) \times [L_e(\lambda_E/\Omega_E)] \tag{1}$$

The TRAC instrument records downwelling solar photosynthetic photon flux density (PPFD) through the canopy during direct light conditions in order to measure foliage clumping at scales larger than the shoot. TRAC is sampled along all five line transects within each quadrat and along one of three sub-plot line transects, dependent on the line transect closest to solar perpendicularity —measurements are made at a solar angle between 30° and 60° [24]. DHP measurements were derived from a Nikon CoolPix 995 digital camera with a Nikon FC-E8 fish-eye converter in diffuse light conditions in order to extract the effective LAI or also known as plant area index (PAI). DHP measurements were made at the 10-, 50-, and 90-m marks along lines A, C, and E within each quadrat and were made during diffuse light conditions (at dawn or dusk). TRAC and DHP measurements were recorded across all four sites for the following dates: Fairystone (1 May 2002, 25 June 2002, 8 July 2002, 4 September 2002), Appomattox (6 March 2002, 23 May 2002, 30 July 2002), Umstead (13 April 2002, 23 April 2002, 21 October 2002), and Hertford (5 March 2002, 9 April 2002, 18 June 2002). Processing of TRAC data using TRACWin 3.7.3 software [24,25] and DHP measurements make use of Gap Light Analyzer (GLA) software [26].

2.3. Soil and Meteorology

Biogeochemical models usually require detailed multilayer soil profiles as simulation inputs. This information was not available for the field study sites, and so we used a representative soil profile at all locations (Supplementary Materials Table S2). Historical weather data from the nearest National Oceanographic Atmospheric Administration's (NOAA) National Centers for Environmental Information (NCEI) Cooperative Observing Station (Coop) to the forest observation site (Stand Site) were selected (Table 1). Figure 3 shows daily time series of maximum temperature, minimum temperature and precipitation at these locations during 2002.

Table 1. Locations of historical weather data sites assigned to each forest site and the distance between the forest and weather data locations.

Stand Site	Lat/Long	Elev. (m)	Coop Site	Lat/Long	Elev. (m)	Dist. (km)
Appomattox	37.22, −78.88	200	Appomattox, VA/51011	37.36, −78.83	277.4	15.8
Hertford	36.38, −77.00	10	Jackson, NC/374456	36.40, −77.42	39.6	37.9
Fairystone	36.77, −80.09	470	Martinsville, VA/515300	36.71, −79.87	231	21
Umstead	35.86, −78.74	94	Raleigh, NC/377079	35.79, −78.70	121.9	8.2

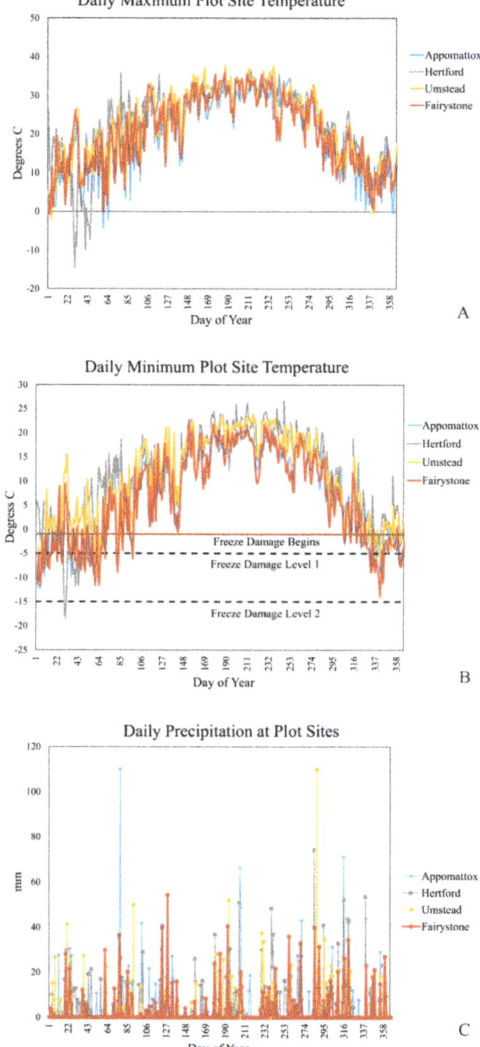

Figure 3. Daily weather at the four NCEI Coop locations representing the year 2002 forest plot sites (**A**) maximum temperature (°C), (**B**) minimum temperature (°C), and (**C**) precipitation (mm). The dashed lines in Figure 3B define two points on the frost damage function curve (see discussion in Section 2.4 and Supplementary Materials Data Part 3).

2.4. The EPIC Model

The simulation of forests in EPIC is adapted from methods pioneered in the Agricultural Land Management Alternatives with the Numerical Assessment Criteria (ALMANAC) model [27]. Changes within this general plant model were developed to better address forest species [28]. Forest regrowth in Canadian Boreal forest ecosystems following disturbances such as forest fires, clear cuts and insect infestations is detailed in a boreal forest version of ALMANAC (ALMANACBF) [29]. Much of this research was performed to improve agro-forest representation in the Soil Water Assessment Tool (SWAT) [30] that characterizes the forest stand as a distinct land unit of a single or consistent combination of tree species of similar age and productivity [31,32]. The EPIC implementation of this research lacks some, and simplifies other ALMANAC processes so that EPIC and ALMANAC simulation results for the same forest ecosystem may be similar, but they will not necessarily be identical.

EPIC is a hybrid, process-based, semi-empirical biogeochemical model characterizing field-sized areas up to about 100 ha [33]. Previous EPIC applications deal primarily with managed agricultural crops, but a few studies consider natural and managed shrubland and forests. EPIC was used to evaluate light interception as a predictor of water use in hedge intercrop, monocrop, and hedge monoculture (*Senna spectabilis cv. Embu*) systems in a semi-arid environment in Kenya [34]. A spatially extended version of EPIC, The Agricultural Policy/Environmental Extender (APEX) was examined to see if it could replicate the effects of silvicultural practices on streamflow and loading of sediments and nutrients in nine small Texas watersheds [35] (http://blackland.tamu.edu/models/apex). In another EPIC application, researchers returned to the same East Texas watersheds to explore N, P and herbicide losses across three undisturbed control watersheds, three conventional clearcut watersheds and three intensive clearcut watersheds [36]. None of these studies fully explores EPIC model performance for forest stand LAI and canopy height. We hypothesize here that the EPIC implementation may be sufficient for the regional-scale air quality applications suggested in the introduction.

The most critical aspects of EPIC for the present study include the calculation of LAI, the characterization of multiple, competing species, and canopy height. For a particular species, potential values of leaf expansion, final LAI, and leaf duration, determined by plant species parameters and temperature, are reduced by unfavorable air temperatures, radiation, soil water, nutrient, aluminum and aeration conditions [33]. A series of s-curve functions provide daily values of (1) maximum potential LAI as a function of plant population, (2) leaf area expansion up to the maximum value for the growing season, and (3) loss of leaf area late in the season (Supplementary Materials Tables S3–S5). While perennials mature within a single growing season, multiple growing seasons are required for trees to reach maturity. In these cases, maximum daily potential LAI adjusts to include plant response to stress, and is scaled on an annual basis to reflect the time to tree species maturity (years). Other EPIC LAI adjustments reflect growth initiation in the spring and leaf freeze damage. Day length triggers leaf initialization. Once triggered, leaf expansion is driven by heat units, calculated using a species-specific base temperature. LAI reductions (i.e., leaf damage) occurs when daily minimum temperatures drop below $-1\ ^\circ$C. A simple exponential function of species age and height at maturity provides annual tree height.

The plant competition component of EPIC adapts expressions developed for the ALMANAC model. Up to 10 plant species can compete for light, water and nutrients. EPIC simulates light interception by leaf canopies using Beer's law [37] and the leaf area index (LAI) of the total canopy. No regular row spacing is assumed in a natural forest stand, and so EPIC assigns the light extinction coefficient (k) a value of 0.685 for all forest stand species. Light competition is then simulated as a function of both LAI and canopy height. This is in contrast to the ALMANAC approach that uses different values of k for different species and for different row spacings [28].

EPIC multi-species simulation of tree growth and decline also responds to water and nutrient stress. The water balance consists of separate transpiration calculations for each species, with each using the water it needs if sufficient water is present in the common/shared rooting zone. The nutrient balance (N and P) allows each species to acquire sufficient nutrients to meet its demands if adequate quantities are available in the common/shared rooting zone. If nutrients or water is limited, the provisioning order rotates each day so that no single species receives a competitive advantage. Table 2 provides an example of water, nitrogen and temperature stress calculations for tree species simulated by EPIC at the Appomattox field site during 2002. In contrast, ALMANAC simulates species-specific competitive advantage for water and nutrients based on physical characteristics such as age, rooting depth and extent.

Table 2. Annual summary of the number of water, nitrogen and temperature stress days for tree species simulated at the experimental plot during 2002 at the Appomattox field site. Multiple stress conditions may exist for each species on a single day. These counts represent the number of days on which the stress factor was dominate. Temperature stress includes both high and low temperature stresses.

Tree Species	Water (Days)	Nitrogen (Days)	Temperature (Days)
Loblolly pine	39	1	125
Sweetgum	39	1	190
White oak	37	0	208
Red maple	41	0	125

3. Results

3.1. Model Calibration

EPIC LAI calibration involves modification of distributed plant parameter values. Some of these parameter values reflect process-level understanding, field data and published literature, while others reflect species similarity. Plant parameter values for tree species added to or modified from the distributed EPIC for this study derive from information contained in the USDA Silvics manual [38] which describes each species geographic distribution, shade tolerance, maximum canopy height and seasonal growth pattern and temperature sensitivity. These initial values were then refined (calibrated) using field observations of stand-level LAI and maximum canopy height while maintaining plausible physical and process driven relationships. Table 3 provides the observation set used to calibrate the model. Section 2.3 describes soil and observed daily temperature and precipitation inputs to EPIC at the four forest sampling locations. We then let EPIC simulate values of daily average wind speed, relative humidity and radiation using statistical moments of climatological data, e.g., a statistical weather generator [33]. Long-term average nitrogen concentrations in precipitation are estimated using NADP data (http://nadp.sws.uiuc.edu). Calibration is complete when there is no further improvement in stand-level LAI correlation with observed LAI, LAI bias and canopy height with further parameter modification.

Table 3. Observed LAI mean and standard deviation used in EPIC calibration and evaluation. Q indicates the field site quadrat from which the sample is taken (Note: * LAI collection dates after mechanic removal of understory, data not used).

Site	Quad	Date	LAI (Mean)	LAI (Std Dev)
Fairystone	Q1	1-May	2.14	0.24
	Q2	1-May	1.75	0.35
	Q1	25-June	2.62	0.32
	Q2	25-June	2.67	0.38
	Q3	8-July	2.33	0.41
	Q4	8-July	2.54	0.35
	Q4	1-September	2.51	0.4
Umstead	Q1	5-April	0.71	0.09
	Q1	13-April	1.34	0.67
	Q1	23-April	2.61	0.49

Table 3. *Cont.*

Site	Quad	Date	LAI (Mean)	LAI (Std Dev)
	Q1	21-October	2.8	0.58
Appomattox	Q1	6-March	1.49	0.13
	Q1	23-May	1.88	0.18
	Q1	30-July	2.5	0.25
	Q1	6-August *	2.17	0.08
Hertford	Q1	5-March	1.73	0.14
	Q1	9-April	1.67	0.17
	Q1	18-June	2.4	0.32
	Q1	25-July	2.33	0.27
	Q1	5-August *	2.11	0.24

3.2. Calibrated Model Results

Table 4 summarizes site species characteristics for the simulation. Species age at biological maturity is a calibrated result, but is constrained to reflect ranges reported in USDA (1990). The letter "C" indicates a species calibration site. The letter "V" indicates a species verification site. Note that data were insufficient to verify the calibration of some species. Original (uncalibrated) and final (calibrated) EPIC tree species parameter values are provided in Supplementary Materials Tables S3–S5. Figure 4 compares calibrated and uncalibrated EPIC simulation results to observed LAI values across all four sites. Calibration improves overall simulated R^2 from 0.005 to 0.83 and reduces mean bias from -0.19 to 0.008. Uncalibrated model behavior at hardwood (Umstead and Fairystone) and Pine dominated (Appomattox and Hertford) sites are distinctly different. This difference relates to an unrealistically large uncalibrated value of maximum LAI, 5.0 for Loblolly Pine. When these metrics are re-calculated just for the hardwood stands, the uncalibrated correlation increases to 0.74 but the uncalibrated mean bias increases to -0.218, indicating the need for additional calibration even for hardwood-dominated stands.

Overall calibrated model performance is important, but we have also identified the importance of accurate, temporally resolved LAI. Figure 4 and the sections that follow examine the calibrated results at each site within the 2002 the growing season. Although no species specific LAI observations are available, simulated seasonal LAI values are provided for each species modeled within the stand to confirm that species seasonality and LAI dynamics are physically plausible for that species as well as relative to other modeled stand species.

Table 4. EPIC species and management information for each simulation location. C = calibration, V = verification, Ch.Oak = Chestnut Oak, Am. Holly = American Holly, Bl Oak = Black Oak, N Red Oak = Northern Red Oak.

	Loblolly Pine	*White Oak*	*Red Maple*	*Sweetgum*	*Chestnut Oak*	*Pignut Hickory*	*American Holly*	*Yellow Poplar*	*Black Oak*
Appomattox	C	C	C	C					
Bio Maturity (years)	55	175	100	60					
Normalized Density (stems)	1246	1655	1805	150					
Species %	25.7	34.1	37.2	3					
Stand Age	19	19	19	19					
Hertford	V		V	V			C		
Bio Maturity (years)	55		100	60			100		
Normalized Density (stems)	1482		2862	668			1626		

Table 4. *Cont.*

	Loblolly Pine	White Oak	Red Maple	Sweetgum	Chestnut Oak	Pignut Hickory	American Holly	Yellow Poplar	Black Oak
Species %	22.3		43.1	10.1			24.5		
Stand Age	20		20	20			20		
Fairystone			V		C	C			
Bio Maturity (years)			100		150	100			
Normalized Density (stems)			120		307	39			
Species %			25.7		65.9	8.4			
Stand Age			80		80	80			
Umstead		V						C *	C
Bio Maturity (years)		175						100	100
Normalized Density (stems)		184						1	86
Species %		30.5						0.1	14.3
Stand Age		80						80	80

* This species was included to capture available species-specific canopy height observations for this site.

3.2.1. Appomattox

Appomattox serves as the calibration site for Loblolly Pine, White Oak (*Quercus alba*), Red Maple (*Acer rubrum*) and Sweetgum (*Liquidamber styraciflua*). Normalizing the observed species densities for the four EPIC species subset upward to the observed total stand density approximates within-stand competition for resources at each experimental location. Although the plot description indicates that Loblolly Pine reaches economic maturity i.e., the point when value increase no longer meets or exceeds net return, in 20 years, biological maturity i.e., maximum merchantable volume, occurs later. Literature suggests an average maturity age of about 40 years [38]. Calibration at this site suggests a slightly longer time to maturity (i.e., 55 years) (Table 4).

Figure 5A summarizes the simulation results for Appomattox. Black squares indicate the mean observed stand-level LAI across samples within the quadrat (Table 3). The whiskers represent one standard deviation (± 1 STD) across samples. Loblolly Pine dominates overall LAI. Comparison of observed and simulated LAI indicates a small negative bias (-0.019), and an R^2 value of 0.91. Simulated canopy height is biased 3% high.

A cold temperature damage function reduces EPIC LAI when the minimum daily temperature lies below $-1\,°C$. White Oak early season LAI losses reflect damage from low temperature events following a period of mean temperatures exceeding the White Oak growth threshold (2 °C). EPIC parameters define two points on a frost damage curve. The calibrated white oak damage function predicts that 10% of LAI is lost for each day the minimum temperature is $-5\,°C$, and 50% LAI loss for each day the minimum temperature is $-15\,°C$. Loss estimates change linearly between these two limits, and are more likely to understate frost damage than to overstate it [33]. In contrast, calibrated Loblolly Pine parameters suggest only a 1% per day LAI loss at $-5\,°C$ and a 3% per day LAI loss at $-15\,°C$. There are several freeze damage events during the month of February, and an isolated single day damaging freeze event on March 23 (Figure 3).

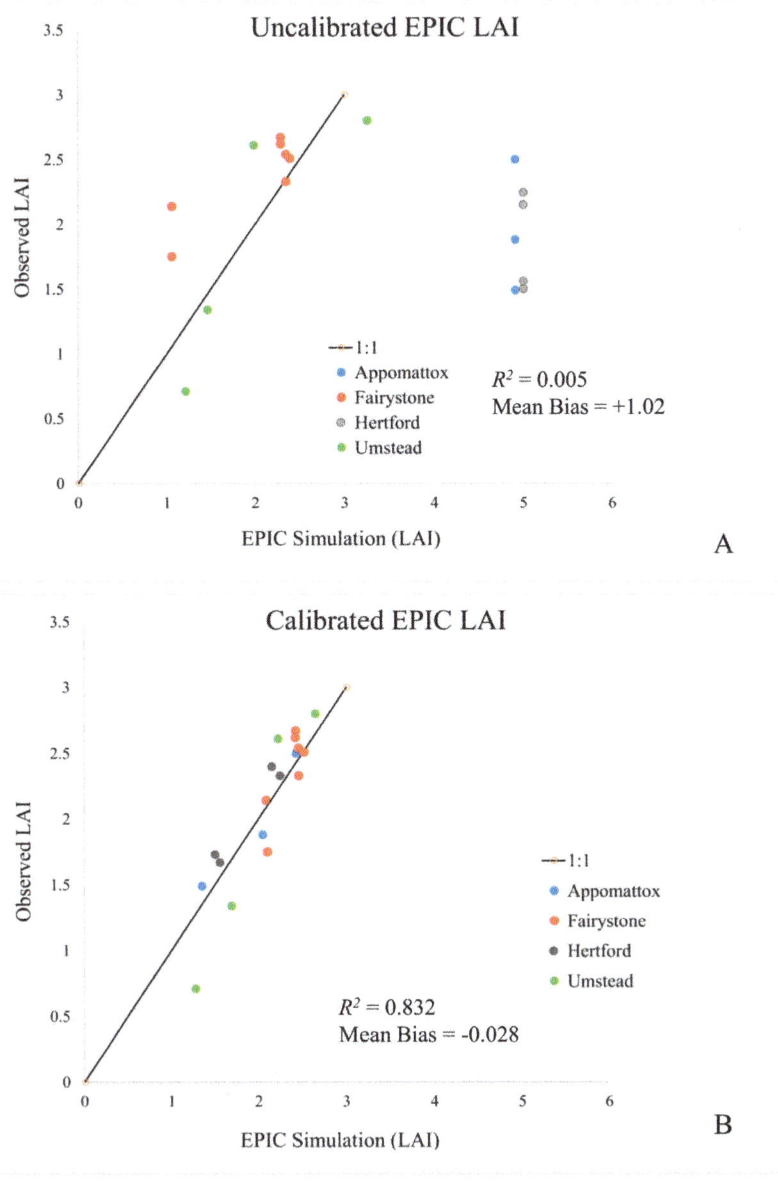

Figure 4. Comparison of (**A**) uncalibrated and (**B**) calibrated EPIC simulated LAI and observation.

3.2.2. Hertford

Hertford (Figure 5B) serves as the calibration site for American Holly (*Ilex opaca*), and a verification site for Loblolly Pine, Red Maple and Sweetgum. Although Hertford appears to experience some of the coldest spring temperatures of the four sites (Figure 3), Loblolly Pine is relatively cold tolerant. The stand level EPIC LAI estimates lie within ±1 STD of the field observations, are biased slightly low (−0.17) and have an R^2 value of 0.96. Dominant species (Pine) canopy height is biased 5% high.

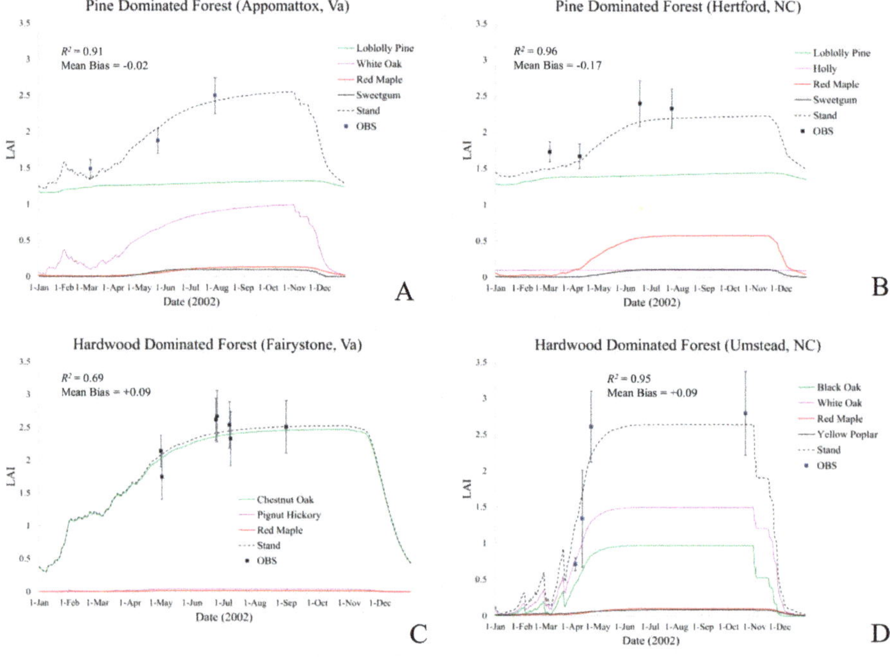

Figure 5. EPIC simulation results at (**A**) Appomattox stand year 19, (**B**) Hertford stand year 20, (**C**) Fairystone stand year 80, and (**D**) Umstead stand year 80. Error bars represent ±1 standard deviations across experimental plot data. R^2 is the simulated and observed data correlation and MB is the mean bias. (Note: 'OBS'—observed)

3.2.3. Fairystone

Fairystone serves as a verification site for Red Maple and a calibration site for Chestnut Oak (*Quercus prinus*) and Pignut Hickory (*Carya glabra*). Calibrated results are shown in Figure 5C. Chestnut Oak completely dominates the stand LAI. Early LAI variability indicate a series of freeze events during February and March. Chestnut Oak is relatively sensitive to low temperatures and, with its extreme dominance as regards stand LAI, there is significant variability in the early stand-level LAI. Model correlation is the poorest here as compared to the other sites, with simulated LAI R^2 of 0.69, but mean bias is quite good at 0.09 and canopy height of 16.33 m falling well within the reported range for this site (14.6–22.1 m). Examination of plot Figure 5C illustrates one shortcoming of this simulated approach which is an inability to capture unusual events beyond heat, cold and excessive or insufficient moisture. In this case, there is an observed anomalous LAI drop between the 25 June and 8 July observations. Further investigation indicates several 2 to 4 cm precipitation events during this period suggesting the potential for localized leaf loss associated with high wind or hail. Our simulation model is unable to account for these kinds of episodic losses.

3.2.4. Umstead

Umstead serves as a verification site for White Oak and a calibration site for Yellow Poplar (*Liriodendron tulipifera*), Black Oak (*Quercus velutina*) and Northern Red Oak (*Quercus rubra*). Calibrated results are shown in Figure 5D. The oak species are the primary contributors to stand level LAI. Like the Appomattox and Fairystone sites, there are indications of late winter, early spring stress that temporarily reduces leaf area.

Stand level EPIC estimates fall within ± 1 STD of the observations with the exception of 23 April, produce a mean bias of 0.09 and an R^2 value of 0.95. Canopy height is biased 9% high.

4. Discussion

While the results presented in Section 3 suggest EPIC has an ability to simulate mixed forest stand LAI at the experimental plot scale, and previous EPIC applications have been performed for small watersheds, regional air quality models require simulations for much larger areas for which detailed information regarding stand characteristics needed by the EPIC model is often lacking. In addition, both experimental sites and SWAT land units often reflect relatively even-aged stands, i.e., the site is cleared at the start of an experiment. This will not be the case on a larger, regional basis and unmanaged or minimally managed stands. The discussion that follows illustrates one means of addressing these challenges.

The EPIC model requires information regarding species, species density and species age. It then uses this information to model a single, homogeneous mixture of competing forest species. Neither characterization is likely to precisely represent any "real world" natural stand, but can serve to bound the range of possible stand-level LAI estimates. Both areal coverage and species density are rarely available for the same experimental site, but the US Forest Service Forest Inventory and Analysis (FIA) plot calculator (http://apps.fs.fed.us/Evalidator/evalidator.jsp) can provide a consistent, reproducible estimate of species area and density information for a ~50 mi^2 (~130 km^2) area (4 mi or 6.4 km radius circle) surrounding the Appomattox field site (Table 5). Approximately 57% of the area within this radius is reported as being forested.

Table 5. FIA plot calculator results for the area surrounding the Appomattox plot site, 2008–2013. TPH = Trees per hectare (percent of stand total).

	Timberland (Ha)	TPH	*Age Class* *1–20 year* %	*Age Class* *21–40 year* %	*Age Class* *41–60 year* %
Loblolly pine	14,755	931 (52)	100		
Virginia Pine	3726	418 (24)	100		
Mixed Oak	12,007	90 (5)			100
Mixed upland hardwoods	14,717	339 (19)		100	
Total Forested	**18,294**				
Total Area	**32,170**				

The FIA summary data are aggregated into three species groupings: (1) Loblolly Pine and Virginia Pine (*Pinus virginiana*) represented by Loblolly Pine; (2) mixed oak species comprised of Chestnut, Black and Red Oaks, represented by Black Oak; and (3) mixed upland hardwoods, represented by Red Maple. Stem densities for the groupings and stem fraction in trees per hectare (TPH) are 1349 (76%), 90 (5%) and 339 (19%). This compares to the Appomattox experimental plot represented by Loblolly Pine (26%), White Oak (34%), Red Maple (37%) and Sweetgum (3%). Stand densities are 1778 TPH for the FIA area stand compared to 4856 TPH for the Appomattox experimental plot. The FIA species age distribution suggests a 60-year simulation period with Black Oaks "planted" in year 1 (1942) of the simulation, Red Maples "planted" in year 20 (1961) of the simulation and Loblolly Pine "planted" in year 40 (1981) of the 60-year simulation. The FIA summary suggests relatively even ages within each species. If this were not the case, a range of ages within a species could be simulated by introducing ("planting") new saplings each year within a desired establishment window. With some simplification, this use of FIA data is similar to that described for boreal forests [29]. For instance, this research introduces a temporally dynamic function to simulate population dynamics that is missing from our example [29] and other research suggests the use of temporally variable growth parameters to represent rapidly developing or "short-rotation" species [39].

Figure 6A shows simulated LAI for the FIA area surrounding the Appomattox field site. For 2002 conditions, the substantial number of Pine provides a relatively constant base LAI of 1.5 throughout the winter months. Black Oak LAI provides the bulk of the stand seasonal signature, with green-up well under-way by April 1 and rapid brown-down (senescence) in late October. Figure 6B compares our plot stand LAI in 2002 to our FIA representation. The seasonal peak LAI value is similar across stands, with the FIA stand's greater maturity compensating for its lower overall stem density. The simulated Pine LAI is similar across the two simulations (plot value of ~1.2 and FIA value of ~1.4). This is, perhaps a bit surprising since pine are introduced into an existing hardwood stand for the FIA simulation but all species are "planted" simultaneously in the plot simulation. This result, however, appears to be supported by [40] who report that hardwoods may facilitate longleaf pine seedling establishment, as opposed to suppressing it, at hardwood densities as high as 1400 stems·ha^{-1}. This simulated species interaction is explored further in Figure 6C, which follows the FIA stand development from 1985 through 2005 (post pine introduction). Changes in stand level LAI magnitude and overall seasonal shape over time reflect simulated stand evolution and weather-driven variability. For instance, mean March minimum temperatures increase slightly between 1985 (~1°C) and 2005 (2.4°C). The simulated LAI seasonal pattern is guided by the accumulation of heat units above a species-specific "base" until a fixed annual total is reached. More rapid heat unit accumulation in the spring means that the annual total may be reached earlier in the year, effectively shifting brown-down initiation earlier as well. Changes such as these could be important for regional simulation of air quality since, as mentioned previously, LAI influences pollutant removal through gas phase deposition, and the magnitude and timing of pollutant precursor emissions can influence subsequent pollutant production and destruction.

An emerging research area to which improved areal LAI estimates could make a significant contribution is episodic land use change associated with wildland fires and prescribed burns used as part of land and resource management (https://www.fs.fed.us/fire/). Figure 7 illustrates the magnitude of the wildfire issue in terms of number of fires and total area burned. Air quality modeling research scientists are responding by explicitly including particulate emissions generated by these fires in their simulations (e.g., [41]). The majority of these fire events are small, but 59 wildfire events, each of which were responsible for burning more than 100,000 acres (247,105 ha) were reported from January 2010 through December 2015. As our ability to simulate these events improves, it becomes important to include associated landscape changes such as surface exposure and forest stand re-growth in our estimates of subsequent pollutant deposition and precursor emission.

EPIC is an attractive option for capturing large-scale interactions between forest land covers and air quality because it is relatively easy to implement and has already been successfully coupled with a regional air quality model [42]. There are, however, other forest ecosystem options that could be considered such as the LANDIS and LANDIS-II models [43,44]. It is always challenging to balance the desire for ecological and biogeochemical realism against the additional resources required for their inclusion. Additional research is needed to determine the level of process detail needed to adequately characterize these complex air and land surface interactions for regional air quality applications.

Finally, modeled LAI has the potential to augment satellite-derived estimates of this same parameter. In a perfect world, satellite estimates would provide reliable LAI in magnitude and without data drops, however this is almost never the case across all forested types and biomes. As an example, the Bigfoot MODIS Validation Project found agreement between validation data and MODIS LAI at low levels of LAI but was problematic at higher biomass levels [45]. Also, they found significant differences dependent on the algorithm pathway chosen, which was a product of atmospheric interference (i.e., cloud contamination) and the number of quality scenes acquired in an 8-day chronosequence. EPIC or any other validated biomass model may work in combination with the satellite feed by constraining or inflating LAI values to reasonable figures based on biases observed in these validation studies [46]. Thus, the modeled LAI could provide bias corrections where the satellite-derived LAI could provide the timing of green-up and senescence and relative seasonal changes in LAI.

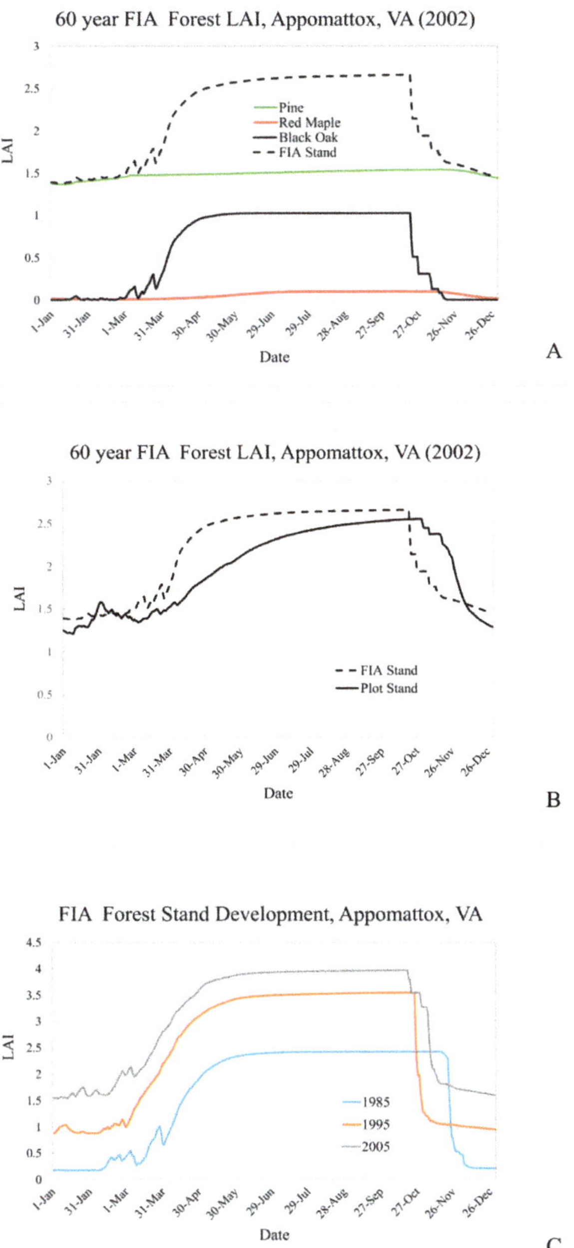

Figure 6. EPIC simulation results for (**A**) FIA stand characterization for the area surrounding the Appomattox, VA experimental plot, (**B**) a comparison of FIA area and plot-level stands and (**C**) FIA stand characterization for the area surrounding Appomattox, VA 1985 through 2006.

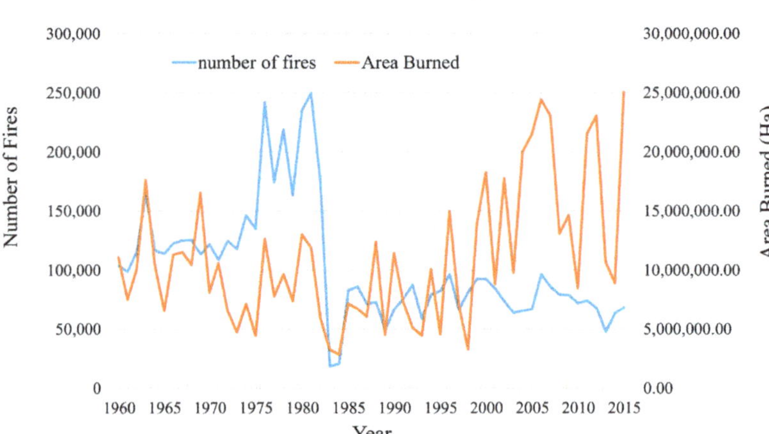

U.S. Wildfire History

Figure 7. U.S. wildfire history, 1960–2014. Source (https://www.nifc.gov/fireInfo/fireInfo_statistics.html).

5. Conclusions

This study has explored the calibration of a semi-empirical, process-based biogeochemical model (EPIC) to estimate stand-level LAI and canopy height in four unfertilized mixed forest sites located in North Carolina and Virginia. Measurements of forest composition (species and number), LAI, diameter (DBH), height and basal area were recorded at each site during the 2002 field season. Calibration and verification results are good at these sites, with modeled to observed LAI correlations ranging from 0.61 to 0.96, correct identification of dominant species evolution, and dominant species (canopy) height estimate within 10% of observation (Table 6). Across all sites and observations, calibration produces an R^2 value of 0.83 and mean bias of +0.008. These results suggest that EPIC offers an LAI estimate that agrees with observations while, at the same time better representing ecosystem processes and dynamic temporal patterns that respond to current and future meteorological conditions. All model calibrations reported here assume a uniform mixture of tree species. This appears to be a reasonable assumption for these stands, but more heavily managed mono-species stands are also easily simulated.

Table 6. Observed and simulated LAI correlation, mean bias and canopy height.

Stand Site	Obs.	LAI R^2	LAI (Mean Bias)	Obs. Dominant Height (m)	EPIC Dominant Species Height (m)
Appomattox	3	0.91	−0.02	15.9	14.9, Loblolly Pine
Hertford	5	0.96	−0.17	14.3	17.1, Loblolly Pine
Fairystone	7	0.69	0.09	14.6–22.1	18.0, Chestnut Oak
Umstead	4	0.95	0.09	12.8	12.3, White Oak

EPIC LAI estimates in regional air quality models such as CMAQ, but additional work is needed to operationalize this approach. For instance, we were able to calibrate and validate only a limited subset of species for a single geographic region. Calibration of additional species and validation across more geographically diverse settings and forest landscapes is needed, e.g., [27]. While application of the EPIC LAI model across the continental U.S., a common CMAQ modeling domain, could require significant resources, this preliminary analysis suggests that it is technically feasible to define general eco-system-level stand composition profiles appropriate for larger geographic or ecological areas. Validation and evaluation of such areal estimates will be needed, but even satellite estimates of areal LAI are prone to error, so that a combination of remotely sensed data and measured atmospheric chemical fluxes may be appropriate. More temporally and spatially detailed characterization of vegetation cover should lead to more realistic atmospheric flux (precursor emission and chemical deposition) and ecosystem exposure estimation. Additional model evaluation is needed in order to better address long-term air quality trends and inter-annual variability that include emerging landscape drivers such as wildland and managed fires.

Supplementary Materials: The following are available online at www.mdpi.com/1999-4907/9/1/26/s1, Figure S1: Appomattox, VA site location. Image on left is a color infrared Ikonos image with plot location (Q1) depicted within the loblolly pine stand (dark red tone). Leaf-off and leaf-on images are shown on the right title, Figure S2: Deciduous forest composition for the Appomattox LAI validation site, dominant-codominant, intermediate, and suppressed canopy, Figure S3: Hertford, NC site location. Image on left is a color infrared Ikonos image with plot location (Q1) depicted within the loblolly pine stand (dark red tone). Leaf-off and leaf-on images are shown on the right, Figure S4: Deciduous forest composition for the Hertford LAI validation site, dominant-codominant, intermediate, and suppressed canopy, Figure S5: Fairystone, VA site location. Image on left is a natural color digital ortho-quarter quadrangle image with plot locations (Q1–Q4) depicted within the oak-hickory hardwood stand. Leaf-off and leaf-on images are shown on the right, Figure S6: Deciduous forest composition for the Fairystone LAI validation site, dominant-codominant, intermediate, and suppressed canopy, Figure S7: Umstead, NC site location. Image on left is a natural color digital ortho-quarter quadrangle image with plot location Q1 depicted within the oak-hickory hardwood stand. Leaf-off and leaf-on images are shown on the right, Figure S8: Deciduous forest composition for the Fairystone LAI validation site, dominant-codominant, intermediate, and suppressed canopy, Table S1: Forest stand structural attributes for 4 sites, Table S2: Initial Cecil (1292NC0018VAC) soil profile used as input to the EPIC model at all forest calibration and verification sites, Table S3: Selected initial EPIC crop parameter values for tree species simulated at the four forest field sample sites. An entry of N/A indicates no initial parameter values were available, Table S4: Selected calibrated EPIC crop parameter values for tree species simulated at the four forest field sample sites, Table S5: EPIC variable Key (https://www.nrcs.usda.gov/Internet/FSE_DOCUMENTS/nrcs143_012924.pdf)

Acknowledgments: The views expressed in this article are those of the author(s) and do not necessarily represent the views or policies of the U.S. Environmental Protection Agency. This research did not receive any specific grant from funding agencies in the public, commercial, or not-for-profit sectors.

Author Contributions: John Iiames and Ellen Cooter conceived and designed this study and contributed equally to the writing of this paper. John Iiames designed the in situ data collection of LAI and the post processing of this data. Ellen Cooter designed the EPIC calibration and ran the model. Donna Schwede and Ellen Cooter provided CMAQ and EPIC model runs for this work. Jimmy Williams contributed background material on EPIC modelled forest applications.

Conflicts of Interest: The authors declare no conflict of interest.

References

1. Cooter, E.J.; Rae, A.; Bruins, R.; Schwede, D.; Dennis, R. The Role of the Atmosphere in the Provision of air-Ecosystem Services. *Sci. Total Environ.* **2013**, *448*, 197–208. [CrossRef] [PubMed]

2. Black, T.A.; Chen, W.J.; Barr, A.G.; Arain, M.A.; Chen, Z.; Nesic, Z.; Hogg, E.H.; Neumann, H.H.; Yang, P.C. Increased carbon sequestration by a boreal deciduous forest in years with a warm spring. *Geophys. Res. Lett.* **2000**, *27*, 1271–1274. [CrossRef]

3. Karl, T.; Harley, P.; Emmons, L.; Thornton, B.; Guenther, A.; Basu, C.; Turnipseed, A.; Jardine, K. Efficient atmospheric cleansing of oxidized organic trace gases by vegetation. *Science* **2010**, *330*, 816–819. [CrossRef] [PubMed]

4. Nichol, J.; Wong, M.S. Estimation of ambient BVOC emissions using remote sensing techniques. *Atmos. Environ.* **2011**, *45*, 2937–2943. [CrossRef]

5. Levis, S.; Foley, J.A.; Pollard, D. Potential high-latitude vegetation feedbacks on CO_2-induced climate change. *Geophys. Res. Lett.* **1999**, *26*, 747–750. [CrossRef]

6. Stull, R.B. *An Introduction to Boundary Layer Meteorology*; Kluwer Academic Publishers: Dordrecht, The Netherlands, 1988; pp. 1–665.

7.   Byun, D.W.; Schere, K.L. Review of the governing equations, computational algorithms, and other components of the models-3 Community Multiscale Air Quality (CMAQ) modeling system. *Appl. Mech. Rev.* **2006**, *59*, 51–77. [CrossRef]

8.   Meyers, T.P.; Finkelstein, P.; Clarke, J.; Ellestad, T.G.; Sims, P.F. Description and evaluation of a multilayer model for inferring dry deposition using standard meteorological measurements. *J. Geophys. Res.* **1998**, *103*, 22645–22661. [CrossRef]

9.   Pleim, J.; Ran, L. Surface Flux Modeling for Air Quality Applications. *Atmosphere* **2011**, *2*, 271–302. [CrossRef]

10.   Sickles, J.E.; Shadwick, D.E. Air quality and atmospheric deposition in the eastern US: 20 years of change. *Atmos. Chem. Phys.* **2015**, *5*, 173–197. [CrossRef]

11.   Dentener, F.; Vet, R.; Dennis, R.L.; Du, E.; Kulshretha, U.C.; Galy-Lacaus, C. Progress in Monitoring and Modelling Estimates of Nitrogen Deposition at Local, Regional and Global Scales. In *Nitrogen Deposition, Critical Loads and Biodiversity*; Sutton, M.A., Mason, K.E., Sheppard, L.J., Sverdrup, H., Haeuber, R., Hicks, K.W., Eds.; Springer: Dordrecht, The Netherlands, 2014; pp. 7–22.

12.   Cooter, E.; Schwede, D. Sensitivity of the National Oceanic and Atmospheric Administration multilayer model to instrument error and parameterization uncertainty. *J. Geophys. Res.* **2000**, *105*, 6695–6707. [CrossRef]

13.   Skamarock, W.C.; Klemp, J.B.; Dudhia, J.; Gill, D.O.; Barker, D.M.; Duda, M.G.; Huang, X.-Y.; Wang, W.; Powers, J.G. *A Description of the Advanced Research WRF Version 3*; National Center for Atmospheric Research: Boulder, CO, USA, 2008; p. 125.

14.   Pleim, J.E.; Xiu, A.; Finkelstein, P.L.; Otte, T.L. A coupled land-surface and dry deposition model and comparison to field measurements of surface heat, moisture and ozone fluxes. *Water Air Soil Pollut. Focus* **2001**, *1*, 243–252. [CrossRef]

15.   Ran, L.; Gilliam, R.; Binkowski, F.S.; Xiu, A.; Pleim, J.; Band, L. Sensitivity of the WRF/CMAQ modeling system to MODIS LAI, FPAR, and albedo. *J. Geophys. Res. Atmos.* **2015**, *120*, 8491–8511. [CrossRef]

16.   Bonan, G.B.; Oleson, K.W.; Fisher, R.A.; Lasslop, G.; Reichstein, M. Reconciling leaf physiological traits and canopy flux data: Use of the TRY and FLUXNET databases in the Community Land Model version 4. *J. Geophys. Res. G Biogeosci.* **2012**, *117*. [CrossRef]

17.   Lloyd, J.; Patiño, S.; Paiva, R.Q.; Nardoto, G.B.; Quesada, C.A.; Santos, A.J.B.; Baker, T.R.; Brand, W.A.; Hilke, I.; Gielmann, H.; et al. Optimization of photosynthetic carbon gain and within-canopy gradients of associated foliar traits for Amazon forest trees. *Biogeosciences* **2010**, *7*, 1833–1859. [CrossRef]

18.   Mercado, L.; Lloyd, J.; Carswell, F.; Malhi, Y.; Meir, P.; Nobre, A.D. Modelling Amazonian forest eddy covariance data: A comparison of big leaf versus sun/shade models for the C-14 tower at Manaus I. Canopy photosynthesis. *Acta Amazon.* **2006**, *36*, 69–82. [CrossRef]

19.   Mercado, L.M.; Lloyd, J.; Dolman, A.J.; Sitch, S.; Patiño, S. Modelling basin-wide variations in Amazon forest productivity—Part 1: Model calibration, evaluation and upscaling functions for canopy photosynthesis. *Biogeosciences*, **2009**, *6*, 1247–1272. [CrossRef]

20.   Goto, Y. Improved Vegetation Characterization and Freeze Statistics in a Regional Spectral Model for the Florida Citrus Farming Region. Ph.D. Thesis, The Florida State University, Tallahassee, FL, USA, 2008.

21.   Ran, L.; Pleim, J.; Gilliam, R.; Binkowski, F.S.; Hogrefe, C.; Band, L. Improved meteorology from an updated WRF/CMAQ modeling system with MODIS vegetation and albedo. *J. Geophys. Res.* **2016**, *121*, 2393–2415. [CrossRef]

22.   Opie, J.E. Predictability of Individual Tree Growth Using Various Definitions of Competing Basal Area. *For. Sci.* **1968**, *14*, 314–323.

23.   Chen, J.M.; Black, T.A. Foliage area and architecture of plant canopies from sunfleck size distributions. *Agric. For. Meteorol.* **1992**, *60*, 249–266. [CrossRef]

24.   Iiames, J.S.; Congalton, R.G.; Pilant, A.N.; Lewis, T.E. Validation of an integrated estimation of Loblolly pine (*Pinus taeda* L.) leaf area index (LAI) utilizing two indirect optical methods in the southeastern United States. *South. J. Appl. For.* **2008**, *32*, 101–110.

25.   Leblanc, S.G. *DHP-TRACWin Manual*; Canada Centre for Remote Sensing, Natural Resources Canada: Ottawa, ON, Canada, 2008.

26. Frazer, G.W.; Canham, C.D.; Lertzman, K.P. *Gap Light Analyzer (GLA), Version 2.0: Imaging Software to Extract Canopy Structure and Gap Light Transmission Indices from True-Color Fisheye Photographs, User's Manual and Program Documentation*; Simon Fraser University: Burnaby, BC, Canada; The Institute of Ecosystem Studies: Millbrook, New York, NY, USA, 1999.

27. Kiniry, J.R.; Williams, J.R.; Gassman, P.W.; Debaeke, P. A general, process-oriented model for two competing plant species. *Trans. ASAE* **1992**, *35*, 801–810. [CrossRef]

28. Kiniry, J.R.; Macdonald, J.D.; Kemanian, A.R.; Watson, B.; Putz, G.; Prepas, E.E. Plant growth simulation for landscape-scale hydrological modelling. *Hydrol. Sci. J.* **2008**, *53*, 1030–1042. [CrossRef]

29. MacDonald, J.D.; Kiniry, J.R.; Putz, G.; Prepas, E.E. A multi-species, process based vegetation simulation module to simulate successional forest regrowth after forest disturbance in daily time step hydrological transport models. *J. Envrion. Eng.* **2008**, *7*, 127–143. [CrossRef]

30. Gassman, P.W.; Reyes, M.R.; Green, C.H.; Arnold, J.G. The soil and water assessment tool: Historical development, applications, and future research directions. *Trans. ASABE* **2007**, *504*, 1211–1250. [CrossRef]

31. Putz, G.; Burke, J.M.; Smith, D.W.; Chanasyk, D.S.; Prepas, E.E.; Mapfuma, E. Modelling the effects of boreal forest landscape management upon streamflow and water quality: Basic concepts and considerations. *J. Environ. Eng. Sci.* **2003**, *2*, S87–S101. [CrossRef]

32. Arnold, J.G.; Fohrer, N. SWAT2000: Current capabilities and research opportunities in applied watershed modelling. *Hydrol. Process.* **2005**, *19*, 563–572. [CrossRef]

33. WWilliams, J.W.; Izaurralde, R.C.; Steglich, E.M. *Agricultural Policy/Environmental eXtender Model Theoretical Documentation Version 0806*; Blackland Research and Extension Center: Temple, TX, USA, 2012; pp. 1–131. Available online: http://epicapex.tamu.edu/files/2014/10/APEX0806-theoretical-documentation.pdf (accessed on 31 October 2017).

34. McIntyre, B.D.; Riha, S.J.; Ong, C.K. Light interception and evapotranspiration in hedgerow agroforestry systems. *Agric. For. Meteorol.* **1996**, *81*, 31–40. [CrossRef]

35. Saleh, A.; Willimas, J.R.; Wood, J.C.; Hauck, L.M.; Blackburn, W.H. Application of APEX for Forestry. *Trans. ASAE* **2004**, *47*, 751–765. [CrossRef]

36. Wang, X.; Saleh, A.; McBroom, M.W.; Williams, J.R.; Yin, L. Test of APEX for Nine Forested Watersheds in East Texas. *J. Environ. Qual.* **2007**, *36*, 983–995. [CrossRef] [PubMed]

37. Monsi, M.; Saeki, T. Uber den lichtfaktor in den pflanzengesellschaften und seine bedeutung fur die stoffproduction. *Jpn. J. Bot.* **1953**, *14*, 22–52.

38. United States Department of Agriculture Forest Service. *Silvics of North America: 1. Conifers; 2. Hardwoods. Agriculture Handbook 654*; U.S. Department of Agriculture, Forest Service: Washington, DC, USA, 1990; Volume 2, pp. 1–877.

39. Guo, T.; Engel, A.; Shao, G.; Arnold, J.G.; Srinivasn, R.; Kiniry, J.R. Functional approach to simulating short-rotation woody crops in process-based models. *BioEnergy Res.* **2015**, *8*, 1598–1613. [CrossRef]

40. Loudermilk, E.L.; Hiers, K.; Pokswinski, S.; O'Brian, J.J.; Barnett, A.; Mitchell, R.J. The path back: Oaks (*Quercus* spp.) facilitate longleaf pine (*Pinus Palustris*) seedling establishment in xeric sites. *Ecosphere* **2016**, *7*, E01361. [CrossRef]

41. Baker, K.; Woody, M.; Tonnesen, G.; Hutzell, W.; Pye, H.; Beaver, M.; Pouliot, G.; Peirce, T. Contribution of regional-scale fire events to ozone and PM 2.5 air quality estimated by photochemical modeling approaches. *Atmos. Environ.* **2016**, *140*, 539–554. [CrossRef]

42. Cooter, E.J.; Bash, J.O.; Benson, V.; Ran, L. Linking agricultural crop management and air quality models for regional national-scale nitrogen assessments. *Biogeosciences* **2012**, *9*, 4023–4035. [CrossRef]

43. Scheller, R.M.; Mladenoff, D.J. A forest growth and biomass module for a landscape simulation model, LANDIS: Design, validation, and application. *Ecol. Model.* **2004**, *180*, 211–229. [CrossRef]

44. Creutzburg, M.K.; Scheller, R.M.; Lucash, M.S.; LeDuc, S.D.; Johnson, M.G. Forest management scenarios in a changing climate: Trade-offs between carbon, timber, and old forest. *Ecol. Appl.* **2017**, *27*, 503–518. [CrossRef] [PubMed]

45. Morisette, J.; Privette, J.L.; Baret, F.; Myneni, R.B.; Nickeson, J.; Garrigues, S.; Shabanov, N.; Fernandes, R.; Leblanc, S.; Kalacska, M.; et al. Validation of global moderate-resolution LAI products: A framework proposed within CEOS Land Product Validation Subgroup. *IEEE Geosci. Remote Sens.* **2006**, *44*, 1804–1817. [CrossRef]

46. Iiames, J.S.; Congalton, R.G.; Lewis, T.E.; Pilant, A.E. Uncertainty analysis in the creation of a fine-resolution leaf area index (LAI) reference map for validation of moderate resolution LAI products. *Remote Sens.* **2015**, *7*, 1397–1421. [CrossRef]

Article

Assessing Forest Cover Dynamics and Forest Perception in the Atlantic Forest of Paraguay, Combining Remote Sensing and Household Level Data

Emmanuel Da Ponte [1,*], Benjamin Mack [2], Christian Wohlfart [2], Oscar Rodas [3], Martina Fleckenstein [4], Natascha Oppelt [1], Stefan Dech [2] and Claudia Kuenzer [2]

[1] Remote Sensing & Environmental Modelling, Department for Geography, Kiel University, 24098 Kiel, Germany; oppelt@geographie.uni-kiel.de

[2] German Remote Sensing Data Center (DFD) of the German Aerospace Center (DLR), Oberpfaffenhofen, 82234 Wessling, Germany; ben8mack@gmail.com (B.M.); Christian.Wohlfart@dlr.de (C.W.); Stefan.Dech@dlr.de (S.D.); Claudia.Kuenzer@dlr.de (C.K.)

[3] World Wide Fund for Nature (WWF) Paraguay, 1001-1925 Asuncion, Paraguay; orodas@wwf.org.py

[4] World Wide Fund for Nature (WWF) Germany, Reinhardtstraße 18, 10117 Berlin, Germany; martina.fleckenstein@wwf.de

* Correspondence: daponte@geographie.uni-kiel.de; Tel.: +49-174-594-2972

Received: 29 August 2017; Accepted: 7 October 2017; Published: 11 October 2017

Abstract: The Upper Parana Atlantic Forest (BAAPA) in Paraguay is one of the most threatened tropical forests in the world. The rapid growth of deforestation has resulted in the loss of 91% of its original cover. Numerous efforts have been made to halt deforestation activities, however farmers' perception towards the forest and its benefits has not been considered either in studies conducted so far or by policy makers. This research provides the first multi-temporal analysis of the dynamics of the forest within the BAAPA region on the one hand, and assesses the way farmers perceive the forest and how this influences forest conservation at the farm level on the other. Remote sensing data acquired from Landsat images from 1999 to 2016 were used to measure the extent of the forest cover and deforestation rates over 17 years. Farmers' influence on the dynamics of the forest was evaluated by combining earth observation data and household survey results conducted in the BAAPA region in 2016. Outcomes obtained in this study demonstrate a total loss in forest cover of 7500 km^2. Deforestation rates in protected areas were determined by management regimes. The combination of household level and remote sensing data demonstrated that forest dynamics at the farm level is influenced by farm type, the level of dependency/use of forest benefits and the level of education of forest owners. An understanding of the social value awarded to the forest is a relevant contribution towards preserving natural resources.

Keywords: BAAPA; remote sensing; household survey; forest; farm types

---

1. Introduction

Deforestation in the tropics today continues inexorably with severe implications for biodiversity conservation, climate regulation and ecosystem services such as carbon storage. The rapid expansion of the agricultural frontier, cattle ranching and illegal logging has converted the world's last remnants of tropical forest into isolated patches endangering their continuity [1]. Between 1999 and 2005, 69 million ha of forest have been lost in Latin America accounting for almost 7% of the forest cover of the continent [2]. Despite the fact that its speed has declined in comparison to previous years [3], deforestation still remains a concern. The latest studies conducted on a global level

identified Paraguay as one of the countries in Latin America with the highest deforestation rates [4,5].
The continuous anthropological pressure on natural resources has led to the loss of 90% of the forest
cover in the eastern region of the country, where the Atlantic Forest is located [6]. The Atlantic
Forest encompasses 15 ecoregions and a total area of 471,204 km^2 [7]. The ecoregion is considered to
be a biodiversity hotspot, due to the presence of numerous endemic species that are unique in the
world [8–11]. Even though the portion of the Atlantic Forest (also known as the Upper Parana Forest
(BAAPA)) within Paraguay only represents a small share of the complete geographic extension of the
ecoregion, it has been recognized as a highly diverse ecosystem [12]. According to Huang et al. [13],
the BAAPA forest cover decreased around 50% of its original cover between 1973 and 2000, in less
than 30 years [1]. Latest studies [6,14] estimated that only 10% of its original cover remains. One of the
major drivers of deforestation in the region is the expansion of mechanized agriculture and a lack of
economic opportunities for forest owners [15,16]. Economic alternatives to service wood production
(e.g., construction wood, fire woods and charcoal) are limited for the local population. As a result,
it is tempting for small-scale farmers to lease their lands to large companies that produce exclusively
monocultural crops such as soy beans and maize [17]. A common perception among farmers in the
region is that one ha of soy crops simply holds a higher economic value than one ha of native forest.
In addition, the low economic compensation that can be obtained for forest products cannot compete
with the high levels of income generated by agricultural exports [18].

Over the past decades, several governmental institutions, e.g., Forest National Institute (INFONA)
and international organizations (e.g., Food and Agriculture Organization of the United Nations (FAO),
the World Wildlife Fund (WWF) and the United States Agency for International Development (USAID)),
have used remote sensing data to assess deforestation in the BAAPA. Nevertheless, despite the existence
of numerous deforestation reports, major parts of the spatial analysis are kept in clusters and some even
considered sensitive information [19]. According to Da Ponte et al. [19], only few scientific studies have
provided a systematic analysis of forest cover change in the BAAPA region [13,20,21]. These studies
estimated the dynamics of the forest cover and forest structure by implementing solely bi-temporal
analysis based on Landsat images spanning the years 1970 to 2001 and 2003 to 2013, respectively. Even
though the discussed studies successfully identified deforestation processes and patterns with remote
sensing techniques, no attempts were made to understand the underlying drivers of change or the
effectiveness of conservation policies. No ground information that could capture local circumstances
(e.g., uses of natural resources, farm types and cultural characteristics) between forest owners has
been included in past analysis. For instance, recent studies conducted in the BAAPA [18,22] have
demonstrated that farmers' perceptions of the importance of the forest vary according to farm types.
Farmers with less economic resources depend more heavily on the forest, whereas larger farmers
consider the forest's main value to be recreational/cultural. Hence, it is to be expected for small-scale
farmers to present a higher percentage of farms exhibiting a decrease in their forest cover.

In order to address these shortcomings, in this study, a dense set of Landsat imagery is applied
on the one hand to provide the first multi temporal analysis of forest cover change in the BAAPA
region (to the knowledge of the authors) between the years 1999 and 2016. On the other, remote
sensing and household level data are combined to understand how farmers' perceptions of the forest
affects conservation practices at the farm level. The goal of this study is to measure the influence
of farmer's educational background on the dynamics of the forest, how changes in deforestation
frequency differ according to farm type (small, medium, and large), how farmers' dependency on
natural forest resources influences changes in the forest cover, and the impact farmers' participation
in conservation programs has on preservation. The outcomes obtained in this study provide useful
information when contemplating the importance of social involvement in land-use planning.

2. Data and Methods

2.1. Study Area

This study was conducted in the Upper Parana Atlantic Forest of Paraguay (BAAPA), located in the eastern region of the country. The ecoregion comprises portions of ten departments, resulting in a total area of 86,000 km^2 (see Figure 1) [23]. Almost 50% (over 3 million inhabitants) of the country's population is located within the boundaries of the BAAPA [23]. The areas of highest population density in the ecoregion are located in the east (Ciudad del Este) and south (Encarnación), whereas in the north the population decreases [23]. The climate in the Atlantic Forest is characterized by frequent rainfalls that fluctuate between 1300 to 1800 mm per year. The temperature in the region varies greatly between seasons. During summer months (December–March), the temperature can increase up to 42 °C, while over winter (May–August), it can decrease down to 0 °C. Most of the diverse biological richness of the BAAPA is distributed in the ecoregions of the Montane Forest in the North (Amambay), the central forest in the south and the Upper Parana forest in the southeast [12]. Although forest cover represents a significant portion of the natural vegetation in the ecoregion, the severe pressure from anthropological activities has degraded the forest with only a few remaining fragmented patches [13,20].

Prior to 1940, the BAAPA forest covered over 55% of the eastern region of the country (accounting for almost 9,000,000 ha). Nevertheless, uninterrupted deforestation practices resulted in the loss of 90% of its original cover [6]. Currently, 90% of the country's soy bean production on 3 million ha is located within the boundaries of the BAAPA region [24]. According to studies such as Huang et al. [13,20], causes of deforestation were related to the long-established perception of the forest as unproductive lands, the rapid expansion of the agricultural frontier and the unsustainable use of natural resources. By the year 2000, almost two-thirds of the Paraguayan Atlantic Forest was lost, with an annual average deforestation rate of 2000 km^2. The government introduced reforestation programs in the late 1990s (incentives to forestation and reforestation law 536/96) to diminish the damage done in the BAAPA, yet unfortunately, these did not obtain remarkable results. The lack of clear regulations and financial support discouraged land owners from introducing further lands into the program [25]. By 2003 at the latest, Paraguay had become the country with the second highest deforestation rate in the world [3]. In response, the Paraguayan government approved in 2004 the "Zero Deforestation Law (2524/04)" for a period of two years, which prohibited the conversion of any parts of the Atlantic forest in eastern Paraguay [14,16]. According to reports from the World Wildlife Fund [3], deforestation rates decreased drastically as a result, slowing by over 90% from 2002 (110,000 ha of forest loss per year) to 2009 (8000 ha of forest loss per year).

Legend

Itaipu watershed		Country boundaries		intersectAreas	
Ñacunday watershed		Administrative boundaries		Parlu Districts	
BAAPA area		Protected areas		Forest cover (2016)	

Figure 1. (**a**) Overview of the study area (Base layer provided by Natural Earth Community and Conservation international [26,27]); (**b**) Paraguay and the Upper Parana Atlantic Forest (BAAPA) location (source: adapted from Natural Earth [26]); (**c**) Household distribution within selected study areas.

2.2. Landsat Image Acquisition and Pre-Processing

For this study, Landsat 5 Thematic Mapper (TM), Landsat 7 Enhanced Thematic Mapper Plus (ETM+) and Landsat 8 Operational Land Imager (OLI) data were acquired between the reference years of 1999 and 2016. Taking into consideration the high temporal and spatial resolution of the Landsat images, the sensor was considered the most suitable for this research. As presented in Table 1, a total of 2775 terrain corrected (L1T) images with less than 30% cloud cover were obtained from the United States Geological Survey (USGS) archives. In order to decrease any possible noise and data gaps resulting from clouds and further atmospheric distortions, the number of satellite images to be used per classification was increased by considering data of two years for each map. This permitted to obtain a denser temporal coverage from the study region. Similar to Wohlfart et al. [28], Knauer et al. [29] and Gebhardt et al. [30], the FMASK (Function of mask) algorithm was applied over the Landsat images to identify and mask pixels classified as clouds shadows or no data (see Figure 2). The FMASK algorithm was developed to automatically detect and mask clouds, cloud shadows and snow from Landsat images by taking the spectral and textural features into consideration based on probabilistic scores [31].

Table 1. Landsat data and number of processed scenes used in this study.

Sensor	Path/Row	Acquisition Dates	Number of Scenes	Total
Landsat 5 TM	224/77	07/1999–11/2011	116	
	224/78	04/1999–11/2011	108	
	224/79	01/1999–11/2011	106	
	225/76	01/1999–11/2011	112	
	225/77	01/1999–11/2011	118	930
	225/78	01/1999–11/2011	121	
	225/79	01/1999–9/2011	127	
	226/76	02/1999–10/2011	122	
Landsat 7 ETM+	224/77	08/1999–07/2016	190	
	224/78	08/1999–08/2016	177	
	224/79	09/1999–08/2016	188	
	225/76	07/1999–07/2016	198	
	225/77	07/1999–08/2016	197	1514
	225/78	07/1999–08/2016	194	
	225/79	10/1999–08/2016	179	
	226/76	08/1999–08/2016	191	
Landsat 8 OLI	224/77	05/2013–08/2016	43	
	224/78	05/2013–08/2016	40	
	224/79	07/2013–08/2016	42	
	225/76	04/2013–07/2016	44	
	225/77	04/2013–07/2016	41	331
	225/78	04/2013–07/2016	45	
	225/79	04/2013–07/2016	43	
	226/76	04/2013–08/2016	33	

Atmospheric corrections were performed with ATCOR-3 [32] for each Landsat scene to obtain physically comparable surface reflectance information, while also integrating topographic corrections by incorporating slope and elevation information from the Shuttle Radar Topography Mission.

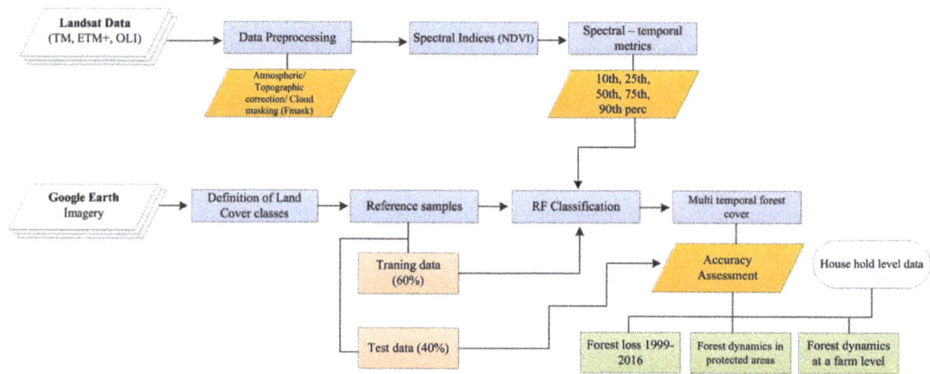

Figure 2. Workflow of remote sensing data processing and classification procedures (source: adapted from Mack et al. [33].

2.3. Spectral-Temporal Landsat Time-Series Metrics

The use of continuous spectral-temporal metrics not only has proven to solve problems related to data gaps (a consequence of clouds), but also has been applied extensively as a reliable approach for separating land cover/use classes [34–37]. A variety of different spectral-temporal metrics were estimated based on bi-annual Landsat stacks, characterizing different land cover classes for the most prominent phenological information. The procedure follows the approach as described in Mack et al. [33] and Wohlfart et al. [28]. The lack of temporal regularity of Landsat acquisitions constrains the direct quantification of phenological metrics. Therefore, several bi-annual spectral-temporal metrics were computed from the Landsat scenes in order to obtain proxies for seasonal information. For this study, several statistical image metrics were derived (percentiles of 10, 25, 50, 75, 90%) from Landsat (TM, ETM+, and OLI) observations based on the reflectances of the five bands (blue, green, red, near infrared and short-wave infrared). For each band and index, multi-year percentile differences (90% minus 10% and 75% minus 25%) were calculated. As described in Mack et al. [33] interannual minimum and maximum were neglected in order to decrease noise and further outliers. In addition, Normalized Vegetation Index (NDVI) percentiles were computed. Hence, a total of 35 multi-temporal spectral features were considered as input variables in the classifications.

2.4. Estimation of Forest Cover between Years 1999–2016

Forest/non-forest maps were produced for every year between 1999 and 2016, employing a random forest (RF) [38] classifier to generate inter-annual thematic change maps based on spectral-temporal metrics (see Figure 2). For each reference year, training samples were randomly collected over the BAAPA area, resulting in a set of at least 100 homogeneous training polygons (as suggested by Congalton and Green [39]) for each of the five land cover/use classes "forest, croplands, grasslands, urban areas, and water". Training and validation samples were well distributed over the study area to obtain the most representative coverage of land cover/uses in the region. Following the procedures of Wohlfart et al. [40], visual interpretation of very-high-resolution images (acquired from the historical imagery function of Google Earth between 1999 and 2016 [41]) was performed to define the classes of the training and validation samples. The interpretation was based not only on the image interpretation but also on local expert knowledge of the area.

The RF algorithm has been increasingly applied to conduct land cover mapping due to its performance, user friendliness and computer proficiency [42–44]. RF is a decision tree algorithm which selects random subsets of learning samples and of variables to build multiple (default value of 500) independent decision trees. Models were built and adjusted using the software R (version 3.3.1,

R Foundation for Statistical Computing, Vienna, Austria) using its random forest package [45,46]. The pixel-wise classification applies the majority vote rule from aggregated decision trees to determine the final category. In this study, RF models with 500 independent trees were built for each two-year composite, resulting in a total of eight individual models. Default values for the mrty parameters were used, which traditionally is $\sqrt{p}$, where the number of predictors in the dataset is represented by p. In order to train the RF classifier, 60% of the reference dataset served as the training input, and the remaining 40% of the samples as the verification set. The quality of each classified image was described through overall accuracies, producers' and users' accuracies, and Kappa coefficients derived from the error matrix [47]. Finally, a non-forest mask was generated by grouping all non-classes. Forest patches with an area smaller than 0.5 ha were excluded from the analysis, considering the forest definition established by FAO [48].

In order to analyze the long-term differences of forest dynamics between protected areas (of different ownership) and among farm types (small, medium and large) a long-term (bi-temporal) analysis of change was conducted by comparing forest classifications results for the reference years of 1999–2000 and 2015–2016.

2.5. Household Survey Data

For this study, a household socio-economic survey (277 households) was conducted in the BAAPA region (see Figure 1c) in January 2016 over a period of one month. Due to the large size of the BAAPA region, three sample areas were chosen to conduct the survey; the ITAIPU watershed dam (10,000 km^2) located in the north, the Ñacunday watershed (2500 km^2), and the Tavapy district (436 km^2) situated in the South (see Figure 2c).

Respondents were stratified according to the size of their farm, following the categorization applied by the Ministry of Agricultural of Paraguay in its rural censuses [49]. Farmers with land size <20 ha represent the small-scale farmers group; farmers with land size of 20–50 ha represent the medium-scale farmers group; and farmers with land size >50 ha represent large-scale farmers. In general, the survey focuses more on aspects of the rural population (e.g., job, income, education level and land size) and their relationship with the forest (e.g., how they define "a forest" their knowledge of its functions, and its importance for their livelihood), their use of forest resources and services (e.g., firewood, construction and forest farming) and conservation programs (e.g., understanding and participation in such programs). For further detailed information on the household surveys methods and results, the interested reader is referred to Da Ponte et al. [18].

2.6. Combining Household and Remote Sensing Data

Using cadastral information, long-term forest cover change results for the reference years 1999–2000 and 2015–2016 were correlated to responses acquired from the field survey. For 106 of the interviewed farmers, cadastral data was obtained from the Paraguayan National Cadastral Service (SNC). Further, information was acquired on site during the field campaigns by measuring the limits of 39 farms while it was feasible to do so; for small-scale farms in particular, the topographic and weather conditions needed to be appropriate for doing so. This resulted in a number of 145 farms where both household survey and cadastral information was available for our comparative study. This sample size can be considered representative of the study region, since according to Yamane´s equation [50], 100 samples are required to achieve a sampling accuracy of approximately 90%. See Equation (1):

$$n = \frac{N}{1 + N(e)^2} \tag{1}$$

where n represents the samples needed; N refers to the sample population; and e the sampling error (0.10).

Changes in forest cover at the farm level were assessed by applying bi-temporal change detection analysis stratifying the changes into three categories: "forest loss", "forest gain" and "no significant changes". A farm was considered to fall into the category "no significant changes" if variations in forest cover occurred between 1 and 4 pixels (0–36 ha).

3. Results

3.1. Forest Classification Accuracy

In general, classification accuracies obtained from Landsat images between the years 1999 and 2016 fluctuated from 85% to 93%, with Kappa coefficients ranging from 0.82 to 0.91 (see Table 2). The Landsat data set from 2001–2002 exhibited the highest accuracy, of which 94% of the pixels were classified correctly as forest. On the other hand, the lowest accuracy values were seen in the 2015–2016 Landsat data sets, obtaining 88% and 87%, respectively. The lower classification values could be attributed to high spectral similarities between forest areas and dense crop fields (e.g., soybean and maize plantations).

Table 2. Classification accuracies for each time step from 1999 to 2016.

Time Period	Overall	KAPPA	Producers Accuracy Forest	Users Accuracy Forest	Producers Accuracy Non-Forest	Users Accuracy Non-Forest
1990/2000	89.04%	0.85	90.15%	89.12%	89.87%	88.10%
2001/2002	93.06%	0.91	94.13%	92.28%	92.86%	93.18%
2003/2004	85.71%	0.82	86.74%	85.95%	84.67%	85.13%
2005/2006	92.86%	0.90	93.75%	92.67%	92.88%	91.43%
2007/2008	91.69%	0.89	93.09%	90.43%	91.15%	92.24%
2009/2010	91.03%	0.87	91.15%	91.08%	89.87%	91.02%
2011/2012	92.35%	0.89	94.88%	93.45%	90.72%	91.23%
2013/2014	92.13%	0.89	92.78%	92.52%	91.36%	91.75%
2015/2016	87.36%	0.83	88.40%	87.78%	87.04%	86.26%

3.2. Forest Loss Rates

In 1999/2000, over 31% (27,000 km^2) of the BAAPA area was covered by forest. As presented in Table 3 and Figure 3, the largest forest areas in the region were located in the departments of Canindeyú and San Pedro, accounting for more than 48% (over 10,000 km^2) of the total forest area in the BAAPA. The lowest levels of forest coverage were found in the departments of Paraguarí, Guairá and Concepción, together accounting for only 9% of the forest cover (around 2500 km^2). In the years 2001/2002, the forest cover in the BAAPA decreased to 29%, equivalent to 630 km^2. The departments of Canindeyú, San Pedro and Alto Parana exhibited the highest relative forest loss of 79%, with more than 500 km^2. In 2003/2004, deforestation rates increased drastically. Almost 9% (2300 km^2) of the forest was being depleted, nearly four times as much as in previous years. Similar to the trends above, the highest rates of deforestation were concentrated in the departments of Canindeyú, San Pedro, and Alto Parana, together totaling over 56% (around 1300 km^2) of the area loss.

Following the year 2004, rates of deforestation gradually decreased from 4.9% (1200 km^2) from 2005/2006 down to 2.5% (549 km^2) between 2011 and 2012, before increasing again slightly in 2015/2016 (2.9%). Overall, by the year 2016, more than 27% (7500 km^2) of forest cover was lost since 1999, at an annual deforestation rate of 1.5% (442 km^2) over the entire BAAPA area. The lowest deforestation rates were shown in the departments of Guairá (12.1%) and Paraguarí (2.4%) accounting for 1.5% (120 km^2) of the total area deforested. In contrast, the departments of San Pedro and Canindeyú consistently evidenced the highest losses, with a total forest cover loss of 41% and 33%, respectively.

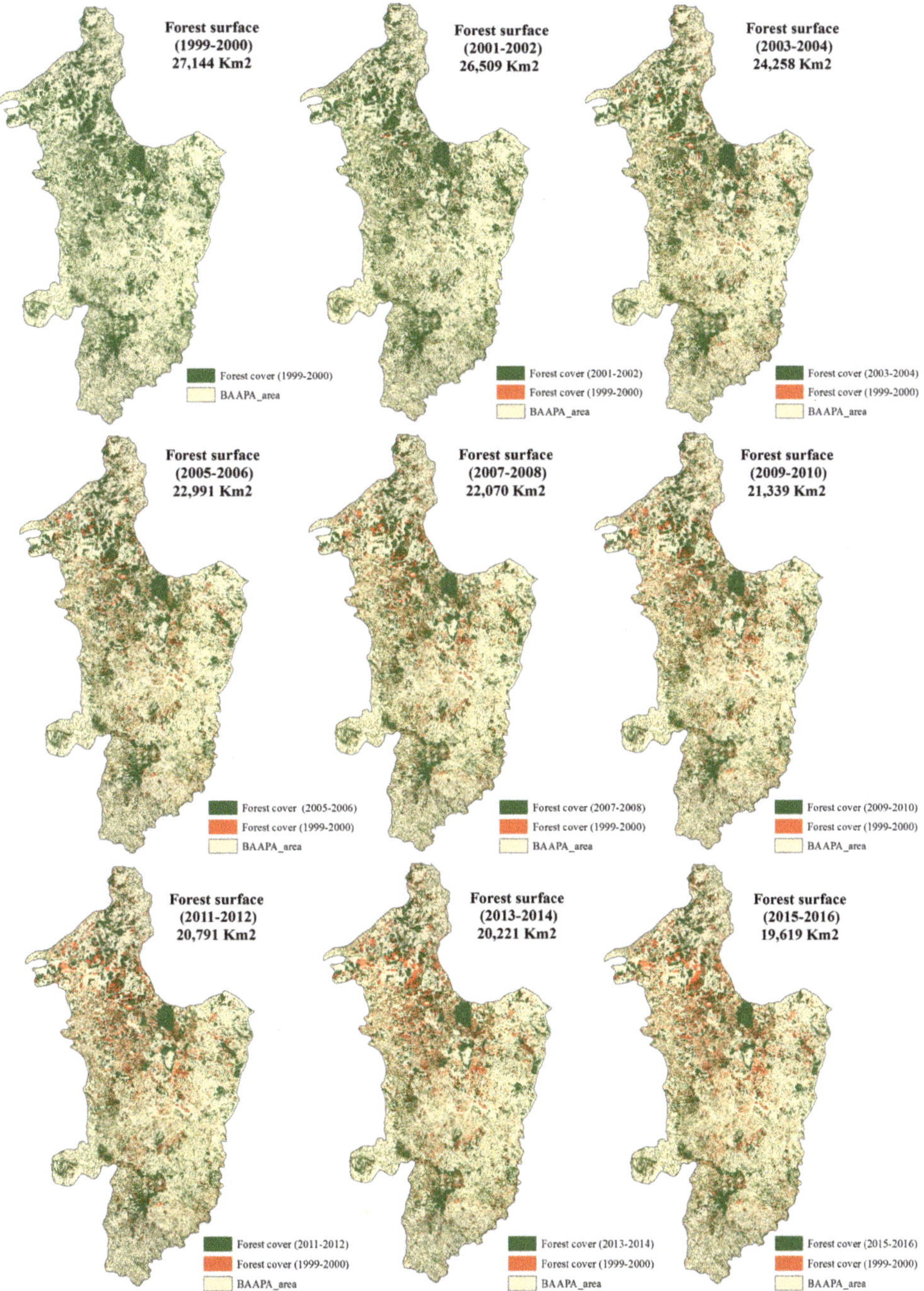

Figure 3. Deforestation results in the BAAPA region between 1999 and 2016.

Table 3. Forest cover and forest cover loss in the BAAPA region from 1999 to 2017.

Department	Forest Cover (1999–2000 km²)	%	Forest Cover (2001–2002 km²)	%	Forest Cover (2003–2004 km²)	%	Forest Cover (2005–2006 km²)	%	Forest Cover (2007–2008 km²)	%
Alto Paraná	3336	12.3	3210	11.8	2856	10.5	2747	10.1	2709	10.0
Amambay	2414	8.9	2371	8.7	2353	8.7	2144	7.9	1998	7.3
Caaguazú	3113	11.5	3069	11.3	2801	10.3	2658	9.8	2649	9.7
Caazapá	2172	8.0	2169	8.0	1901	7.0	1801	6.6	1787	6.6
Canindeyú	5812	21.4	5602	20.6	5036	18.5	4889	18.0	4692	17.3
Concepción	1246	4.6	1236	4.5	1084	4.0	1027	3.8	940	3.5
Guairá	916	3.4	891	3.3	862	3.2	860	3.2	856	3.1
Itapúa	3086	11.4	3084	11.3	2833	10.4	2802	10.3	2730	10.0
Paraguarí	357	1.3	324	1.2	347	1.3	335	1.2	327	1.2
San Pedro	4735	17.4	4570	16.8	4124	15.2	3728	13.7	3383	12.4
Total	27,187	100	26,526		24,197		22,991		22,071	

Department	Forest Cover (2009–2010 km²)	%	Forest Cover (2011–2012 km²)	%	Forest Cover (2013–2014 km²)	%	Forest Cover (2015–2016 km²)	%	Total Forest Loss (km²)	%
Alto Paraná	2664	9.8	2609	9.6	2598	9.6	2528	9.3	808	24.2
Amambay	1946	7.2	1911	7.0	1827	6.7	1808	6.7	606	25.1
Caaguazú	2548	9.4	2487	9.1	2434	9.0	2322	8.5	791	25.4
Caazapá	1768	6.5	1768	6.5	1732	6.4	1639	6.0	533	24.5
Canindeyú	4427	16.3	4278	15.7	4091	15.0	3904	14	1908	32.8
Concepción	916	3.4	901	3.3	890	3.3	876	3.2	370	29.7
Guairá	834	3.1	816	3.0	806	3.0	805	3.0	111	12.1
Itapúa	2705	9.9	2700	9.9	2678	9.9	2634	9.7	452	14.6
Paraguarí	328	1.2	358	1.3	360	1.3	348	1.3	9	2.4
San Pedro	3204	11.8	2963	10.9	2804	10.3	2754	10	1981	41.8
Total	21,340		20,791		20,220		19,618		7569	27.8

Figure 4 reveals a clear pattern concerning the effectiveness of the protecting reserves based on their style of governance (ownership). For instance, each protected area owned by a governmental entity (Gov) showed a decrease in total forest cover. The highest deforestation rates were found in the National Parks of Cerro Corá (4.5%), Ybytyryzú (3%) and San Rafael (2.9%), totaling almost 30 km². In contrast, each natural reserve under ITAIPU-IT (binational hydroelectric dam (partially owned by the government)) management exhibited increments on their forest cover, with natural restoration rates (natural reforestation) varying between 1% (in the Yvytyrokai) and 69% (Biological Reserve Mbaracayú). As for protected areas privately owned (Prv) (e.g., Mbaracayú and Morombí), no clear trend was found. While the Mbaracayú reserve exhibited a small increase in forest cover, (0.8%), the Morombí reserve, by contrast, presented the highest deforestation rates (4.7%) among all the protected areas in the BAAPA region.

Forest cover dynamics within protected areas

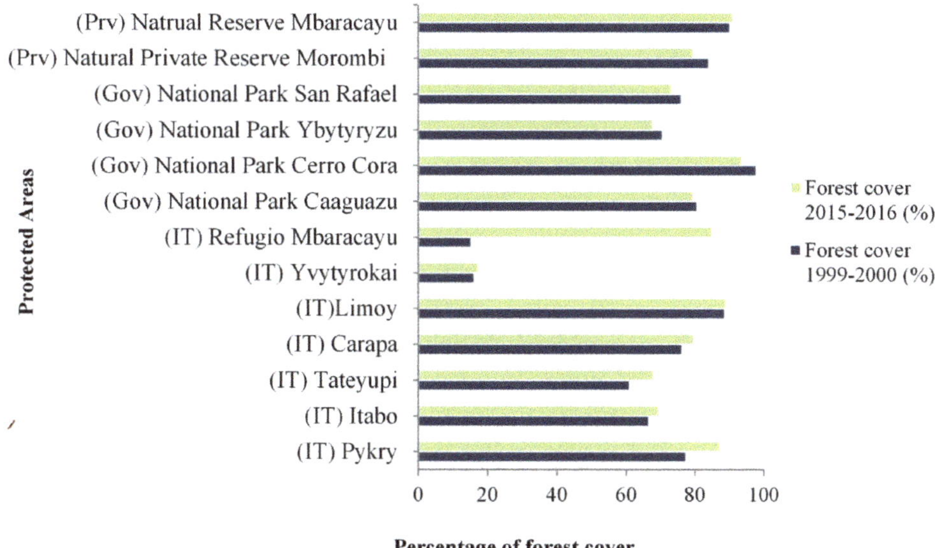

Figure 4. Comparison of forest dynamics between protected areas based on the type of ownership: Prv. (Private ownership), Gov. (Governmental ownership) and IT (Owned by the ITAIPU hydroelectric dam).

3.3. Forest Cover Change and Household Survey

3.3.1. Demography and Influence on Forest Dynamics

All interviews were conducted with the designated head (by the families) of each household. The vast majority of respondents were males (around 85%) with an age between 30 and 62 years. Ownership was mixed between Paraguayans and Colons (Brazilians), with the Paraguayans tending to own the smaller farms (82% Paraguayans), and Colons larger ones (83% Colons). The principal occupation of 91% interviewees was farmer, while a small share (9%) occupied positions in governmental institutions in addition to farming activities. Respondents' main agricultural activities were soy bean production (mainly large-scale farmers), cattle ranching (mostly medium-scale and small-scale farmers) and subsistence agriculture (small-scale farmers in particular). When analyzing the dynamics of the forest at the farm level, Figure 5 shows that forest loss/gain are closely related to farm size. For instance, the majority of forest loss (62%) occurred on small-scale farmers' properties. The percentage of farmers experiencing deforestation gradually decreases with an increased farm size, declining from 50% for medium-scale to 38% for large-scale farmers groups. Forest gain, on the other hand, is more common among large-scale farmers, accounting for 48% of the interviewees. On the contrary, small scale farmers exhibited the lowest percentage (23%) of respondents with an increment in their forest cover. Hence, when farm sizes increase, the percentage of farms showing forest cover gain increases as well.

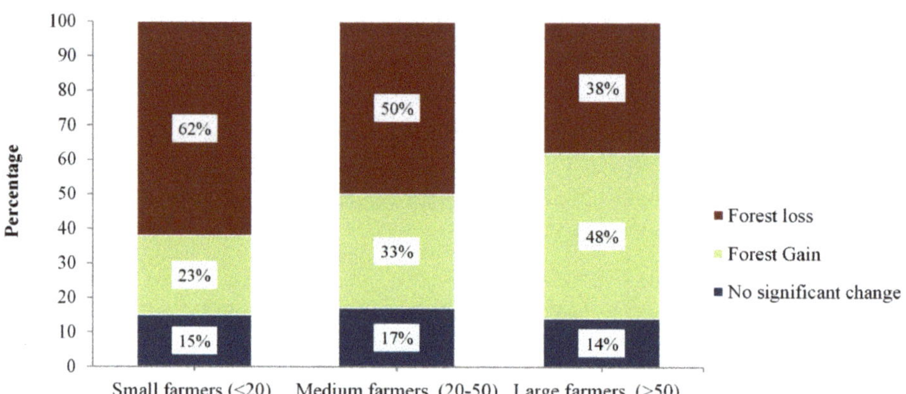

Figure 5. Dynamics of the forest stratified by farm type.

Figure 6 presents the distribution of the education level between farm types. Basically, education level increases slightly as farm sizes increases. The majority of farmers with higher education (a university degree) is found among the large-scale farmers group (32%). Small-scale farmers more frequently reveal lower levels of education, with 22% of the respondents having no school degree. Notwithstanding, a primary school education remains the most common level of education among all farm types with 65% (small-scale farmers), 61% (medium-scale farmers) and 40% (large-scale farmers) of respondents, respectively. Figure 7 presents strong tendency between a farmers' education level and the dynamics of the forest on their parameters. The highest percentage of farmers exhibiting forest loss is found in the group with no school degree.

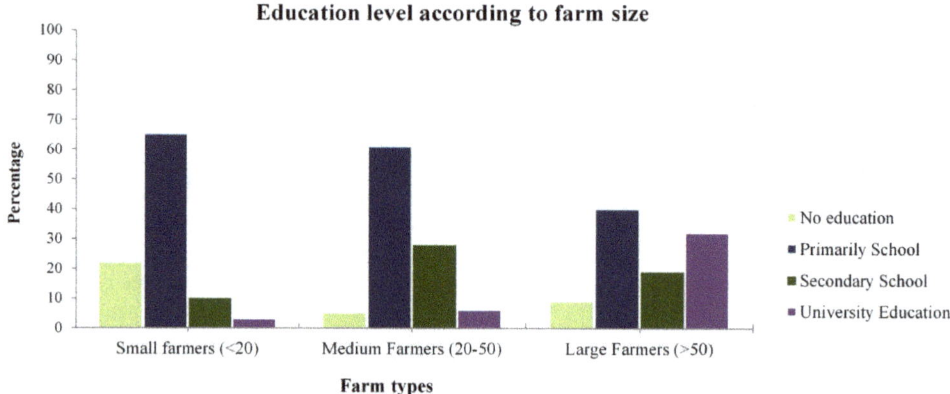

Figure 6. Education level according to farm size of the respondents.

Forest dynamics according to education level

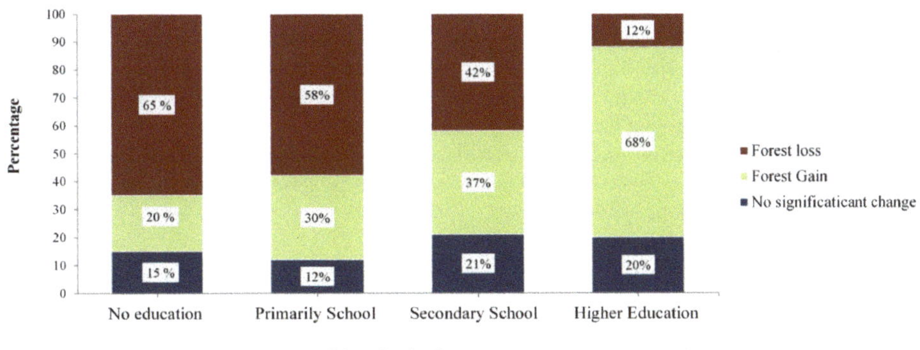

Figure 7. Dynamics of the forest based on the education level of the respondents.

This tendency gradually decreases as education level increases, down from 65% in the group with no formal education to 12% in the group with a college education. Furthermore, farmers with higher education (University degree) most commonly saw gains of forest cover on their property. This tendency decreases with decreasing education level, dropping from 68% (higher education) to 20% (no education).

3.3.2. Uses of Forest Benefits and Influence on the Forest Cover

To capture the level of dependency on forest benefits by different farm types, interviewees were asked which products they obtained from the forest and how important they were to them. A total of 68% of the farmers remarked that they frequently benefited from the forest, and 92% stated that forests are very important for their livelihood. A deeper analysis of the results revealed a higher dependency of forest products among small (97%) and medium-scale (78%) farmers, whereas large-scale farmers (44%) stated that they made use of the forest but not as intensively. The high reliance on forest products and services, in particular among the small-scale farmers group, can be attributed to a lack of other sufficient financial resources. Figure 8 presents the different uses of the forest according to farm type. Over 88% of the small-scale farmers group admit to collecting firewood from the forest. A total of 94% of this group stated that their main use was for subsistence, in particular cooking. On the other hand, only 40% of medium and 18% large-scale farmers claimed a certain level of dependency, in clear contrast to the above. Small (73%) and medium-scale (44%) farmers were more reliant on forest wood for construction (e.g., households, barns and fences construction) than large-scale farmers (16%).

In rural areas, the vast majority of small households (in particular within the group of small-scale farmers) own houses that are built with wood from the forest, while medium- and large-scale farms often present permanent homes. When asked about the cultural value of the forest, around 55% of large-scale farmers considered the forest's main value to be recreational. However, this inclination is less frequent among medium- and small-scale farmers, of which only 22% and 4%, respectively, held the same opinion.

Figure 9 presents how the forest cover of each farm group is affected by the use of forest benefits and products. For this analysis, we considered the percentage of farmers that acknowledged the use of the forest for any purpose (e.g., construction, firewood, agroforestry and recreation) and the spatial information from the forest cover of each farm. When analyzing the small-scale farmers group, 60% of the respondents presented forest loss, whereas only 26% and 14% showed forest gain or non-significant changes. Similar trends were observed among medium-scale farmers, among whom a high percentage of respondents (58%) presented a decrease in their forest cover, in comparison to the ones showing increments (35%) or no changes (7%). In contrast, a much higher proportion of the large-scale farmer's group (53%) revealed forest gain. Overall trends disclose a correlation between the uses of forest by different farm groups and their influence on the forest cover. Whereas small- and medium-scale farms evidenced high rates of deforestation, large-scale farmers demonstrated more sustainable use of the natural resources. However, it is important to mention that main uses of forest benefits differed among farm groups, which could have an impact on the forest cover itself. As described previously, small- and medium-scale farmers exhibited a higher tendency to use the forest as a source of firewood and construction-wood, whereas large-scale farmers were more inclined to use the forest for recreational proposes.

Main uses of forest benefits

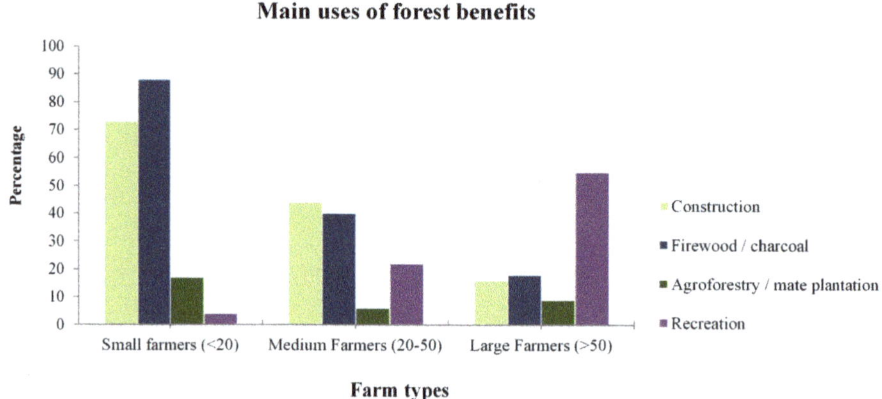

Figure 8. Main uses of forest related products based on farm type.

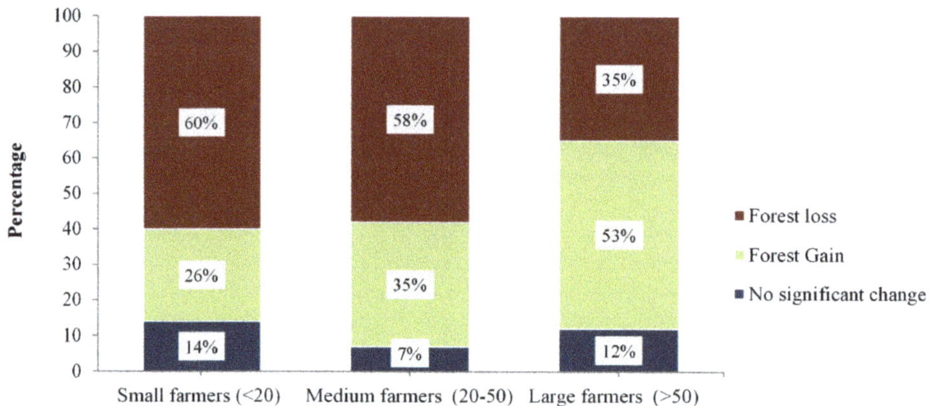

Figure 9. Dynamics of the forest cover based on the level of dependency.

3.3.3. Conservation Programs and Forest Cover Change

To comprehend landowner's perception about the willingness to conserve forests, farmers were asked if natural areas should be protected. In general, positive responses were obtained, with 99% of farmers expressing their support for protecting forests. Additionally, 88% of the farmers remarked on the importance of the forest and the negative effects of a disappearing forest, in particular on the prevailing natural resources, flora and fauna and water reservoirs. Overall, the interviewed farmers exhibited high degrees of environmental awareness. Furthermore, 54% of respondents reported participating in environmental programs such as the Payment for Ecosystem Service (PES) program, reforestation programs and water courses protection programs. Particularly large-scale farmers (65%) participated in such programs followed by medium (47%) and small-scale (45%) farmers' groups.

To understand the influence of conservation programs on forest dynamics, respondents' participation in environmental programs/workshops was reflected by the variations of the forest cover in each farm. Overall, results also highlighted that sustainable use of the forest was related with program participation. Of the group that reported not participating in any environmental workshops or conservation programs, 73% evidenced forest loss and only 9% an increase in forest cover. On the contrary, the fraction of properties experiencing forest loss decreased to 48% among participants in environmental programs, and the fraction of farmers with forest gains increased to 41%. When analyzing the results at a farm type level, small-scale farmers presented similar tendencies to the ones described above (see Figure 10). The percentage of famers exhibiting forest loss decreased from 73% to 48% for respondents involved in conservation/workshops programs.

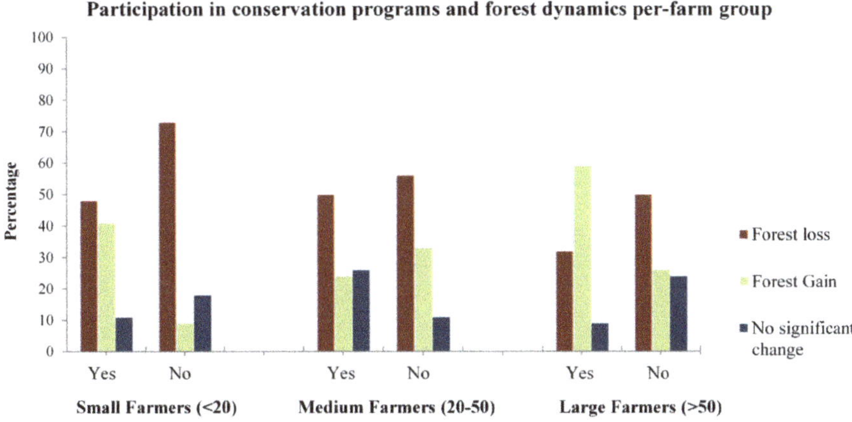

Figure 10. Relationship between conservation programs and forest cover change per-farm type.

Additionally, forest cover increase was found among farmers with environmental education, accounting for 41% of the respondents in this group. Large-scale farmers revealed different patterns from the ones described previously. The group of respondents which attended workshops was higher among farmers with increases in their forest cover (59%) in comparison to the ones exhibiting forest loss (32%) in the same group. Trends among the non-participant group resemble the ones shown for small- and medium-scale farmers, where farmers exhibiting forest loss (50%) dominated. Overall, results revealed a clear pattern where environmental education has an influence on the dynamics of the forest. Farmers with environmental knowledge tend to experience less forest loss.

4. Discussion

4.1. Forest Cover Change Analysis

The dynamics of the BAAPA forest and the perception of its benefits is a topic of great importance which has not been adequately studied so far. When analyzing changes of the forest cover, most studies were based solely on remote sensing [4,5,20] or field survey data [18]. To the knowledge of the authors, no research has combined both information types in the region. If the objective is to preserve natural resources, it is important to understand not only the historical distribution of the forest, but also the landowners' perceptions of it [51,52]. In a first part, the present study focuses on how the forest cover has been changing over the last 17 years, using a multi-temporal analysis approach. For this purpose, changes in the forest cover were assessed by applying a RF classifier using spectral temporal metrics derived from a dense set of Landsat Imagery (TM, ETM+ and OLI/TIRS). In a second part, this study examines how different farmer groups perceive the forest and how their perceptions influence its dynamics, integrating remote sensing and household level data.

Overall, change detection results revealed a total forest cover loss of 7500 km^2 (almost 28% of its original cover) between 1999 and 2016. However, in comparison with studies conducted for the previous years, a decreasing trend in deforestation rates is observed. Huang et al. [13,20] demonstrated that between 1989 and 2000, the BAAPA forest lost almost 40% (over 13,500 km^2) of its original cover, a figure nearly two times higher than deforestation figures found in this research. Inconsistencies with other studies regarding deforestation rates were found. For example, Da Ponte et al. [21] analyzed changes in the BAAPA forest between the years 2003 and 2013 based on Landsat images (OLI and ETM+). The study reported a total forest cover loss of 37% (around 33,039 km^2) by the year 2003 and 30% (over 26,966 km^2) in 2013. The present research, however, revealed lower forest coverage for the same period, of 28% (24,197 km^2) and 23% (20,221 km^2), respectively. Differences between findings could be related to sensors applied, input data (e.g., percentiles, vegetation indexes) and definition of forest. Similar trends were observed when comparing classification results with the ones obtained by the National University of Asuncion—UNA. For the year 2011, the UNA found natural forest coverage of 20% in the BAAPA, or almost 17,500 km^2 [53]. Results from this study presented higher forest coverage values (around 23% (20,000 km^2)) for the same period. It is worth noting that both studies assessed changes in the forest cover by applying bi-temporal-approaches considering only single scenes. Further spectral features such as percentiles derivation, vegetation indices, and a dense set of Landsat imagery were excluded from the analysis. The described data was already proven to be essential to achieving higher accuracy in land cover predictions and change detection assessment [28,37].

A detailed analysis of the time series indicates that deforestation rates seem to increase abruptly between the years 2002–2004 (almost four times higher than previous years) and gradually decrease again until the years 2015–2016, where slight increase is once again observed. This trend can be attributed to the Zero Deforestation Law (No. 2524/04) established in the year 2004, which banned the conversion of forest lands for other purposes [14]. According to farmers interviewed during the field survey, the Zero Deforestation Law was anticipated by rural communities and large landowners, who increased their deforestation activities before the deforestation law took effect.

A clear difference in the effectiveness of conservation programs under different forest management regimes is observed. For instance, every natural reserve owned by the Government-Gov presented a certain degree of forest loss within their boundaries. Reserves managed by the ITAIPU-IT hydroelectric, on the other hand, showed increases in forest cover, in some cases up to 71% of its total area (e.g., Mbaracayu reserve). ITAIPU is a binational hydroelectric company owned by the Paraguayan and Brazilian government. They invest economic resources to protect natural areas, but solely in areas located directly in their watersheds (as a part of their environmental mitigation program). Subsidies given directly by the government, in contrast, are rather scarce. According to several national environmental institutions (e.g., Guyra Paraguay, WWF, Fundacion Moises Bertoni), further support to protecting natural forest areas is needed. Until today, there still remains a strong weakness in the enforcement of environmental laws, policies and proper criteria to include stakeholder's needs and concerns when implementing conservation programs [25].

4.2. Household Survey and Remote Sensing Data

The analysis of forest cover change at the farm level in combination with results derived from field surveys permits comparing variations in forest surfaces along with the influence of change by different farm groups. Overall, results revealed a clear difference in the dynamics of the forest cover between farm types. In general, a higher proportion of small-scale farms presented forest loss compared with medium- and large-scale farmers. Similar tendencies were found in several reports from Peruvian environmental agencies (e.g., Ministry of Environment and Ministry of Agriculture and Irrigation) that tag small-scale farmers as the principal responsible group for over 90% of the deforestation activities that occurred in the Peruvian Amazon [54]. However, it is important to clarify that the total area of forest loss between farm types was not considered in this study. Therefore, even though small-scale farmers were shown to be the farm group most responsible for deforestation activities, the total forest area loss might not be as high compared to medium- and large-scale farmers. This study also analyzed the influence of education level on deforestation tendencies in the BAAPA. The percentage of farmers exhibiting forest loss decreases as education levels increase. Comparable trends were found among the Amerindian farmers in Honduras, where forest clearings tended to decline as education levels increased [55]. Additionally, according to Turner II et al. [56], a higher level of education could imply a better management of natural resources and a decrease in pressure on the forest.

A clear correlation was observed between the various farm types and differing ways of using the forest. In general, results show a higher percentage of small- and medium-scale farmers demonstrating forest loss. The present trend is consistent with the results obtained in the 2016 field survey, which revealed a high dependency on forest benefits (in particular construction wood and fire-wood) by the same farm types. The extraction of forest goods without considering any concept of sustainability or management plans could greatly influence its continuity [14]. Similar to the trends discussed above, a correlation between the tendencies found among large-scale farmers and results obtained in the 2016 survey was observed. The majority of large-scale farmers stated that their main use of the forest was recreational; correspondingly, their properties did not show evidence of intense harvesting. Results in our study demonstrated not only that large-scale farms presented the lowest percentage of respondents exhibiting forest loss, but also that the same group showed the highest percentage of respondents with increases in their forest cover. However, it is important to remark that the level of use of forest goods is highly associated with the level of income and daily subsidence needs (housing and cooking in particular), from each farm group. According to the main results obtained in the survey, the level of reliance on forest products varies with farm size; dependency on the forest tends to increase as farm size decreases. Whereas medium- and large-scale farmers are inclined to see the forest as an additional source of income, small-scale farmers, on the contrary, rely directly on forest products for subsistence.

Lastly, this study related the influence of environmental education on the variation in forest cover between farm types. The properties of farmers that participated in environmental programs/workshops were less likely to exhibit forest loss. However, at least for small- and medium-scale farmers, the percentage of farms showing forest loss was overwhelming. Large-scale farmers presented different trends, the majority of farms demonstrating increases in forest cover, which can be related to their increased participation in environmental programs.

Finally, it is important to take notice of certain biases on the input data that might have influenced the outcomes of this study. Firstly, the responses of the survey could have been influenced by the background of the interviewer, problems with environmental authorities and personal thoughts over the nature of the research itself. The present study combined remote sensing data with the available cadastral information for the area as far as possible. However, since most of the cadastral data was not accessible (in particular for small-scale farmers), field measurements relied on the knowledge of the farmer regarding the boundaries from their farm. Lastly, the distribution of the samples was dependent on the cadastral information; therefore, the representation for the study area could have been biased.

5. Conclusions

The results of this study provided a description of deforestation trends over the BAAPA region between 1999 and 2016. The correlation of household and remote sensing data permitted the obtainment of relevant information with regards to farmers' influence on the dynamics of the forest at the farm level. Based on the major findings and discussions in this study, the main conclusions are described as follows:

- Results of the forest change detection analysis based on Landsat imagery revealed a total forest cover loss of almost 7500 km^2 between the years 1999 and 2016, which represents almost 27% of its cover.

- The outcomes of the time series analysis presented a drastic increase in deforestation rates between the years 2001–2002 and 2003–2004, almost four times the deforestation rates observed for previous years (2300 km^2). According to local farmers, the present trend could be attributed to the upcoming Zero Deforestation Law in the country, which influenced the rapid deforestation before the law was applied.

- Forest cover change analysis in protected areas demonstrated a clear difference between their effectiveness. Whereas protected areas under the ITAIPU hydroelectric management regime presented increases in forest cover, protected areas managed by the Government, on the contrary, showed a decrease in their forest cover in each of the reserves.

- According to the 145 households interviewed, forest dynamics at the farm level is related to farm types. While the frequency of farmers presenting forest loss increases as farm sizes decreases, forest gains, on the contrary, increase as farm sizes increases as well.

- Education level has been shown to have an influence on the dynamics of the forest at the farm level. Overall, results demonstrated that, as education level increases, the percentage of famers exhibiting forest loss decreases. When considering forest gain, on the other hand, a higher percentage of farms with increases in forest cover can be found among the group with higher education.

- The level of dependency on forest products by different farm groups affects the status of their forest. Higher levels of dependency resulted in a higher percentage of farmers presenting forest cover loss.

- Environmental programs provide a certain degree of influence on changes in the forest cover at the farm level. Among the groups participating in environmental programs and workshops, a lower percentage of respondents showed forest loss on their properties than for comparable groups that did not attend the workshops.

Further studies could make use of higher resolution imagery to increase the accuracy of the results, in particular when considering an assessment of forest cover change at the farm level. In addition, absolute values of deforestation between farm types should be addressed in future studies to assess what the actual impact of different farm types on the forest cover is. Moreover, it would be interesting to consider additional dynamic information on the state of the forest (such as yearly forest degradation and regeneration rates) which would add more information with regards to the pressure exerted by different farm types on forest resources. The use of multi-temporal information, along with ground data, are key components to designing and supporting conservation strategies and policies. It is crucial to consider not only the outlook of rural population but their influence on the behavior of natural resources over time, as well.

Acknowledgments: This work was conducted under the Paraguay Land Use (PARLU) 'Protecting Forest for the Benefit of Climate, People and Nature in Paraguay' executed by WWF Paraguay and supported by the German Federal Ministry of the Environment, Nature Conservation, Building and Nuclear Safety (BMUB).

Author Contributions: D.P.E. conducted the field surveys, processed and validated the remote sensing data, and wrote the majority of the manuscript. K.C. was involved in research design, the design of the household survey questionnaire, methodological guidance for data processing and a critical revision of the manuscript. B.M. and C.W. contributed to the processing of remote sensing data. R.O. contributed to conducting the field surveys. N.O., S.D. and M.F. supported the manuscript with discussion and critical revision.

Conflicts of Interest: The authors declare no conflict of interest.

References

1. Carlson, M.J.R.; Mitchell, L.R. Scenario analysis to identify viable conservation strategies in Paraguay's imperiled Atlantic forest. *Ecol. Soc.* **2011**, *16*, 8.
2. Tejaswi, G. *FAO Manual on Deforestation, Degradation, and Fragmentation Using Remote Sensing and GIS*; Forestry Department, Food and Agriculture Organization of the United Nations: Rome, Italy, 2007. Available online: www.fao.org/forestry/18222-045c26b711a976bb9d0d17386ee8f0e37.pdf (accessed on 29 August 2017).
3. WWF Paraguay Extends Zero Deforestation Law to 2018. Available online: http://wwf.panda.org/?210224/Paraguay-extends-Zero-Deforestation-Law-to-2018 (accessed on 29 August 2017).
4. Hansen, M.C.; Stehman, S.V.; Potapov, P.V. Quantification of global gross forest cover loss. *Proc. Natl. Acad. Sci. USA* **2010**, *107*, 8650–8655. [CrossRef] [PubMed]
5. Hansen, M.C.; Potapov, P.V.; Moore, R.; Hancher, M.; Turubanova, S.A.; Tyukavina, A.; Thau, D.; Stehman, S.V.; Goetz, S.J.; Loveland, T.R.; et al. High-Resolution Global Maps of 21st-Century Forest Cover Change. *Science* **2013**, *342*, 850–853. [CrossRef] [PubMed]
6. Fleytas, C. Cambios en el paisaje. Evolución de la cobertura vegetal en la Región Oriental del Paraguay. In *Biodiversidad del Paraguay: Una imacion a Sus Realidades*, 1st ed.; Salas-Deunas, D., Facetti, J., Eds.; Fundacion Moises Bertoni, USAID, GEF/BM.: Asuncion, Paraguay, 2007; pp. 77–88. Available online: https://de.scribd.com/document/225612694/Cambios-en-El-Paisaje-Evolucion-de-La-Cobertura-Vegetal-en-La-Region-Oriental-Del-Paraguay (accessed on 4 April 2017).
7. Bitetti Di, M.S.; Placci, G.; Alves, D. *Vision de la Biodiversidad de la Ecorregión Bosque Atlántico del Alto Paraná: Diseño de un Paisaje para la Conservación de la Biodiversidad y prioridades para las acciones de conservación*; World Wildlife Fund: Washington, DC, USA, 2003; pp. 29–135. Available online: http://assets.wwf.org.br/downloads/altoparana_version_completa.pdf (accessed on 15 April 2017).
8. Mayers, N. Threatened biotas: "Hot spots" in tropical forests. *Environmentalist* **1988**, *8*, 187–208. [CrossRef]
9. Mittermeier, R.A.; Myers, N.; Mittermeier, C.G. *Hotspots: Earth's Biologically Richest and Most Endangered Terrestrial Ecoregions*; Graphic Arts Center Publishing Company: Washington, DC, USA, 1999; pp. 1–431.
10. Myers, N.; Mittermeier, R.A.; Mittermeier, C.G.; Da Fonseca, G.A.; Kent, J. Biodiversity hotspots for conservation priorities. *Nature* **2000**, *403*, 853–858. [CrossRef] [PubMed]
11. Olson, D.M.; Dinerstein, E. The Global 200: Priority ecoregions for global conservation. *Ann. Missouri Bot. Gard.* **2002**, *89*, 199–224. [CrossRef]
12. Catterson, T.M.; Fragano, F.V. Tropical Forestry and Biodiversity Conservation in Paraguay: A Report to USAID/Paraguay. 2004. Available online: https://rmportal.net/library/content/1/118_paraguay/at_download/file (accessed on 29 August 2017).
13. Huang, C.; Kim, S.; Altstatt, A.; Townshend, J.R.G.; Davis, P.; Song, K.; Tucker, C.J.; Rodas, O.; Yanosky, A.; Clay, R.; et al. Rapid loss of Paraguay's Atlantic forest and the status of protected areas–A Landsat assessment. *Remote Sens. Environ.* **2007**, *106*, 460–466. [CrossRef]
14. Hutchison, S.; Aquino, L. Making a Pact to Tackle Deforestation in Paraguay. International Tree Foundation, 2011. Available online: internationaltreefoundation.org/wp-content/.../Paraguay-FINAL-30-march-2011.pdf (accessed on 29 August 2017).
15. Grossman, J.J. A case study of smallholder eucalyptus plantation silviculture in Eastern Paraguay. *For. Chron.* **2012**, *88*, 528–534. [CrossRef]
16. Kernan, B.S.; Cordero, W.; Macedo Sienra, A.M.; Marín, J.V. *Report on Biodiversity and Tropical Forests in Paraguay*; USAID: Asuncion, Paraguay, 2010; pp. 1–99. Available online: https://www.usaid.gov/sites/default/files/documents/1862/paraguay_biodiversity_tropical_forest_report.pdf (accessed on 29 August 2017).
17. Valiente, E.; Gerard, F. Fortalecimiento de las políticas agroambientales en los países de américa latina y el caribe. Diagnostico Nacional de Politica Agroambiental Paraguay, 2016. Available online: http://www.fao.org/in-action/programa-brasil-fao/proyectos/politicas-agroambientales/es/ (accessed on 29 August 2017).
18. Da Ponte, E.; Kuenzer, C.; Parker, A.; Rodas, O.; Oppelt, N.; Fleckenstein, M. Forest cover loss in Paraguay and perception of ecosystem services: A case study of the Upper Parana Forest. *Ecosyst. Serv.* **2017**, *24*, 200–212. [CrossRef]

19. Da Ponte, E.; Fleckenstein, M.; Leinenkugel, P.; Parker, A.; Oppelt, N.; Kuenzer, C. Tropical forest cover dynamics for Latin America using Earth observation data: A review covering the continental, regional, and local scale. *Int. J. Remote Sens.* **2015**, *36*, 37–41. [CrossRef]

20. Huang, C.; Kim, S.; Song, K.; Townshend, J.R.G.; Davis, P.; Altstatt, A.; Rodas, O.; Yanosky, A.; Clay, R.; Tucker, C.J.; et al. Assessment of Paraguay's forest cover change using Landsat observations. *Glob. Planet. Chang.* **2009**, *67*, 1–12. [CrossRef]

21. Da Ponte, E.; Roch, M.; Leinenkugel, P.; Dech, S.; Kuenzer, C. Paraguay's Atlantic Forest cover loss—Satellite-based change detection and fragmentation analysis between 2003 and 2013. *Appl. Geogr.* **2017**, *79*, 37–49. [CrossRef]

22. Bragayrac, E. *Proyecto Conservación de Bosques del Paraguay—REDD BAAPA*; Monitoreo Social de la Colonia Amistad: Asuncion, Paraguay, 2014; pp. 1–46. Available online: http://www.worldlandtrust.org/documents/annexe_1_-_social_monitoring_results_in_spanish.pdf (accessed on 29 August 2017).

23. DGEEC (Dirección Nacional de Censos y Estadisticas), Atlas censal del Paraguay. Available online: http://www.dgeec.gov.py/ (accessed on 29 August 2017).

24. MAG (Ministerio de Agricultura y Ganaderia). Informer sector Agropecuario ISA. Available online: http://www.mag.gov.py/index.php/noticias/logros-del-sector-agropecuario-en-el-pais (accessed on 29 August 2017).

25. Gonzales, R.; Rivadeneira, M. *Estudio de tendencias y perspectivas del Sector Forestal en America Latina Documento de Trabajo. Informe nacional Paraguay*; FAO: Rome, Italy, 2004; Volume 9, pp. 1–98. Available online: http://www.fao.org/forestry/outlook/2406/es/ (accessed on 29 August 2017).

26. Patterson, T.; Kelso, V.N. Earth Natural Free Vector and Raster Map Data. Available online: http://www.naturalearthdata.com/downloads (accessed on 10 March 2017).

27. Mittermeier, R.A.; Robles-Gil, P.; Hoffman, M.; Pilgrim, J.D.; Brooks, T.B.; Mitterneier, C.G.; Lamoreux, J.L.; Fonseca, G.A. Biodiversity Hotspots Revisited, Conservation International 2004. Databasin, 2004. Available online: http://www.biodiversityhotspots.org/xp/Hotspots/resources/maps.xml (accessed on 3 April 2017).

28. Wohlfart, C.; Mack, B.; Liu, G.; Kuenzer, C. Multi-faceted land cover and land use change analyses in the Yellow River Basin based on dense Landsat time series: Exemplary analysis in mining, agriculture, forest, and urban areas. *Appl. Geogr.* **2017**, *85*, 73–88. [CrossRef]

29. Knauer, K.; Gessner, U.; Fensholt, U.; Kuenzer, C. Agricultural Expansion in Burkina Faso over 14 Years with 30 m Resolution Time Series: The Role of Population Growth and Implications for the Environment. *Remote Sens.* **2017**, *9*, 132. [CrossRef]

30. Gebhardt, S.; Wehrmann, T.; Ruiz, M.A.M.; Maeda, P.; Bishop, J.; Schramm, M.; Kopeinig, R.; Cartus, O.; Kellndorfer, J.; Ressl, R.; et al. MAD-MEX: Automatic wall-to-wall land cover monitoring for the mexican REDD-MRV program using all landsat data. *Remote Sens.* **2014**, *6*, 3923–3943. [CrossRef]

31. Zhu, Z.; Woodcock, C.E. Object-based cloud and cloud shadow detection in Landsat imagery. *Remote Sens. Environ.* **2012**, *118*, 83–94. [CrossRef]

32. Richter, R.; Schläpfer, D. *Atmospheric/Topographic Correction for Satellite Imagery (ATCOR-2/3 User Guide)*; ReSe Applications Schläpfer: Langeggweg, Switzerland, 2016; pp. 13–250. Available online: www.rese.ch/pdf/atcor3_manual.pdf (accessed on 29 August 2017).

33. Mack, B.; Leinenkugel, P.; Kuenzer, C.; Dech, S. A semi-automated approach for the generation of a new land use and land cover product for Germany based on Landsat time-series and Lucas in-situ data. *Remote Sens. Lett.* **2017**, *8*, 244–253. [CrossRef]

34. Franklin, S.E.; Ahmed, O.S.; Wulder, M.A.; White, J.C.; Hermosilla, T.; Coops, N.C. Large area mapping of annual land cover dynamics using multitemporal change detection and classification of Landsat time series data. *Can. J. Remote Sens.* **2015**, *41*, 293–314. [CrossRef]

35. Griffiths, P.; Kuemmerle, T.; Baumann, M.; Radeloff, V.C.; Abrudan, I.V.; Lieskovsky, J.; Munteanu, C.; Ostapowicz, K.; Hostert, P. Forest disturbances, forest recovery, and changes in forest types across the Carpathian ecoregion from 1985 to 2010 based on Landsat image composites. *Remote Sens. Environ.* **2014**, *151*, 72–88. [CrossRef]

36. Hansen, M.C.; Egorov, A.; Potapov, P.V.; Stehman, S.V.; Tyukavina, A.; Turubanova, S.A.; Roy, D.P.; Goetz, S.J.; Loveland, T.R.; Ju, J.; et al. Monitoring conterminous United States (CONUS) land cover change with Web-Enabled Landsat Data (WELD). *Remote Sens. Environ.* **2014**, *140*, 466–484. [CrossRef]

37. Müller, H.; Griffiths, P.; Hostert, P. Long-term deforestation dynamics in the Brazilian Amazon—Uncovering historic frontier development along the Cuiabá–Santarém highway. *Int. J. Appl. Earth Obs. Geoinf.* **2016**, *44*, 61–69. [CrossRef]

38. Breiman, L. Random forests. *Mach. Learn.* **2001**, *45*, 5–32. [CrossRef]

39. Congalton, R.G.; Green, K. *Assessing the Accuracy of Remotely Sensed Data: Principles and Practices*; Second Edition (Mapping Science); CRC Press: Boca Raton, FL, USA, 2008.

40. Wohlfart, C.; Liu, G.; Huang, C.; Kuenzer, C. A River Basin over the course of time: Multi-temporal analyses of land surface dynamics in the Yellow River Basin (China) based on medium resolution remote sensing data. *Remote Sens.* **2016**, *8*, 186. [CrossRef]

41. Paraguayan Atlantic Forest. 24°58′31.38″ S and 55°03′27.73″ W. Google Earth. 1999–2016. Available online: https://earth.google.com/web/@-24.58598683,-55.12783014,432.7883121a,308634.30743907d,35y,11.83507172h,7.24129773t,0r (accessed on 12 April 2017).

42. Pal, M. Random forest classifier for remote sensing classification. *Int. J. Remote Sens.* **2005**, *26*, 217–222. [CrossRef]

43. Gislason, P.O.; Benediktsson, J.A.; Sveinsson, J.R. Random Forests for land cover classification. *Pattern Recognit. Lett.* **2006**, *27*, 294–300. [CrossRef]

44. Schneider, A. Monitoring land cover change in urban and peri-urban areas using dense time stacks of Landsat satellite data and a data mining approach. *Remote Sens. Environ.* **2012**, *124*, 689–704. [CrossRef]

45. R Core Team. *R, 3.3.1*; R Foundation for Statistical Computing: Vienna, Austria, 2016.

46. Liaw, A.; Wiener, M. Classification and regression by randomForest. *R News* **2002**, *2*, 18–22.

47. Lillesand, T.; Kiefer, R.W.; Chipman, J. *Remote Sensing and Image Interpretation*, 6th ed.; Flahive, R., Kelleher, L., Hong, S., Eds.; Wiley: New York, NY, USA, 2008; Volume 6.

48. FAO. *Global Forest Resources Assessment 2010*; FAO: Rome, Italy, 2010; pp. 1–193.

49. *PNUD Sector rural paraguayo: una visión general para un dialogo informado*; Mercurio S.A.: Asuncion, Paraguay, 2010; Volume 7, pp. 8–109.

50. Yamane, T. *Statistics: An Introductory Analysis*; Harper and Row: New York, NY, USA, 1967.

51. Sodhi, N.S.; Lee, T.M.; Sekercioglu, C.H.; Webb, E.L.; Prawiradilaga, D.M.; Lohman, D.J.; Pierce, N.E.; Diesmos, A.C.; Rao, M.; Ehrlich, P.R. Local people value environmental services provided by forested parks. *Biodivers. Conserv.* **2009**, *19*, 1175–1188. [CrossRef]

52. Casado-Arzuaga, I.; Madariaga, I.; Onaindia, M. Perception, demand and user contribution to ecosystem services in the Bilbao Metropolitan Greenbelt. *J. Environ. Manag.* **2013**, *129*, 33–43. [CrossRef] [PubMed]

53. Rajelaga, L.; Guerrero, N.; Cabrera, L.; Hirata, Y.; Takahashi, M.; Vega, L. Mapa de Cobertura de la Tierra Paraguay 2011. In *Facultad de Ciencias Agrarias y Forestry and Forest Products Research Institute (Japon)*; SIG/CIF/FCA/UNA: San Lorenzo, Paraguay, 2013; pp. 1–24.

54. Ravikumar, A.; Sears, R.R.; Cronkleton, P.; Menton, M.; Pérez-Ojeda del Arco, M. Is small-scale agriculture really the main driver of deforestation in the Peruvian Amazon? Moving beyond the prevailing narrative. *Conserv. Lett.* **2017**, *10*, 170–177. [CrossRef]

55. Godoy, R.; Groff, S.; O'Neill, K. The role of education in neotropical deforestation: Household evidence from Amerindians in Honduras. *Hum. Ecol.* **1998**, *26*, 649–675. [CrossRef]

56. Turner, B.L., II; Geoghegan, J.D. *Integrated Land-Change Science and Tropical Deforestation in the Southern Yucatan*; OUP Oxford: Oxford, UK, 2004.

 **forests** 

Article

Validation and Application of European Beech Phenological Metrics Derived from MODIS Data along an Altitudinal Gradient

Veronika Lukasová [1,*], Tomáš Bucha [2], Jana Škvareninová [3] and Jaroslav Škvarenina [1]

[1] Faculty of Forestry, Technical University in Zvolen, T. G. Masaryka 24, 960 53 Zvolen, Slovakia; jaroslav.skvarenina@tuzvo.sk

[2] National Forestry Center, T. G. Masaryka 22, 960 92 Zvolen, Slovakia; bucha@nlcsk.org

[3] Faculty of Ecology and Environmental Sciences, Technical University in Zvolen, T. G. Masaryka 24, 960 53 Zvolen, Slovakia; skvareninova@tuzvo.sk

* Correspondence: veronika.lukasova@tuzvo.sk; Tel.: +421-902-117-277

Received: 16 November 2018; Accepted: 9 January 2019; Published: 14 January 2019

Abstract: Monitoring plant phenology is one of the means of detecting the response of vegetation to changing environmental conditions. One approach for the study of vegetation phenology from local to global scales is to apply satellite-based indices. We investigated the potential of phenological metrics from moderate resolution remotely sensed data to monitor the altitudinal variations in phenological phases of European beech (*Fagus sylvatica* L.). Phenological metrics were derived from the NDVI annual trajectories fitted with double sigmoid logistic function. Validation of the satellite-derived phenological metrics was necessary, thus the multiple-year ground observations of phenological phases from twelve beech stands along the altitudinal gradient were employed. In five stands, the validation process was supported with annual (in 2011) phenological observations of the undergrowth and understory vegetation, measurements of the leaf area index (LAI), and with laboratory spectral analyses of forest components reflecting the red and near-infrared radiation. Non-significant differences between the satellite-derived phenological metrics and the in situ observed phenological phases of the beginning of leaf onset (LO_10); end of leaf onset (LO_100); and 80% leaf coloring (LC_80) were detected. Next, the altitude dependent variations of the phenological metrics were investigated in all beech-dominated pixels over the area between latitudes 47°44′ N and 49°37′ N, and longitudes 16°50′ E and 22°34′ E (Slovakia, Central Europe). In all cases, this large-scale regression revealed non-linear relationships. Since spring phenological metrics showed strong dependence on altitude, only a weak relationship was detected between autumn phenological metric and altitude. The effect of altitude was evaluated through differences in local climatic conditions, especially temperature and precipitation. We used normal values from the last 30 years to evaluate the altitude-conditioned differences in the growing season length in 12 study stands. The approach presented in this paper contributes to a more explicit understanding of satellite data-based beech phenology along the altitudinal gradient, and will be useful for determining the optimal distribution range of European beech under changing climate conditions.

Keywords: validation; phenology; NDVI; LAI; spectral analyses; European beech; altitude

1. Introduction

Recently, the Intergovernmental Panel on Climate Change has placed high importance on the gathering and interpretation of phenology observations to improve understanding of the ecological impacts of climate change [1]. Trends in the timing of plant development can have major impacts on plant productivity, competition between plant species, and interactions with heterotrophic organisms [2]. These have prompted further species-specific studies for inclusion in biodiversity and ecosystem adaptation assessments [3]. Forest ecosystems where the phenology variability can be observed along the broad temperature gradient associated with the large altitudinal gradient are particularly vulnerable to global warming. The consequences can already be observed on the altitudinal range of species [4].

This study focused on the phenology of European beech (*Fagus sylvatica* L.), whose spatial distribution covers most of the continent of Europe. With regard to the ecological amplitude of European beech, we can assume that forests growing at low altitudes will suffer from drought [5]. Although beech is resistant to fairly low winter temperatures, it is sensitive to late frosts, which limit its occurrence at lower altitudes where the cold air can accumulate [6]. Beech can either adapt to extreme conditions via their phenotype plasticity through shifts in the timing of phenological phases or migrate to altitudes with milder climate forcing.

Understanding the links between shifts in phenological phases and climatic variability and changes from a local to global scale requires more detailed in situ observations that underpin the remote sensing data. Remote sensing and optical technologies, as well as spectral analyses, bring new dimensions to phenological monitoring, while validation still plays a key role in the successful implementation of satellite data-based phenological metrics. Previously, considerable effort has been spent on satellite data processing [4,7,8], finding the best indicator of the changing green leaf area [4,9–11], and the best fitting procedures [8,12,13]. Timing of the phenological phases of forest ecosystems from remote sensing data is commonly based on the seasonal trajectories of the vegetation indices of single pixels. The limits of matching the in situ observed phenological phases to the phenological derived from different fitting procedures and data sources were revealed [9,14,15]. The effect of the background of forests was less investigated. In the vertical structure of forest ecosystems, there are several components which, similar to the canopy leaves, reflect a considerable amount of radiation in the spectra usually used in the calculations of vegetation indices. The effect of these components is significant, especially in the period before and during canopy greening, the same as during and after leaf fall [16,17]. The mixture of species in deciduous forests differing in phenological behavior makes it difficult to understand the link between satellite information and ground phenology, and it is not easy to clearly extract the phenological response of a single species [2]. Therefore, the utilization of more detailed phenological scales could bring better consistence between the satellite based phenological metrics and ground phenological observations [18,19]. Coupling the field measurements and reflectance data on the species-specific level studys, along with one dominant species, is suitable for analysis in moderate to small spatial resolution of satellite data.

The aim of the current study was to investigate the potential of phenological metrics from moderate resolution remotely sensed data to monitor the phenological phases of beech-dominated stands and to estimate the impact of the altitude on the timing of the phenological phases. This study was based on the 250 m Moderate Resolution Imaging Spectroradiometer (MODIS) data time series acquired over the area of the Western Carpathians during the years 2000–2012. Under laboratory conditions, we quantified the spectral reflectance of forest components that could participate in the overall beech-stand reflectance, and calculated their normalized difference vegetation index (NDVI) ($NDVI_{LAB}$). This was helpful in the reconstruction of satellite-derived NDVI ($NDVI_{MOD}$) seasonal trajectories and in choosing suitable phenological metrics for pairing with in situ observed phenological phases. A good match between the remote sensing and ground phenological data was the supposition to large-scale analyses of the impact of the altitude-associated environmental conditions on the beech phenology.

2. Materials and Methods

2.1. Study Stands and In Situ Phenological Observations

The study area was located at latitudes between 47°44′ N and 49°37′ N and at longitudes between 16°50′ E and 22°34′ E in Slovakia (Central Europe). For this region, a beech mask was created in three steps:

1. The representation of beech in a 250 × 250 m pixel (corresponding to MO09GQ resolution) was derived from the classification of the tree species composition in Slovakia [20] from Landsat satellite images in 30 m spatial resolution using the ArcGIS Aggregate function. The pixels with presence of beech 60% and higher were included in the mask.
2. In order to eliminate a possible tree classification error from Landsat, the Forestry Information System database was used. Here, the actual data of tree species composition are available for each forest compartment (generally, compartment area varies from 1 to 15 ha). The presence of beech from the database was assigned to the pixels selected in the first step. Pixels with beech 60% and higher remained in the beech mask.
3. Due to the possible contamination of DN values with non-forest land cover classes (meadows, fields, water areas, etc.), pixels at the edges of the forest were removed from the derived beech mask.

The created beech mask of Slovakia was composed of 16,025 beech-dominated pixels (Figure 1), where the presence of European beech trees was 60% and higher. These beech-dominated pixels were located at altitudes from 156 to 1331 m above sea level (a.s.l.), with the dominant incidence in the altitudinal range between 400–600 m a.s.l. where 66% of all beech pixels were located.

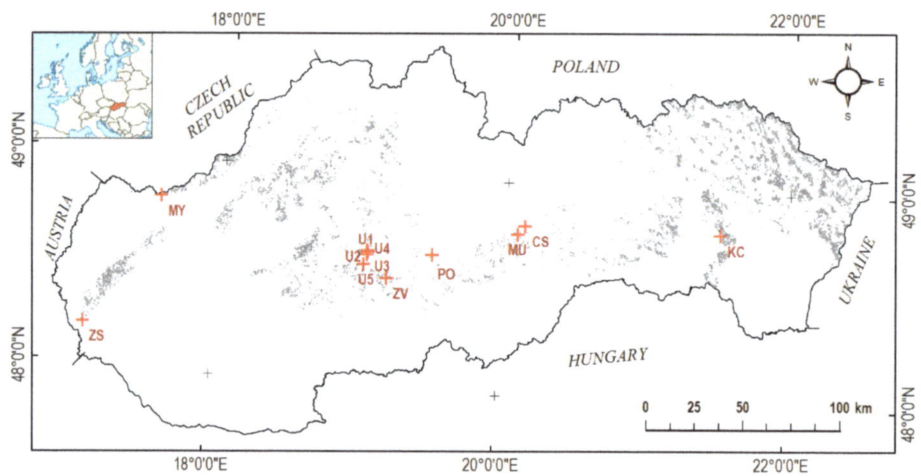

Figure 1. Map of the beech stands with the presence of European beech at 60% and higher (grey color). The locations of the study stands with in situ phenological observations are marked with red crosses.

To assess the ability of phenological metrics to capture interannual variability in spring and autumn phenology, we used a 12 beech stands from the beech mask where the phenological observations had been realized. The stands covered three main climatic areas—warm, moderate, and cold—in an altitudinal range from 304 to 1051 m a.s.l. This allowed us to track phenology in different altitude-dependent climate conditions (Table 1).

Table 1. The basic characteristics of beech study stands.

Stand Number	Identifier	Altitude	Aspect	Presence of Beech (%)	Climatic Region	Climate Normal (1981–2010)	
						T [7] (°C)	P [8] (mm)
1	ZS	304	NW	60	W7 [1]	10.1	675
2	MY	457	W	65	M3 [2]	9	729
3	U1	490	W	85	M6 [3]	8.6	691
4	U5	512	N	70	M3	8.3	700
5	U2	522	W	65	M6	8.3	700
6	ZV	566	SW	65	M6	8	745
7	KC	570	SW	90	M3	8.1	644
8	MU	579	N	100	M7 [4]	7.9	689
9	U4	607	E	100	M6	7.8	760
10	U3	615	W	65	M6	7.8	760
11	CS	1003	N	90	C1 [5]	5.2	857
12	PO	1051	SE	80	C2 [6]	5.7	863

[1] Warm, moderate moist with mild winters, [2] Moderate warm, moderate moist, upland to highland, [3] Moderate warm, moist, highland, [4] Moderate warm, very moist, highland, [5] Moderate cold, [6] Cold mountain, [7] Temperature, [8] Precipitation.

The phenological data of 7 beech-dominated stands were adopted from the phenological network of Slovak Hydrometeorological Institute (SHMI). We focused on the period 2000–2012, which corresponds to the availability of MODIS data (from the year 2000). The ground measurements we used here consist of visual observations of phenological phases performed always by the same observer, on 10 selected trees from the main canopy. The group of observed beech trees was situated in the 6.25 ha beech-dominated pixel. The main phenological phases observed in all of the study stands are presented in Table 2 and are also marked by the international BBCH codes [21]. BBCH codes are commonly used for observations in the International Phenological Gardens (IPG) [22]; in the monitoring of plant species reactions to climate changes [23] and environmental changes in urban areas [24]; and in creation of phenological calendars [25], etc.

Table 2. Definition of the phenological phases observed in all study stands.

Phenological Phase	BBCH Codes	Definition
Beginning of Bud Bursting (BB_10)	BBCH07	When bud scales opened and green top of leaf was sticking out
Beginning of Leaf Onset (LO_10)	BBCH11	When 10% of leaves have final shape, but not final size and color
General Leaf Onset (LO_50)	BBCH13	When 50% of leaves have final shape, but not final size and color
Beginning of Leaf Coloring (LC_10)	-	When 10% of leaves changed their color from green to yellow, red or brown
General Leaf Coloring (LC_50)	BBCH94	When 50% of leaves changed their color from green to yellow, red or brown
Beginning of Leaf Fall (LF_10)	BBCH93	When 10% of leaves fell down from trees to the ground
End of Leaf Fall (LF_100)	BBCH97	When 100% of leaves fell down from trees to the ground

In the five stands, U1–U5, located within the area of the University Forest Enterprise of the Technical University in Zvolen, the phenological observations were performed in two years—2011 and 2012. Here, we employed intensive field observations specifically designed to monitor the vegetation phenology from the ground, through the undergrowth, to the canopy, and record the changes in the canopy structure on hemispherical photographs in 2011. Photographs were taken seven times in each stand: 3–4 May (day of the year–DOY 123), 12–13 May (DOY 133), 19 Jun (DOY 170), 10–11 August (DOY 222), 28–29 September (DOY 272), 17–18 October (DOY 290), and 15–21 November (DOY 319), and the leaf area index (LAI) was calculated as a structural characteristic of the beech canopies. The LAI values over the season were used to investigate the responsiveness of NDVI to the structural changes in the canopies. The details of the LAI calculations were published in our previous study that aimed at comparing different methods of leaf area index calculation [26]. Cover of ground vegetation was observed visually as a percentage on five square sample plots in each beech stand U1–U5. The distance between the sample plots was 50 m and the area of a single plot was 100 m^2. The estimated percentage cover was rounded to 5%.

In the U1–U5 stands, the same phenological scale as in the seven SHMI stands was used, but extra was noted every 10% of the developmental stage of each phenological phase. We also captured the phenological phase final leaf onset, when all leaves had their final shape, size, and color. The phenological phases of the under-growing beech trees in these stands were evaluated by the same methodology as on the canopy trees.

2.2. Validation Supporting Laboratory Spectral Analyses

When analyzing beech-dominated stands from the 250 m MODIS pixel, several forest components participated with their reflectance on the single pixel $NDVI_{MOD}$. Therefore, simultaneously with the in situ observations in stands U1–U5, individual samples of fresh and fallen beech leaves, fresh green undergrowth and understory vegetation (spring heliophytes), and beech bark were collected in 2011. At first, the samples of fallen beech leaves were collected at the time before the beginning of the phenological activity (DOY97). On DOY110, samples of two spring heliophytes dominating in the stands were collected: coral-root (*Cardamine bulbifera* L.) and dog's mercury (*Mercurialis perennis* L.). The samples of fresh beech leaves were gathered twice during the season. The leaf collection time was subjected to a specific phenological phase of beech: the end of leaf onset (DOY127), and final leaf onset (DOY159). The samples of beech bark were cut from two parts of the currently harvested trees: from the stems and branches (DOY159).

All samples collected in the field were placed in closed plastic bags and transported to the laboratory for measurements not later than two hours to minimize the wilting. Under laboratory conditions, the spectral reflectance was measured with a Li-1800 spectroradiometer with an integrating sphere 1800-12 equipped with a light source (LI-COR, Lincoln, NE, USA). The measured wavelength range covered the red and near-infrared band. The measurements of each sample were repeated three times and the values were averaged. The $NDVI_{LAB}$ was calculated from the reflectance in the red (RED: 620–670 nm) and the infrared spectra (IRED: 841–876 nm):

$$NDVI = \frac{IRED - RED}{IRED + RED} \tag{1}$$

The cover and visibility of forest components through the canopies were considered when evaluating their participation on the satellite-derived NDVI.

2.3. Deriving MODIS NDVI Phenological Metrics

In this study, the satellite data from a MODIS spectroradiometer, MOD09 and MYD09 products (Collection 5) were used. The images of the MOD09/MYD09 products represent the spectral reflectance of the Earth in a spatial resolution of 250 m (MOD09GQ and MYD09GQ product) and 500 m (MOD09GA, MYD09GA) with radiometric and atmospheric corrections at a daily time step [27]. This product contains the spectral channels needed for NDVI calculations at a 250 m resolution, the red channel (RED: 620–670 nm), and the infrared channel (IRED: 841–876 nm). We used the MOD/MYD09GQ products because of the spatial resolution and the necessity to capture vegetation dynamics on a daily level, despite the potentially adverse effect of anisotropic reflectance of the vegetation [28]. Our choice was supported by the findings of Franch et al. [29] that the effect of surface anisotropy in the red and infrared band of MODIS data barely influenced NDVI estimation, obtaining an RMS of around 1%. The value of the spectral reflectance of each pixel is partly influenced by the spectral properties of adjacent pixels [30]. Therefore, pixels on the boundary between the forest stands and other land cover classes were excluded from the analyses.

Furthermore, we focused on the elimination of the images and pixels affected by clouds and cloud shadows, as well as the impact of image distortions. All MODIS images covering the territory of Slovakia were reviewed at the NASA archive from 2000 to 2012, and their quality was visually assessed. Only cloudless or partly cloudy imagery were utilized for the analyses. Another criterion for the selection of appropriate images was the satellite position in relation to Slovakia during data collection. Since the MODIS tracks are stable with 16-day repetitions, we could identify that there were six images in the in-nadir position related to Slovakia; four images in close-to-nadir position, and six images in the off-nadir position. Off-nadir images were completely excluded from the download. Thus, we reduced the potential effect of anisotropic reflectance and achieved a spatial resolution close to 250 m. In addition to the visual inspection of image quality, we analyzed the 1 km Reflectance Data State QA layer present in the MOD09GA/MYD09GA products. The dataset enabled us to check the quality of the recorded reflectance at the pixel level. The data qualities of the individual pixels are encoded as the values of individual bits in a 16-bit integer. Based on the data from the quality dataset, we selected pixels with the following bit values: 8, 72, 76, 136, 140, 200, and 8200. Other pixel values were replaced by 0.

Reflectance values (DN values) of the selected pixels underwent a final quality analysis. The arithmetic mean of the DN values and standard deviation (SD) were calculated for each day of an image selected from the database. A pixel was included in the analysis if its value ranged within $x \pm 2\,SD$. The pixels outside this range were assigned a bit value of 0.

The MODIS data from the DOY 1 to 80 were excluded from the analyses. At that time, no phenological activity of any vegetation in the beech stands was recorded and there was a high probability of signal contamination by snow cover. Over the examined 13-year period, after applying the rules for image and pixel selection, we collected a folder of 433 images from Terra and Aqua satellites. A detailed description of the procedure of selecting MODIS data for fitting the model is described in [31].

The time series of the NDVI derived from the satellite data reflect the seasonal changes in the amount of the green leaf biomass. European beech is a species with one annual growing season. To fit the seasonal phenological model from the NDVI time series for the single beech-dominated pixels, we applied the sigmoid logistic function [9]. This model estimated the phenological phases as a function of day of year (DOY). To determine the transition dates of the phenological phases–phenological metrics, inflection points of the first derivative and local minima and maxima of the second derivative of the function were calculated. Six key phenological metrics were derived, three in the spring (S1, S2, S3) and autumn periods (A1, A2, A3), respectively (Figure 2). Phenological metrics were characterized as follow:

- S1—the acceleration of leafing in forest stand.
- S2—the leafing in forest stand reaches the half-maximum.
- S3—the deceleration of leafing in forest stand.
- A1—the acceleration of leaf coloring in forest stand.
- A2—the leaf coloring in forest stand reaches the half-maximum.
- A3—the deceleration of leaf coloring in forest stand.

First, the phenological metrics for the study stands were derived to validate their applicability in phenological monitoring. The phenological metrics obtained from $NDVI_{MOD}$ fits were compared against in situ observed phenological phases. The statistics used here were from the paired sample *t*-test. With the *t*-test, the significance of differences between the in situ observed phenological phases and satellite data-based phenological metrics was tested. The *t*-statistic was compared to the critical value $T_\alpha(n-1)$. The RMSE was used to evaluate the average prediction uncertainty relative to the in situ observations. The Pearson's correlation coefficient was calculated to quantify the relationship between the onset days derived by the two above-mentioned methods. The strength of the correlation values was described according to the guide suggested for the absolute value of r [32]: very weak—0.00–0.19; weak—0.20–0.39; moderate—0.40–0.59; strong—0.60–0.79; very strong—0.80–1.0.

Second, after the successful validation process, the sigmoid logistic function was fitted for all beech-dominated pixels from the beech mask of Slovakia, and the phenological metrics were derived. Phenological metrics from all of these pixels were used in the large-scale altitudinal study.

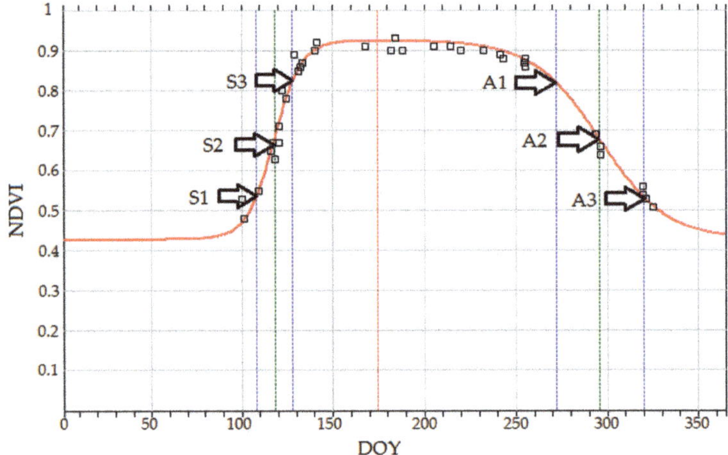

Figure 2. Graphical illustration of NDVI$_{MOD}$ seasonal profile fitted by a sigmoid logistic function with the inflection points of the first derivative (phenological metrics S2 and A2), and the local minima and maxima of the second derivative (phenological metrics S1, S3 and A1, A3 for the spring and autumn period, respectively). NDVI, normalized difference vegetation index. DOY, day of year.

2.4. Altitudinal Study

As previous studies have indicated, the timing of the onset of phenological phases depends on temperature [33–36], light [37,38], rainfall and humidity [5], stand characteristics, and on the internal plant periodicity [39]. Other factors include the stand location [40], altitude [4,41–43] latitude [44], and oceanic or continental climate [9]. In the mild climate, temperature is the most significant limiting factor affecting a variety of physiological and biochemical processes in plants [39]. The air temperature decreases with increasing altitude; in the troposphere it decreases by about 0.65 °C per 100 m of altitude increase (average lapse rate) [45]. The next significant meteorological element is atmospheric rainfall. Under the conditions of Central Europe, it can generally increase by 50–60 mm per 100 m of altitude increase and depends on the geographical location, altitude, and exposure to wind bringing moist air masses and frontal systems [46]. The Climate Normals (1981–2010)—normal temperature and normal precipitation corresponding to the study stands—are included in Table 1.

The altitudinal study was performed at two scales: the stand scale and a large scale. Validated phenological markers S2, S3, A2, and the growing season length (GSL) (GSL$_i$ = A2$_i$ − S2$_i$) were analyzed as multi-year averages. First, the regression analyses were used to test the onset days' variations of phenological metrics and GSL in relation to altitude (days 100 m^{-1}), normal temperature (days °C^{-1}), and normal precipitation at the level of the twelve study stands. Second, the impact of altitude on the onset of phenological metrics and GSL derived for all beech-dominated stands from the above-mentioned beech mask in the large-scale altitudinal study was investigated.

3. Results

3.1. What Is Hidden behind the NDVI Value of Beech-Dominated Stands?

In the laboratory, we analyzed the spectral reflectance of beech leaves fallen in the previous growing season which covered the ground in beech stands before the beginning of the growing season. The NDVI$_{LAB}$ of the fallen leaves was lower than the satellite data-based NDVI$_{MOD}$ (Table 3). This suggested the presence of other forest components enhancing NDVI$_{MOD}$ at this time. Analyses revealed that a bark from beech trees reflected the considerable amount of the red and near infrared radiation (Figure 3a). It was interesting to find the differences between the NDVI$_{LAB}$ sampled from the bark of the branches and $NDVI_{LAB}$ of the bark from the stems (Table 3). These parts of trees are visible for satellites through the canopies up to the end of leaf onset (LO_100). The spectral analyses of the collected samples of green leaves revealed that the $NDVI_{LAB}$ of the leaves from the under-growing beech trees was lower than the $NDVI_{LAB}$ of the beech leaves from the canopy (Table 3). The spectral reflectance of these leaves differ in the red spectra (Figure 3a). Next, at the time of LO_100, the $NDVI_{LAB}$ of the canopy beech leaves was lower than the $NDVI_{LAB}$ at the time of final leaf onset (Table 3). This stemmed from the differences in spectral reflectance which at LO_100 was higher in the red spectrum (RED) and lower in the near-infrared spectrum (IRED) than at FLO (Figure 3b). The beech leaves collected when these two phenological phases started differed in their color, size, and structure (Figure 4).

Table 3. $NDVI_{LAB}$ of the samples collected on the particular DOY and the satellite-based $NDVI_{MOD}$ derived tor the same DOY: mean $\pm$ standard deviation, and the number of samples and values respectively, in the brackets.

Time of Sampling–DOY	97	110	127	159
Date	7.4	20.4	7.5	8.6
Canopy beech leaves	-	-	0.71 ± 0.023 (10)	0.79 ± 0.015 (8)
Under-growing beech leaves	-	-	0.66 ± 0.011 (15)	-
Fallen leaves	0.34 ± 0.067 (8)		-	-
Bark of branches	-	-	-	0.63 ± 0.028 (5)
Bark of stems	-	-	-	0.46 ± 0.042 (14)
Coral-root	-	0.71 ± 0.012 (4)	-	-
Dogs mercury	-	0.70 ± 0.003 (3)	-	-
$NDVI_{MOD}$	0.48 ± 0.062 (7)	0.58 ± 0.017 (10)	0.87 ± 0.010 (10)	0.91 ± 0.008 (5)

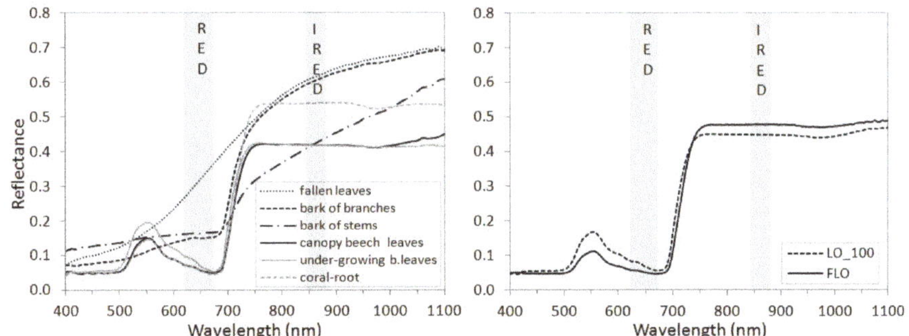

Figure 3. Spectral reflectance of all beech forest components measured with a Li 1800 integration sphere (**a**) and beech leaves from the phenological phases LO_100 (end of leaf onset) and final leaf onset (FLO) (**b**). The grey columns marked RED and IRED highlight the wavelengths used in the $NDVI_{LAB}$ calculations.

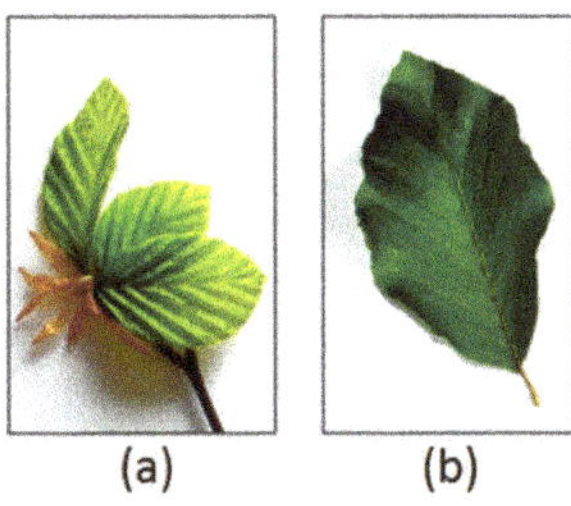

Figure 4. Difference in the beech leaves between the phenological phases of full leaf onset LO_100 (**a**) and final leaf onset (FLO) (**b**).

Based on data from the laboratory spectral analyses and in situ observations, it was concluded that the low values of $NDVI_{MOD}$ from the time with no green vegetation in the beech-dominated stands was caused together by presence of fallen leaves and the above-ground parts of the trees. This state lasted until DOY 93, when the $NDVI_{MOD}$ was 0.45 on average.

The slight increase of the $NDVI_{MOD}$ after DOY 98 caused flushing of the spring heliophytes—coral-root and dog's mercury. $NDVI_{LAB}$ of them reached 0.71 and 0.70 for coral-root and dog's mercury, respectively, and they covered from 5% (stands U1, U2, U4, U5) to 14% (stand U3) of the forest floor (Table S1). Subsequently, following the spring heliophytes, the under-growing beeches started their phenological phases. The $NDVI_{MOD}$ was around 0.58 (Table 3) when the canopy green leaves finished only the bud bursting (BB_100). The beech trees from the understory ended the leaf onset around DOY 110, when the cover of the ground vegetation in study stands was from 7 to 20% (Table S1). The most rapid increase of $NDVI_{MOD}$ was recorded from 0.58 (DOY 110) to 0.87 (DOY 124) from the end of canopy beech bud bursting to the end of leaf onset (from BB_100 to LO_100). After LO_100, both the $NDVI_{MOD}$ and $NDVI_{LAB}$ showed the continuing development of the canopy beech leaves. This development was also captured by the hemispherical photography, when the LAI values increased after LO_100 (Figure 5).

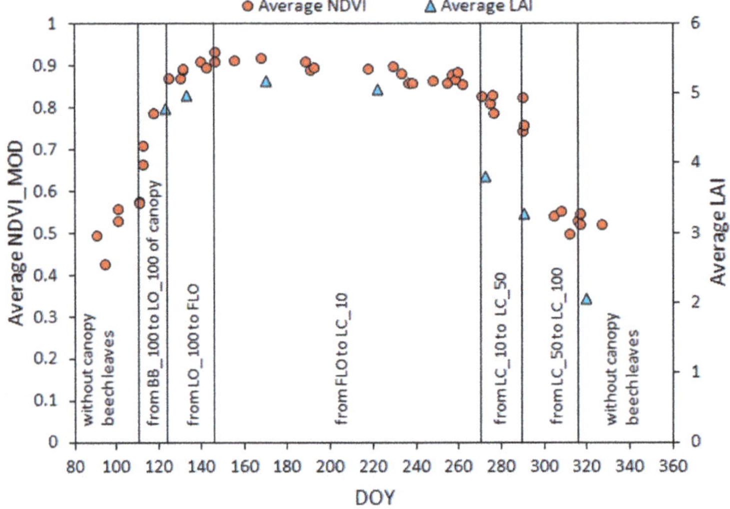

Figure 5. Seasonal changes in the $NDVI_{MOD}$ and LAI averaged over the test stands of U1–U5 in 2011. The year was divided into sections according to the phenological phases of canopy beech trees. NDVI_MOD, normalized difference vegetation index from MODIS data, LAI, leaf area index. DOY, day of year. BB_100, end of bud bursting. LO_100, end of leaf onset. LC_10, beginning of leaf coloring. LC_50, general leaf coloring. LC_100, end of leaf coloring.

The maximum values of $NDVI_{MOD}$ equal to 0.93 were achieved on DOY 146. At this time, the period of spring phenological phases of the canopy beech trees finished. The foliage was fully developed and this was the time of the final leaf onset. The maximum LAI values of the growing season were observed on DOY 170, but could appear earlier between the measurements on DOY 133 and 170. After reaching the final leaf onset, $NDVI_{MOD}$ started to slowly decrease. This gradual decrease from the average $NDVI_{MOD}$ of 0.93 to 0.86 lasted until DOY260.

The beginning of leaf coloring (LC_10) was observed on DOY 270. From this time, the $NDVI_{MOD}$ started to decrease faster, and this reduction was also recorded by the hemispherical images (Figure 5). After the leaf fall started on average on DOY 290, the decreasing LAI captured this loss of leaves in the canopy. The $NDVI_{MOD}$ continued decreasing and when the general leaf coloring (LC_50) was reached around DOY 290, the average $NDVI_{MOD}$ value was 0.76. Over the next two weeks, the $NDVI_{MOD}$ dropped to 0.54. The last $NDVI_{MOD}$ value recorded on DOY 319 was 0.52. At that time, all leaves were discolored and the $NDVI_{MOD}$ represented the reflectance of the forest components without green leaves and vegetation–litter, above-ground parts of trees, and brown leaves lasting on branches.

3.2. Assigning the Phenological Metrics to the In Situ Observed Phenological Phases

The phenological metrics were assigned to the in situ observed phenological phases according to the closest onset day and the differences were tested. In the first step of validation, the phenological data from the twelve study stands and the single years were compared against the corresponding data from phenological metrics. If the validations could not be realized due to the missing phenological phases that were not observed in the phenological network of the Slovak Hydrometeorological Institute, we validated only the phenological data from the detailed phenological observations from stands U1–U5.

3.2.1. Spring Phenological Phases

- In situ observed phenological phases such as bud swelling and bud bursting were not possible to observe in the satellite data because the increasing $NDVI_{MOD}$ at the time of their occurrence was caused by other forest components, and not by the canopy beech leaves. Therefore, phenological metric S1 was not paired with any in situ observed phenological phases.
- Phenological metric S2 was assigned to the beginning of leaf onset (LO_10). The paired sample *t*-test revealed non-significant differences between S2 and LO_10 ($t = 0.03 < T_{0.01}(90) = 1.96$). The correlation coefficient indicated very strong correlation between LO_10 and S2 (Figure 6). The average $NDVI_{MOD}$ of the S2 was 0.67.

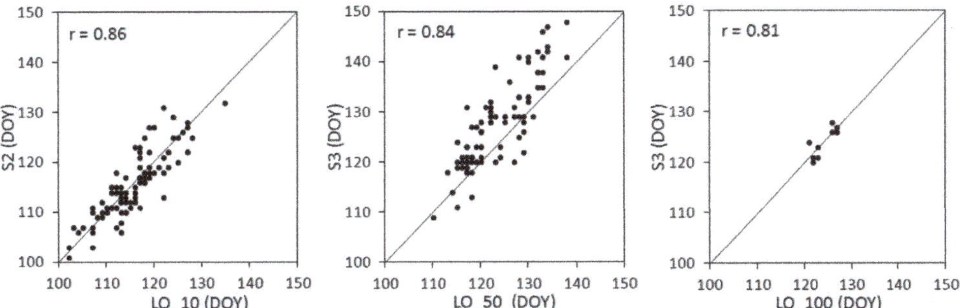

Figure 6. Correlation between the in situ observed spring phenological phases and assigned spring phenological metrics: S2 and beginning of leaf onset LO_10 (**a**), S3 and general leaf onset LO_50 (**b**), S3 and end of leaf onset LO_100 (**c**). The high values of the correlation coefficient indicated very strong correlation between the phenological phases and phenological metrics. DOY, day of year.

- Phenological metric S3 was assigned to the phenological phase LO_50, and although the r = 0.84 (Figure 6) suggested a very strong correlation between S3 and LO_50, the *t*-test revealed significant differences with a RMSE of 6.24. The phenological metric S3 was overestimated. The phenological data from the SHMI, which were the base data for the validation, did not contain information on the phenological phase LO_100, which, according to our detailed phenological monitoring, occurred on average four days after LO_50. No significant differences were revealed between S3 and LO_100 ($t = 0.10 < T_{0.01}(9) = 3.25$) in our U1–U5 stands, although the correlation coefficient was smaller than the one between S3 and LO_50 (Figure 6). A control test by adding four days (average difference between LO_50 and LO_100) to the days of LO_50 from the SHMI observations was performed. The differences between the estimated LO_100 and S3 were non-significant ($t = 0.54 < T_{0.01}(90) = 1.96$) with the RMSE of 4.57. The average $NDVI_{MOD}$ of S3 was 0.81.

3.2.2. Autumn Phenological Phases

- Phenological metric A1 occurred between the two autumn phenological phases: beginning of leaf coloring (LC_10) and general leaf coloring (LC_50), on average eight days after LC_10. Thus, we paired A1 with LC_10 and LC_50 from all twelve study stands and single years, but the differences in both cases were significant. Then, using data from detailed phenological monitoring on the U1–U5 stands, the pairs of A1 with LC_20, LC_30, and LC_40 were tested, but the differences in all of the tested pairs were significant.

- The paired test of the onset days of A2 and LC_50 revealed an average time lag of 13 days for the A2 phenological metric. Based on that, the pairs of A2 with LC_60, LC_70, LC_80, LC_90, and LC_100 recorded in the U1–U5 stands were tested. Only the paired t-test with LC_80 revealed non-significant differences ($t = 1.51 < T_{0.01}(9) = 3.25$) between the in situ LC_80 and satellite data-based A2.

- A3 occurred after the end of leaf coloring (LC_100) when only the understory vegetation was green, but the beech trees had no green leaves. Therefore, this phenological metric was not paired with any autumn in situ observed phenological phase.

- The correlation between the onset days of the autumn phenological phases and phenological metrics was weak, with the correlation coefficient not exceeding 0.30 (Figure 7).

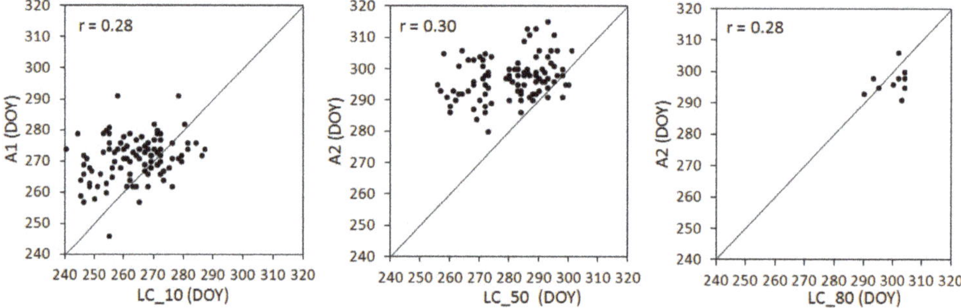

Figure 7. Correlation between the in situ observed autumn phenological phases and assigned autumn phenological metrics: A1 and beginning of leaf coloring LC_10 (**a**), A2 and general leaf coloring LC_50 (**b**), A2 and 80% leaf coloring LC_80 (**c**). The correlation was weak in all cases. DOY, day of year.

3.3. Phenological Metrics along the Altitudinal Gradient

Phenological metrics validated in the previous sections were used in the large-scale analyses of the effect of altitude on the onset days of phenological phases. We evaluated the phenological metrics S2, S3, A2, and GSL corresponding to the phenological phases beginning of leaf onset, end of leaf onset, 80% leaf coloring, and the growing season length, respectively. The representative onset dates of a particular phenological phase expressed by the phenological metrics were calculated for each beech-dominated pixel from the beech mask and averaged over a temporal series of 13 years (2000–2012). In all cases, the large-scale regression analyses revealed non-linear dependences of the phenological metrics on altitude. The second order polynomial function was used to fit the datasets (Figure 8a–d). The altitudinal gradients derived from regression functions are shown in Table 4.

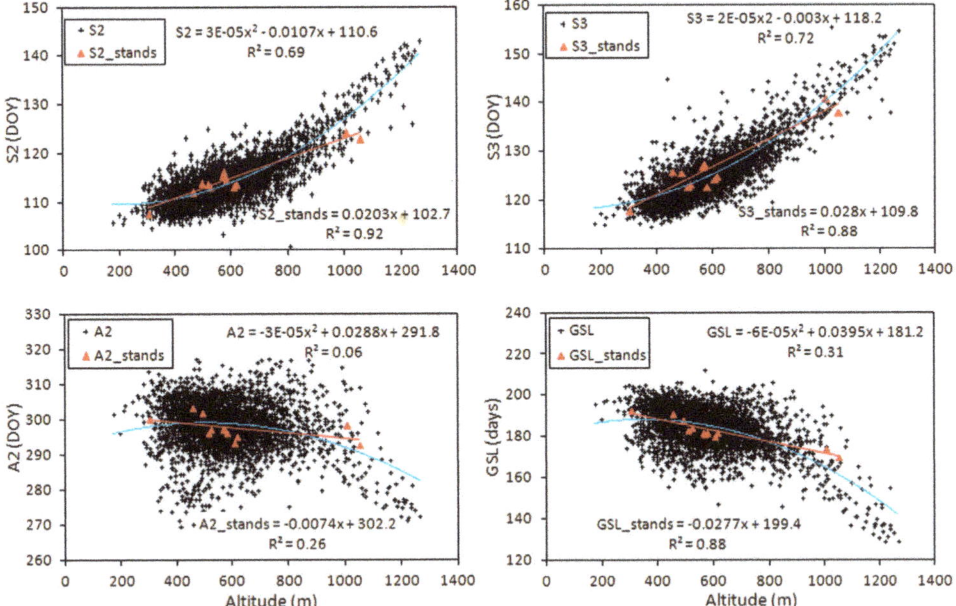

Figure 8. Satellite-based phenological metrics S2 (**a**), S3 (**b**), A2 (**c**), and growing season length, GSL (**d**) as a function of altitude at a stand-level and large-scale level. The individual phenological metrics of 12 study stands are indicated by red triangles and the altitudinal trend is the linear red line. The phenological metrics derived from all beech-dominated stands (60% of beech and higher) in the area of Slovakia are indicated by black crosses and the altitude trend is the polynomial blue line. DOY, day of year.

Table 4. The altitudinal gradient of the phenological metrics S2, S3, A2, and GSL calculated from regression functions for each 100 m of the vertical distribution range of beech-dominated pixels (from 156 to 1331 m a.s.l.). Values in the table were rounded to the whole day.

Altitude	Altitudinal Gradient of Large-Scale Phenology				Difference on Altitudinal Gradient			
	S2	S3	A2	GSL	S2 [1]	S3 [1]	A2 [2]	GSL [3]
156	109	118	296	186	0	-	−3	−2
200	108	118	296	187	0	0	−2	−1
300	108	119	298	188	-	1	−1	-
400	109	120	299	187	0	2	0	0
500	110	122	299	186	1	3	-	−2
600	111	124	298	183	3	5	0	−4
700	113	126	297	179	5	8	−1	−8
800	116	129	296	174	8	10	−3	−13
900	120	132	293	168	11	13	−5	−20
1000	124	135	291	161	15	17	−8	−27
1100	128	139	287	152	20	21	−12	−36
1200	133	143	283	142	25	25	−16	−45
1300	139	148	279	131	31	30	−20	−57
1331	141	150	277	127	33	31	−22	−60

[1] Differences from the earliest onset of S2 and S3, respectively, [2] Differences from the latest onset of A2, [3] Differences from the longest growing season.

The average beginning of leaf onset expressed by the S2 metric increased with altitude. At low altitudes (200–300 m a.s.l.), the beginning of leaf onset occurred earliest on DOY 108 and delayed with the increasing altitude up to DOY 141 (Figure 8a). Strong dependence was revealed between S2 and altitude, since 68% of the total variation in the onset days of S2 was caused by their dependence on the altitude. The difference between the earliest and the latest S2 days estimated from the regression function was 33 days.

The polynomial function, which characterized the altitudinal gradient of average end of leaf onset expressed by the S3 (Figure 8b), had a very similar course to the one for S2 with a 31-day difference between the earliest onset at lowest altitudes (DOY 118) and the latest onset of S3 at the highest altitudes (DOY 150). Dependence of S3 on the altitude was strong, as 72% of the total variation in the onset days of S3 could be explained by the applied second order polynomial function.

The altitudinal gradient of the average onset days of 80% leaf coloring, expressed by the phenological metric A2, differed from those of the spring phenological phases. The latest onset of A2 was generally revealed in the altitudinal range of 400–600 m a.s.l. Only a weak relationship was found between A2 and altitude as the polynomial function could explain only 6% of the A2 variation. According to our datasets, the variability in the onset of this autumn phenological phase was greater in the comparison with leaf onset. In particular, in the altitudinal range 300–700 m a.s.l., an unexpected very early onset of A2 from DOY 270 in more pixels was recorded (Figure 8c).

The length of the growing season was quite varying from 127 days at the highest altitudes to 188 days at the lowest altitudes. Using the second order polynomial function, 27% of the total variation in the growing season length of beech pixels by its dependence on altitude could be explained. Used regression function indicated that the differences in growing season length in the altitudinal range from 156 to 500 m a.s.l. varied within two days and was shortened with increasing altitude.

When considering only the phenological metrics from our 12 study stands, the regression analyses revealed that a linear function fit the data better. The changes at a stand scale estimated by the linear regression along the altitudinal gradient were smaller than those revealed at a large-scale. The S2 of stands delayed by two days at 100 m^{-1} and S3 delayed by 2.8 days at 100 m^{-1}. The onset of S3 along the altitudinal gradient changed a little bit more gradually than that of S2. The onset of A2 derived from the regression model advanced only 0.8 days at 100 m^{-1} and GSL was shortened by about 2.5 days at 100 m^{-1}.

The average GSL of the 12 study stands was considered from the climatic aspect using the normal temperatures and precipitation. The longest growing season (188–193 days) with the early onset of LO_10 (S2) and the latest LC_80 (A2) was at the lowest altitudes <500 m a.s.l. (Table S2). The stands situated from 500 to 700 m a.s.l. had the GSL between 181–185 days. The shortest growing seasons (170–174 days) with the latest LO_10 (S2) were found in the stands located at the highest altitudes >1000 m a.s.l. (Table S2). The normal temperature of stands increased linearly with altitude by 0.61 °C for every 100 m increase in altitude. The $R^2 = 0.97$ indicated a very strong correlation between altitude and normal temperature (Figure S1). The absolute difference of normal temperature between the beech stands at the lowest and the highest altitude was 4.9 °C. The length of the growing season extended along the temperature gradient by 4.5 days for every 1 °C of normal temperature increase and a very strong relationship between these two variables was found (Figure 9a). The rate of change in the normal precipitation with increasing altitude was 28 mm with 100 m of the lift. Study stands are situated in moderate moist to moist climatic areas and the absolute difference in precipitation between the beech stands at the lowest and the highest altitude was only 188 mm. The relationship between altitude and precipitation was strong with $R^2 = 0.77$ (Figure S1). GSL shortened by −6.9 days with every 100 mm of precipitation increase. The relationship between GSL and normal of precipitation expressed by R^2 was 0.55 (Figure 9b).

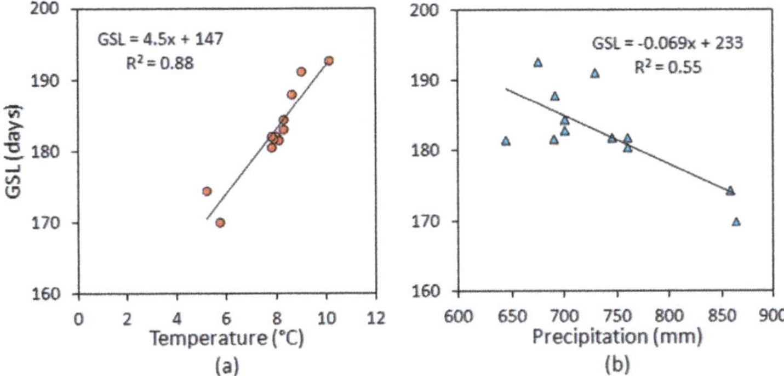

Figure 9. Growing season length in the twelve study stands in relation with normal temperature (**a**) and precipitation (**b**).

4. Discussion

The results presented in this study were based on the daily MODIS products called MOD09 and MYD09, which had good capability of detecting gradual changes in green leaves at the time of leaf onset and to date it accurately. Derived single-pixel $NDVI_{MOD}$ values represented the reflectance of all forest components. The effect of different forest components on the satellite data-based vegetation indices had already been outlined in previous studies. Reed et al. [16] pointed out that the reflectance at every pixel was the conglomeration of a variety of vegetation and background features, which may obscure observations of initial bud burst. During spring, the understory generally starts earlier than the main tree species. Testa et al. [17] found out that when the understory in beech stands reached the 90% level of leaf onset, the main species had only reached its 10% level. These tendencies were also revealed in our beech study stands and raised questions about what is hidden behind the NDVI value of the beech forests, especially at the time before the canopy closing. The laboratory spectral analyses revealed that not only did the leaves of the understory and canopy participate in the pixel reflectance, but other parts of trees also reflected the red and near-infrared radiation important for the NDVI calculation. The chlorophyll content in the bark of beech stems and branches [47] absorbed a considerable amount of radiation in the red spectra. This resulted in $NDVI_{LAB}$ values similar to that of the fallen leaves (Figure 3). The presence of fallen leaves, stems, and branches in the beech pixels kept the $NDVI_{MOD}$ on a relatively stable level (around 0.42) when there was no snow cover on the forest floor. $NDVI_{MOD}$ started to increase from that level after the incidence of the first spring heliophytes and the undergrowth flushing, but the canopy beech trees were only in the phenological phase of 10% bud bursting, and therefore had no green leaves. The reflectance signal of bud burst was not sufficient to alter the reflectance signal from a larger mixed pixel, while field observers may be able to note bud swelling and bursting [13,14]. With regard to previous studies [8,17,48–50], it is probable that the incidence of spring heliophytes and the undergrowth strategy of phenological escape is a common occurrence in forests. Therefore, the first phenological metric occurring simultaneously with the beginning of the bud bursting of the canopy (BB_10), usually situated at the beginning of NDVI increase on the sigmoid function (in this study phenological metric S1), characterized the leafing of the understory and thus was not suitable to track phenological phases of the canopy trees. Liang et al. [14], who assigned the first point on the NDVI sigmoid function (S1) to the end of bud bursting (BB_100) of the canopy trees, revealed an underestimation of S1 of more than two weeks. In most cases, this point on the NDVI profile could be interpreted as a start of the vegetation period in forests in general [12], but not as a tree specific phenological phase. Different results may be obtained when the enhanced vegetation index (EVI) is used to track the phenological phases because of its insensitivity to the forest background. Significant agreement was found between the first point on the EVI piecewise sigmoid function and BB_10 [14]. According to these results, we concluded that the pairing of phenological metrics from EVI and NDVI fitted with the in situ observed phenological phases needed two different approaches. However, MODIS spectral bands needed for EVI calculations were available from the spatial resolution 500 m, we preferred to analyze NDVI time series calculated from the bands available from the resolution 250 m.

The variables derived from the in situ observations provided information that could be used for the interpretation of the variables derived from the satellite data. The leaf area index is one of the main variables affecting the vegetation reflectance and is also an indicator of changing phenological phases. The LAI measurements were used to analyze the phenological changes, especially after the leaf onset and during the autumn leaf coloring and leaf fall. The analyses of $NDVI_{MOD}$ seasonal profiles revealed a continuous increase after the in situ observed end of leaf onset (DOY124), similar to the findings with EVI [13]. Our findings were supported by the measurements of the LAI and $NDVI_{LAB}$ of the beech leaf samples. $NDVI_{LAB}$ increased from 0.71 on DOY 127 to 0.79 on DOY 159. This increase was caused by the increased reflectance in the near-infrared spectrum and decreased reflectance in the red spectrum related to the increasing chlorophyll content between LO_100 and final leaf onset (FLO) [51]. In FLO, the leaves achieved their final shape, size, and structure, which were also recognized by the LAI measurements based on hemispherical photography. At that time, the foliage was fully developed. FLO is visually hardly observable because of the minimal structural changes of canopy leaves that are barely visible to the human eye, but is recordable by optical and spectral measurement based indices. Subsequently, after reaching the maximum leaf area, the LAI and $NDVI_{MOD}$ started to slightly decrease. Similar results have been revealed in beech and oak stands in France, where the NDVI, measured by an in situ downward looking spectroradiometer, recorded a slight monotonic decrease during the spring period [52,53]. In the summer months, there is no biosynthesis of the chlorophyll in beech leaves, thus the photo-degraded chlorophyll is not replaced by the new [54]. This phenomenon of NDVI decrease measured at the leaf-level was marked as "summer greendown", to which NDVI is sensitive [51]. The reduction of LAI during the summer period can be explained by a drop of turgor in the leaf cells due to the lack of moisture. In autumn, the indices reacted to the different phenological phases. While the decreasing LAI was caused by the ongoing leaf fall, the NDVI reflected the changes in leaf color and structure.

The next step in the validation process was to assign the satellite data-based phenological metrics to the ground observed phenological phases. The key problem in pairing these two data sources presented by previous authors was the insufficient frequency of the ground phenological observations, resulting in greater differences between the paired values [9]. In this study, very detailed phenological monitoring was performed to exclude this deficiency, and significant matches were found. According to our results, the first spring vegetative phenological phase that could be reliably deduced from the $NDVI_{MOD}$ phenology metrics used in this study was the LO_10 of canopy beech trees, which corresponded to S2 (inflection point) with a bias of 0 days and an average $NDVI_{MOD}$ of 0.67. The phenological phase LO_10 was observed, on average, three days after full bud bursting, which is usually paired with the inflection point of the NDVI sigmoid function (S2) [48,55]. Nevertheless, Fisher and Mustard [9] also revealed that the inflection point of the sigmoid function occurred between the observed BB_50 and LO_50, closer to LO_50. In this study, the next spring phenological phase LO_100 was linked to S3, with an average $NDVI_{MOD}$ equal to 0.81. The $NDVI_{MOD}$ values of S2 and S3 corresponded to those measured in the field spectral experiment published by Nagai et al. [50], who concluded that NDVI of 0.6 indicated the start of leaf expansion, a NDVI of 0.7 indicated that about 50% of the canopy leaves had expanded, and a NDVI of 0.8 meant that all of the canopy leaves had expanded.

Determination of the autumn phenological phases from the metrics used in this study was more complicated when compared to the spring phenological phases. The main reason for this is the development of the European beech phenological phases, especially the order of beech leaves coloring. The phenological phase of the autumn senescence of beech leaves start from the upper and edge parts of the canopy [56]. These leaves are directly exposed to sunlight and photodegrade first. The satellite data scan forests from above and thus are able to record these first changes of leaf senescence, while the in situ observers, looking from the ground concurrently, do not recognize any progress. Hence, there is a question of which beginning of leaf coloring is more reliable as they do not appear to be comparable. In this study, A1 was paired with different stages of observed leaf coloring and no matches were found, although this phenological metric appeared between the observed LC_10 and LC_50. For further validation of the first autumn phenological metric, we proposed the use of cameras placed above the canopy [57] or drones [58], which could identify the phenological phases from the same direction as the satellites. The next phenological metric, A2, which is commonly used in other studies to identify leaf senescence, was the inflection point of the sigmoid function. Some other authors have marked this point as the beginning of yellowing [58] or paired it with the leaf fall phase [15,50,59]. This phenological metric (A2) appeared after LC_50 was observed in all of the study stands. Using the data from the detailed monitoring in U1–U5, it was compared to all phenological stages of leaf coloring observed after LC_50. Significant agreement was found with LC_80. The last point on the sigmoid function, A3, occurred after LC_100 and therefore was not interpreted in relation to the beech phenological phases. In a coarser context, it could be interpreted as the beginning of dormancy in these stands. Liu et al. [60] marked the last point as the full leaf coloring with differences of 0–12 days, but also concluded that the MODIS VI data consistently estimated later dormancy onset relative to the ground observed full leaf coloration. In this study, the leaf fall phenological phase was not evaluated using $NDVI_{MOD}$ based phenological metrics because it is related to the position of leaves, and not their reflectance.

Altitude is the most significant factor when considering the effect of topography on the onset of phenological phases, and it is linked with the reduction of air temperature and increase of precipitation [61]. Averaging the phenological metrics from 13 years into a single representative value made it possible to investigate the altitude variations in seasonal remote sensing responses over a large range of altitudes (from 156 to 1331 m a.s.l.). Regression analyses were utilized to discover the dependence of the onset of European beech phenological phases on the altitude and climatic conditions. The analyses were performed at two levels: at the stand scale and at a large-scale, which demonstrated different results. While the stand scale regressions revealed linear relationships with correlations (R^2) of 0.92, 0.88, 0.26, and 0.88 for S2, S3, A2, and GSL, respectively, the large-scale analyses found non-linear regressions between the phenology and altitude. In previous studies, linear regression trends have been demonstrated in stand scale analyses, which were similar to our results. The studies discovered a lag of 1.1 days 100 m^{-1} with R^2 = 0.63 (10 stands, France) [43], and a lag of 1 day 100 m^{-1} with R^2 = 0.51 (22 stands, France) [4] in the oceanic climate of Western Europe; a lag of 2.6 days 100 m^{-1} with R^2 = 0.73 in the south of Central Europe (47 stands, Slovenia) [42]; and a lag of 1.7–3.9 days 100 m^{-1} with R^2 = 0.22–0.81 to the west of Central Europe (97 stands divided into six regions, Germany) [41]. In the above-mentioned studies, regression analyses of the onset of the autumn coloring of European beech revealed only a weak dependence on altitude with a R^2 less than 0.20 [41,42].

Missing the ground observed data in the highest altitude could lead into the linear dependence of the spring phenological phases on altitude. The study stands observed in situ are located in the ecological optimum or suboptimum of the distributional range of European beech, where the effect of the temperature could be dominant. In the highest altitudes, where the ecological pessimum for growing occurs, the phenology onset reflects the interaction of extreme ecological conditions such as lower temperature; longer snow cover; stronger winds, etc. The trees in such conditions react by the much later onset of leafing, bringing the non-linearity in the shape of regression. When a large amount of data along the whole altitudinal gradient is available, the regression curve may acquire a variety of shapes. Hwang et al. [61] applied a non-linear model fitting procedure to derive the local patterns of topography-mediated vegetation phenology from MODIS vegetation indices. Except for midday of the green-up period, all variables showed quadratic associations with altitude, reflecting an interaction between the topoclimatic patterns of temperature and water availability. Gyon et al. [4]

used linear regression for the leaf onset days along the altitudinal gradient, but the results revealed that for low altitudes (<500 m a.s.l.), the delay of leaf onset was less than the expected one: 1.5 days $100\ m^{-1}$ compared to 2.3 days $100\ m^{-1}$ when applied to the whole altitudinal range. The second order polynomial function best fitted our phenological data in relation to altitude. When comparing the spring and autumn phases, the onset days of the S2 and S3 metrics were less variable, with the data concentrated around the approximated value.

The onset of LC_80 had already started in some beech forests at the end of the summer. This early onset may result from a variety of causes. One source of discrepancies could be the lack of $NDVI_{MOD}$ data, hence, the distorted fitting procedure. This may appear during the autumn months, when the incidence of fog and low clouds is more probable in the mountainous landscape. The next cause of a decline in $NDVI_{MOD}$ could be due to some kind of leaf damage. During the growing season, beech trees and leaves are threatened by several biotic (insects) and abiotic (hot temperatures, drought stress, incidence of spring and summer frost, etc.) factors, and trees could also be harvested in the managed forests. Regardless of these anomalies, the onsets of the autumn phenological phases were very variable and only partially depended on the altitude.

The capability of detecting altitude-mediated phenology offers the potential to detect vegetation responses to local-climate conditions. In this study, we used the normal values of temperature and precipitation from the last 30 years to assess the altitude-conditioned differences in growing season length at the study stand scale. The rate of temperature change along the altitudinal gradient ($0.61\ ^\circ C$ $100\ m^{-1}$) corresponded with the generally valid moist adiabatic lapse rate. The temperature conditions with a sufficient amount of precipitation ensured the longest growing season (188–193 days) at the lowest altitudes (<500 m a.s.l.). This is related to both the early beginning of leaf onset LO_10 and the late onset of 80% leaf coloring (LC_80). However, the temperature and precipitation conditions seemed to be the most suitable for European beech growth; the beech trees at these altitudes could be most endangered by the late frost damage [62,63]. Fresh new leaves are especially vulnerable. The average length of the growing season in the stands situated from 500 to 700 m a.s.l. varied between 181–185 days. Here, the normal precipitation differed only slightly between the study stands, while the temperature decrease resulted in shorter growing seasons with both later onset of LO_10 and the earlier onset of LC_80 in comparison to stands at the lowest altitudes. The shortest growing seasons (170–174 days) were found in the stands located at the highest altitudes (>1000 m a.s.l.). The lowest temperature in the moist environment lead to the late onset of LO_10, but the end of the growing season in these climate conditions appeared on similar days as at the lower altitudes. From the point of beech production, this is an important finding since canopy duration affects the carbon assimilation rates and the energy budget of forest ecosystems from the stand-level to the global scale [64]. It has been proven that the length of the growing season correlates with tree height and diameter growth while this effect is associated with the timing of growth cessation rather than with leaf onset [65]. It follows that beech stands at higher altitudes are similarly productive as those situated lower, while the probability of late frost damage is much smaller. When the sensitivity of leaf onset days to temperature from our study were compared with the data from Western Europe, there was a much lower sensitivity of European beech to temperature decrease (-2.1 days $^\circ C^{-1}$) in Western Europe [43]. The differences in the sensitivity of leaf onset days to temperature between the data from Western (France) and Central Europe (Slovakia) may result from the genetically conditioned internal periodicity of different populations of European beech. Furthermore, the data from the west of Central Europe (Germany) showed differences in the sensitivity of the beech growing season length on altitude-conditioned climatic factors and population locations [41]. The onset of phenological phases and growing season length could vary around the presented averaged data with regard to other climatic or topographic factors, which were not considered in this study.

5. Conclusions

The remote sensing-based approaches for monitoring phenology have been the focus of research over the past decades. The validation of the phenological metrics against the in situ observation data is still crucial. Here, the data retrieved from the in situ phenological observations, detailed monitoring of the forest stands, laboratory spectral analyses, and LAI measurements were used to detect what the satellite data represent. Next, we described and tested a method to detect the timing of spring and autumn phenological phases for beech-dominated stands. Our results show that phenological metrics applied here correspond closely to phenological observations collected on the ground in 12 beech-dominated stands. Significant compliance was found especially in the spring period. The onset of autumn phenological phases needs to be further explored due to the observation discrepancy between downward looking satellites and upward looking observers. In the future, the utilization of phenological cameras or drones could fill the gaps in the data validation.

The phenological metrics we described in this paper have the potential to improve the knowledge of phenological responses to a variety of factors. The large-scale altitudinal study presented here showed the advantage of validated satellite data-based phenological metrics. However, the moderate spatial resolution data source MODIS was used, so the results suggest that it should be possible to generate multi-year observations of phenological phases at 250 m spatial resolution over large areas. The large-scale analysis can provide new opportunities to study tree-species phenology in the context of climate change.

Supplementary Materials: The following are available online at http://www.mdpi.com/1999-4907/10/1/60/s1, Figure S1: The relationship between altitude and normal temperature and precipitation on the study stands, Table S1: Average cover of ground vegetation in stands U1–U5, Table S2: Average transition days of the phenological metrics S2 and A2, and the growing season length, with the temperature and precipitation normal.

Author Contributions: V.L., J.Š. (Jaroslav Škvarenina), and J.Š. (Jana Škvareninová) conceived and designed the experiments; V.L. and J.Š. (Jana Škvareninová) performed the experiments; V.L. and T.B. analyzed the data; T.B. and J.Š. (Jaroslav Škvarenina) contributed reagents/materials/analysis tools; and V.L. wrote the paper.

Acknowledgments: This work was accomplished as part of the VEGA projects No.: 1/0589/15, 1/0111/18 and 1/0500/19 of the Ministry of Education, Science, Research, and Sport of the Slovak Republic and the Slovak Academy of Science; and the projects of the Slovak Research and Development Agency No.: APVV-15-0425 and APVV-15-0497. The authors thank these agencies for their support. The authors are notably thankful to two researchers: Ing. Ivana Vasiľová and Ing. Jozef Zverko for their participation in detailed phenological monitoring; and to Ing. Vladimír Kriššák for his technical support during the terrestrial works.

Conflicts of Interest: The authors declare no conflict of interest. The founding sponsors had no role in the design of the study; in the collection, analyses, or interpretation of data, in the writing of the manuscript, and in the decision to publish the results.

References

1. Ciais, P.; Sabine, C.; Bala, G.; Bopp, L.; Brovkin, V.; Canadell, J.; Chhabra, A.; DeFries, R.; Galloway, J.; Heimann, M.; et al. Carbon and Other Biogeochemical Cycles. In *Climate Change 2013: The Physical Science Basis. Contribution of Working Group I to the Fifth Assessment Report of the Intergovernmental Panel on Climate Change*; Stocker, T.F., Qin, D., Plattner, G.-K., Tignor, M., Allen, S.K., Boschung, J., Nauels, A., Xia, Y., Bex, V., Midgley, P.M., Eds.; Cambridge University Press: Cambridge, UK; New York, NY, USA, 2013; pp. 465–570. [CrossRef]

2. Badeck, F.W.; Bondeau, A.; Böttcher, K.; Doktor, D.; Lucht, W.; Schaber, J.; Sitch, S. Responses of spring phenology to climate change. *New Phytol.* **2004**, *162*, 295–309. [CrossRef]

3. Polgar, C.; Gallinat, A.; Primack, R.B. Drivers of leaf-out phenology and their implications for species invasions: Insights from Thoreau's Concord. *New Phytol.* **2014**, *202*, 106–115. [CrossRef]

4. Guyon, D.; Guillot, M.; Vitasse, Y.; Cardot, H.; Hagolle, O.; Delzon, S.; Wigneron, J.P. Monitoring elevation variations in leaf phenology of deciduous broadleaf forests from SPOT/VEGETATION time-series. *Remote Sens. Environ.* **2011**, *115*, 615–627. [CrossRef]

5. Mátyás, C.; Berki, I.; Czúcz, B.; Gálos, B.; Móricz, N.; Rastovits, E. Future of Beech in Southeast Europe from the Perspective of Evolutionary Ecology. *Acta Silv. Lignaria Hung.* **2010**, *6*, 91–110.

6. Becker, M. Taxonomie et characteres botaniques. In *Le Hetre*; Teissier du Cros, E., Ed.; INRA: Paris, France, 1981; pp. 35–49. ISBN 978-2-7592-0512-7.

7. Myneni, R.B.; Hoffman, S.; Knyazikhin, Y.; Privette, J.L.; Glassy, J.; Tian, Y.; Wang, Y.; Song, X.; Zhang, Y.; Smith, G.R.; et al. Global products of vegetation leaf area and fraction absorbed PAR from year one of MODIS data. *Remote Sens. Environ.* **2002**, *83*, 214–231. [CrossRef]

8. Fisher, J.I.; Mustard, J.F.; Vadeboncoeur, M.A. Green leaf phenology at Landsat resolution: Scaling from the field to the satelite. *Remote Sens. Environ.* **2006**, *100*, 265–279. [CrossRef]

9. Fisher, J.I.; Mustard, J.F. Cross-scalar satellite phenology from ground, Landsat and MODIS data. *Remote Sens. Environ.* **2007**, *109*, 261–273. [CrossRef]

10. Beck, P.S.A.; Atzberger, C.; Høgda, K.A.; Johansen, B.; Skidmore, A.K. Improved monitoring of vegetation dynamics at very high latitudes: A new method using MODIS NDVI. *Remote Sens. Environ.* **2006**, *100*, 321–334. [CrossRef]

11. Jönsson, A.M.; Hellström, M.; Bärring, L.; Jönsson, P. Annual changes in MODIS vegetation indices of Swedish coniferous forests in relation to snow dynamics and tree phenology. *Remote Sens. Environ.* **2010**, *114*, 2719–2730. [CrossRef]

12. Zhang, X.; Friedl, M.A.; Schaaf, C.B.; Strahler, A.H.; Hodges, J.C.F.; Gao, F.; Reed, B.C.; Heute, A. Monitoring vegetation phenology using MODIS. *Remote Sens. Environ.* **2003**, *84*, 471–475. [CrossRef]

13. White, K.; Pontius, J.; Schaberg, P. Remote sensing of spring phenology in northeastern forests: A comparison of methods, field metrics and sources of uncertainty. *Remote Sens. Environ.* **2014**, *148*, 97–107. [CrossRef]

14. Liang, L.; Schwartz, M.D.; Fei, S. Validating satellite phenology through intensive ground observation and landscape scaling in a mixed seasonal forest. *Remote Sens. Environ.* **2011**, *115*, 143–157. [CrossRef]

15. Chen, X.; Luo, X.; Xu, L. Comparison of spatial patterns of satellite-derived and ground-based phenology for the deciduous broadleaf forest of China. *Remote Sens. Lett.* **2013**, *4*, 532–541. [CrossRef]

16. Reed, B.C.; Brown, J.F.; VanderZee, R.; Merchant, J.W.; Ohlen, D.O. Measuring phenological variability from satellite imagery. *Int. Assoc. Veg. Sci.* **1994**, *5*, 703–714. [CrossRef]

17. Testa, S.; Soudani, K.; Boschetti, L.; Borgogno Mondino, E. MODIS-derived EVI, NDVI and WDRVI time series to estimate phenological metrics in French deciduous forests. *Int. J. Appl. Earth Obs. Geoinf.* **2018**, *64*, 132–144. [CrossRef]

18. Škvarerninová, J.; Snopková, Z. The Development of Phenological Stages of European Beach (*Fagus sylvatica* L.) in Slovakia During the Period of 1996–2010. In Proceedings of the Bioclimate: Source and Limit of Social Development, Topoľčianky, Slovakia, 6–9 September 2011; pp. 87–90.

19. Škvareninová, J. *Vplyv Zmeny Klimatických Podmienok na Fenologickú Odozvu Ekosystémov*; Vydavateľstvo Technickej Univerzity vo Zvolene: Zvolen, Slovakia, 2013; 132p, ISBN 978-80-228-2598-6.

20. Bucha, T. Classification of tree species composition in Slovakia from satellite images as a part of monitoring forest ecosystems biodiversity. *Acta Inst. For. Zvolen* **1999**, *9*, 65–84.

21. Meier, U. *Growth Stages of Mono- and Dicotyledonous Plants*, 2nd ed.; Federal Biological Research Centre for Agriculture and Forestry: Berlin, Germany; Braunschweig, Germany, 2001; 158p.

22. Bruns, E.; Chmielewski, F.M.; van Vliet, A.J.H. The Global Phenological Monitoring Concept. In *Phenology: An Integrative Environmental Science*; Schwartz, M., Ed.; Springer: Dordrecht, The Netherlands, 2003; pp. 93–104. ISBN 978-94-007-0632-3.

23. Chmielewski, F.M.; Rötzer, T. Phenological trends in Europe in relation to climatic changes. *Agrometeorol. Schr.* **2000**, *7*, 1–15.

24. Škvareninová, J.; Tuhárska, M.; Škvarenina, J.; Babálová, D.; Slobodníková, L.; Slobodník, B.; Středová, H.; Minďaš, J. Effects of light pollution on tree phenology in the urban environment. *Morav. Geogr. Rep.* **2017**, *25*, 282–290. [CrossRef]

25. Babálová, D.; Škvareninová, J.; Fazekaš, J.; Vyskot, I. The dynamics of the phenological development of four woody species in south-west and central Slovakia. *Sustainability* **2018**, *10*, 1497. [CrossRef]

26. Lukasova, V.; Lang, M.; Skvarenina, J. Seasonal changes in NDVI in relation to phenological phases, LAI and PAI of beech forests. *Balt. For.* **2014**, *20*, 248–262.

27. Land Processes Distributed Active Archive Center. Available online: https://lpdaac.usgs.gov/dataset_discovery/modis/modis_products_table/mod09gq_v006 (accessed on 13 March 2018).

28. Ju, J.; Roy, D.P.; Shuai, Y.; Schaaf, C. Development of an approach for generation of temporally complete daily nadir MODIS reflectance time series. *Remote Sens. Environ.* **2011**, *114*, 1–20. [CrossRef]

29. Franch, B.; Vermote, E.F.; Sobrino, J.A.; Fédèle, E. Analysis of directional effect on atmospheric correction. *Remote Sens. Environ.* **2013**, *128*, 276–288. [CrossRef]

30. Townshend, J.R.G.; Huang, S.N.; Kalluri, V.; Defries, R.S.; Liang, S. Beware of the per-pixel characterization of land cover. *Int. J. Remote Sens.* **2000**, *21*, 839–843. [CrossRef]

31. Bucha, T.; Koreň, M. Phenology of the beech forests in the Western Carpathians from MODIS for 2000–2015. *iForest Biogeosci. For.* **2017**, *10*, 537–546. [CrossRef]

32. Evans, J.D. *Straightforward Statistics for the Behavioral Sciences*; Brooks/Cole Publishing: Pacific Grove, CA, USA, 1996; 600p.

33. Kramer, K. Phenotypic plasticity of the phenology of seven European tree species in relation to climatic warming. *Plant Cell Environ.* **1995**, *18*, 93–104. [CrossRef]

34. Matsumoto, K.; Ohta, T.; Irasawa, M.; Nakamura, T. Climate change and extension of the *Ginkgo biloba* L. growing season in Japan. *Glob. Chang. Biol.* **2003**, *9*, 1634–1642. [CrossRef]

35. Estrella, N.; Menzel, A. Response of leaf colouring in four deciduous tree species to climate and weather in Germany. *Clim. Res.* **2006**, *32*, 253–267. [CrossRef]

36. Wang, T.; Ottle, C.; Peng, S.; Janssens, I.A.; Lin, X.; Poulter, B.; Yue, C.; Ciais, P. The influence of local spring temperature variation on temperature sensitivity of spring phenology. *Global Chang. Biol.* **2014**, *20*, 1473–1480. [CrossRef]

37. Koike, T. Autumn coloring, photosynthetic performance and leaf development of deciduous broad-leaved trees in relation to forest succession. *Tree Physiol.* **1990**, *7*, 21–32. [CrossRef]

38. Lee, D.W.; O'Keefe, J.; Holbrook, N.M.; Feild, T.S. Pigment dynamics and autumn leaf senescence in a New England deciduous forest, eastern USA. *Ecol. Res.* **2003**, *18*, 677–694. [CrossRef]

39. Menzel, A. Phenology: Its importance to the global change community. *Clim. Chang.* **2002**, *54*, 379–385. [CrossRef]

40. Broich, M.; Huete, A.; Tulbure, M.G.; Ma, X.; Xin, Q.; Paget, M.; Restrepo-Coupe, N.; Davies, K.; Devadas, R.; Held, A. Land surface phenological response to decadal climate variability across Australia using satellite remote sensing. *BioGeoSci. Discuss.* **2014**, *11*, 7685–7719. [CrossRef]

41. Dittmar, C.; Elling, W. Phenological phases of common beech (*Fagus sylvatica* L.) and their dependence on region and altitude in Southern Germany. *Eur. J. Forest Res.* **2006**, *125*, 181–188. [CrossRef]

42. Čufar, K.; Luis, M.D.; Saz, M.A.; Črepinšek, Z.; Kajfež-Bogataj, L. Temporal shifts in leaf phenology of beech (*Fagus sylvatica*) depend on elevation. *Trees* **2012**, *26*, 1091–1100. [CrossRef]

43. Vitasse, Y.; Delzon, S.; Dufrene, E.; Pontiller, J.Y.; Louvet, J.M.; Kremer, A.; Michalet, R. Leaf phenology sensitivity to temperature in European trees: Do within-species populations exhibit similar responses? *Agric. Forest Meteorol.* **2009**, *149*, 735–744. [CrossRef]

44. Dunn, A.H.; Beurs, K.M. Land surface phenology of North American mountain environments using moderate resolution imaging spectroradiometer data. *Remote Sens. Environ.* **2011**, *115*, 1220–1233. [CrossRef]

45. Ahrens, C.D.; Henson, R. *Meteorology today: An Introduction to Weather, Climate, and the Environment*, 11th ed.; Cengage Learning: Boston, MA, USA, 2015; 656p, ISBN 978-1-305-26500-4.

46. Climatic Conditions of Slovak Republic. Available online: http://www.shmu.sk/sk/?page=1064 (accessed on 12 March 2018).

47. Pfanz, H.; Aschan, G. The Existence of Bark and Stem Photosynthesis in Woody Plants and Its Significance for the Overall Carbon Gain. An Eco-Physiological and Ecological Approach. In *Progress in Botany*; Springer: Berlin, Germany, 2001; Volume 62, pp. 477–510. [CrossRef]

48. Richardson, A.D.; O'Keefe, J. Phenological differences between understory and overstory: A case study using the long-therm Harvard forest records. In *Phenology of Ecosystem Processes*; Noormets, A., Ed.; Springer: New York, NY, USA, 2009; pp. 88–117. [CrossRef]

49. Ahl, D.E.; Gower, S.T.; Burrows, S.N.; Shabanov, N.V.; Myneni, R.B.; Knyazikhin, Y. Monitoring spring canopy phenology of a deciduous broadleaf forest using MODIS. *Remote Sens. Environ.* **2006**, *104*, 88–95. [CrossRef]

50. Nagai, S.; Nasahara, K.N.; Muraoka, H.; Akiyama, T.; Tsuchida, S. Field experiments to test the use of the normalized-difference vegetation index for phenology detection. *Agric. Forest Meteorol.* **2010**, *150*, 152–160. [CrossRef]

51. Yang, X.; Tang, J.; Mustard, J.F. Beyond leaf color: Comparing camera-based phenological metrics with leaf biochemical, biophysical, and spectral properties throughout the growing season of a temperate deciduous forest. *J. Geophys. Res. Biogeosci.* **2014**, *119*, 181–191. [CrossRef]

52. Soudani, K.; Hmimina, G.; Delpierre, N.; Pontailler, J.Y.; Aubinet, M.; Bonal, D.; Caquet, B.; de Grandcourt, A.; Burban, B.; Flechard, C.; et al. Ground-based Network of NDVI measurements for tracking temporal dynamics of canopy structure and vegetation phenology in different biomes. *Remote Sens. Environ.* **2012**, *123*, 234–245. [CrossRef]

53. Hmimina, G.; Dufrêne, E.; Pontailler, J.Y.; Delpierre, N.; Aubinet, M.; Caquet, B.; de Grandcourt, A.; Burban, B.; Flechard, C.; Granier, A.; et al. Evaluation of the potential of MODIS satellite data to predict vegetation phenology in different biomes: An investigation using ground-based NDVI measurements. *Remote Sens. Environ.* **2013**, *132*, 145–158. [CrossRef]

54. García-Plazaola, J.I.; Becerril, J.M. Seasonal changes in photosynthetic pigments and antioxidants in beech (*Fagus sylvatica*) in a Mediterranean climate: Implications for tree decline diagnosis. *Aust. J. Plant Physiol.* **2001**, *28*, 225–232. [CrossRef]

55. Soudani, K.; Maire, G.; Dufrêne, E.; François, C.H.; Delpierre, N.; Ulrich, E.; Cecchini, S. Evaluation of the onset of green-up in temperate deciduous broadleaf forests derived from Moderate Resolution Imaging Spectroradiometer (MODIS) data. *Remote Sens. Environ.* **2008**, *112*, 2643–2655. [CrossRef]

56. Chalupa, V. Počátek, trvání a ukončení vegetační činnosti u lesních dřevin. *Práce VÚLHM* **1969**, *37*, 41–68.

57. Toda, M.; Richardson, A.D. Estimation of plant area index and phenological transition dates from digital repeat photography and radiometric approaches in a hardwood forest in the Northeastern United States. *Agric. Forest Meteorol.* **2018**, *249*, 457–466. [CrossRef]

58. Klosterman, S.; Richardson, A.D. Observing spring and fall phenology in a deciduous forest with aerial drone imagery. *Sensors* **2017**, *17*, 2852. [CrossRef]

59. Nagai, S.; Inoue, T.; Ohtsuka, T.; Kobayashi, H.; Kurumado, K.; Muraoka, H.; Nasahara, K.N. Relationship between spatio-temporal characteristics of leaf-fall phenology and seasonal variations in near surface- and satellite-observed vegetation indices in cool-temperate deciduous broad-leaved forest in Japan. *Int. J. Remote Sens.* **2014**, *35*, 3520–3536. [CrossRef]

60. Liu, L.; Liang, L.; Schwartz, M.D.; Donnelly, A.; Wang, Z.; Schaaf, C.B. Evaluating the potential of MODIS satellite data to track temporal dynamics of autumn phenology in a temperate mixed forest. *Remote Sens. Environ.* **2015**, *160*, 156–165. [CrossRef]

61. Hwang, T.; Song, C.; Vose, J.M. Topography-mediated controls on local vegetation phenology estimated from MODIS vegetation index. *Landsc. Ecol.* **2011**, *26*, 541–556. [CrossRef]

62. Dittmar, C.; Fricke, W.; Elling, W. Impact of late frost events on radial growth of common beech (*Fagus sylvatica* L.) in Southern Germany. *Eur. J. Forest Res.* **2006**, *125*, 249–259. [CrossRef]

63. Kreyling, J.; Thiel, D.; Nagy, L.; Jentsch, A.; Huber, G.; Konnert, M.; Beierkuhnlein, C. Late frost sensitivity of juvenile *Fagus sylvatica* L. differs between southern Germany and Bulgaria and depends on preceding air temperature. *Eur. J. Forest Res.* **2012**, *131*, 717–725. [CrossRef]

64. Kindermann, J.; Wurth, G.; Kohlmaier, G.H.; Badeck, F.W. Interannual variation of carbon exchange fluxes in terrestrial ecosystems. *Global Biogeochem. Cycles* **1996**, *10*, 737–755. [CrossRef]

65. Gomory, D.; Paule, L. Trade-of between height growth and spring flushing in common beech (*Fagus sylvatica* L.). *Ann. Forest Sci.* **2011**, *68*, 975–984. [CrossRef]

 *forests*

Article

Assessment of the Response of Photosynthetic Activity of Mediterranean Evergreen Oaks to Enhanced Drought Stress and Recovery by Using PRI and R690/R630

Chao Zhang [1,2,*], Catherine Preece [1,2], Iolanda Filella [1,2], Gerard Farré-Armengol [1,2] and Josep Peñuelas [1,2]

[1]   CSIC, Global Ecology Unit, CREAF-CSIC-UAB, Bellaterra, 08193 Barcelona, Catalonia, Spain; catherine.preece09@gmail.com (C.P.); iola@creaf.uab.cat (I.F.); g.farre@creaf.uab.cat (G.F.-A.); josep.penuelas@uab.cat (J.P.)
[2]   CREAF, Cerdanyola del Vallès, 08193 Barcelona, Catalonia, Spain
*   Correspondence: c.zhang@creaf.uab.cat; Tel.: +34935813355

Received: 31 July 2017; Accepted: 7 October 2017; Published: 10 October 2017

Abstract: The photochemical reflectance index (PRI) and red-edge region of the spectrum are known to be sensitive to plant physiological processes, and through measurement of these optical signals it is possible to use non-invasive remote sensing to monitor the plant photosynthetic status in response to environmental stresses such as drought. We conducted a greenhouse experiment using *Quercus ilex*, a Mediterranean evergreen oak species, to investigate the links between leaf-level PRI and the red-edge based reflectance ratio (R690/R630) with CO_2 assimilation rates (A), and photochemical efficiency (F_V/F_M and Yield) in response to a gradient of mild to extreme drought treatments (nine progressively enhanced drought levels) and corresponding recovery. PRI and R690/R630 both decreased under enhanced drought stress, and had significant correlations with A, F_V/F_M and Yield. The differential values between recovery and drought treatments of PRI ($\Delta PRI_{recovery}$) and R690/R630 ($\Delta R690/R630_{recovery}$) increased with the enhanced drought levels, and significantly correlated with the increases of $\Delta A_{recovery}$, $\Delta F_V/F_{M recovery}$ and $\Delta Yield_{recovery}$. We concluded that both PRI and R690/R630 were not only sensitive to enhanced drought stresses, but also highly sensitive to photosynthetic recovery. Our study makes important progress for remotely monitoring the effect of drought and recovery on photosynthetic regulation using the simple physiological indices of PRI and R690/R630.

Keywords: chlorophyll fluorescence (ChlF); drought; Mediterranean; photochemical reflectance index (PRI); photosynthesis; R690/R630; recovery

1. Introduction

The increasing occurrence of drought in the Mediterranean region is widely reported [1–3] and during the last 20 years observations show that it has affected ecosystem functioning and structure [4,5], decreased plant growth [6] and primary production [7,8], and triggered vegetation mortality [9]. The increase of drought has also enhanced water scarcity [10], and elicited ecological damage [11] and crop failures [12]. Faced with these negative impacts, monitoring the timing of drought onset and the extent of the effects on vegetation ecosystems is being increasingly warranted.

Summer drought, one of the key constraints of Mediterranean ecosystems [4], elicits a water deficit in the leaf tissue that can down-regulate photosynthesis [13]. This reduction in photosynthesis for Mediterranean species such as *Quercus ilex* L. is generally caused by stomatal closure in response to drought stress [14–16]. *Quercus ilex* is a broadleaved evergreen tree or shrub that is widely distributed from semi-arid to humid areas of the Mediterranean region [17]. The sclerophyllous characteristics of *Q. ilex* make it possible to reduce its transpiration and to withstand the effect of long summer drought events. *Quercus ilex* also has the ability to reactivate its photosynthetic machinery and increase photosynthesis rates when the soil water content (SWC) is high enough to again support water transport through the plant [18].

When plants suffer water stress, the reducing power from photosynthesis becomes excessive, but then the increases in zeaxanthin pigments from xanthophyll de-epoxidation can safely dissipate excessive energy and prevent the photosystems from potential damage by accumulating excitation energy [19,20]. The photochemical reflectance index (PRI) was originally defined based on changes in zeaxanthin at the wavelength of 531 nm to assess the efficiency of absorbed photosynthetic active radiation (APAR) for photosynthesis (light use efficiency; LUE) [21,22]. A great number of studies have found that PRI is able to track LUE changes at diurnal and seasonal scales from leaf to canopy to ecosystem levels [23–25], because apart from xanthophyll pigments, PRI was also associated with changes of carotenoid/chlorophyll ratios [26–29], which play a key role in long-term dynamics of photosynthesis [20,30]. Increasing attention has been focused on using PRI to detect the effect of environmental stress (e.g., intense irradiance, water shortage, nutrition deficit, and high and low temperature stresses) on photosynthetic changes [8,14,31–36].

The changes in photosynthesis are synchronously accompanied by the emission of chlorophyll *a* fluorescence (ChlF), which is regulated by the photochemical conversion and heat dissipation of excitation energy [37]. The contribution of the ChlF emission is only a small part of the total radiation reflected from vegetation, however, ChlF is highly sensitive to the variability of plant physiological processes in response to various environmental conditions, and can provide a direct approach to detect the functional status of photosynthetic machinery [38,39]. The range of the ChlF emission spectrum is from the red to near-infrared regions, with two peaks at around 690 and 740 nm [38]. The active ChlF, mainly obtained from a pulse-amplitude modulated (PAM) fluorometer, has been extensively used for monitoring foliar photosynthetic apparatus [40]. Recently, the passive ChlF (solar-induced fluorescence, SIF), derived from such methods as the Fraunhofer Line Depth technique or laser induction, has scaled up the measurements and the applications for detecting photosynthesis from leaf level to canopy, ecosystem and regional scales [31,41–44].

Some studies have illustrated that reflectance ratios such as R690/R600, R690/R630, R690/R655, and R740/R800 were sensitive to the changes of ChlF, because the ChlF signal is superimposed on the red-edge of leaf reflectance [38,45,46]. These reflectance ratios have been shown to track the changes in fluorescence, photochemical efficiency and carbon assimilation rates for healthy and stressed plants due to high co-variation of red-edge reflectance/absorption, pigment concentration, and leaf physiology [31,47–53].

Numerous studies had shown PRI is sensitive to water stress [33,49,54–56], but no studies have illustrated that PRI was sensitive enough to track recovery after drought stress, at least to the best of our knowledge. Few studies so far have either exploited the applicability of the reflectance ratios based on the red-edge region of the spectrum for detecting the photosynthetic response to water stress [31,48,49]. Only the study by Dobrowski et al. [31] monitored photosynthetic response to water stress and recovery with only a four-day-long experiment.

Along with the increasing occurrence of drought, it is important to find an easy, non-destructive and efficient method to monitor both the effect of drought on plant physiological and functional variability and the subsequent recovery of the photosynthetic machinery. This is particularly important for regions confronted with progressively more extreme summer droughts such as the Mediterranean basin. Based on the previous studies, PRI and reflectance ratio based on the red-edge region of the spectrum (R690/R630) are thus promising methods to test such photosynthetic responses. In this study, we conducted a two-month long greenhouse experiment using the typical Mediterranean species *Q. ilex* by exposing the plants to nine different progressive drought treatments through controlling the SWC, and then re-watering for five weeks to explore the levels of plant recovery. We simultaneously measured foliar reflectance to obtain PRI and R690/R630, PAM ChlF to get maximum (F_V/F_M) and actual (Yield) photochemical efficiency of photosystem II (ΦPSII), and CO_2 assimilation rates (A). We hypothesized that both PRI and R690/R630 would be sensitive to the progressively increasing drought, and could therefore be used to monitor photosynthetic dynamics. We also hypothesized that both PRI and R690/R630 would detect the photosynthetic recovery after experiencing different levels of drought stresses.

2. Material and Methods

2.1. Experimental Design

Three-year-old *Quercus ilex* saplings of approximately 60 cm height were obtained in May 2015 (Forestal Catalana, Barcelona, Spain) and were re-potted in 3.5 litre pots, with a substrate consisting of 45% sand, 45% autoclaved peat, and 10% natural soil inoculum. The soil was collected from a natural holm oak forest on a south-facing slope (25% slope) in the Prades Mountains in Northeastern Spain (41°13′ N, 0°55′ E). There were 162 saplings in total, divided into six blocks, and plants were grown in the greenhouse of the Autonomous University of Barcelona (Barcelona, Spain) with a six-week period of adequate watering, to allow them to acclimate to greenhouse conditions. During the experiment, in the greenhouse the mean air temperature was 26.7 °C (measured with EL-USB-2 data logger, Lascar Electronics, Wiltshire, UK), and the soil temperature monitored at a fine scale in five pots across the different soil types was 27.0 °C on average (measured with Decagon Em50 data logger with 5TM soil probes, Decagon Devices, Pullman, WA, USA). The saplings were then exposed to water stress by withholding water for nine different drought treatments with 18 plants each. The length of time without water ranged from 0 to 21 days, which means the first drought level, with 0-day withholding water, was effectively the control treatment. At the end of each treatment, the foliar photochemical efficiency of photosystem II (ΦPSII) and the reflectance were synchronously measured on half of the pots (nine per drought level) inside the greenhouse under clear skies within one hour of solar noon (12:00 to 14:00). The remaining pots then had six weeks of recovery, which involved re-watering at optimal levels, and then identical leaf measurements were collected for these plants.

The water content of the substrate was determined in each pot at the start of the study and at the end of its drought period, and recovery period, if relevant (using ML3 Theta Probe connected to a HH2 Moisture Meter from Delta-T Devices, Cambridge, UK). Mean soil moisture at the start of the experiment was 22.6% and it decreased exponentially to 0.3% at the end of the most extreme drought treatment. Soil moisture recovered quickly to ca. 20% within one week of re-watering and was 24.7% on average during the recovery phase. The differential SWC (ΔSWC$_{recovery}$) was calculated by subtracting the nine different drought treatments values from corresponding recovery treatments.

2.2. Leaf Photosynthesis Measurements

In the greenhouse, the CO_2 assimilation rates (A) of leaves were measured using an ADC pro (LCpro1Portable Photosynthesis System; ADC BioScientific Ltd., Hoddesdon, Herts, UK) gas exchange system. The measurements were conducted under a quantum flux density of 1000 μmol m^{-2} s^{-1} and ambient CO_2 concentration of 395 μmol mol^{-1}. Five measurements were recorded after the values were stabilized and had reached a steady state for each plant. After each measurement, the leaf area that was enclosed in the cuvette was marked on the leaves; leaves were later photographed and the leaf areas for which A was measured were estimated using ImageJ 1.46r (NIH, Bethesda, MD, USA) to standardize all A measurements. The differential A ($\Delta A_{recovery}$) was calculated by subtracting the nine different drought treatments values from corresponding recovery treatments.

2.3. Chlorophyll Fluorescence Measurements

Chlorophyll fluorescence measurements were performed by a pulse-amplitude-modulated photosynthesis yield analyzer (PAM-2000; Walz, Effeltrich, Germany). The measurements were conducted on three healthy and mature leaves at the top of each plant. A saturating light pulse (SP) was applied to dark-adapted (at least 20 min) leaves for the determination of minimum (F_o) and maximum (F_M) fluorescence. The maximum photochemical efficiency of photosystem II (PSII) (F_V/F_M) was then calculated according to Genty et al. [57]:

$$F_V/F_M = (F_M - F_o)/F_M \tag{1}$$

In parallel with F_V/F_M, the actual photochemical efficiency of PSII (Yield) was also calculated based on the rapid measurements of steady-state (F_S) and the maximal (F_M') fluorescence yield during the full closing of the PSII center in light-adapted leaves under ambient light and full sun conditions around noon:

$$\text{Yield} = (F_M' - F_S)/F_M' \tag{2}$$

The differential F_V/F_M ($\Delta F_V/F_{Mrecovery}$) and Yield ($\Delta \text{Yield}_{recovery}$) between the nine different drought treatments values and recovery treatments were also calculated.

2.4. Reflectance Measurements

Leaf spectral reflectance measurements were collected using a broad range mini spectroradiometer (LR1; ASEQ, Vancouver, BC, Canada) with a fiber-optic of 25° field of view. The instrument measures spectral reflectance between 300 and 1000 nm with a sampling interval of less than 1 nm. To reduce atmospheric condition changes, the spectroradiometer was calibrated using a white Spectralon reference panel (Labsphere, North Sutton, NH, USA), which can be regarded as a Lambertian reflector. Incident solar irradiance was immediately determined using the same white reference panel prior to the radiance measurements. All spectral measurements were carried out at a nadir view angle ca. 1 cm above the leaf. The ground-projected instantaneous field of view (GIFOV) was thus about 4.4 mm. In each plant, we measured three different leaves at the top of the canopy as replicates. The photochemical reflectance index (PRI, [21,22]) was calculated from the reflectance data (Rx implies reflectance at x nm) as:

$$PRI = (R531 - R570)/(R531 + R570) \tag{3}$$

Additionally, we also retrieved the reflectance ratio R690/R630 based on the red-edge region of the spectrum. Because changes in the reflectance at 690 nm (R690) can be attributed to the absorptions of pigments, plant physiological variations and also ChlF which are associated with photosynthetic activity, and R630 is relatively less sensitive to photosynthesis and fluorescence emission [49].

The differential PRI ($\Delta PRI_{recovery}$) and R690/R630 ($\Delta R690/R630_{recovery}$) were obtained through nine different drought treatments values subtracted from corresponding recovery treatments.

2.5. Statistical Analysis

Differences of soil water content, CO_2 assimilation rate, photochemical efficiency of photosystem II (ΦPSII) and reflectance indices between the nine drought levels and corresponding plant recovery were analyzed using repeated-measures analyses of variance (ANOVAs). The differences were considered statistically significant at $p < 0.05$. The applicability of PRI and R690/R630 for assessing CO_2 assimilation, ΦPSII responses to drought stress and recovery were analyzed using standardized major-axis regression to identify correlations between the variables. All analyses were conducted with R version 3.3.2 (R Core Development Team, Vienna, Austria, 2016).

3. Results

3.1. Responses to Enhanced Drought Stress

In our experiment, the *Q. ilex* seedlings were treated by withholding watering from 0 to 21 days in nine different drought levels treatments. Along with the increasing days of drought and decreasing soil water content (i.e., enhanced drought stress), CO_2 assimilation rates (A) and maximum (F_V/F_M) and actual (Yield) photosystem II efficiency (ΦPSII) decreased from highest values at the beginning of the experiment to lowest values in severe drought conditions (Figure 1). The physiological indices of the photochemical reflectance index (PRI) (Figure 1a) and reflectance ratio (R690/R630) (Figure 1b) also decreased gradually. Both changes in PRI and R690/R630 were highly consistent with the decreases in soil water content (SWC), A, F_V/F_M and Yield. However, in the last three extreme drought level treatments, PRI did not change significantly.

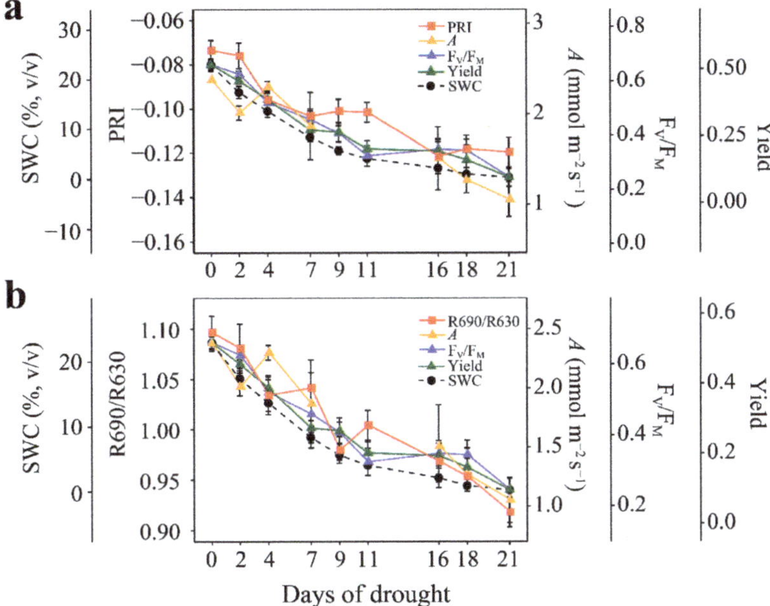

Figure 1. Changes in CO_2 assimilation rates (A), maximum (F_V/F_M) and actual (Yield) photochemical efficiency of photosystem II (ΦPSII), and soil water content (SWC) with the photochemical reflectance index (PRI) (**a**) and with reflectance ratio R690/R630 based on the red-edge region of the spectrum (**b**) in response to nine different drought levels treatment for *Quercus ilex*. Error bars denote the standard errors of the mean ($n = 9$).

3.2. Relationships of PRI and R690/R630 with A, F_V/F_M, and Yield under Enhanced Drought Stress

PRI and R690/R630 were used to assess the photosynthetic response to drought stress. Both PRI (Figure 2a) and R690/R630 (Figure 2b) were significantly correlated with A and explained 82% and 86% of variance of A, respectively. PRI had strong relationships with F_V/F_M and Yield ($R^2 \geq 0.85$ and $p < 0.001$ for both, Figure 2c). Similar significant correlations were found between R690/R630 and F_V/F_M and Yield ($R^2 = 0.89$ and $p < 0.001$ for both, Figure 2d). In contrast, the normalized difference vegetation index (NDVI, calculated by (R900 − R680)/(R900 + R680)) showed no significant correlation with A under enhanced drought conditions ($R^2 = 0.08$ and $p = 0.58$).

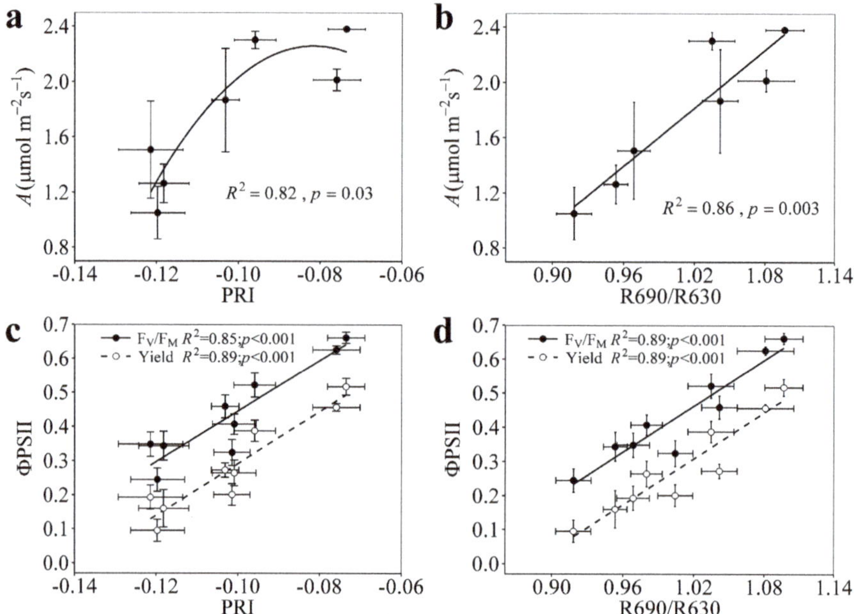

Figure 2. Relationships of *A* with PRI (**a**) and R690/R630 (**b**), and photochemical efficiency of PSII (ΦPSII) (F_V/F_M and Yield) with PRI (**c**) and R690/R630 (**d**) for *Quercus ilex*. All the values are from Figure 1.

3.3. Responses to Recovery

The SWC (Figure 3a) clearly recovered to the pre-drought (0-day drought level, namely, control treatment) condition in all drought treatments after five weeks of re-watering. They all reached identically high values, significantly higher than in the drought treatments from the 4-day drought level treatment onwards ($p < 0.001$). *A* (Figure 3b) was higher and remained similar in the recovery treatment, particularly in the last three severe drought treatments. The mean of F_V/F_M (Figure 3c) in the recovery treatment was significantly increased from the 7-day drought level treatment ($p < 0.001$). Yield (Figure 3d) presented analogous changes with SWC, *A*, and F_V/F_M and increased from the 9-day drought level treatment ($p < 0.001$). Both PRI (Figure 3e) and R690/R630 (Figure 3f) had no significant difference in the first four drought treatments with corresponding recovery treatments, but significantly increased from the recovery treatment of the 9-day drought level.

The differential values of variables were calculated by subtracting nine drought levels treatments from corresponding recovery treatments to assess the changes of photosynthesis after one month of re-watering the plants of the drought treatment, and to evaluate the applicability of PRI and R690/R630 in detecting the recovery. The values of $\Delta A_{recovery}$, $\Delta F_V/F_{Mrecovery}$, and $\Delta Yield_{recovery}$ increased in conjunction with increasing ΔSWC and increasing drought levels (Figure 4). Both $\Delta F_V/F_{Mrecovery}$ and $\Delta Yield_{recovery}$ increased slowly in the last three drought levels. Changes in $\Delta PRI_{recovery}$ (Figure 4a) were consistent with $\Delta A_{recovery}$, $\Delta F_V/F_{Mrecovery}$, and $\Delta Yield_{recovery}$. $\Delta R690/R630_{recovery}$ (Figure 4b) also showed a gradual increase as with the other variables, with a high value in the 9-day drought level treatment.

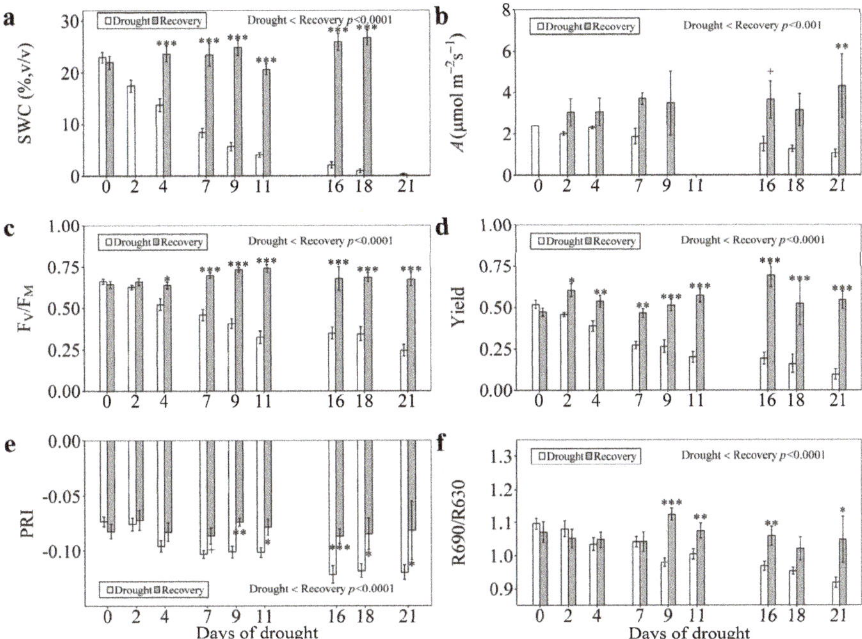

Figure 3. Changes in SWC (**a**), A (**b**), F_V/F_M (**c**), Yield (**d**), PRI (**e**) and R690/R630 (**f**) in response to nine different drought levels and corresponding recovery treatments for *Quercus ilex*. The significances were denoted as $^+ p < 0.1$, $^* p < 0.05$, $^{**} p < 0.01$ and $^{***} p < 0.001$.

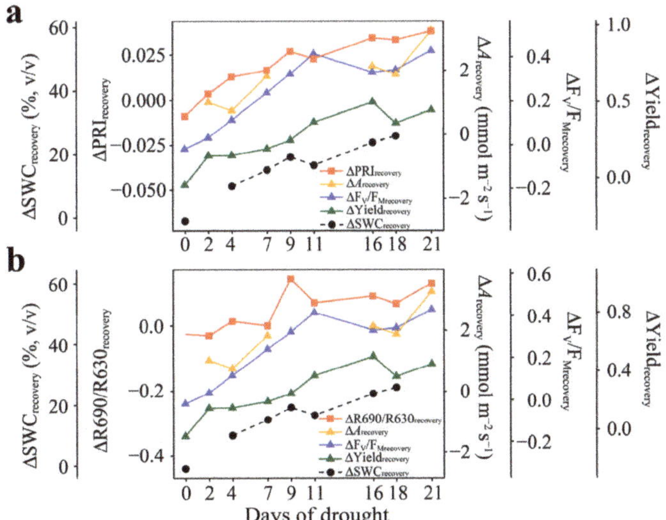

Figure 4. Changes in differential A (ΔA), $\Delta F_V/F_M$, ΔYield, and ΔSWC with ΔPRI (**a**) and with ΔR690/R630 (**b**) in response to nine different drought levels for *Quercus ilex*. The differential values of variables calculated by subtracting the mean of nine drought levels treatment from corresponding mean of recovery treatments in Figure 3.

3.4. Relationships of $\Delta PRI_{recovery}$ and $\Delta R690/R630_{recovery}$ with $\Delta A_{recovery}$, $\Delta F_V/F_{Mrecovery}$, and $\Delta Yield_{recovery}$

Both $\Delta PRI_{recovery}$ and $\Delta R690/R630_{recovery}$ accounted for large proportions of the variability of $\Delta A_{recovery}$ ($R^2 \geq 0.70$ and $p < 0.05$ for both, Figure 5a,b). $\Delta PRI_{recovery}$ was highly significantly correlated with $\Delta F_V/F_{Mrecovery}$ and $\Delta Yield_{recovery}$ ($R^2 \geq 0.85$ and $p < 0.001$ for both, Figure 5c). The relationships of $\Delta R690/R630_{recovery}$ with $\Delta F_V/F_{Mrecovery}$ and $\Delta Yield_{recovery}$ were also significant, but they were not as good as those of $\Delta PRI_{recovery}$ with $\Delta F_V/F_{Mrecovery}$ and $\Delta Yield_{recovery}$ (Figure 5d).

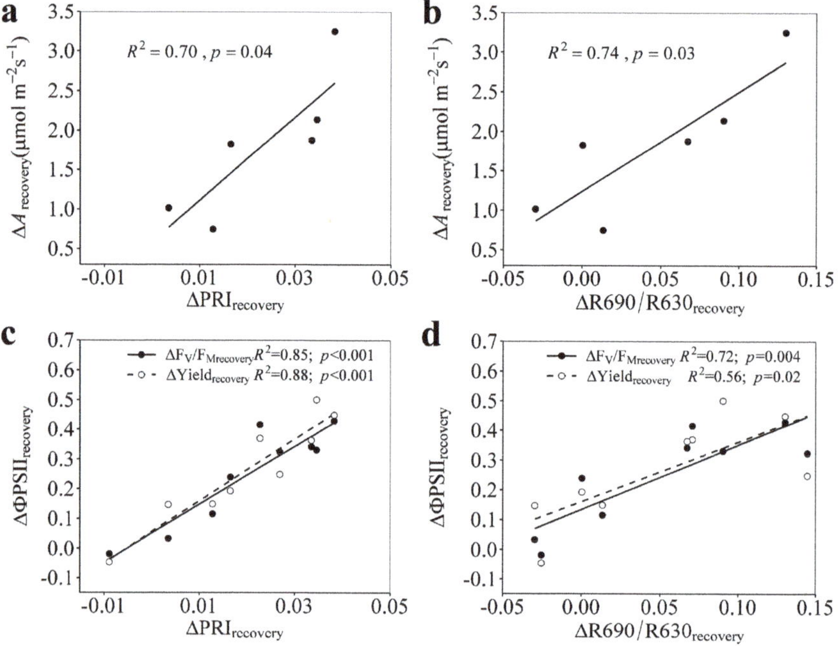

Figure 5. Relationships of differential A ($\Delta A_{recovery}$) with $\Delta PRI_{recovery}$ (**a**) and $\Delta R690/R630_{recovery}$ (**b**), and photochemical efficiency of PSII ($\Delta\Phi PSII$) (F_V/F_M and Yield) with $\Delta PRI_{recovery}$ (**c**) and $\Delta R690/R630_{recovery}$ (**d**) for *Quercus ilex*. All the differential values are from Figure 4.

4. Discussion

4.1. PRI and R690/R630 Tracked the Photosynthetic Changes under Enhanced Drought Levels

The enhanced levels of drought caused reductions in CO_2 assimilation rates (A) and photochemical efficiency of photosystem II ($\Phi PSII$) (F_V/F_M and Yield) in Mediterranean evergreen leaves (Figure 1). Similar trends were also recorded for the photochemical reflectance index (PRI, Figure 1a) and reflectance ratio based on the red-edge region of the spectrum (R690/R630, Figure 1b), indicating the sensitivity of PRI and R690/R630 to gradually increased drought stress. The close correspondence of PRI and R690/R630 with photosynthetic activity (Figure 3a) and $\Phi PSII$ (Figure 3b) demonstrated the promise of pigment- or red-edge reflectance-based approaches to remote monitoring of evergreen photosynthetic activity under gradually enhanced drought.

PRI is used to reveal the facultative xanthophyll cycle activity at short-term scale (hours to few days) [21,22,28], but also to estimate the changes in the constitutive pigment pool size at long-term scale (weeks to months) [27,29,58,59]. In our study, PRI showed promise as an index of photosynthetic down-regulation and illustrated the activation of photo-protective carotenoid pigments under photosynthetic down-regulation in response to gradually enhanced drought stress. During the summer drying period, leaves in evergreen vegetation of semi-arid Mediterranean region experience decreases in pigment contents and increases in carotenoids/chlorophyll (Car/Chl) ratio [60]. This change in Car/Chl ratios plays a key role in photosynthetic down-regulation, and is associated with variability in PRI [27,58]. Decreases in PRI and photosynthesis presented in this study thus indirectly suggested that changes in Car/Chl had a significant effect on photosynthetic down-regulation. Therefore, PRI detected the effect of the progressive drought on photosynthesis, consistent with many previous studies that have demonstrated that PRI is correlated with photosynthesis and Yield under water stress conditions [14,34].

Additionally, PRI was quite stable in the last three extreme drought level treatments (plants without water for 16–21 days), unlike the slight decrease of A, F_V/F_M, Yield, and R690/R630, reflecting the insensitivity of PRI to severe drought, at intensities where there is observable leaf fall. Previous leaf-level studies have reported that PRI was unable to assess the negative light use efficiency (LUE) of severely damaged plants [34]. Under severe stress conditions, evergreen leaves could retain zeaxanthin pigments to protect the photosystem from damage, which might contribute to such stability in PRI [61]. At canopy level, changes in structure, heterogeneity in irradiance, and differences in sun angles within the crown have hindered PRI interpretation and assessment of LUE [24]. At larger spatial scales, satellite-based (Moderate Resolution Imaging Spectroradiometer, MODIS) PRI has tracked ecosystem LUE, even during severe water limitation of summer for *Q. ilex* [62]. The normalized PRI by absorbed light detected the drought effect on gross primary production (GPP) in a deciduous forest and an evergreen broadleaf forest in France, but did not capture the reduction of GPP in a semi-arid grassland in Hungary [8]. Guarini et al. [36] demonstrated that MODIS-based PRI should be used with care under severe water stress, because the disturbances from canopy structure, illumination and viewing angles, and so forth, could increase the uncertainty of PRI in tracking LUE. These studies presented the complications of the applicability of PRI under severe drought conditions at different spatial scales, and other factors, such as soil backgrounds, vegetation functional types, atmospheric interference and instrument characteristics, can also complicate the applications of PRI in assessing LUE [24]. However, the strong correlations between PRI and A and ΦPSII (F_V/F_M and Yield) under enhanced drought levels demonstrate that PRI can be applied to detect the effect of continuously increased drought on photosynthesis for Mediterranean evergreen sclerophylls, which generally experience summer drought.

Interestingly, the reflectance ratio based on the red-edge region of the spectrum R690/R630 presented significant correlations with A, F_V/F_M, and Yield, and efficiently tracked the photosynthetic changes to enhanced drought levels (Figure 2b,d). Previous studies have shown reflectance ratios such as R685/R630 and R690/R630 are sensitive to changes in foliar F_V/F_M. Dobrowski et al. [31] used similar reflectance ratios to quantify the chlorophyll fluorescence (ChlF) emissions because of their superimposition at the red reflectance region, with a maximum emission in the near red-edge region of 690 and 740 nm. The reference band (e.g., R630 or R655) was chosen due to its insensitivity to ChlF emission and sensitivity to chlorophyll pigments [47]. These simple reflectance ratios have tracked the foliar ChlF under both diurnal natural, stress and recovery conditions [31,47,49]. The links between R690/R630 and ChlF emissions probably contribute to interpreting the relationships between R690/R630 and photosynthesis in this study. In addition, ChlF can increase under severely low quantum yields [63], however, continuous decreases in R690/R630 during extreme drought periods suggested that, apart from ChlF emission at 690 nm, other plant physiological functions also control the changes in R690/R630. The reabsorption of red ChlF (particularly at ChlF emission peak near 690 nm) by chlorophyll pigments can also affect the variability of R690/R630 [45]. Further, the low reflectance in the red-edge of the spectrum generally resulted from the high absorption of chlorophyll pigments. Zarco-Tejada et al. [49] showed the large effect of chlorophyll content on similar reflectance ratios such as R685/R630 and R680/R630. However, NDVI, which could indirectly indicate the chlorophyll pigment changes, was quite stable and did not track the photosynthetic activity during the study period. The chlorophyll content thus probably had low effects on R690/R630 and its correlation with photosynthesis for this Mediterranean evergreen species.

Additionally, the down-regulation of maximum photochemical efficiency of PSII (F_V/F_M) is associated with changes of non-photochemical quenching (NPQ) under environmental stress [61,64,65]. The significant correlation of R690/R630 with F_V/F_M in this and previous diurnal studies [47,48] indicated that NPQ probably plays a key role in linking R690/R630 with changes in photosynthetic activity. Although PRI was also highly correlated with F_V/F_M, the rather constant PRI in the extreme drought conditions (the last three drought treatments) was decoupled with the strong depression of F_V/F_M. Such decoupling of PRI with LUE and NPQ was also found in early spring when Scots pine needles demonstrated large down-regulation due to the effect of severe stress [27]. In contrast, in our study, R690/R630 presented consistent changes with F_V/F_M, Yield and A even under extreme water stress conditions.

4.2. PRI and R690/R630 Tracked the Photosynthetic Recovery from Progressively Enhanced Drought Stresses

The CO_2 assimilation rates and ΦPSII of *Q. ilex* rapidly recovered to normal values that were similar with control (0-day drought level treatment) after five weeks of re-watering (Figure 3). The values of PRI and R690/R630 were increased to similar values after re-watering, indicating that both PRI and R690/R630 were also sensitive to the recovery of the plant. The significant correlations of differential PRI (ΔPRI$_{recovery}$) and R690/R630 (ΔR690/R630$_{recovery}$) with $\Delta A_{recovery}$, $\Delta F_V/F_{Mrecovery}$ and ΔYield$_{recovery}$ (Figure 5) illustrated their potential to monitor photosynthetic recovery response to drought stress.

PRI was originally defined to detect the variability of zeaxanthin changes at 531 nm [21,22]. However, it should be noted that 531 nm was the band associated with carotenoid pigment changes, including zeaxanthin, and also other carotenoid pigments such as lutein and neoxanthin [66]. These carotenoids both play key roles in energy dissipation of photosynthetic down- and up-regulation [67,68]. However, studies have demonstrated that constitutive long-term changes of PRI were greatly attributable to carotenoid pigment pool sizes [27,29,58,59]. Thus, the mechanism of PRI detecting photosynthetic recovery was probably due to the carotenoid pigments acting on the photochemical process. Also, during water stress and recovery, PRI was significantly correlated with carotenoid pigments in olive saplings [32].

In contrast with PRI, changes in R690/R630 were probably related with ChlF emissions but not with chlorophyll pigments for evergreen broadleaves in this study. ChlF is the energy of absorbed photosynthetically active radiation (APAR) that is lost during the first steps of photosynthesis which involves the emission of red to far-red light (ca. 660 to 800 nm) [39]. ChlF is therefore associated with the fraction of APAR and also with light use efficiency (LUE), giving the possibility of using R690/R630 to monitor photosynthetic recovery after experiencing different drought stresses. Additionally, ChlF is mainly emitted from PSII, so the recovery of photochemistry could increase the ChlF emission, particularly in red wavelengths at 690 nm, which is close to the absorption peak of the pigments of PSII [39,45]. Consequently, R690/R630 is a good indicator of photosynthetic recovery in response to progressive drought stress for evergreen leaves.

5. Conclusions and Final Remarks

This study makes progress in assessing mild to extreme drought stresses in Mediterranean evergreen species with the PRI and reflectance ratio R690/R630. Both PRI and R690/R630 were not only sensitive to progressive drought, but also to plant recovery, significantly tracking the photosynthetic response to enhanced drought levels, and also detecting the photosynthetic recovery after re-watering. Carotenoid/chlorophyll pigment ratios probably control the correlation between PRI and photosynthesis, and ChlF at 690 nm probably plays a role in the relationship of R690/R630 with photosynthesis. Both PRI and R690/R630 can be used for remotely monitoring the effect of drought on the carbon uptake and productivity of Mediterranean species.

Our work also promotes the possibility of parameterizing LUE models based on the PRI and the ratio R690/R630 for evergreen species. This is because of links between PRI is linked with short-term changes in xanthophyll pigments proportions and long-term shifts in pigment pools, and R690/R630 with chlorophyll content and fluorescence [69]. For other non-evergreen vegetation such as annual or deciduous plants, the potential uses of PRI and R690/R630 in detecting the progressive drought and recovery effects on photosynthetic activity should be further tested given their different physiological traits, but these results open a promising window for them too. The increasing use of hyperspectral spectroradiometers carried on unmanned aerial vehicles (UAVs) [70] and satellites (such as Metop and Sentinel series) provides the possibility of testing the LUE model using PRI and red-edge reflectance indices for multiple vegetation types and at different spatiotemporal scales.

Acknowledgments: This work was supported by the European Research Council Synergy grant SyG-2013-610028 IMBALANCE-P, the Spanish Government project CGL2016-79835-P, and the Catalan Government project SGR 2014-274. Chao Zhang gratefully acknowledges the support from the Chinese Scholarship Council.

Author Contributions: Josep Peñuelas, Catherine Preece and Iolanda Filella conceived and designed the experiments; Chao Zhang, Catherine Preece, Iolanda Filella and Gerard Farré-Armengol performed the experiments; Chao Zhang analyzed the data and wrote the initial draft and figures. All of the authors contributed to the discussion of the results and to the writing of the manuscript.

Conflicts of Interest: The authors declare no conflict of interest.

References

1. Hoerling, M.; Eischeid, J.; Perlwitz, J.; Quan, X.; Zhang, T.; Pegion, P. On the increased frequency of Mediterranean drought. *J. Clim.* **2012**, *25*, 2146–2161. [CrossRef]
2. Giorgi, F.; Lionello, P. Climate change projections for the Mediterranean region. *Glob. Planet. Chang.* **2008**, *63*, 90–104. [CrossRef]
3. Nicault, A.; Alleaume, S.; Brewer, S.; Carrer, M.; Nola, P.; Guiot, J. Mediterranean drought fluctuation during the last 500 years based on tree-ring data. *Clim. Dyn.* **2008**, *31*, 227–245. [CrossRef]
4. Peñuelas, J.; Sardans, J.; Filella, I.; Estiarte, M.; Llusià, J.; Ogaya, R.; Carnicer, J.; Bartrons, M.; Rivas-Ubach, A.; Grau, O.; et al. Assessment of the impacts of climate change on Mediterranean terrestrial ecosystems based on data from field experiments and long-term monitored field gradients in Catalonia. *Environ. Exp. Bot.* **2017**. [CrossRef]
5. Liu, D.; Llusia, J.; Ogaya, R.; Estiarte, M.; Llorens, L.; Yang, X.; Peñuelas, J. Physiological adjustments of a Mediterranean shrub to long-term experimental warming and drought treatments. *Plant Sci.* **2016**, *252*, 53–61. [CrossRef] [PubMed]
6. Granda, E.; Camarero, J.J.; Gimeno, T.E.; Martínez-Fernández, J.; Valladares, F. Intensity and timing of warming and drought differentially affect growth patterns of co-occurring Mediterranean tree species. *Eur. J. For. Res.* **2013**, *132*, 469–480. [CrossRef]
7. Liu, D.; Ogaya, R.; Barbeta, A.; Yang, X.; Peñuelas, J. Contrasting impacts of continuous moderate drought and episodic severe droughts on the aboveground-biomass increment and litterfall of three coexisting Mediterranean woody species. *Glob. Chang. Biol.* **2015**, *21*, 4196–4209. [CrossRef] [PubMed]
8. Vicca, S.; Balzarolo, M.; Filella, I.; Granier, A.; Herbst, M.; Knohl, A.; Longdoz, B.; Mund, M.; Nagy, Z.; Pintér, K.; et al. Remotely-sensed detection of effects of extreme droughts on gross primary production. *Sci. Rep.* **2016**, *6*, 1–13. [CrossRef] [PubMed]
9. Barbeta, A.; Ogaya, R.; Peñuelas, J. Dampening effects of long-term experimental drought on growth and mortality rates of a Holm oak forest. *Glob. Chang. Biol.* **2013**, *19*, 3133–3144. [CrossRef] [PubMed]
10. Iglesias, A.; Garrote, L.; Flores, F.; Moneo, M. Challenges to manage the risk of water scarcity and climate change in the Mediterranean. *Water Resour. Manag.* **2007**, *21*, 775–788. [CrossRef]
11. Lloret, F.; Siscart, D.; Dalmases, C. Canopy recovery after drought dieback in holm-oak Mediterranean forests of Catalonia (NE Spain). *Glob. Chang. Biol.* **2004**, *10*, 2092–2099. [CrossRef]
12. Turner, N.C. Sustainable production of crops and pastures under drought in a Mediterranean environment. *Ann. Appl. Biol.* **2004**, *144*, 139–147. [CrossRef]

13. Farquhar, G.D.; Sharkey, T.D. Stomatal conductance and photosynthesis. *Annu. Rev. Plant Physiol.* **1982**, *33*, 317–345. [CrossRef]

14. Peñuelas, J.; Filella, I.; Llusia, J.; Siscart, D.; Pinol, J. Comparative field study of spring and summer leaf gas exchange and photobiology of the mediterranean trees *Quercus ilex* and *Phillyrea latifolia*. *J. Exp. Bot.* **1998**, *49*, 229–238. [CrossRef]

15. Peña-Rojas, K.; Aranda, X.; Fleck, I. Stomatal limitation to CO_2 assimilation and down-regulation of photosynthesis in *Quercus ilex* resprouts in response to slowly imposed drought. *Tree Physiol.* **2004**, *24*, 813–822. [CrossRef] [PubMed]

16. Ogaya, R.; Llusià, J.; Barbeta, A.; Asensio, D.; Liu, D.; Alessio, G.A.; Peñuelas, J. Foliar CO_2 in a holm oak forest subjected to 15 years of climate change simulation. *Plant Sci.* **2014**, *226*, 101–107. [CrossRef] [PubMed]

17. De Rigo, D.; Caudullo, G. *Quercus ilex* in Europe: Distribution, habitat, usage and threats. In *European Atlas of Forest Tree Species*; European Union: Luxembourg, 2016; pp. 130–131.

18. Vaz, M.; Pereira, J.S.; Gazarini, L.C.; David, T.S.; David, J.S.; Rodrigues, A.; Maroco, J.; Chaves, M.M. Drought-induced photosynthetic inhibition and autumn recovery in two Mediterranean oak species (*Quercus ilex* and *Quercus suber*). *Tree Physiol.* **2010**, *30*, 946–956. [CrossRef] [PubMed]

19. Demmig-Adams, B.; Adams, W.W. The role of xanthophyll cycle carotenoids in the protection of photosynthesis. *Trends Plant Sci.* **1996**, *1*, 21–26. [CrossRef]

20. Adams, W.W.; Demmig-Adams, B. Carotenoid composition and down regulation of photosystem II in three conifer species during the winter. *Physiol. Plant.* **1994**, *92*, 451–458. [CrossRef]

21. Gamon, J.A.; Peñuelas, J.; Field, C. A narrow-waveband spectral index that tracks diurnal changes in photosunthetic efficiency. *Remote Sens. Environ.* **1992**, *41*, 35–44. [CrossRef]

22. Peñuelas, J.; Filella, I.; Gamon, J.A. Assessment of photosynthetic radiation use efficiency with spectral reflectance. *New Phytol.* **1995**, *131*, 291–296. [CrossRef]

23. Garbulsky, M.F.; Peñuelas, J.; Gamon, J.; Inoue, Y.; Filella, I. The photochemical reflectance index (PRI) and the remote sensing of leaf, canopy and ecosystem radiation use efficiencies. A review and meta-analysis. *Remote Sens. Environ.* **2011**, *115*, 281–297. [CrossRef]

24. Zhang, C.; Filella, I.; Garbulsky, M.; Peñuelas, J. Affecting factors and recent improvements of the photochemical reflectance index (PRI) for remotely sensing foliar, canopy and ecosystemic radiation-use efficiencies. *Remote Sens.* **2016**, *8*, 677. [CrossRef]

25. Peñuelas, J.; Garbulsky, M.F.; Filella, I.; Papp, T. Photochemical reflectance index (PRI) and remote sensing of plant CO_2 uptake. *New Phytol.* **2011**, *191*, 596–599. [CrossRef] [PubMed]

26. Garbulsky, M.F.; Peñuelas, J.; Papale, D.; Ardö, J.; Goulden, M.L.; Kiely, G.; Richardson, A.D.; Rotenberg, E.; Veenendaal, E.M.; Filella, I. Patterns and controls of the variability of radiation use efficiency and primary productivity across terrestrial ecosystems. *Glob. Ecol. Biogeogr.* **2010**, *19*, 253–267. [CrossRef]

27. Porcar-Castell, A.; Garcia-Plazaola, J.I.; Nichol, C.J.; Kolari, P.; Olascoaga, B.; Kuusinen, N.; Fernández-Marín, B.; Pulkkinen, M.; Juurola, E.; Nikinmaa, E. Physiology of the seasonal relationship between the photochemical reflectance index and photosynthetic light use efficiency. *Oecologia* **2012**, *170*, 313–323. [CrossRef] [PubMed]

28. Peñuelas, J.; Gamon, J.A.; Fredeen, A.L.; Merino, J.; Field, C.B. Reflectance indices associated with physiological changes in nitrogen- and water-limited sunflower leaves. *Remote Sens. Environ.* **1994**, *48*, 135–146. [CrossRef]

29. Wong, C.Y.S.; Gamon, J.A. Three causes of variation in the photochemical reflectance index (PRI) in evergreen conifers. *New Phytol.* **2015**, *206*, 187–195. [CrossRef] [PubMed]

30. Frank, H.A.; Cogdell, R.J. Carotenoids in photosynthesis. *Photochem. Photobiol.* **1996**, *63*, 257–264. [CrossRef] [PubMed]

31. Dobrowski, S.Z.; Pushnik, J.C.; Zarco-Tejada, P.J.; Ustin, S.L. Simple reflectance indices track heat and water stress-induced changes in steady-state chlorophyll fluorescence at the canopy scale. *Remote Sens. Environ.* **2005**, *97*, 403–414. [CrossRef]

32. Sun, P.; Wahbi, S.; Tsonev, T.; Haworth, M.; Liu, S.; Centritto, M. On the use of leaf spectral indices to assess water status and photosynthetic limitations in *Olea europaea* L. during water-stress and recovery. *PLoS ONE* **2014**, *9*, e105165. [CrossRef] [PubMed]

33. Moreno, A.; Maselli, F.; Gilabert, M.A.; Chiesi, M.; Martínez, B.; Seufert, G. Assessment of MODIS imagery to track light-use efficiency in a water-limited Mediterranean pine forest. *Remote Sens. Environ.* **2012**, *123*, 359–367. [CrossRef]

34. Peñuelas, J.; Llusia, J.; Pinol, J.; Filella, I.; Penuelas, J.; Llusia, J.; Pinol, J.; Filella, I. Photochemical reflectance index and leaf photosynthetic radiation-use-efficiency assessment in Mediterranean trees. *Int. J. Remote Sens.* **1997**, *18*, 2863–2868. [CrossRef]

35. Filella, I.; Amaro, T.; Araus, J.L.; Peñuelas, J. Relationship between photosynthetic radiation-use efficiency of barley canopies and the photochemical reflectance index (PRI). *Physiol. Plant.* **1996**, *96*, 211–216. [CrossRef]

36. Guarini, R.; Nichol, C.; Clement, R.; Loizzo, R.; Grace, J.; Borghetti, M. The utility of MODIS-sPRI for investigating the photosynthetic light-use efficiency in a Mediterranean deciduous forest. *Int. J. Remote Sens.* **2014**, *35*, 6157–6172. [CrossRef]

37. Krause, G.; Weis, E. Chlorophyll fluorescence and photosynthesis: The basics. *Annu. Rev. Plant Physiol. Plant Mol. Biol.* **1991**, *42*, 313–349. [CrossRef]

38. Lichtenthaler, H.K.; Miehé, J.A. Fluorescence imaging as a diagnostic tool for plant stress. *Trends Plant Sci.* **1997**, *2*, 316–320. [CrossRef]

39. Porcar-Castell, A.; Tyystjärvi, E.; Atherton, J.; Van Der Tol, C.; Flexas, J.; Pfündel, E.E.; Moreno, J.; Frankenberg, C.; Berry, J.A. Linking chlorophyll a fluorescence to photosynthesis for remote sensing applications: Mechanisms and challenges. *J. Exp. Bot.* **2014**, *65*, 4065–4095. [CrossRef] [PubMed]

40. Ač, A.; Malenovský, Z.; Olejníčková, J.; Gallé, A.; Rascher, U.; Mohammed, G. Meta-analysis assessing potential of steady-state chlorophyll fluorescence for remote sensing detection of plant water, temperature and nitrogen stress. *Remote Sens. Environ.* **2015**, *168*, 420–436. [CrossRef]

41. Guanter, L.; Zhang, Y.; Jung, M.; Joiner, J.; Voigt, M.; Berry, J.A.; Frankenberg, C.; Huete, A.R.; Zarco-Tejada, P.; Lee, J.-E.; et al. Global and time-resolved monitoring of crop photosynthesis with chlorophyll fluorescence. *Proc. Natl. Acad. Sci. USA* **2014**, *111*, E1327–E1333. [CrossRef] [PubMed]

42. Carter, G.A.; Jones, J.H.; Mitchell, R.J.; Brewer, C.H. Detection of solar-excited chlorophyll a fluorescence and leaf photosynthetic capacity using a Fraunhofer Line Radiometer. *Remote Sens. Environ.* **1996**, *55*, 89–92. [CrossRef]

43. Zarco-Tejada, P.J.; González-Dugo, M.V.; Fereres, E. Seasonal stability of chlorophyll fluorescence quantified from airborne hyperspectral imagery as an indicator of net photosynthesis in the context of precision agriculture. *Remote Sens. Environ.* **2016**, *179*, 89–103. [CrossRef]

44. Freedman, A.; Cavender-Bares, J.; Kebabian, P.L.; Bhaskar, R.; Scott, H.; Bazzaz, F.A. Remote sensing of solar-excited plant fluorescence as a measure of photosynthetic rate. *Photosynthetica* **2002**, *40*, 127–132. [CrossRef]

45. Buschmann, C.; Lichtenthaler, H.K. Principles and characteristics of multi-colour fluorescence imaging of plants. *J. Plant Physiol.* **1998**, *152*, 297–314. [CrossRef]

46. Meroni, M.; Rossini, M.; Guanter, L.; Alonso, L.; Rascher, U.; Colombo, R.; Moreno, J. Remote sensing of solar-induced chlorophyll fluorescence: Review of methods and applications. *Remote Sens. Environ.* **2009**, *113*, 2037–2051. [CrossRef]

47. Zarco-Tejada, P.J.; Miller, J.R.; Mohammed, G.H.; Nolan, T.L. Chlorophyll fluoresence effects on vegetation apparent reflectanc: I. Leaf-level measurements and model simulation. *Remote Sens. Environ.* **2000**, *74*, 582–592. [CrossRef]

48. Zarco-Tejada, P.J.; Miller, J.R.; Mohammed, G.H.; Noland, T.L.; Sampson, P.H. Chlorophyll fluorescence effects on vegetation apparent reflectance: II. Laboratory and Airborne canopy-level measurements with hyperspectral data. *Remote Sens. Environ.* **2000**, *74*, 596–608. [CrossRef]

49. Zarco-Tejada, P.J.; Berni, J.A.J.; Suárez, L.; Sepulcre-Cantó, G.; Morales, F.; Miller, J.R. Imaging chlorophyll fluorescence with an airborne narrow-band multispectral camera for vegetation stress detection. *Remote Sens. Environ.* **2009**, *113*, 1262–1275. [CrossRef]

50. Ač, A.; Olejníčková, J.; Mishra, K.B.; Malenovský, Z.; Hanuš, J.; Trtílek, M.; Nedbal, L.; Marek, M.V. Towards remote sensing of vegetation processes. In Proceedings of the Workshop Sensing a Changing World 2008, Wageningen, The Netherlands, 19–21 Novemebr 2008; pp. 19–23.

51. Ač, A.; Malenovský, Z.; Hanuš, J.; Tomášková, I.; Urban, O.; Marek, M.V. Near-distance imaging spectroscopy investigating chlorophyll fluorescence and photosynthetic activity of grassland in the daily course. *Funct. Plant Biol.* **2009**, *36*, 1006–1015. [CrossRef]

52. Furuuchi, H.; Jenkins, M.W.; Senock, R.S.; Houpis, J.L.J.; Pushnik, J.C. Estimating plant crown transpiration and water use efficiency by vegetative reflectance indices associated with chlorophyll fluorescence. *Open J. Ecol.* **2013**, *3*, 122–132. [CrossRef]

53. Ni, Z.Y.; Liu, Z.G.; Li, Z.L.; Nerry, F.; Huo, H.Y.; Li, X.W. Estimation of solar-induced fluorescence using the canopy reflectance index. *Int. J. Remote Sens.* **2015**, *36*, 5239–5256. [CrossRef]

54. Panigada, C.; Rossini, M.; Meroni, M.; Cilia, C.; Busetto, L.; Amaducci, S.; Boschetti, M.; Cogliati, S.; Picchi, V.; Pinto, F.; et al. Fluorescence, PRI and canopy temperature for water stress detection in cereal crops. *Int. J. Appl. Earth Obs. Geoinf.* **2014**, *30*, 167–178. [CrossRef]

55. Zarco-Tejada, P.J.; González-Dugo, V.; Williams, L.E.; Suárez, L.; Berni, J.A.J.; Goldhamer, D.; Fereres, E. A PRI-based water stress index combining structural and chlorophyll effects: Assessment using diurnal narrow-band airborne imagery and the CWSI thermal index. *Remote Sens. Environ.* **2013**, *138*, 38–50. [CrossRef]

56. Rossini, M.; Fava, F.; Cogliati, S.; Meroni, M.; Marchesi, A.; Panigada, C.; Giardino, C.; Busetto, L.; Migliavacca, M.; Amaducci, S.; et al. Assessing canopy PRI from airborne imagery to map water stress in maize. *ISPRS J. Photogramm. Remote Sens.* **2013**, *86*, 168–177. [CrossRef]

57. Genty, B.; Briantais, J.-M.; Baker, N.R. The relationship between the quantum yield of photosynthetic electron transport and quenching of chlorophyll fluorescence. *Biochim. Biophys. Acta Gen. Subj.* **1989**, *990*, 87–92. [CrossRef]

58. Filella, I.; Porcar-Castell, A.; Munné-Bosch, S.; Bäck, J.; Garbulsky, M.F.; Peñuelas, J. PRI assessment of long-term changes in carotenoids/chlorophyll ratio and short-term changes in de-epoxidation state of the xanthophyll cycle. *Int. J. Remote Sens.* **2009**, *30*, 4443–4455. [CrossRef]

59. Gamon, J.A.; Berry, J.A. Facultative and constitutive pigment effects on the photochemical reflectance index (PRI) in sun and shade conifer needles. *Isr. J. Plant Sci.* **2012**, *60*, 85–95. [CrossRef]

60. Baquedano, F.J.; Castillo, F.J. Comparative ecophysiological effects of drought on seedlings of the Mediterranean water-saver *Pinus halepensis* and water-spenders *Quercus coccifera* and *Quercus ilex*. *Trees Struct. Funct.* **2006**, *20*, 689–700. [CrossRef]

61. Demmig-Adams, B.; Adams, W.W. Photoprotection in an ecological context: The remarkable complexity of thermal energy dissipation. *New Phytol.* **2006**, *172*, 11–21. [CrossRef] [PubMed]

62. Goerner, A.; Reichstein, M.; Rambal, S. Tracking seasonal drought effects on ecosystem light use efficiency with satellite-based PRI in a Mediterranean forest. *Remote Sens. Environ.* **2009**, *113*, 1101–1111. [CrossRef]

63. Van Der Tol, C.; Berry, J.A.; Campbell, P.K.E.; Rascher, U. Models of fluorescence and photosynthesis for interpreting measurements of solar-induced chlorophyll fluorescence. *J. Geophys. Res. Biogeosci.* **2014**, *119*, 2312–2327. [CrossRef] [PubMed]

64. Porcar-Castell, A. A high-resolution portrait of the annual dynamics of photochemical and non-photochemical quenching in needles of *Pinus sylvestris*. *Physiol. Plant.* **2011**, *143*, 139–153. [CrossRef] [PubMed]

65. Verhoeven, A. Sustained energy dissipation in winter evergreens. *New Phytol.* **2014**, *201*, 57–65. [CrossRef]

66. Gitelson, A.A.; Zur, Y.; Chivkunova, O.B.; Merzlyak, M.N. Assessing carotenoid content in plant leaves with reflectance spectroscopy. *Photochem. Photobiol.* **2002**, *75*, 272–281. [CrossRef]

67. Ruban, A.V.; Berera, R.; Ilioaia, C.; van Stokkum, I.H.M.; Kennis, J.T.M.; Pascal, A.A.; van Amerongen, H.; Robert, B.; Horton, P.; van Grondelle, R. Identification of a mechanism of photoprotective energy dissipation in higher plants. *Nature* **2007**, *450*, 575–578. [CrossRef] [PubMed]

68. Jahns, P.; Holzwarth, A.R. The role of the xanthophyll cycle and of lutein in photoprotection of photosystem II. *Biochim. Biophys. Acta Bioenerg.* **2012**, *1817*, 182–193. [CrossRef] [PubMed]

69. Gamon, J.A. Reviews and Syntheses: Optical sampling of the flux tower footprint. *Biogeosciences* **2015**, *12*, 4509–4523. [CrossRef]

70. Gago, J.; Douthe, C.; Coopman, R.E.; Gallego, P.P.; Ribas-Carbo, M.; Flexas, J.; Escalona, J.; Medrano, H. UAVs challenge to assess water stress for sustainable agriculture. *Agric. Water Manag.* **2015**, *153*, 9–19. [CrossRef]

MDPI

St. Alban-Anlage 66

4052 Basel

Switzerland

Tel. +41 61 683 77 34

Fax +41 61 302 89 18

www.mdpi.com

Forests Editorial Office

E-mail: forests@mdpi.com

www.mdpi.com/journal/forests